FOURTH EUROPEAN MARINE BIOLOGY SYMPOSIUM

FOURTH EUROPEAN MARINE BIOLOGY SYMPOSIUM

Edited by

D. J. CRISP

CAMBRIDGE

AT THE UNIVERSITY PRESS

1971

CAMBRIDGE UNIVERSITY PRESS
Cambridge, New York, Melbourne, Madrid, Cape Town,
Singapore, São Paulo, Delhi, Tokyo, Mexico City

Cambridge University Press
The Edinburgh Building, Cambridge CB2 8RU, UK

Published in the United States of America by Cambridge University Press, New York

www.cambridge.org
Information on this title: www.cambridge.org/9780521178259

First published 1971
First paperback edition 2011

A catalogue record for this publication is available from the British Library

Library of Congress Catalogue Card Number: 71–173829

ISBN 978-0-521-08101-6 Hardback
ISBN 978-0-521-17825-9 Paperback

Additional resources for this publication at www.cambridge.org/9780521178259

CONTENTS

[v]

2. *The responses of marine animals to light*

FOREWORD

The IVth European Symposium on Marine Biology (E.M.B.S.) took place at Bangor between 14 and 20 September 1969 by invitation of the Marine Science Laboratories of the University College of North Wales. It was attended by some two hundred participants from twenty countries. Financial support for the Symposium was received from the Council of Europe and the Royal Society of London; the Bangor City Council, the University College of North Wales and the University of Wales provided hospitality to the guests.

Previous symposia of this series have been held at the Biologische Anstalt, Helgoland, German Federal Republic (1966); at the Biological Station, Espegrend, Norway (1967), and at the Station Biologique d'Arcachon, France (1968). The aim of these meetings is to foster personal contact between scientists working in marine institutes and university departments of European countries and to offset the tendency for marine biologists to lose touch with the subject as a whole, through the necessity to specialise on particular lines. Topics of wide interest to ecologists, physiologists and taxonomists are therefore chosen as themes around which the programme of each symposium is built.

The two themes of the IVth Symposium were, 'Larval biology' and 'Light in the marine environment'. Forty-two contributions were delivered and discussed; the majority of them have been included in this volume.

My task of editing the proceedings has been made much lighter through the willing assistance given me by Professor J. M. Dodd, Professor E. W. Knight-Jones, Professor Robert Weill, Dr J. H. S. Blaxter as well as by the staff of my own department. Mrs Marian Jones checked the references and Mrs. M. Flowerdew undertook the secretarial work in connection with editing the volume.

OPENING ADDRESS

D. J. CRISP

Marine Sciences Laboratories, Menai Bridge, Anglesey

When the invitation was issued to European marine biologists to hold their Fourth Symposium in Bangor, the College had in mind not only the merits of North Wales as a venue for such a meeting, but also the honour that the Symposium would do us in meeting here on the 21st birthday of the Marine Science Laboratories. It seems fitting, therefore, that in my address of welcome, I should give a brief outline of the history of our institute.

For the benefit of those who are not acquainted with the politics of the University of Wales, I should first explain the relationship between the Marine Science Laboratories, the University, and the University College at Bangor. The essential point to be remembered is that in Wales it is the constituent colleges that are responsible for all the teaching and research, each college being in effect a separate university, whereas the 'University of Wales' is an administrative structure which serves to link the colleges loosely together. The Marine Science Laboratories contain two independent university departments, Marine Biology and Physical Oceanography, belonging to and administered by the University College of North Wales, Bangor.

A journey to these parts of the Principality is no longer a major expedition to a wild and remote area, as it was in the days when the naturalists John Ray and Thomas Pennant visited North Wales. Nevertheless, the mountains of Snowdonia still create a sense of isolation from the rest of the country. It is perhaps owing to this happy illusion, as well as to the vigilance of conservationists, that North Wales has, at least until recently, resisted the despoilation that usually follows in the wake of development of communications and growth of population.

Accessibility, combined with an unspoilt and natural environment, has been the major asset of the University College situated at Bangor and accounts for its success in many fields of environmental biology. The marine life is exceptionally rich and easily available for other reasons also. The Menai Strait is unique in combining shelter from storms with a great cleansing flush of tidal water. The coastline within some twenty miles of Menai Bridge is both extensive and varied and includes an important faunal disjunction at Carmel Head, North-west Anglesey. There are all grades of deposits from mud to coarse gravel and boulders, coasts of smooth and fissured rock. Within easy reach to the east are estuaries and shores exhibiting all types and conditions of pollution.

Not surprisingly, the affinity that marine biologists have felt for North Wales predates by half a century the interest now shown by the University College.

William Herdman, following his appointment in 1881 as the first Derby Professor of Natural History at Liverpool, attempted to set up a Marine Biological Station within easy reach of his University. He was perhaps the originator of that unfortunate tradition that the devotees of marine biology should reside on an island and work in dilapidated buildings. His first attempt to found a station was at Hilbre Island, accessible only across the sands of Dee at low water. His second attempt was made on Puffin Island, now a rocky and exposed bird sanctuary at the extreme easterly tip of Anglesey. Here, in 1889, he and his staff landed by a small boat, entered a disused semaphore telegraph station and restored it as a laboratory. But after several strandings of ships on the mainland and of scientists on the island, the enterprise was abandoned in 1891. Herdman then turned his attention to Port Erin on the Isle of Man, where the laboratory associated with Liverpool University was eventually established.

The first permanent marine laboratory in North Wales was established at Conway and arose through an unfortunate occurrence of typhoid poisoning, allegedly caused by the proximity of the mussel beds to the town's sewage outfall. The Ministry's Shellfish Experimental Station at Conway was initially headed by the bacteriologist R. W. Dodgson, whose task was to establish a cleansing procedure to safeguard the important Conway mussel industry. The location of this laboratory is perhaps unique in having been determined by proximity to pollution. Yet, despite this disadvantage, its programme now encompasses the whole range of shellfish biology and is prominent in the development of shellfish cultivation in association with the White Fish Authority's oyster hatchery.

The foundations of the interest of the University College of North Wales in marine biology were laid by the late Professor F. W. Rogers Brambell, F.R.S., who was made Lloyd Roberts Professor of Zoology at Bangor in 1930. Shortly after his appointment he instituted a vacation course for undergraduate students which was, I think, the first of its kind offered by a British university department. This not only gave a strong impetus to marine studies at Bangor, but also familiarized many zoology students of other universities with the faunal diversity of the area.

In 1942, Rogers Brambell and Sir D. Emrys Evans, the then College Principal, suggested that a Marine Biology Station should be set up at

Bangor as part of the post-war plans of the University. In 1947 the University Grants Committee agreed in principle to this development, on the understanding with the University of Wales, that it would meet the needs of all four constituent colleges. The following year marked the first meeting of the University Advisory Committee which was set up to steer the station's development. Plans were made for the appointment of a Director, the purchase of a boat and the acquisition of land on the Caernarvonshire side of the Menai Straits. In so far as an institute can be said to have been born, rather than merely conceived, the year 1948 should be regarded as the birthday of the University's Marine Station.

The station suffered a tragic setback when the first Director, Dr Fabius Gross, died within a year or so of his appointment. By then no financial commitment had been made to erect a laboratory, though Gross and the College architects had planned a large building of the character required to stand on the Caernarvonshire side of the Straits, close to Bangor, at Nantporth.

By 1951, when I joined Dr E. W. Knight-Jones, then acting Director, and Dr C. P. Spencer, who was training under H. W. Harvey, F.R.S., at Plymouth, the political situation had deteriorated and any priority accorded to the Marine Station had been dropped in favour of new science buildings and halls of residence. It became necessary, therefore, to buy time. The laboratory had to be set up in temporary accommodation. By good fortune, a house built in spacious Victorian style, known as Westbury Mount, came on the market. Admittedly, it had defects characteristic of old buildings; an unsuitable layout, unstable floors, dry rot and, it is believed, a poltergeist; nevertheless its situation was ideal. It commanded fine views of the Straits, it was centrally placed in relation to the best collecting areas, and its proximity to the public pier was very convenient for boat work and for seawater supplies. It rightly became the permanent location of the laboratory.

Curiously, despite the vigorous growth of the Department of Marine Biology during a period of great university expansion, it never acquired a major science building of its own, but continued to lean on the massive rubble walls of Westbury Mount. Indeed, the Department's achievements would not have been possible, but for its having adopted the way of life of the hermit crab. Having outgrown its original home, it has repeatedly moved into another vacated by the previous occupant, whenever the opportunity arose. But unlike the old shells of the hermit crab, those of the Marine Biology Department continue to be occupied.

In 1962, a metamorphosis began; the 'Marine Biology Station' was renamed 'The Marine Science Laboratories' to accommodate two new

developments; a Department of Physical Oceanography of which Professor
J. Darbyshire is now the Head, and a Unit of Marine Invertebrate Biology
which was set up by the Development Commissioners, a body which has
now been absorbed into the Natural Environment Research Council. This
Unit, which is closely linked with the Department of Marine Biology, is
concerned at present with studies of marine invertebrate larvae. These two
new ventures, together with a Research Vessel, the Prince Madog, and
a laboratory for fish physiology generously given by the Nuffield Foundation, have led to a considerable broadening of the disciplines studied.

One of the themes chosen for the Symposium, 'Larval biology', is
especially pertinent to this occasion, since so many marine Bangorians, past
and present, have made significant contributions to the subject. Some years
before the Marine Station was founded, Dr Cole and Dr Knight-Jones, both
old students of the college, studied the setting behaviour of oysters at the
Conway laboratory; Dr Walne has since extended this work into the practical
world of oyster culture, and Dr Knight-Jones, now Professor of Zoology
at University College Swansea, has continued his interest in gregarious
settlement. It is indeed a great pleasure to see at this Symposium so many
colleagues and past students who have contributed to this field of research.

The second theme, 'Light in the marine environment', is less obviously
related to the recent programmes of the laboratory. Nevertheless, it would
have gratified its first Director, since Dr Fabius Gross had intended that
the Marine Station should become a focus for the study of the microbiology
of primary production in the sea, and Dr Spencer, with his collaborators,
has continued to give emphasis to this subject.

To see so many European countries, not to mention some outside Europe,
represented among the participants, encourages those who initiated these
Symposia to believe that, as they become better known year by year, they
are fulfilling their purpose more successfully. In welcoming you all to this,
the Fourth Symposium, I would like to express my own view that the
scientific sessions, important as they are, serve only as a framework. The
more important side of such a gathering is the opportunity it offers, through
informal discussion, to forge good personal relationships and to get to grips
with fundamental scientific problems. I hope, therefore, that the domestic
arrangements and the social events will provide sufficient time for such
activities as well as allowing you to see something of our beautiful
countryside.

Finally, I would like to thank all those who have made this Symposium
possible. We have no formal organisation, and therefore must rely entirely

on the assistance so generously given by members of the host institution. There are too many to name individually, but I would like to say that not only have many of the staff, students and their wives from my own department given willingly of their time, but we have also had generous help from many others, including the College administration, the women's hall of residence, the Department of Zoology and our colleagues from the laboratory at Conway.

THE DISPERSAL OF THE LARVAE OF SHOAL-WATER BENTHIC INVERTEBRATE SPECIES OVER LONG DISTANCES BY OCEAN CURRENTS*

R. S. SCHELTEMA

Woods Hole Oceanographic Institution, Woods Hole, Massachusetts 02543

INTRODUCTION

Planktonic larvae are the dispersal stages of marine benthic invertebrate organisms. They provide a means whereby bottom-dwelling species can attain a wide geographical distribution, and they serve as the links that maintain genetic continuity between populations spatially isolated from one another.

Dispersal by means of pelagic larvae is particularly characteristic of the benthos in warm-temperate and tropical seas because it is here that a large proportion of all species have pelagic planktotrophic larvae (Thorson, 1946). Indeed, the importance of free-swimming larval stages to the geographical distribution of tropical marine organisms was already recognized by Alfred Russel Wallace (1876) who suggested that univalve and bivalve molluscs with pelagic larval stages '... have a powerful means of dispersal, and are carried by tides and currents so as ultimately to spread over every shore and shoal that offers conditions favourable for development' (vol. 1, p. 30).

Thorson (1961) has assembled data from the biological literature on the duration of pelagic larval life of 195 species of bottom invertebrates and concludes that in most instances the period of larval development is too short to account for larval transport across major ocean basins. However, most of the 195 species tabulated are cold-water forms, which as a rule have a shorter pelagic larval development than that of most tropical and warm-temperate species (Thorson, 1961, p. 473; Mileikovsky, 1966, p. 400 in translation). Because these data are unavoidably biased toward species of short planktonic duration, it seems better that they not be extrapolated to make conclusions on the probability of larval dispersal across tropical seas.

Mileikovsky (1960, 1966, 1968a, 1968b) has recorded the distribution of larvae from shoal-water benthic invertebrates in the open waters of the

---

* Contribution No. 2341 from the Woods Hole Oceanographic Institution, Woods Hole, Massachusetts 02543.

Norwegian and Barents Sea. Larvae of bivalve molluscs and polychaetes are dispersed as far as 1000 km. From this, Mileikovsky concludes that (1) '... long-range larval drift may most frequently give rise to "pseudo populations" [expatriate, non-reproducing populations] rather than to true populations of the species...' and (2) because the duration of larval development is so short, the dispersal of larvae across ocean basins, when it occurs, must be a stepwise progression from island to island or from one shoal region to another. Stepwise dispersal is possible in the Norwegian Sea because it has extensive shoal water regions of less than 200 m depth (Stock, 1950), but 'island hopping' must be less important in the warm-temperate and tropical Atlantic where islands are very scattered and few in number.

Three lines of evidence are needed to demonstrate that the long-distance larval dispersal of a particular species is possible. (1) The larvae must be found throughout the length of the major surface currents required to carry them from their 'origin' to their subsequent 'destination'. If reproduction of a species is restricted to a short season, the larvae may occur in various portions of the currents at different times. It would also be expected that morphological development will proceed, and that the larvae will increase in size with increasing distance from the point of their origin. The larvae must also tolerate the conditions likely to be encountered en route. (2) The velocity of the surface currents of the ocean should be known so that the rate of transport can be estimated and the time required for dispersal between points computed. (3) The length of time required for larval development must be approximately known, or at least it should be demonstrated that the larvae remain pelagic for a time sufficient to account for their transport across the ocean.

Most evidence either for or against long-distance dispersal of larvae, up until now, has been chiefly concerned with the duration of planktonic development and its relationship to the velocity of surface currents (Gardiner, 1904; Thorson, 1961). No systematic observations have been made over large areas of the oceans to determine whether in fact the larvae are found throughout the entire length of major ocean currents. It is this latter evidence in particular which I wish to present here. Further data on the durations of pelagic life will also be offered.

METHODS

The conclusions to be presented are restricted to the warm-water shelf fauna of the North and tropical Atlantic Ocean and are based upon approximately 800 plankton tows throughout this region (Fig. 1). Plankton samples were obtained from the R.V. *Atlantis II* and R.V. *Chain* by taking oblique tows with three-quarter metre plankton nets having mesh sizes of from 223 to 316 μm (No. 2 and No. 6 netting). Two hundred metres of wire was payed out; the depth to which the tow extended depended on the wire angle (30–70°) and differed for each station (70–170 m, usually near 100 m). The

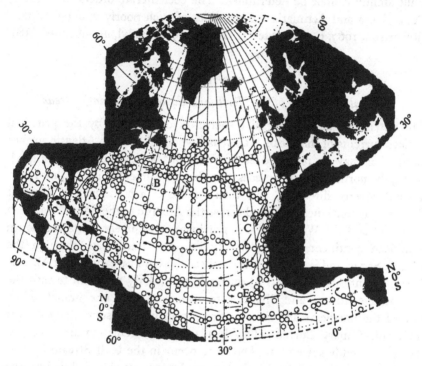

Fig. 1. North Atlantic Ocean showing the distribution of surface plankton stations and the general circulation of surface waters. Some of the station locations represent more than one collection. A = Gulf Stream; B = North Atlantic Drift; C = Canary Current; D = North Equatorial Current; E = Equatorial Counter Current; F = South Equatorial Current.

net was towed for 20 min at approximately 4·5 km/h. Preliminary sorting was done at sea, where as many larval specimens as possible were removed alive. Certain species of larvae were selected for rearing to settlement. These were maintained in 1 l polyethylene boxes and were fed either on phytoplankton cultures or on brine shrimp nauplii.

Because very few previous taxonomic investigations have been made on the larvae of the tropical benthos, the adult affinities of each larval species must be determined individually. Identifications can be made either by comparing hard parts of the larvae with those of the adults (e.g. the proto-

conchs of molluscs or the setae of polychaetes) or by rearing the larvae in
the laboratory to settlement and through post-larval development until
their adult morphology is attained. There are many groups of tropical
invertebrates, however, in which the systematics of the adults is so poorly
known that, even when the larvae are reared to near sexual maturity, the
adult affinity cannot be determined. The coelenterate orders Zoanthidia,
Ceriantharia and Actiniaria are examples of such poorly understood taxa
(Robertson, 1967, p. 248, note 14; Leloup, 1964; Bzdyl, unpublished MS).

RESULTS

General circulation of the North and Equatorial Atlantic Ocean

The extent or range of larval dispersal is determined by the principal
surface currents of the ocean. Taken as a whole, the North Atlantic surface
circulation may be regarded as an enormous anticyclonic gyre (Fig. 1).
Along the northwest African coast the Canary Current (Fig. 1 C) moves in
a southeasterly direction, and then as the North Equatorial Current
(Fig. 1 D) it continues westward across the Atlantic to the South American
continent and the West Indies. A northwesterly current skirts the South
American continent, passing mostly into the Caribbean, through the
Florida Strait, and issuing northeasterly as the Gulf Stream (Fig. 1 A). The
North Atlantic Drift (Fig. 1 B) moves in an easterly direction toward the
European continent and divides into two parts with one branch going
toward the northeast and the remaining arm passing southeast where it
meets the Canary Current. The details of this circulation are complex.
Temporary eddy systems are known to occur in the Gulf Stream System
(Fuglister, 1963), and the North Atlantic Drift is much simplified by our
description (*vide* Worthington, 1962; Mann, 1967). Doubtless other portions of the gyre will prove to be equally complex when further studied.
Notwithstanding, the generalized description above allows prediction of
the transport that may be expected of an object drifting on or near the sea
surface in the North Atlantic Ocean.

In the equatorial region, the North and South Equatorial Currents
(Fig. 1 D, F) are separated from each other by the rather weakly developed
Equatorial Counter Current (Fig. 1 E) flowing eastward. Unlike currents on
either side of it, the countercurrent transports a relatively small volume of
water and usually extends only part way across the western Atlantic.
Monthly charts of the North and South Atlantic surface circulation both
by Schumaker (1940, 1943) and by the U.S. Hydrographic Office show the
Equatorial Counter Current as diffuse and seasonal. However, directly

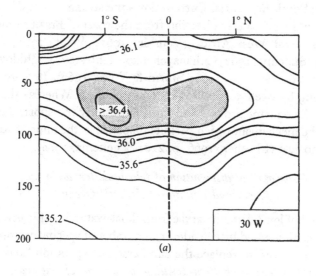

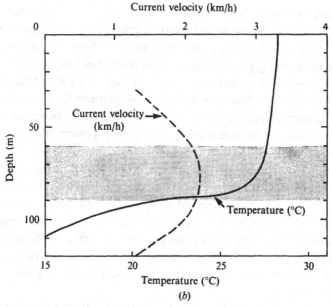

Fig. 2. The Equatorial Undercurrent is a subsurface core of saline water moving in an easterly direction at a depth of approximately between 60 and 90 m. It flows directly under the equator between Brazil and the island of São Tomé in the Gulf of Guinea. (A) Vertical cross section at 30° W latitude showing central core of saline water (from Scheltema, 1968 modified after Khanaychenko, Khlystov & Zhidov, 1965); (B) Temperature and current velocity in the Equatorial Undercurrent in the region between 25 and 32° W latitude. Temperature data (solid line) from bathythermograph lowerings. Velocity data (broken line) are average current velocity computed from Stalcup & Metcalf (1966) (from Scheltema, 1968).

beneath the South Equatorial Current flows the strong Atlantic Equatorial Undercurrent in an easterly direction from the coast of Brazil to the Island of São Tomé off West Africa (Voorhis, 1961; Voigt, 1961; Metcalf, Voorhis & Stalcup, 1962; Khanaychenko, Khlystov & Zhidov, 1965; Rinkel, Sund & Newman, 1966; Sturm & Voigt, 1966). Its core is quite shallow, lying between 50 and 100 m depth (Fig. 2). Whereas the surface currents at the equator provide a means for westerly transport, the undercurrent is the mechanism which permits dispersal in the opposite direction from west to east (Chesher, 1966, p. 210; Scheltema, 1968).

Geographical distribution of teleplanic larvae in the North and Equatorial Atlantic Ocean

The dispersal of long-distance larvae from shoal-water, benthic invertebrate species is best illustrated by considering a number of specific examples. It is useful first, however, to replace the cumbersome expression '*long-distance larvae from shoal-water bottom-dwelling organisms of the shelf*' by the term *teleplanic* larvae derived from the Greek 'teleplanos' meaning far-wandering (Aesculus, Prometheus bound, l. 576). A teleplanic larva must (1) originate from shoal-water, continental shelf benthos, (2) be regularly found in the open sea, (3) have a larval development of long duration, and (4) serve as a means of dispersal over long distances. Many teleplanic larvae are easily recognized by their morphological adaptations for long-distance dispersal. It is possible to give examples of teleplanic larvae from any of the major invertebrate phyla (Scheltema, 1964); the present discussion is restricted to a few examples.

Many of the Mollusca are particularly well represented in the teleplanos (Scheltema, 1966a). Among the gastropods, the families Cymatiidae, Architectonicidae, Cypraeidae, Tonnidae, Thaiidae, Triphoridae, Ovulidae, Magilidae, and Naticidae are all commonly found in the open sea plankton; indeed, approximately 75% of all plankton samples taken in the open sea contain teleplanic gastropod veliger larvae. Some species are particularly well adapted to drift on the sea-surface for long periods and over great distances (Plates 1, 2).

The Cymatiidae (tritons) is a family of predatory gastropods amply represented in the tropics. There are over 100 species, many of which are considered circumtropical. The group is known from the fossil record, particularly from the Tertiary, and it has been suggested that dispersal into the Atlantic from the Indo–Pacific was by way of the Tethys Sea (Clench & Turner, 1957). Most tropical species have teleplanic larvae. None has yet been reared through its complete development to settlement in the laboratory, though larvae taken from plankton samples can be held to metamorphosis. I shall take but one example from the many species of this family whose larvae have been found in the plankton of the North Atlantic.

Cymatium parthenopeum (von Salis) is one of a number of species with a wide geographical distribution throughout the North Atlantic. It is known in the Western Atlantic from the West Indies, Gulf of Mexico, Florida, and northward up the coast of the United States as far as Cape Hatteras; and in the Eastern Atlantic from the Azores, the Mediterranean and along the West African coast southward to Senegal. Its vertical distribution does not exceed 100 m.

Larvae of *C. parthenopeum* are found in all the major currents throughout the North Atlantic gyre (Plate 2). Veligers occur in the Gulf Stream at all seasons of the year and are also continuously distributed along the North Atlantic Drift to the Azores, southward along the northwest coast of Africa, in the North Equatorial Current and throughout the Caribbean region (Fig. 3). When captured, many of the larvae are heavily fouled with algae and folliculinid protozoa, indicating that they have been at sea for a very long time. Larvae of *C. parthenopeum* are unquestionably dispersed over long distances, and from their geographical distribution on the sea surface it seems certain that they are frequently transported across the Atlantic on the North Equatorial Current and the North Atlantic Drift.

The Architectonicidae (sun dials) is a family of gastropods known from tropical waters throughout the world. Bathymetrically they are distributed from a depth of a few meters to the edge of the continental shelves. The adults are known to feed upon colonial zoantharian coelenterates belonging to the family Zoanthidae and specifically on the genera *Palythoa* and *Zoanthus* (Robertson, 1967). Most species of Architectonicidae have a wide geographical distribution and some Atlantic forms have analogues in the Indo–Pacific. Of the ten known species from off West Africa (Senegal), six are also recorded from the West Indies while two others are considered analogues of Western Atlantic species (Marche-Marchad, 1969; R. Robertson, personal communication.)

The occurrence of teleplanic larvae among members of the Architectonicidae is widespread. Their veligers are known from throughout the North Atlantic and have been found also along the Brazilian coast as far south as 18° (Fig. 4). More than 15 species are recognizable from the plankton, and the adult affinities of about half of these are already established.

Philippia krebsii is a North Atlantic species infrequently found in museum collections. Its known geographical distribution in the western Atlantic includes the Barbados, Curaçao, Yucatan north to Cape Hatteras, and Bermuda. In the eastern Atlantic it is recorded only from the Canary Islands. The larvae of *P. krebsii* (Plate 1*a*; *vide* Robertson, Scheltema & Adams, 1970, for description) is found distributed in the major currents along the entire North Atlantic gyre (Robertson, 1964; Scheltema, 1968). Both the adults and larvae of *P. krebsii* are apparently eurythermal; the

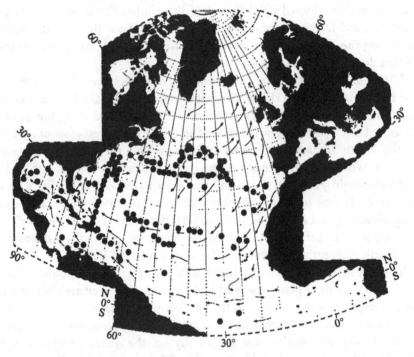

Fig. 3. Distribution of veliger larvae of *Cymatium parthenopeum* (von Salis). The number of larvae generally decreased with increasing distance from the coast.

former are sometimes found in the cooler waters of the continental shelf, whereas the latter are taken in surface plankton samples between temperatures of 15 and 29 °C. The larvae are sufficiently eurythermal to survive transport across the North Atlantic Drift.

To judge from the distribution of *P. krebsii* larvae, it is probable that the geographical range of this species is very much greater than is presently known.

Among the bivalves, a number of families have teleplanic veligers (e.g. the Teredinidae, Pinnidae and Mytilidae). Forty-eight % of all plankton collections made in the North Atlantic have bivalve larvae. Particularly surprising is the dispersal of the larvae of Teredinidae (shipworm) which may be found distributed in the surface waters throughout the North and Equatorial Atlantic, the Caribbean, the Gulf of Mexico and the Mediterranean (Scheltema, In Press).

Teleplanic larvae are known among a number of polychaete families, including the Spionidae, Chaetopteridae, Sternaspididae, Sabellariidae,

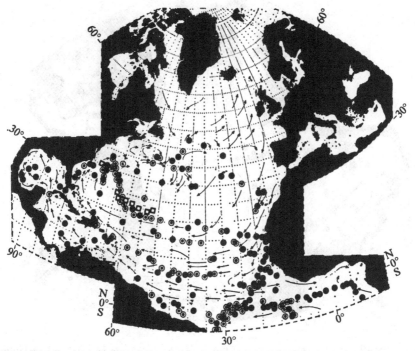

Fig. 4. Distribution of veliger larvae belonging to the gastropod family Architectonicidae in the surface waters of the North Atlantic Ocean. Circles enclosing dots are locations from which larvae of *Philippia krebsii* were collected. Squares show locations from which veliger larvae of *P. krebsii* were collected by the Dana Expedition (*vide* Robertson, 1964). Larger uncircled dots are stations where other species of Architectonicidae larvae were found.

Oweniidae and Euphrosynidae. Only a single species belonging to the Spionids will be considered here.

Laonice cirrata (Sars) is a species of very wide geographical distribution. It is known from the North and South Atlantic, the Indo–Pacific, the Mediterranean, and the Arctic and Antarctic Oceans (Wesenberg-Lund, 1951; Ushakov, 1955; Pettibone, 1956; Kirkegaard, 1959; Hartman, 1966). In the North Atlantic it is known from North America, Europe and West Africa. Its vertical distribution extends down to 2000 m (Hartman, 1966), but it is principally restricted to the continental shelf.

Häcker (1898) described and figured from samples taken by the 'Plankton Expedition' a polychaete larva which he named 'Chaetosphaera' and placed in the family Spionidae. This larva was characterized by setae of a very special type, being flattened, slightly curved, and unilaterally serrated. Hannerz (1956) has assigned the 'Chaetosphaera' larva to the genus

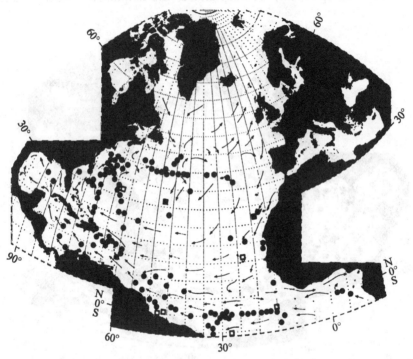

Fig. 5. Distribution of 'Chaetosphaera' larvae of *Laonice cirrata* in the North Atlantic Ocean. Circles indicate locations where *L. cirrata* was collected by R.V. *Atlantis II* and R.V. *Chain* in the present study. Open squares denote stations taken by the 'Plankton Expedition' which included 'Chaetosphaera' larvae. Closed squares are stations where 'Chaetosphaera' larvae were taken by the *Princesse Alice* and reported by Fauvel (1916).

Laonice and has shown that the description of Häcker agrees very well with that of *Laonice cirrata*. The 'Chaetosphaera' larvae of *L. cirrata* are found in the entire North Atlantic gyre (Fig. 5). Consequently, the great geographical distribution of the species is not surprising.

The Sipunculida represent a small but ubiquitous phylum of marine worms found intertidally, on the continental shelf, and in the deep sea. In the tropics they are invariably found on coral reefs. The larval development within the group is known in detail for only nine species which have been grown in the laboratory; three of these were recently described by Rice

(1967). Only two of the nine species have a pelagic planktotrophic development. Though the pelagosphaera larvae of sipuncuids are seldom described from plankton collections, they are common in the surface waters of the North Atlantic Ocean (Jägersten, 1963; Scheltema, 1963, 1964; Scheltema & Hall, 1965). About 75% of all plankton collections made by the R.V. *Atlantis II* and R.V. *Chain* in the open waters of the warm-temperate and tropical North Atlantic contain larvae of sipunculids; at least 15 larval species are recognizable.

The occurrence of sipunculid larvae in the open sea was first recorded by Häcker (1898) in the volume on the Polychaeta from the 'Plankton Expedition'. Three species were recognized, but it was not possible to determine their adult affinities. Even from the inadequate illustrations it is possible to recognize one of these larval forms (*Baccaria citrinella*) as a particularly common species in the surface waters of the North Atlantic and Mediterranean. However, even when *B. citrinella* is held in the laboratory for 1 year after settlement, it has not been identifiable by specialists of Sipunculida. This is because the tropical sipunculid fauna is still so poorly known and because the intraspecific variability within the sipunculids is not yet well understood. Indeed, *B. citrinella* changes genus during the course of its ontogony!

The very wide capacity for dispersal of pelagosphaera larvae can be illustrated by the geographical distribution of a species found throughout the surface waters of the North Atlantic (Fig. 6) and probably identical to one of several forms termed 'smooth' (Plate 1 *b*) by Jägersten (1963). The distribution of this larva is typical of a large number of other pelagosphaera species.

Zoogeographical evidence confirms the fact that many sipunculids have a high capacity for long-distance dispersal. Of the 21 known tropical West African forms, 30% also occur in the West Indies (Wesenberg-Lund, 1959). Moreover, oceanic islands in the Atlantic are populated principally by eastern and western Atlantic sipunculid species rather than by endemic forms.

Among the coelenterates there are three orders of Zoantharians commonly dispersed by means of teleplanic larvae. These are the Ceriantharia, the Actiniaria and the Zoanthidia.

Cerianthula larvae are frequently encountered throughout the warm-temperate and tropical Atlantic Ocean. A large number have been described from expeditions (namely Van Beneden, 1898, the 'Plankton Expedition'; Carlgren, 1946, the *Michael Sars* Expedition; Leloup, 1964, the *Discovery* Expedition), but the adult affinities of none of these have ever been determined. I have found that some species are easily maintained alive and can be induced to metamorphose, but it is still quite impossible to determine the adult affinities of the post-larvae.

Larvae of Actiniaria also are found in the warm waters of the North Atlantic. R. Bzdyl (unpublished MS) examined larvae belonging to the

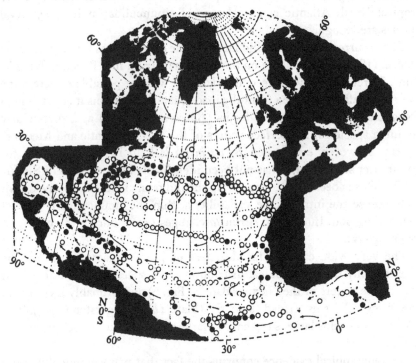

Fig. 6. Distribution of sipunculid larvae (pelagosphaera) from surface plankton collections made in the North Atlantic Ocean. Material taken by the research vessels *Petula* and *Yakutak* have been added. Filled circles indicate positions at which the larval species 'smooth' was encountered. Open circles indicate locations where any of about 15 other species of pelagosphaera larvae were taken. The species 'smooth' was taken throughout the North Atlantic gyre.

genus *Telmatactis*. From material collected by the R.V. *Atlantis II* and the R.V. *Chain*, he found that one larval species belonging to this genus was disseminated throughout all the warm-temperate and tropical waters of the North Atlantic Ocean. Again, although held until after settlement, it was not possible to determine the adult affinities of this larval species even by specialists of this taxa (*vide* R. Bzdyl, unpublished MS).

In the order Zoanthidia, two types of Semper's larvae may be distinguished, the so-called *Zoanthella* and *Zoanthina*. Both have been commonly found in the open waters of the tropical Atlantic, but only the former will be considered here. Several species of the larval genus *Zoanthella* have been described (namely *Z. semperi* and *Z. henseni*, Van Beneden, 1898; *Z. galapagoensis*, Heath, 1906), but most of the differences between these 'species' are trivial and are probably the result of preservation. Conklin

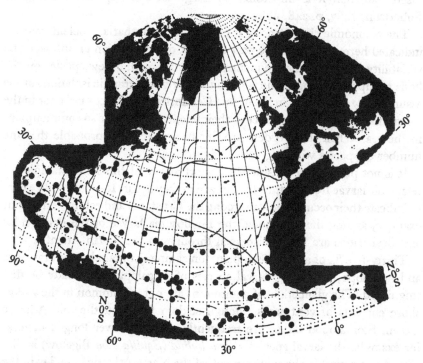

Fig. 7. Distribution in the North Atlantic Ocean of Semper's larvae belonging to *Palythoa* sp. This species appears to be stenothermal and restricted to sea surface temperatures exceeding 21 °C. Summer (northernly) and winter (southernly) 21° isotherms are indicated by continuous lines. Records of specimens north of the winter 21° isotherm were all collected during the summer months.

(1909) concluded that the specimens of *Zoanthella* which he examined from the Bahamas were identical to those described as *Z. henseni* by Van Beneden from the Guinea Current off West Africa (Plate 1c). Menon (1926) held *Z. galapagoensis* collected from near Madras in the laboratory to settlement and concluded that the post-larvae belonged to the genus *Palythoa*.

The Semper's larvae of *Zoanthella* or *Palythoa* are very common throughout the tropical Atlantic (Fig. 7), but their distribution is strictly limited to waters exceeding 21 °C. The specific adult affinities of the larval species *Z. henseni* is still questionable. Pax & Müller (1956) recently examined the

Zoanthidia from tropical West Africa and recorded only one of the total 15 species as amphi-Atlantic in distribution. This was *Palythoa variabilis*, a form also known from Jamaica, Puerto Rico, and Tortugas. Conjecture might lead one to suggest that *Zoanthella henseni* is the larva of *Palythoa variabilis*: however, the taxonomy of the order Zoanthidia is in such a confused state that it is impossible to assign an adult specific name (*vide* Robertson, 1967, p. 248, note 14).

The taxonomic chaos among the orders of Zoantharian coelenterates as indicated here is due in large part to a lack of information on intraspecific variability and upon the frequent practice of erecting new species on the basis of supposed geographic isolation. Clearly, geographic isolation can be assumed *a priori* neither in the orders Ceriantharia and Actiniaria nor in the Zoanthidia since the occurrence of their teleplanic larvae is so commonplace in the sea. When these orders are finally revised, it is probable that the number of species will be greatly reduced.

It is not possible to consider in a short article of this kind examples of teleplanic larvae from all the phyla in which they occur. It should suffice here to indicate their occurrence also among the Crustacea and echinoderms. For example, among the Decapoda, the Phyllosoma larvae of the Palinuridae and Scyllaridae are found dispersed throughout the North Atlantic.

Thirty-five % of all open-sea plankton samples contain brachyuran zoea and megalopa; at least 30 larval species of megalopa can already be distinguished. The glaucothöe of anomurans are not uncommon in the alongshore currents of continents, though seldom found in the mid-Atlantic Ocean. Stomatopod larvae are frequently dispersed over long distances; for example, the larval species *Alima hylina* (*Squilla alba* Bigelow) is dispersed along most of the southern half of the North Atlantic gyre, i.e. in the Canary Current off West Africa, in the South Equatorial Current, off the coast of northeastern South America, in the Caribbean, and along the Gulf Stream to about 38 °N.

Echinoderm larvae are found in 38% of all open ocean samples. Other commonplace forms that I will not discuss here are the tornaria larvae of the Enteropneusta and the Amphioxides larvae of the Cephalochordata.

The relationship between the geographical distribution of teleplanic larvae and the circulation of the North and Equatorial Atlantic Ocean

Two kinds of geographical distribution are distinguishable for the teleplanic larvae in the surface waters of the North Atlantic gyre. The first is that found among eurythermal tropical and warm-temperate forms, and is well represented by a variety of species belonging to the gastropod genera *Cymatium, Tonna, Coralliophila, Pedicularia* and *Thais* and by some members of the family Architectonicidae.

Cymatium parthenopeum is such a eurythermal tropical and warm-temperate species and its larvae as already noted are found throughout the entire North Atlantic gyre (Fig. 3). Theoretically it is possible for the larvae of this species to be carried between any two points along the edge of the gyre, and for adult populations to occur on the continental shelf anywhere near its periphery. The only limitation to the distance through which a particular veliger of *C. parthenopeum* can be transported on the currents of the gyre is the length of its pelagic development, and the only constraints to the occurrence of adult populations in the shelf benthos around the gyre are the physical and biological requirements for survival (e.g. temperature, bottom type, competition, occurrence of prey and predator species, etc.). It may therefore be expected that other factors being favourable, *C. parthenopeum* will be found in the Azores, along the northwestern coast of Africa and the northeastern coast of South America, in the Caribbean, and along the North American continent to approximately Cape Hatteras. This, in fact, describes the geographical distribution of adult populations of *C. pathenopeum* and parallels that of many other species of warm-water eurythermal forms.

The second type of distribution is that found among tropical stenothermal species. It is represented by such forms as the Semper's larva *Zoanthella henseni*, belonging to the coelenterate genus *Palythoa*, and by the larval species *Alima hylina*, belonging to the stomatopod crustacean *Squilla alba*.

The Semper's larvae of *Palythoa* are found only in the southern portion of the North Atlantic gyre (Fig. 7), and consequently their trans-Atlantic dispersal by surface currents appears to be limited to transport from east to west along the North and South Equatorial Current. A mechanism which permits an eastward dispersal across the North Atlantic is the warm Equatorial Undercurrent, with temperatures at its core between 24 and 28 °C (Fig. 2B). Closing net samples have shown that larvae of *Palythoa* and numerous other tropical species are indeed found within the Equatorial Undercurrent, and that dispersal by this means is available to *Zoanthella henseni*. The Equatorial Undercurrent provides a way whereby teleplanic larvae may be dispersed from the South American to the African continent. The reciprocal larval exchange of stenothermal, tropical, amphi-Atlantic species is thereby made possible between the two continents.

There are, in addition to the above two categories, eurythermal species which live throughout an extreme temperature range, cold boreal to tropical. Their larvae have a geographical distribution on the sea surface similar to tropical and warm-temperate eurythermal species and, like them, are also distributed around the entire North Atlantic gyre. Examples of such forms are *Laonice cirrata*, a number of chaetopterid polychaete species, and the veligers of the prosobranch gastropod, *Triphora*.

Current velocities in the North and Equatorial Atlantic Ocean

The rate of larval dispersal is determined by the velocities of the ocean currents. Three sources of data are available, namely (1) Pilot Charts which are based upon information from ships' logs, (2) direct measurements made from ships at anchor or from buoys, and (3) drift-bottle data.

There are two independent sources for velocity data of the North Equatorial Current. Dickson & Evans (1956) regularly made careful, direct current measurements during the cruise of the *Petula* between Cape Verde and the Barbados and found the current velocity to be between 1.0 and 1.2 km/h. Currents shown on the US Pilot Charts of the North Atlantic for January and June show a current velocity of 0.5 knots or about 0.9 km/h. To traverse the distance of 3700 km at these rates would take between 120 and 150 days (17–21 weeks).

Data for the South Equatorial Current is available from drift-bottle data (Guppy, 1917). The velocity may be computed from his data to be 2.3 km/h. At this rate the 4625 km is crossed in 84 days (12 weeks). Pilot Chart data give velocities of from 1.3 to 2.2 km/h, making the 4625 km crossing between the Gulf of Guinea and Brazil in 88–148 days (12–21 weeks). Guppy records one crossing of 60 days or about 9 weeks.

The velocity of the Equatorial Undercurrent has been measured directly from buoys by Stalcup & Metcalf (1966) and Rinkel *et al.* (1966) on either end of the current. These measurements showed velocities of 50–150 cm/s or 1.8–5.4 km/h, but the average velocity at the core of the undercurrent is around 2 km/h (Fig. 1 B). The distance between the Brazilian coast and São Tomé could be crossed in from 36 to 109 days (5–15 weeks).

An average current velocity of the North Atlantic Drift from Pilot Charts is about 0.7 knots or 1.3 km/h. The distance of 4000 km between Cape Hatteras and the Azores could be crossed in about 132 days (19 weeks) at this rate. An integrated value from drift-bottle data is considerably longer (Scheltema, 1966b); the maximum rate is about 0.8 km/h. The shortest time recorded was 120 days (17 weeks), but the modal time was 300 days (43 weeks).

The time required for trans-Atlantic transport can now be summarized:

North Equatorial Current 120–150 days, 17–21 weeks.

South Equatorial Current 60–148 days, 9–21 weeks.

Equatorial Undercurrent 36–109 days, 5–15 weeks.

North Atlantic Drift (Cape Hatteras to Azores) 120–300 days, 17–43 weeks.

The relationship between the duration of larval development and current velocities of the North and Equatorial Atlantic Ocean

The duration of larval development and its relation to current velocity determines the maximum distance which larvae can be dispersed. Mileikovsky (1966) believes that '...the larval drift of all benthic invertebrates in the warmer waters of currents in the tropical regions is more prolonged and lengthy than that of their counterparts in the waters of temperate and high latitude currents'. But it must be admitted that the knowledge of larval development of tropical bottom invertebrates is still very rudimentary, although there are certain families of tropical invertebrates that characteristically do seem to have teleplanic larvae.

Experimental studies show that the duration of larval development is dependent upon such obvious physical and biological conditions as temperature and the availability of the proper food organisms, etc. Perhaps more important is the ability of many larvae to delay settlement until they encounter a favourable environment for post-larval development. This settlement response was first strikingly pointed out by D. P. Wilson for the polychaetes and comprehensively reviewed by him in a long article on the influence of substratum on metamorphosis (Wilson, 1952). Subsequent research has shown examples of delay in settlement in almost every other major invertebrate phylum. It is not possible to discuss in detail here the large amount of recent experimental work on this problem. The present discussion is restricted to some new empirical evidence concerning the duration of pelagic development of teleplanic larvae taken from the open sea.

Cymatium parthenopeum has a very long pelagic development. Larvae captured when already near the stage for settlement have delayed metamorphosis for 138 days (Scheltema, 1966b), long enough to permit the transport in either direction both across the equatorial Atlantic and eastwardly across the North Atlantic Drift under favourable conditions. Most likely, larvae were already several months old when captured. Delay in settlement has been demonstrated in neritic gastropod species (Scheltema, 1956, 1961; Thompson, 1958; Kiseleva, 1966, 1967) and probably it will also be found among teleplanic gastropod veligers when they are studied experimentally.

Many polychaetes have a remarkable capacity for delay in settlement. Recently Wilson (1968) has shown that *Sabellaria alveolata*, known to complete its development in 6 weeks under optimum conditions, can delay settlement up to 8 months. *Chaetopterus variopedatus* has a larval development of over 3 months (Allen & Nelson, 1911), and probably much longer. Its dispersal throughout the North Atlantic is thereby assured. Another species, captured from the plankton and belonging to the genus *Spiochaetopterus*, has been held in my laboratory for over 1 year.

Rice (1967) has shown that the development of the sipunculid *Phascolosoma agassizii*, an amphi-Atlantic species occurring both in tropical West Africa and in the West Indies, has a pelagic planktotrophic development exceeding 7 months. Sipunculid larvae captured in the open sea can be held for long periods. A species belonging to the genus *Aspidosiphon* was held for 133 days before settlement occurred. The median length of delay for 22 individuals was 9 weeks, enough for transport across the equatorial Atlantic in either direction, even though when captured these larvae had essentially completed their morphological development.

The very long periods of larval development (sometimes over 1 year) among the decapod *Palinuridae* and *Scyllaridae* phyllosma larvae is well known (Robertson, 1968).

Finally, some very recent results give direct evidence for a long pelagic larval development among coelenterates. The larvae of a species of Actiniaria belonging to the genus *Telmatactis* that was taken from the plankton of the open sea was held as a pelagic stage for 146 days at 20 °C. Larvae held at 15 °C delayed metamorphosis 163 days (R. Bzdyl, unpublished MS). The Semper's larvae of the zoanthid *Zoanthella henseni* taken from plankton samples in the South Equatorial Current and held in my laboratory remained planktonic for 13 weeks.

Though enough is now known to demonstrate that many species have teleplanic larvae, further studies on the larval development of tropical marine invertebrate species are urgent to an understanding of the systematics and ecology of these forms.

CONCLUSIONS

Larvae from the tropical and warm-temperate invertebrate shelf fauna are widely distributed in the surface waters of the open sea. Present knowledge of their distribution, the length of pelagic larval development and the velocities of ocean currents leaves no doubt that there is a reciprocal exchange of larvae between the eastern and western Atlantic. Such an exchange implies that the larvae act as genetic links between disjunct, geographically isolated populations of benthic marine species.

I am indebted to my research assistants Mr John R. Hall and Mrs Janet Moller for their industry in sorting samples and assistance at sea. Mr William Sargent and Mr Robert Simons also accompanied me at sea. I also wish to acknowledge help from a number of colleagues who have made collections available to me. These include Dr John Allen, Dove Marine Laboratory, University of Newcastle, England; Mr George C. Grant, Narragansett Marine Laboratory, University of Rhode Island; Dr Allen Bé, Lamont Geophysical Observatory, Columbia University; and particularly Dr Robert L. McMaster, also from the University of Rhode Island; and Dr Richard H. Backus and Dr V. T. Bowen of the Woods Hole Oceanographic Institution. I am grateful to the officers and men of the research vessels R.V. *Atlantis II* and R.V. *Chain*. Finally to my wife, Amelie, for assistance in innumerable ways, I am as always indebted. This research was supported by grants from the National Science Foundation.

REFERENCES

ALLEN, E. J. & NELSON, E. W. (1911). On the artificial culture of marine organisms. *J. mar. biol. Ass. U.K.* **8**, 421–74.

BZDYL, R. (1969). Evidence for trans-Atlantic transport of planulae larvae probably belonging to the genus *Telmatactis*. B.A. Thesis, Kalamazoo College, Kalamazoo, Mich., USA, unpublished MS, 32 pp.

CARLGREN, O. (1946). Ceriantharia, Zoantharia, Actiniaria. *Michael Sars* North Atlantic Deep-sea Expedition, 1910 **5**, 3–27.

CHESHER, R. H. (1966). The R.V. *Pillsbury* deep-sea biological expedition to the Gulf of Guinea, 1964–65. 10. Report on the Echinoidea collected by R.V. *Pillsbury* in the Gulf of Guinea. *Stud. trop. Oceanogr. Miami* **4** (pt. 1), 209–23.

CLENCH, W. J. & TURNER, R. D. (1957). The family Cymatiidae in the Western Atlantic. *Johnsonia* **3**, 189–244.

CONKLIN, E. G. (1909). Two peculiar actinian larvae from Tortugas, Florida. *Carnegie Instn Washington Pub.* **103**, 173–86.

DICKSON, C. N. & EVANS, F. (1956). The *Petula's* meteorological logbook. *Mar. Obsr* pp. 215–18.

FAUVEL, P. (1916). Annelides polychètes pelagiques provenant des campagnes de l'*Hirondella* et la *Princesse Alice* (1885–1910). *Camp. Sci. Monaco, Res.* **48**, 1–152.

FUGLISTER, F. C. (1963). Gulf Stream '60'. In *Progress in Oceanography*. **1**, 263–373.

GARDINER, J. S. (1904). Notes and observations on the distribution of the larvae of marine animals. *Ann. Mag. nat. Hist.*, Ser. 7, **14**, 403–10.

GUPPY, H. B. (1917). *Plants, Seeds and Currents in the East Indies and Azores.* London: William Norgate.

HÄCKER, V. (1898). Die pelagischen Polychaeten und Acheaten Larven der Plankton-Expedition. *Ergebn. Plankton Exped.* II, H, d, 50 p.

HANNERZ, L. (1956). Larval development of the Polychaete families Spionidae Sars, Disomidae Nesnil, and Poecilochaetidae, N. Fam. in Gullmar Fjord (Sweden). *Zool. Bidr. Upps.* **31**, 1–204.

HARTMAN, O. (1966). Polychaeta Myzostomidae and Sedentaria of Antarctica. *Antarct. Res. Ser.* 7 (Am. Geophys. Union) 1–158.

HEATH, H. (1906). A new species of Semper's larva from the Galapagos Islands. *Zool. Anz.* **30**, 171–5.

JÄGERSTEN, G. (1963). On the morphology and behavior of pelagosphaera larvae (Sipunculoidea). *Zool. Bidr. Upps.* **36**, 27–35.

KHANAYCHENKO, H. K., KHLYSTOV, N. Z. & ZHIDOV, V. G. (1965). The system of equatorial countercurrents in the Atlantic Ocean. *Okeanologiya* **5**, 222–9. (Translation: *Acad. Sci. USSR Oceanology*, Scripta Technica, pp. 24–32.)

KIRKEGAARD, J. B. (1959). The polychaeta of West Africa. Part I. Sedentary species. *Atlantide Rep.* No. 5, 7–117.

KISELEVA, G. A. (1966). Investigations on the ecology of larvae of certain larger forms of benthic animals of the Black Sea. Abstracts of dissertations for the candidate's degree in Biological Sciences. Odessa, Gosudarstvenuyy Universitet I. I. Mechnikova, Odessa, 20 pp.

KISELEVA, G. A. (1967). The effect of the substrate on settling and metamorphosis of larvae of benthic animals. In *Donnyye Biotsendsy i Biologiya Bentoshykh Organismnov Chernogo Morya* (Bottom Biocenoses and Biology of Benthic Organisms of the Black Sea) pp. 71–84. (Respublikansky Mezhedom – Stvennyy Sbornik, Ser. 'Biologiya Morya' Kiev.)

LELOUP, E. (1964). Larves de Cerianthaires. '*Discovery*' *Rep.* **33**, 251–307.

MANN, C. R. (1967). The termination of the Gulf Stream and the beginning of the North Atlantic Current. *Deep Sea Res.* **14**, 337–59.

MARCHE-MARCHAD, I. (1969). Les Architectonicidae (Gastropoda Prosobranches) de la côte occidentale d'Afrique. *Bull. Inst. fondam. Afr. noire Ser. A* **31**, 461–86.

MENON, K. R. (1926). On the adults of *Zoanthella* and *Zoanthina. Rec. Indian Mus.* **28**, 61–4.

METCALF, W. G., VOORHIS, A. D. & STALCUP, M. C. (1962). The Atlantic Equatorial Undercurrent. *J. geophys. Res.* **67**, 2499–508.

MILEIKOVSKY, S. A. (1960). On the distance of dispersal with marine currents of pelagic larvae of bottom invertebrates on the example of *Limapontia capitata* Müll (Gastropoda opisthobranchia) from the Norwegian and Barents Seas (Russ.) *Dokl. Akad. Nauk SSSR* **135** (4), 965–7.

MILEIKOVSKY, S. A. (1966). Range of dispersion of the pelagic larvae of benthic invertebrates by currents and the migratory role of this dispersion taking *Gastropoda* and *Lamellibranchia* as examples. *Okeanologiya* **6**, 482–92. (Translation: *Acad. Sci. USSR Oceanology, Scripta Technica*, pp. 396–404.)

MILEIKOVSKY, S. A. (1968*a*). Some common features in the drift of pelagic larvae and juvenile stages of bottom invertebrates with marine currents in temperate regions. *Sarsia* **34**, 209–16.

MILEIKOVSKY, S. A. (1968*b*). Distribution of pelagic larvae of bottom invertebrates of the Norwegian and Barents Seas. *Mar. Biol.* **1** (3), 161–7.

PAX, F. & MÜLLER, I. (1956). Zoantharien aus Franzosisch Westafrika. *Bull. Inst. fr. Afr. noire Ser.* A **28**, 418–58.

PETTIBONE, M. (1956). Marine polychaete worms from Labrador. *Proc. US natn. Mus.* **105**, 531–84.

PILOT CHARTS (US Naval Hydrographic Office, 1967).

RICE, M. E. (1967). A comparative study of the development of *Phascolosoma agassizii, Golfingia pugettensis* and *Themiste pyroides* with a discussion of developmental patterns in the Sipunculida. *Ophelia* **4**, 143–71.

RINKEL, M. O., SUND, S. & NEUMAN, G. (1966). The location of the termination area of the equatorial undercurrent in the Gulf of Guinea based on observations during Equalant III. *J. geophys. Res.* **71**, 3893–901.

ROBERTSON, P. B. (1968). The complete larval development of the sand lobster, *Scyllarus americanus* (Smith), (Decapoda, Scyllaridae) in the laboratory, with notes on larvae from the plankton. *Bull. mar. Sci. (US)* **18**, 294–342.

ROBERTSON, R. (1964). Dispersal and wastage of larval *Philippia krebsii* (Gastropoda: Architectonicidae) in the North Atlantic. *Proc. Acad. nat. Sci. Philad.* **116**, 1–27.

ROBERTSON, R. (1967). *Heliacus* (Gastropoda: Architectonicidae) symbiotic with Zoanthiniaria (Coelenterata). *Science, N.Y.* **156**, 246–8.

ROBERTSON, R., SCHELTEMA, R. S. & ADAMS, F. W. (1970). The feeding, larval dispersal and metamorphosis of *Philippia* (Gastropoda: Architectonicidae). *Pacif. Sci.* **24**, 55–65.

SCHELTEMA, R. S. (1956). The effect of substrate on the length of planktonic existence of *Nassarius obsoletus*. *Biol. Bull. mar. biol. Lab., Woods Hole* **111**, 312.

SCHELTEMA, R. S. (1961). Metamorphosis of the veliger larvae of *Nassarius obsoletus* (Gastropoda) in response to bottom sediment. *Biol. Bull. mar. biol. Lab., Woods Hole* **120**, 92–109.

SCHELTEMA, R. S. (1963). Larvae in the open sea. *Oceanus* **9** (3), 2–9.

SCHELTEMA, R. S. (1964). Origin and dispersal of invertebrate larvae in the North Atlantic. *Am. Zool.* **4** (3), 299–300.

SCHELTEMA, R. S. & HALL, J. R. (1965). Trans-oceanic transport of sipunculid larvae belonging to the genus *Phascolosoma*. *Am. Zool.* **5**, 216.

SCHELTEMA, R. S. (1966*a*). Trans-Atlantic dispersal of veliger larvae from shoal-water benthic mollusca. *Second Int. oceanogr. Congr. (Moscow) Abstr.* No. 375, 320.

SCHELTEMA, R. S. (1966*b*). Evidence for trans-Atlantic transport of gastropod larvae belonging to the genus *Cymatium*. *Deep Sea Res.* **13**, 83–95.

SCHELTEMA, R. S. (1968). Dispersal of larvae by equatorial ocean currents and its importance to the zoogeography of shoal-water tropical species. *Nature, Lond.* **217**, 1159–62.

SCHELTEMA, R. S. (In press). Dispersal of phytoplanktotrophic shipworm larvae over long distances by ocean currents (Bivalvia: Teredinidae). *Mar. Biol.*

SCHUMACHER, A. (1940). Monatskarten der Oberflächenströmungen im Nord Atlantischen Ozean (5 °S bis 50 °N). *Annln. Hydrogr. Berl.* **68**, 109–23.

SCHUMACHER, A. (1943). Monatskarten der Oberflächenströmungen im äquatorialen und Südlichen Atlantischen Ozean. *Annln. Hydrogr. Berl.* **71**, 209–19.

STALCUP, M. C. & METCALF, W. G. (1966). Direct measurements of the Atlantic Equatorial Undercurrent. *J. mar. Res.* **24**, 44–55.

STOCK, T. (1950). Die Tiefenverhältnisse des Europäischen Nordmeers. *Dt. hydrogr. Z.* **3** (1/2), 93–100.

STURM, M. & VOIGT, K. (1966). Observations on the structure of the equatorial undercurrent in the Gulf of Guinea in 1964. *J. geophys. Res.* **71**, 3105–8.

THOMPSON, T. E. (1958). The natural history, embryology, larval biology and post-larval development of *Adalaria proxima* (Alder and Hancock) (Gastropoda Opisthobranchia). *Phil. Trans. R. Soc.* B **242**, 1–58.

THORSON, G. (1946). Reproduction and larval development of Danish marine bottom invertebrates. *Meddr. Kommn. Danms Fisk.-og Havunders.*, Ser. *Plankton* **4** (1), 1–523.

THORSON, G. (1961). Length of pelagic life in marine invertebrates as related to larval transport by ocean currents. In *Oceanography* Pub. 67, A.A.A.S. (ed. M. Sears), pp. 455–74.

USHAKOV, P. V. (1955). Polychaeta of the far eastern seas of the USSR. Keys to the fauna of the USSR. *Zool. Instn, Akad. Nauk. SSSR.* No. 56, 419 pp. (Translated from the Russian, Office of Technical Services, US Department of Commerce.)

VAN BENEDEN, E. (1898). Les anthozoaires de la 'Plankton-Expedition'. *Ergebn. Plankton-Exped.* II K, e, 222 pp.

VOIGT, K. (1961). Äquatoriale Unterströmung auch im Atlantik (Ergebnisse von Strömungsmessungen auf einer Atlantischen Ankerstation der 'Michail Lomonossov' an Äquator im Mai 1959). *Beitr. Meersk. dt. Akad. Wiss. Berl.* **1**, 56–60.

VOORHIS, A. D. (1961). Evidence of an eastward equatorial undercurrent in the Atlantic from measurements of current shear. *Nature, Lond.* **191**, 157–8.

WALLACE, A. R. (1876). The geographical distribution of animals, vol. I. New York: Harpers.

WILSON, D. P. (1952). The influence of the substratum on the metamorphosis of the larvae of marine animals, especially the larvae of *Ophelia bicornis* Savigny. *Annls. Inst. oceanogr., Monaco* **27**, 49–156.

WILSON, D. P. (1968). Some aspects of the development of eggs and larvae of *Sabellaria alveolata* (L.). *J. mar. biol. Ass. U.K.* **48**, 367–86.

WESENBERG-LUND, E. (1951). The Zoology of East Greenland Polychaeta. *Meddr. Grønland* **122**, 1–169.

WESENBERG-LUND, E. (1959). Sipunculoidea and Echiuroidea from tropical West Africa. *Atlantide Rep.* **5**, 177–210.

WORTHINGTON, L. V. (1962). Evidence for a two gyre circulation system in the North Atlantic. *Deep Sea Res.* **9**, 51–67.

TRANSPORT OF BIVALVE LARVAE
IN A TIDAL ESTUARY*

L. WOOD† AND W. J. HARGIS, JR.

*Virginia Institute of Marine Science, Gloucester Point,
Virginia 23062 USA*

INTRODUCTION

It has long been recognized that oyster larvae are retained within the estuary of spawning and may in fact settle in areas upstream from the major spawning populations. This is not to say that no larvae are lost from the estuary by way of its seaward-moving upper waters; indeed, the net loss to the sea may be enormous. But in many estuaries of the world, and in particular that of the James River (Virginia), sedentary molluscan populations replace themselves annually despite – as is the case of *Crassostrea virginica* Gmelin – intensive harvesting. Further, the existence in the James estuary of a shell stratum many metres thick suggests that such periodic replenishment has been going on for millenia.

These observations lead finally to the logical conclusion that estuaries with longstanding and relatively high rates of bivalve production became so because paths of geologic, hydrographic, behavioural and developmental evolution intersected to produce mechanisms of larval retention.

The point at issue is not whether such retention occurs, but whether evolved patterns of larval behaviour contribute significantly to the process. As Bousfield (1955) said in his study of barnacle larval transport, the question is not one of direct contribution to horizontal movement. In the James River, horizontal tidal currents may exceed 80 cm/s, nearly two orders of magnitude faster than the most rapid swimming speed recorded for *C. virginica* larvae. In the vertical axis, on the other hand, water movements are slow enough (0.01 cm/s estimated by Pritchard, 1953) so that an oyster larva moving at its fastest rate – about 1 cm/s (author's observations) – could negotiate a 10 m water column in about a quarter of an hour.

Even such relatively rapid movements could contribute little to the process of larval retention unless they are selective: simply stated, the larvae must spend at least as much time in the lower, landward-moving water

* Contribution Number 328 from the Virginia Institute of Marine Science.
† Present address: Department of Zoology, Spaulding Building, University of New Hampshire, Durham, N.H. 03824 USA.

strata as in the upper, seaward-moving strata if they are to remain in the estuary. The difficulty with this notion of selective residence is chiefly that no sensory system has yet been described which could detect the *direction* of water movement when the system is itself a part of the moving mass.

An alternative kind of vertical selection is possible, as first suggested by J. Nelson (1912) and supported by T. Nelson (1931), Carriker (1951), Kunkle (1957) and Haskin (1964). According to these investigators, oyster larvae swim with the flood and rest on the bottom during ebb. The question of a sensing mechanism was thought by T. Nelson to have been resolved by the probability that larvae could detect the increase in salinity which would accompany a flooding tide. This hypothesis was supported by experimental evidence from the laboratory of Haskin (1964), but was questioned by Korringa (1952) and Verwey (1966).

Other mechanisms whereby larvae might determine their position in a vertical water column have been suggested with varying experimental support, but such considerations should follow, rather than anticipate, the presentation of unequivocal evidence from nature that bivalve larvae swim selectively. Otherwise the absence of really firm supporting evidence renders academic any discussion of a mechanism.

Hence the primary purpose of this paper will be to describe an intensive study of the distribution through time of bivalve larvae in the lower James estuary. We offer it in the hope that it will lay to rest at least the first part of the controversy, by showing that bivalve larval transport is not entirely passive, and in the further hope that studies of the physiological and sensory bases of selective swimming behaviour will now be further stimulated.

The estuary

The James River (Fig. 1) is the third largest and the southernmost of the major tributaries entering Chesapeake Bay. Throughout most of its winding, 156 km length it is narrow and features a 10 m channel that permits ships to reach the inland port of Richmond. The river broadens below Jamestown Island to form an estuary with an average depth of 3–4 m. The shipping channel which runs through the estuarine portion has an average depth of 10–12 m.

For many years, the oyster seed areas of this estuary (marked in black in Fig. 1) have produced a major portion of the young oysters that are transplanted annually to growing areas along the US East Coast.

During the 3 years prior to the time of sampling, a drought in the James River drainage area had reduced the freshwater discharge at Richmond

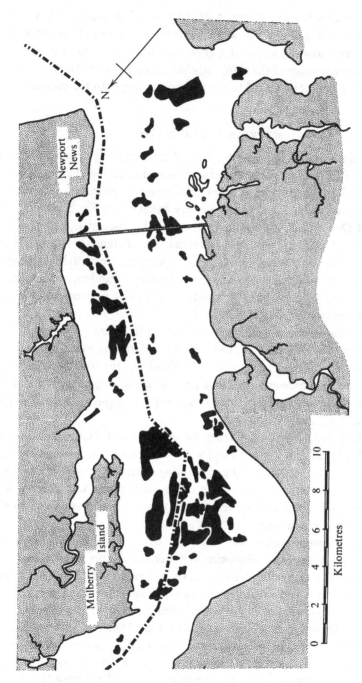

Fig. 1. Map of the James River showing areas occupied by oyster seed (in black).

from the seasonal norm of about 80 m³/s to a low during the sampling period of about 34 m³/s. An accompanying reduction of oyster spatfall created grave concern about the effects of river flow and the effects of a proposed deepening of the shipping channel upon oyster larval transport.

It was this concern that led to our decision to carry out the present study.

MATERIALS AND METHODS

Detailed descriptions appear elsewhere of the sampling operations (Wood & Hargis, In Press) and methods of sorting and counting plankton samples (Norcross, Wood & Hargis, In Press). Suffice it here to state that five vessels were deployed in the lower James estuary as shown in Fig. 2. Stations CD (Channel Down), MC (Mid-Channel) and CU (Channel Up) followed the gently curved axis of the shipping channel. The distance between CU and MC was about 2.5 km, while MC and CD were approximately 2 km apart. From the decks of the vessels were suspended standard assemblies of plankton pumps (c. 20 l/min each) whose depths were constant at 1, 4, 7 and 10 m below the surface. CU and CD each had one pump assembly, while MC had two, one at the bow and one at the stern (20 m apart). The vessels were double anchored, fore and aft, to prevent swinging.

Stations SS (Shoal South) and SN (Shoal North) were located in the areas known as Naseway Shoal and Brown Shoal, respectively. Each was equipped with a single assembly of two pumps, at depths of 1 and 3 m (SS) and 0.5 and 2.5 m (SN).

Sampling was carried out during the latter part of August and the first part of September of 1965. Samples were taken on the hour every hour throughout the day and night. Sampling duration was exactly 900 s. Immediately after the samples were taken, each pump was calibrated by determining the number of seconds to the overflow point of a metal bucket whose volume was known.

Ideally, we would have liked to have begun sampling when mature oyster larvae were most abundant. But because of the sheer size of the exercise, it was logistically necessary to set a beginning date well in advance of the operation. On the basis of plankton records from previous years and an intensive larval monitoring programme carried out during July and early August (Andrews, personal communication), the start was set at 30 August. As this date drew nearer, we became increasingly aware that the larval season was delayed and that there might not be an abundance of larvae in time, but our commitment to the starting date could not then be altered.

The actual number of days sampled during the latter part of August and

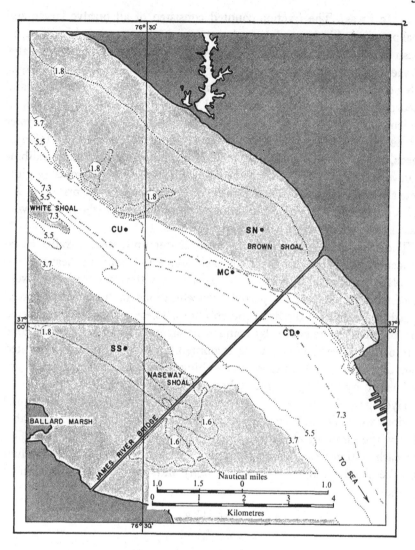

Fig. 2. Sampling stations used in survey (see text).

the first part of September 1965 was seven, but for various reasons the samples for only one 24 h period (1 September 1965) will be presented here.

Synoptic information was obtained of current speed and direction, salinity, temperature, oxygen concentration, light, wind and other meteorological variables. Dye patches released at the surface were followed visually and by fluormetric analysis of water samples. The procedure for sorting and counting plankton samples has been described elsewhere (Norcross *et al.* In Press). The fraction counted consisted of all bivalve larvae and coal particles which passed through a 210 μm mesh but were held on a 44 μm mesh screen. This included straight-hinge and umbo stages of bivalve larvae, but did not include mature stages, which were found in only negligible concentrations. The larvae of *Crassostrea virginica* were identified and recorded separately from those of all other bivalve mollusc species, of which approximately 16 were noted. Counts of three categories (oyster larvae, other bivalve larvae, and coal particles) were punched on Hollerith cards, and in the VIMS Computer Center were factored for pump calibration volumes. These volume-corrected counts were checked against original data sheets for clerical errors which were adjusted in a second print-out. All subsequent analyses, including graphic presentations in this paper, were based on these corrected computer print-outs.

The correlation coefficients between oyster and other bivalve larval concentrations were calculated for each pump and hour by means of the standard regression equation $Y = A + Bx$. Net transport estimates were obtained by first plotting particle concentrations multiplied by current speeds separately for the flood and ebb tides. These graphs were integrated for 'area under the curve' by manual planimetry. Differences in the total numbers of larvae and coal were accounted for by reducing the integrated data to a ratio ($\times$ 100); hence the resulting net transport estimates are not influenced by the large disparity between coal and larval totals.

Hydrographic profiles (Fig. 3) of the James River Estuary at channel stations were calculated on the basis of all complete tidal cycles available during the sampling operation (usually seven). Similarly, net flow profiles for each station were plotted from algebraic sums of planimetered values for the flood (assigned a positive sign) and ebb (negative sign) tidal currents.

RESULTS

Circulation

As a basis for understanding circulation patterns in the lower James, summary profiles of salinity, temperature, density and net flow for both the channel and shoal stations are presented in Fig. 3. It can be seen that while there is a general gradient of salinity in both the vertical and axial planes along the channel, the absolute differences are slight, falling within a range between slightly more than 20‰ at CU (1 m), to a little less than 22.5‰ at CD (10 m). Temperature profiles for the three stations indicate that the upriver stations are slightly cooler than the downriver stations, but that in each case the water column is nearly isothermal. Nearly homogeneous distribution of densities is confirmed by the calculations of σ_t shown in the third figure.

Calculations of net flow at the channel stations suggest that the 'level of no net motion' bisects the vertical profiles of the upstream (CU) and downstream (CD) stations, but not that of the midchannel (MC) station.

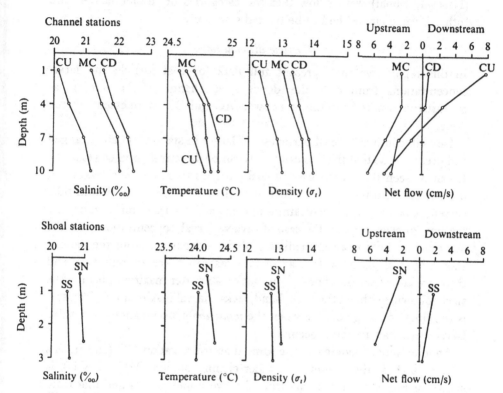

Fig. 3. Salinity, temperature, density and net flow profiles for stations shown in Fig. 2.

Vertical profiles of the three variables for the shoal stations indicate that the northeast shoal (SN) water is saltier, warmer and denser than that of the southwest shoal (SS) station. Further, flow calculations indicate that the net movement of the water at SN is upstream at both surface and bottom, and downstream at both the surface and bottom of SS. These shoal observations, taken together with the fact that the 'level of no net motion' does not intersect the MC vertical profile, suggests that this part of the lower James Estuary has a counter-clockwise circulation pattern, and, further, that station MC was inadvertently located slightly too far to the northeast, toward the shoal.

Pooling larval categories

Regression analysis of the distribution patterns of the two classes of bivalve larvae in the channel stations showed that the two were sufficiently similar to permit the data to be pooled. However, the correlation for station SS (Naseway Shoal) was so low that the categories of 'oyster larvae' and 'other bivalve larvae' had to be treated separately.

Particle distribution

In this report, we shall present full data for coal particle and larval concentrations from only the down river channel (CD) station, as quantitative results from the other two, MC and CU, were essentially the same.

Isopleth contours of coal particles and larvae in station CD are arranged and presented so that the two categories can be directly compared (Plate 1). It can be seen that in the case of coal particles there are *four* maxima of particle concentration; these occur at or near the time of maximum tidal current. Conversely, concentrations of coal particles are minimal at or near the time of slack tide. In the case of bivalve larval concentrations, on the other hand, there is an immediately obvious departure from the pattern shown by coal particles. Instead of four maxima there are only two, and these are at or near the times of the slack water after maximum flood. The slack following ebb, on the other hand, lacks a larval maximum: in fact, this is the time and stage of tide when the *minimum* concentrations of bivalve larvae (less than 10/100 l) occur.

Another way of looking at the same data from station CD (and in this case we include the results from other channel stations MC and CU), is shown in Fig. 4. Here the data from the two upper pumps and that from the two lower pumps are averaged. In each block of the figure the upper graph represents the speed and general direction of tidal current, the graph below it represents salinity, and the lowermost graph depicts the changing concentrations of total bivalve larvae (full lines) and coal particles (dotted lines). In most, but not all, of the six cases, fluctuations in concentrations of coal particles coincide with variations in tidal current speed, while those of total bivalve larvae coincide with increases in salinity. The exceptions to this statement are found in the results from station CU. Here, in the upper part of the water column the early morning maxima of both coal particles and larvae nearly coincided; in the lower part of the water column the temporal relationship was reversed. In both cases, however, the rise in larval concentrations appears to coincide with an increase in salinity. At this

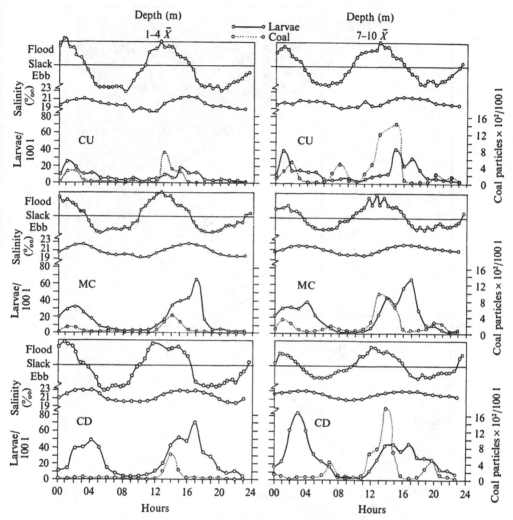

Fig. 4. Hourly measurements of flow (uppermost curve), salinity (middle curve) and density of larvae and particles (two lowest curves) averaged for each of the channel stations over depths ($\bar{X}$) of 1–4 m (left) and 7–10 m (right).

upriver station the salinity variations are not as definite as they appear to be in the two stations farther down the river.

In Fig. 5, we present graphically our calculations of per cent net transport of bivalve larvae. At all but two of the 12 channel sampling points, a strong net transport upriver is indicated. The two exceptions are the surface point at station CU, which shows a slight but statistically significant downriver transport, and the 7 m point at station CD, where the net trans-

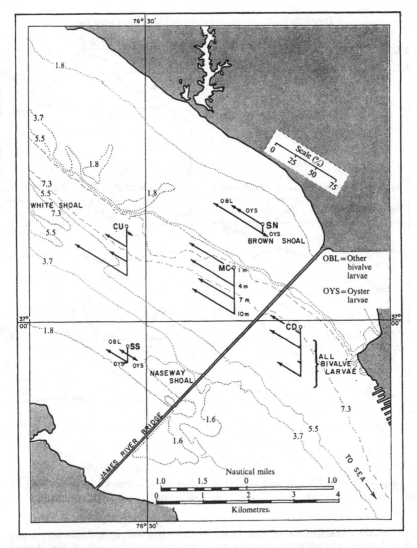

Fig. 5. Calculated transport of oyster and other bivalve larvae at each station. The arrows represent as percentage of total larvae the net transport in the direction shown. At the channel stations the distribution of oyster and other bivalve larvae were not significantly different, and so have been pooled.

port is upriver but is not statistically significant. In any event, the general direction of transport in the channel is away from the sea.

On the shoals the situation is more complex. On the northeast side of the river (station SN), the net transport of both categories of larvae is upriver at the surface, while on the bottom it is strongly upriver in the case of 'other bivalve larvae' but slightly downriver in the case of oysters. This apparent seaward transport of oyster larvae on Brown Shoal is barely significant. At station SS on the southwest side of the river, despite the fact that net tidal flows at both the surface and bottom are seaward, there is a strong indication of net transport in the opposite, upriver, direction in the case of both surface and bottom populations of 'other bivalve larvae'. For oyster larvae, the calculated per cent net transport is downriver at the surface and upriver on the bottom, though the latter is of a greater magnitude than the former.

In general summary of these results, we note that passive coal particles and bivalve larvae differed markedly in their patterns of distribution in time and space, despite the fact that they were of approximately the same density (M. Nichols, personal communication) and fell within the same size range (44–210 μm). Further, larval maxima coincided in most cases with the salinity increase that accompanies flood tide, while coal maxima usually coincided with current speed maxima, regardless of the direction taken by the current. Finally, calculations of per cent net transport of bivalve larvae generally confirm our hydrographic observations and indicate that larvae in the deeper channel and northeast shoal waters are carried upriver, while those in southwest shoal waters are transported seaward.

DISCUSSION

Before proceeding to a summary of our interpretations of these results, it is appropriate at this point to mention the limitations that must be placed on the data.

Initially, we attempted to identify all bivalve larvae present in the samples. While the larvae of *C. virginica* and a few others could be identified reliably, the notorious similarity of bivalve molluscan larvae prevented identification of the other species to the degree of statistical reliability required. Therefore, we decided to train the counting staff to identify only *C. virginica*, and to enumerate all the other bivalve species together. By so doing, we were able to train counters to a criterion of less than 10 % error in tests which involved samples of known species composition and number (Norcross *et al.* In Press). Although we were aware of the loss of resolution consequent to this decision, and were further cautioned on the point by Bousfield (personal communication), subsequent events and our analysis of the data have convinced us that our decision was correct.

Station locations were selected on the basis of our previous knowledge of current speeds in, and the geometry of, the lower James River Basin, and the locations of shoals containing large populations of oysters. In retrospect, it would have been desirable to have added stations both upriver and downriver in the channel. But this increase in information would necessarily have been purchased at the high cost of many additional samples, which, because of our insistence upon the highest accuracy of enumeration obtainable, would have delayed longer the reporting of these results. The 24 h portion presented here required nearly a year of counting time, *after* we had arrived at a procedure and tested it to our satisfaction. The other samples obtained during the other 4 days of the operation were counted by less accurate methods. While the results could not be used for the more precise analysis we have attempted in this paper, they presented the same general picture.

Temporal resolution in this study would have been improved if we had had high-capacity pumps which could have been operated more briefly and more often. If our 300 l samples could have been taken in 5 min instead of 15, it would have been possible to take samples at 30 min intervals, with a consequent improvement in resolution.

While the current meters used during the investigation were the best we could obtain at the time, their directional accuracy ($\pm 17°$) was not sufficient to enable us to carry out one of the analyses originally intended. Instead, we were forced to operate on the assumption that those tidal currents faster than 20 cm/s were moving up and down the river parallel to the axis of the channel.

Finally, there is the general qualification that, because of its very nature, a study of this kind will yield only indirect evidence. Considering the extremely large size of natural populations, it is not practicable to tag and release larvae, and the small size of individuals precludes any possibility of direct observation in nature.

With the above qualifications in mind, we can advance the following interpretations of the results.

The coal particles found in nearly all of our field samples presumably originated at the coal-loading docks on Newport News Point southeast of Station CD (see Fig. 2). Their transport upriver undoubtedly results from their generally higher concentrations near the bottom; in short, they are passively carried back and forth with the tide, their concentrations correlated with the turbulence resulting from increased current speeds, as suggested for oyster larvae by Pritchard (1953). Bivalve larvae, on the other hand, show a quite different temporal pattern in their concentration maxima: First, the concentration maxima of coal particles associated with the ebb tide are entirely lacking in the case of the larvae. Secondly, the larval peaks generally lag behind the flood-tide coal particle peaks and occur more nearly at the time of the high slack water following flood. This marked difference between the behaviour of coal particles (which obviously cannot swim), and that of bivalve larvae which are known to have the potential to control their vertical distribution, provides firm evidence that some kind of selective behaviour must have occurred.

If only one field station had been occupied at a time, it would be possible to interpret these results in a different way. While such a coincidence would probably have a fairly low coefficient of probability, it is possible that a discrete cloud, or several such clouds, of bivalve larvae could have passed by the plankton pumps at the same stage of two successive tidal cycles. But that such a coincidence could have occurred at three stations simultaneously, in synchrony with the tide, would be so unlikely that the idea must be rejected.

In summary, we conclude that the bivalve larvae sampled during this investigation were not being transported passively, but by a process of selective swimming which contributed actively to their upriver movement on the northeast shoal and in the main channel. Furthermore, the data suggests that this behaviour is correlated not with an increasing current speed but with the increases in salinity that accompany a flood tide. The concept of selective swimming in synchrony with tidal cycles was first proposed by J. Nelson (1912) and later confirmed and extended by T. Nelson (1931, 1954). In its original form, the hypothesis stated that oyster larvae rise from the bottom during flood tide, stimulated either by current or by an increase in salinity or both. A rather severe modification of the Nelsons' hypothesis was provided by Prytherch (1928), who reported evidence that oyster larvae swam up from the bottom *only* at times of slack water or very slow tidal currents. His data resembles our own in basic outline; had he been able to occupy several stations simultaneously, and to measure current speed as a basis for net transport calculations, Prytherch might have been able to draw a conclusion similar to the one presented here.

Carriker (1951) found his data to be in agreement with the Nelsons' hypothesis. He added a refinement concerning the distribution of various larval stages in the water column, disclosing that while the younger stages had a fairly uniform vertical disposition, the older stages tended to remain near the bottom. In fact he found that large numbers of mature and eyed larvae were actually on the bottom during ebb tide. This is particularly noteworthy because it is perhaps the only source of evidence from a field investigation that bivalve larvae spend a specific part of the tidal cycle resting on the bottom. In the present study, no bottom samples were taken.

Manning & Whaley (1954) proposed an upriver transport and retention system which was relatively free of behavioural considerations, depending as it did on prevailing southerly winds. Results of the extensive sampling for oyster larvae conducted by Kunkle (1957) were essentially in accord with those of Carriker (1951).

The earliest experiments of which we are aware came from the laboratory of T. Nelson (1931), and showed that oyster larvae in a finger bowl could be stimulated to swim up from the bottom either by a jet of current from a pipette or by a similar jet of higher salinity sea water. Many years later, Haskin performed similar experiments under more rigorous conditions, and reported (1964) that results from both field and laboratory studies confirmed the relationship between increases in swimming activity and increases in salinity.

Bousfield (1955), in his extraordinary study of the transport of barnacle larvae in the Miramichi estuary of Canada, concluded that the early stage larvae of *C. virginica*, like those of *Balanus improvisus*, were near the surface, while the late stages were near the bottom. He concluded that the oyster larvae were transported landward by residual currents along the bottom. He observed further that adult oysters were found in upriver areas where salinities were probably not sufficient to permit the oysters to reproduce, and concluded that this upriver population was maintained by mature oyster larvae drifting in the bottom waters. In most respects, Bousfield's findings appeared to agree with those of Carriker (1951), but Bousfield did not bring forth any data to suggest that oyster larvae were swimming selectively.

Of the several reviews which have touched upon the subject of selective swimming in bivalve larvae, in only two have the authors indicated a general disinclination to accept the evidence put forth in favour of the Nelsons' hypothesis. The first of these was by Korringa (1952), who thought that such a complex mechanism seemed unnecessary, as the spread of oyster larvae upriver could be explained as easily in terms of passive transport. There can be no substantial argument with this point of view, and the present study offers no evidence that selective swimming on the part of oyster larvae is necessary. Indeed, the fact that coal particles are carried up the James River suggests that passive transport alone might be sufficient in this specific estuary. But whether or not an adaptive value has been demonstrated for this behaviour, our data shows that it exists.

In the other review paper, Verwey (1966) simply expressed the opinion that the evidence put forward had not convinced him that the hypothesis was sound; but he gave no details. His specific criticism of the field data given by Haskin (1964), to wit, that the increase of larvae at high water might be due to their proximity to the center of production, does not apply to the design of the present investigation, simply because three stations, 2 and 2.5 km apart, respectively, were occupied simultaneously.

A major task lies ahead: the experimental analysis of physiological mechanisms that may trigger selective swimming in bivalve larvae. Many sensory modalities have been proposed by various authors, but evidence of a clear-cut mechanism which can be experimentally demonstrated and confirmed has yet to appear.

A force of some 50 men and women, all members of the staff of the Virginia Institute of Marine Science in the summer of 1965, contributed in many ways and in various degrees to this work. While it is not possible to mention all of them, we would like to express our appreciation to a few whose contributions were outstanding. Dr J. D. Andrews gave freely of his time and knowledge in monitoring the James River bivalve populations both during the summer preceding and at the time of sampling operations. Mr Paul Chanley helped with identification of larval species. The people chiefly involved with the development and execution of larval counting procedures were Mrs Martha Ebel, Mrs Lynne Davis and Miss Angela Bond. Mr John Norcross contributed substantially to the solution of statistical problems, and Miss Janet Olmon handled the necessary chore of tabulation and calculations. Mrs Jane Davis was responsible for the design and execution of the graphic presentations. To these and the many others who co-operated go our thanks.

REFERENCES

BOUSFIELD, E. L. (1955). Ecological control of the occurrence of barnacles in the Miramichi estuary. *Bull. natn. Mus. Can. Biol. Ser.* 46, **137**, 1–65.

CARRIKER, M. R. (1951). Ecological observations on the distribution of oyster larvae in New Jersey estuaries. *Ecol. Monogr.* **21**, 19–38.

HASKIN, H. H. (1964). The distribution of oyster larvae. In *Symp. Exp. Marine Ecol.* Occasional Pub. 2, pp. 76–80. Naragansett Marine Laboratory.

KORRINGA, P. (1952). Recent advances in oyster biology. *Q. Rev. Biol.* **27**, 274–308, 339–65.

KUNKLE, D. E. (1957). The vertical distribution of oyster larvae in Delaware Bay. *Proc. natn. Shellfish Assoc.* **48**, 90–1.

MANNING, J. H. & WHALEY, H. H. (1954). Distribution of oyster larvae and spat in relation to some environmental factors in a tidal estuary. *Proc. natn. Shellfish Assoc.* **48**, 56–65.

NELSON, J. (1912). Report of the Biological Department of the New Jersey Agricultural Experiment Station for the year 1911.

NELSON, T. C. (1931). Annual Report of the Department of Biology, 1 July 1929 to 30 June 1930. N.J. Agricultural Experiment Station.

NELSON, T. C. (1954). Observations of the behavior and distribution of oyster larvae. *Proc. natn. Shellfish Assoc.* **45**, 23–8.

NORCROSS, J., WOOD, L. & HARGIS, W. J., JR. (In Press). Statistical evaluation of two methods of sorting and counting planktonic bivalve larvae. In *The James: Portrait of an Estuary* (ed. W. J. Hargis, Jr.). Gloucester Point: Va. Inst. Marine Sci.

PRITCHARD, D. W. (1953). Distribution of oyster larvae in relation to hydrographic conditions. *Proc. Gulf Caribb. Fish Inst.* 1952, 123–32.

PRYTHERCH, H. F. (1928). Investigation of the physical conditions controlling spawning of oysters and the occurrence, distribution and setting of oyster larvae in Milford Harbor, Conn. *Bull. Bur. Fish., Wash.* **54**, 429–503.

VERWEY, J. (1966). The role of some external factors in the vertical migration of marine animals. *Neth. J. Sea Res.* **3**, 245–66.

WOOD, L. & HARGIS, W. J., JR. (In Press). Transport and retention of bivalve larvae. In *The James: Portrait of an Estuary* (ed. W. J. Hargis, Jr.). Gloucester Point: Va. Inst. Marine Sci.

OBSERVATIONS ON THE OCCURRENCE OF PLANKTONIC LARVAE OF SEVERAL BIVALVES IN THE NORTHERN ADRIATIC SEA

MIRJANA HRS-BRENKO

Centar za istraživanje mora 'Ruder Bošković', Rovinj, Yugoslavia

INTRODUCTION

Comprehensive investigations of bottom communities in the Adriatic Sea were initiated many years ago (Vatova, 1928, 1935, 1949; Gamulin-Brida, 1967; Gamulin-Brida, Požar & Zavodnik, 1968). Recently these observations were extended to the study of bivalve populations. Thorson (1946) indicated that complementary research is necessary to reach a better understanding of the ecology of the sea bottom. He placed prime emphasis on the study of life histories of species that belong to taxonomic groups which have a wide geographic distribution. He stated that detailed knowledge of the biology of each single species of invertebrate living in a community must be known before the relation between a species and the biocenosis can be understood. In particular, one must know their requirements regarding food, temperature, salinity, substratum, surrounding animals, and, above all, their breeding habits and mode of larval development.

The present study was initiated in order to ascertain the breeding time for some of the more important representative bivalve species in bottom communities of the northern Adriatic Sea. Plankton samples were, therefore, collected over a 2 year period for determination of the time of appearance, duration of planktonic life and disappearance of the larvae.

METHODS

Plankton samples were collected from several stations in the northern Adriatic Sea by vertical hauls. Samples were taken approximately every 10 days at a station 1 mile offshore from Banjole Island in 1967 and 1968. In 1968, monthly samples were taken in addition at two offshore stations which were 11 and 20 miles west of Rovinj. Some observations were also made of the plankton in Pula harbour in 1968 and in the Kvarnerić area in August 1968 (Fig. 1).

The bivalve larvae from each sample were counted and the more advanced larvae were measured; their hinge structures were studied by the method previously described by Rees (1950). The larvae were identified by their shape, hinge structure and dimensions (Odhner, 1914; Lebour, 1938; Jørgensen, 1946; Rees, 1950; Zahvatkina, 1959; Loosanoff, Davis & Chanley, 1966; Poggiani, 1968).

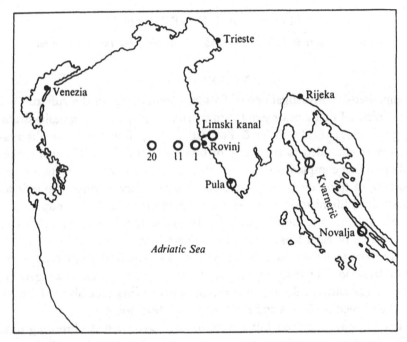

Fig. 1. Distribution of the sample stations in the northern Adriatic Sea, indicated by rings. Those numbered are: 1 mile from Banjole Island and 11 and 20 miles offshore from Rovinj.

The data presented are the results of 2 years of field observations only. To provide corroborative evidence as to the breeding season of the various important bivalves in this area it will be necessary to study gonad changes throughout the year and to rear the larvae of known species under laboratory conditions.

RESULTS AND DISCUSSION

Bivalve larvae were present in the Adriatic Sea throughout the year as shown by samples taken near Banjole Island in 1967 and 1968 (Fig. 2). Bivalve larvae began to increase in late February 1967, but not until mid-March in 1968. A comparatively high variable number of larvae were present in the plankton samples throughout the spring and summer months of both years, followed by a striking decrease in numbers during September. Very few larvae were found in the October and November samples. Thorson (1946) reported an abundance of larvae in the Oresund in October, and Rees (1951) reported similar findings in the North Sea. Thus it appears that

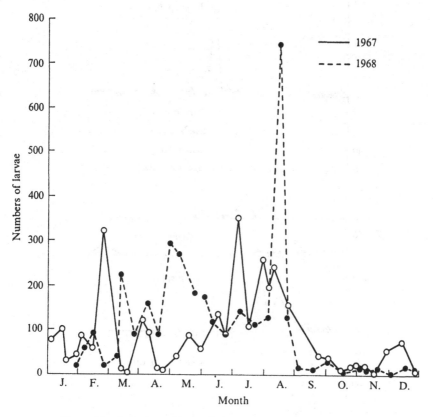

Fig. 2. Total number of larvae from each sample at the station off Banjole Island throughout 1967 and 1968.

many bivalves in the Adriatic Sea have earlier spawning seasons or have less intensive spawning than have bivalve in more northern waters.

The early increase in the number of bivalve in late February 1967 and mid-March 1968 was due to the appearance of many *Chlamys* larvae in the samples (Fig. 3a). It was not possible to identify the *Chlamys* larvae to species but because of the abundance of adult *C. opercularis* L. in the sample areas, especially at the station 11 miles offshore, it was assumed that

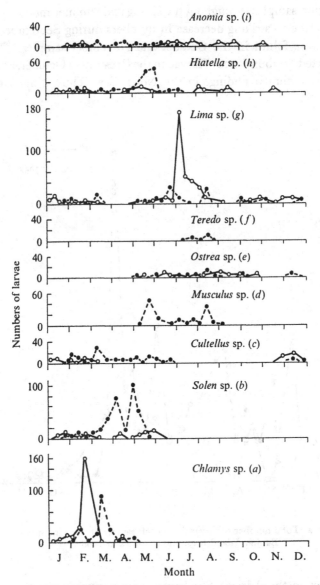

Fig. 3. Frequency of various bivalve larvae in the plankton at the
station off Banjole Island in 1967 (—) and 1968 (---).

the majority were *C. opercularis*. According to Lebour (1938), near Plymouth *C. opercularis* spawns from January to June, when the water temperature is below 11 °C, which is similar to the spawning temperature of *Chlamys* in the Adriatic Sea. It appears probable that adult *Chlamys* sp.

become sexually mature in the autumn at temperatures above 11 °C, spawning being induced by sharp falls in temperature.

Solenacea larvae were observed in plankton samples from January to June, but were abundant in the plankton at Banjole station only during late March to mid-May in 1968 (Fig. 3 b). The lengths ranged from 229 to 365 μm. It was not possible to identify Solenacea larvae to species and not much is known of their distribution and life history. Vatova (1949) found sexually mature *Ensis ensis* L. in the northern and mid-Adriatic Sea during March and Poggiani (1968) found Solenacea larvae in the plankton from January to May in the mid-Adriatic Sea.

Cultellus adriaticus Coen, another species of Solenacea, was present in the plankton from December to the following June (Fig. 3 c). The lengths ranged from 229 to 389 μm. The appearance of the more advanced larvae in early December indicated that the adults were ripe in November. The larvae appeared in the plankton when the temperature of the bottom water dropped below 17 °C, and disappeared in June when the temperature exceeded this level. According to Kinne (1963) spawning may be initiated by rapid change in temperature, therefore, it is possible that in case of *C. adriaticus* a fall in temperature stimulated spawning. A second spawning in March, however, appeared to be stimulated by rising temperature. It was difficult to establish the number of spawnings during spring, but it was apparent that spawning continued over several weeks.

Almost all Mytilacea larvae, observed from early February to early June, appeared to be *Mytilus galloprovincialis* Lmk. Gonadal samples indicated that an intensive spawning occurred in late January and several less intensive spawnings later. The more advanced larvae were rare in February and March but their number increased in April, especially in Pula harbour. Small numbers of Mytilacea larvae were later observed again in this area in December and January after the initial onset of spawning in the autumn.

From May to September, when a variety of larvae were present in the plankton, it was possible to identify another genus of Mytilacea with a shape different than that of *M. galloprovincialis*. The shape and dimensions of these Mytilacea larvae resembled those of *Musculus* sp. (Jørgensen, 1946). Two peaks of abundance occured, one in May and another in August with fewer larvae present during the intervening months (Fig. 3 d). The largest larvae were 357 μm in length, which was similar to the average length (340–370 μm) of *Musculus* larvae (Jørgensen, 1946). *Musculus* larvae were present in the summer and autumn, both at Plymouth (Lebour, 1938) and in the Oresund (Jørgensen, 1946). It appeared that the earlier increase in the water temperature in the spring in the Adriatic Sea initiated spawning of *Musculus* sp., as well as some other bivalves, a few months earlier than in more northern waters.

Some bivalve larvae prefer higher temperature for their early development; therefore, they are more abundant in the warmer months of the year. Larvae of *Teredo* for example, were present in Pula harbour from June to November but were most abundant during July and August (Fig. 4). At other stations in the northern Adriatic Sea they were rare and were observed only in July and August (Fig. 3*f*). None of these larvae were found in the Kvarnerić area, except in Novalja harbour and it appeared that their distribution was associated with the presence of ships or other wooden structures in the harbours. The length of *Teredo* larvae ranged from 171 to 243 μm. Jørgensen (1946) also found *Teredo* sp. in the Oresund in summer and autumn as did Zahvatkina (1959) in Black Sea.

Because larvae of *Teredo* were found in early June, spawning must have commenced in late May when the water temperature increased from 16 to 20 °C. Another less intensive spawning occurred in August and the last larvae were observed in November when the temperature had fallen to about 18 °C. Loosanoff & Davis (1963) found in laboratory experiments that *Teredo* sp. of New England, USA, spawned at a temperature of 14 °C and higher and that the larvae were released at temperatures ranging from 16 to 20 °C. This data also appears to apply to *Teredo* sp. in the northern Adriatic Sea.

Ostrea edulis larvae appeared in the plankton from April to December, and abundantly in May and July in Pula harbour (Fig. 4), at lengths ranging from 200 to 271 μm. Adult *O. edulis* commenced spawning in late March, as was observed by the appearance of white larvae inside the mantle cavity of several specimens in the Limski canal in 1966 (Hrs-Brenko, 1969). It was apparent that a sudden increase of water temperature during the first warm days of spring was enough to induce ripe oysters to spawn. Davis & Calabrese (1969) found that although a high percentage (81.2%) of *O. edulis* larvae survived at 12.5 °C, they grew very slowly at temperatures less than 17.5 °C; therefore, slow growth of brooding larvae can be expected with a prolongation of the incubation period of more than 15 days. This would tend to corroborate the finding of Hrs-Brenko (1969) that larvae first appeared in the plankton in the second part of April in the Limski Canal. In Pula harbour the first larvae were observed somewhat later, in early May. *O. edulis* larvae were less numerous at other stations (Fig. 3*e*).

The bivalves considered thus far do not spawn throughout the year but have a spawning season restricted to the colder or warmer months. Plankton samples provided evidence that some genera of larvae were present in the water almost throughout the year. The presence of larvae of a particular genus in the plankton for such a long time may have resulted either from successive spawnings of different generation of animals of one physiological race or species or of different spawning seasons of the different physiological races or species.

Lima hians larvae, for example, were present throughout the year, with the exception of the months of April and September. An abundance of these

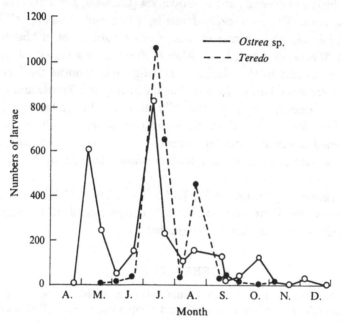

Fig. 4. Frequency of *Ostrea* larvae (—) and *Teredo* (- - -) larvae in the plankton of Pula harbour in 1968.

larvae was present in June and August with the maximum number appearing in July (Fig. 3g). The lengths ranged from 214 to 357 μm. Poggiani (1968) also found larvae from January to April, in June and July, and from October to December in the mid-Adriatic.

Anomia larvae were found occasionally throughout the year (Fig. 3i). These larvae were the most numerous of the bivalve larvae in a plankton sample taken near Plavnik Island (Kvarnerić area) in August 1968. The lengths ranged from 229 to 286 μm.

Hiatella larvae were frequently observed almost throughout the year in the plankton samples, but were numerous in May and June (Fig. 3h) ranging in size from 214 to 357 μm. Poggiani (1968), in his plankton studies of the mid-Adriatic Sea, found larvae of *Hiatella arctica* from January to June.

SUMMARY

Plankton samples were collected at several stations in the northern Adriatic Sea (Bay of Venice and Kvarneric Region) in order to ascertain the planktonic succession of the larvae of the more important bivalve species of this area. Near Banjole Island, bivalve larvae were present in the plankton throughout the year. Detailed observations were conducted on the more advanced larvae of several genera of bivalves (*Chlamys, Mytilus, Musculus, Solen, Cultellus, Teredo, Ostrea, Hiatella, Lima* and *Anomia*). *Chlamys* larvae were dominant in winter months, *Cultellus* abundant in March, and *Solen* and *Mytilus* in April. From May to September a variety of bivalve larvae were present in the plankton. During these months the larvae of *Musculus* were abundant in May and again in August. *Teredo* and *Ostrea* were most numerous in July especially at Pula. The total number of bivalve larvae decreased in September and November though one peak of undetermined larvae appeared in December. The larvae of *Hiatella, Lima* and *Anomia* were present in the plankton almost throughout the year.

I wish to express my thanks to Mr H. C. Davis, Dr A. Calabrese and Mr P. E. Chanley for their suggestions and comments of this paper and Mr G. Sosić for preparation of the graphs.

REFERENCES

DAVIS, H. C. & CALABRESE, A. (1969). Survival and growth of larvae of the European oyster (*Ostrea edulis* L.) at different temperatures. *Biol. Bull. mar. biol. Lab. Wood's Hole* 136 (2), 193–9.

GAMULIN-BRIDA, H. (1967). Biocenološka istraživanja pomičnog norskog dna sjevernog Jadrana kod Rovinja. *Thalassia jugosl.* 3, 23–33.

GAMULIN-BRIDA, H. POŽAR, A. & ZAVODNIK, D. (1968). Contribution aux recherches sur la bionomie des fonds meubles de l'Adriatic du Nord – II. *Biol. Glasn.* (In Press.)

HRS-BRENKO, M. (1969). Observations sur l'Huître (*Ostrea edulis*) du canal de Lim (Adriatique du nord). *Rapp P.-V. Réun. Commn. int. Explor. scient. Mer. Méditerr.*, 19 (5), 855–7.

JØRGENSEN, C. B. (1946). Reproduction and larval development of Danish marine bottom invertebrates. 9. Lamellibranchia. *Meddr. Kommn. Havunders.* Kbh Ser. (*d*), *Plankton* 4, 277–311.

KINNE, O. (1963). The effects of temperature and salinity on marine and brackish water animals. I. Temperature. *Oceanogr. mar. Biol. A. Rev.* 1, 301–40.

LEBOUR, M. V. (1938). Notes on the breeding of some labmellibranchs from Plymouth and their larvae. *J. mar. biol. Ass. U.K.* 23, 119–45.

LOOSANOFF, V. L. & DAVIS, H. C. (1963): Rearing of bivalve molusks. *Adv. mar. Biol.* London and New York **1**, 1–136.

LOOSANOFF, V. L., DAVIS, H. C. & CHANLEY, P. E. (1966). Dimensions and shapes of larvae of some marine bivalve mollusks. *Malacologia* **4** (2), 351–435.

ODHNER, N. H. (1914). Notizen über die fauna der Adria bei Rovigno. Beiträge zur kenntnis der marinen molluskenfauna von Rovigno in Istrien. *Zool. Anz.* **44**, 156–70.

POGGIANI, L. (1968). Note sulle parve planctoniche di alcuni Molluschi dell'-Adriatico medio-occidentale e sviluppo post-larvale di alcuni di essi. *Not. Lab. Biol. mar. pesca-Fano.* **2** (8), 137–80.

REES, C. B. (1950). The identification and classification of lamellibranch larvae. *Hull Bull. mar. Ecol.* **3**, 73–104.

REES, C. B. (1951). First report on the distribution of lamellibranch larvae in the North Sea. *Hull Bull. mar. Ecol.* **3** (20), 105–34.

THORSON, G. (1946). Reproduction and larval development of Danish marine bottom invertebrates, with special reference to the planktonic larvae in the Sound (Oresund). *Meddr. Kommn. Havunders.* Kbh Ser. (d), *Plankton* **4**, 1–523.

VATOVA, A. (1928). Compendio della flora e fauna del mare Adriatico presso Rovigno. *Memorie. R. Com. taalassogr. ital.* **143**, 1–614.

VATOVA, A. (1935). Ricerce preliminari sulle biocenosi del Golfo di Rovigno. *Thalassia* **2** (2), 1–30.

VATOVA, A. (1949). La fauna bentonica dell'alto e medio Adriatico. *Nova Thalassia* **1** (3), 1–110.

ZAHVATKINA, K. A. (1959). Ličinki dvuhstvorčatyn molljuskov Sevastopolj-skogo rajona Cernogo morja. *Trudy Sevastopol'. biol. Sta.* AN SSSR **11**, 108–51.

COMPARATIVE INVESTIGATIONS OF THE DEVELOPMENT OF EPIBENTHIC COMMUNITIES FROM GLOUCESTER MASSACHUSETTS TO St THOMAS VIRGIN ISLANDS

J. B. PEARCE and J. R. CHESS

Sandy Hook Marine Laboratory, Highlands, New Jersey 07732, USA

INTRODUCTION

Quantitative sampling of sessile epibenthic organisms to determine their temporal and latitudinal distribution, colonization, survival, growth and succession presents difficulties not found in infaunal studies. This is reflected in the lack of discussion of epibenthic sampling techniques in recent review articles which concern benthic ecology (Hedgpeth, 1957; Thorson, 1957; Holme, 1964; Hopkins, 1965).

To determine the effects of artificial substrata on the productivity of marine and estuarine environments we have developed a new apparatus (the Multiple Disc Sampling Apparatus, or MDSA) for sampling associations of epibenthic organisms. The use of a standard technique will allow valid comparison of data from different latitudes.

For our purposes the MDSA is a functional tool for determining the relative values for various substrates as artificial reef material and provides a method of determining the optimum time for establishing such structures. We are also using the MDSA technique to monitor and compare similar polluted and pristine ecosystems as well as to study the development and succession of communities, settling of larvae, growth rates of individual species, and reproductive behaviour.

APPARATUS AND PROCEDURE

The MDSA consists of an array of discs 0.05 m² in upper surface area (24.6 cm in diameter). Discs were fastened to channel iron supports which were attached parallel to each other on a square frame made of 2 in (O.D.) galvanized pipe (Plate 1a). Two basic frames have been used. One was 3 m square and supported 108 discs (Plate 1b). Each corner of this frame was supported and anchored by a 2.5 in (O.D.) pipe embedded in a concrete-filled drum weighing approximately 600 kg. The second frame was 2 m square and supports 36 discs. Each corner was supported by a 2 in (O.D.) pipe which was attached to a concrete-filled car tyre weighing 75 kg.

To determine settling preference of invertebrates for artificial substrata, discs made from a variety of substances were used. The discs included fibre reinforced rubber, concrete, untreated commercial grade steel, untreated aluminium, glass and wood (pine). In the prototype 108 disc apparatus, at least 12 discs of each material were arranged linearly along each support. Because rubber tyres, concrete rubble and steel are the most readily available scrap substances for large-scale construction of artificial reefs, two rows of 12 discs each were allowed for these materials. On the smaller frames only 12 discs each of rubber, concrete, and steel were attached. The discs were fastened to the channel iron supports by stainless steel wire which was passed through the discs and secured around the support.

Divers equipped with SCUBA recover the discs after predetermined periods of submersion. As suggested by Fager *et al.* (1966), diving operations have been standardized and kept as simple as possible. By wiring rather than bolting the discs to their supports, the problem of removing corroded fastenings has been avoided. The diver simply cuts the wires to free the discs. This operation can be performed in near zero visibility and economizes the diver's time on the bottom.

One disc of each material was removed approximately every 4 weeks over a 1-year period. As duplicate sets of rubber, concrete and steel were attached to the larger frame, these substances were sampled every 4 weeks over a 2-year period. In addition, the discs were examined *in situ* by divers weekly or fortnightly in order to monitor reef communities.

The original discs provide information on growth, competition, and succession over extended periods. In order to obtain synoptic data concerning larval settling during monthly intervals, independently of the established epifauna, new concrete and rubber discs were attached to the frame each time one of the original series of discs was collected. These new discs were collected on the next sampling date.

We have devised an efficient sample collecting technique for use in the field. A special, tightly woven cloth bag is drawn over a disc prior to cutting the wire fastening. The bag is closed around each disc by a draw-string and secured with a final half-hitch knot. Each bag and enclosed disc is returned to the surface, placed in narcotizing solution (7.5% $MgCl_2$ mixed 1:1 with sea water) for 1 h and then transferred to approximately 10% formalin.

Following overnight fixation in the laboratory each disc was gently rinsed free of sediments and most of the unattached epifauna. The rinsed material consisted of motile organisms as well as those attached forms which had been freed. A graded series of screens (1.00, 0.500 and 0.125 mm mesh) was used to separate the organisms from fine sediments and detritus. The organisms washed from each disc were preserved in 70% ethanol containing 5% glycerin, and subsequently identified and counted.

After rinsing, the upper and lower surfaces of each disc was photographed with black and white (Kodak Plus-X) and colour (Kodachrome II, type A) film. Each disc was then placed in an individual plastic bag containing 70% ethanol and 5% glycerin.

Sessile organisms attached to the discs were counted as follows: A transparent plastic disc (0.05 m²) from which a thirty-six degree sector containing one-tenth of the area (0.005 m²) had been removed was oriented at random on the substrate disc (Plate 1 b). All organisms occurring in the wedge were counted under a dissecting microscope. Two sectors were examined from the upper surface of each disc and one from the lower surface.

Rinsing removes a number of organisms from the discs. To estimate the number of these to be assigned to upper and lower surfaces, extrapolations are made for individuals of the same species which were present in the rinsed portions of the disc sample as well as attached to the discs themselves. For instance, if 75% of the total attached *Mytilus* were found on the upper surface and the remainder on the lower, then 75% of the total number of *Mytilus* rinsed free were added to the count of *Mytilus* relating to the upper surface.

Those unattached species associated with the attached epifauna, such as amphipods, many decapod crustaceans and certain finfish, were counted and distributed equally between the upper and lower surface counts unless a specific relationship was known to exist between such motile forms and an attached species. We have observed such relationships between certain nudibranchs and caprellid amphipods and hydroids of the genus *Tubularia*. Polyclad flatworms were most frequently found with living *Balanus improvisus* or inside empty barnacle shells.

The determination of surface preferences is an important part of the present study. *In situ* observations of the disc assemblages for the determination of surface preferences is valuable for the macroinvertebrates and finfish. Preferences of the meiofauna and microfauna must be deduced either from preserved materials or laboratory experimentation with discs transferred from natural environments to laboratory aquaria.

DISCUSSION OF PRELIMINARY RESULTS

A MDSA bearing 108 discs was placed approximately 3.2 km off the coast of Sea Bright, New Jersey (40° 20' N, 72° 56' W) in 18 m of water on 1 July 1967. The apparatus rested approximately 8 m from an artificial reef consisting of 16 automobile bodies, about 1000 m south of Shrewsbury Rocks, a natural outcrop extending 5 km from shore in an easterly direction.

The MDSA placed off the New Jersey coast has been exposed to several intense storms and has not moved or become distorted during the first 24 months of submergence. The supporting concrete-filled drums resting on compacted sands have not settled appreciably into the bottom. The frame and discs are still supported 1.5 m above the substrate and have thus avoided abrasion by sand. This clearance also permits diver inspection of the underside and ensures easy access for removal of individual discs.

Smaller, 36 disc samplers have been located at three other sites along the Atlantic coast. The first of these was placed in 15 m of water approximately 27 km SSW of Charleston Harbour (32° 29' N, 79° 59' W) on 6 December 1967. On 9 February 1968 a second MDSA was located near Cow and Calf Reef, Jersey Bay, St Thomas, Virgin Islands (18° 18' 30" N, 64° 51' 00" W) in water 10 m in depth. A third MDSA installation was made on 28 May 1968 near the Gloucester Harbour, Massachusetts breakwater (42° 35' 07" N, 70° 40' 18" W) in water 13 m deep.

The analysis of data collected has resulted in considerable information regarding periods and intensity of larval settlement, competition for available space, rate of growth, and community development. In addition some of the techniques used in this study have provided information not available from previous studies of epibenthic communities. For instance, our collecting of discs *in situ* has allowed us to obtain motile forms not ordinarily taken in quantitative epibenthic or fouling studies. Investigators using fouling plates or blocks hauled from considerable depth to the surface do not record quantitatively the Amphipoda, which we found abundant both in numbers of individuals and species. Undoubtedly, motile forms living among hydroids and worm tubes are disturbed by unusual motions and easily lost from hauled or scraped samples. In addition to the smaller Peracarida many of our disc samples have contained highly motile forms such as juvenile finfish and shrimps, as well as larval and juvenile crabs.

Other advantages are inherent to the technique besides reduced losses and greater selectivity in sampling. The standardized disc size of 0.05 m² permits direct comparison of epifaunal standing crops with the results of earlier infaunal studies in which 0.05 and 0.10 m² Petersen, Smith-

McIntyre and Van Veen grabs were used. The surface areas of our discs are larger than those used in most previous fouling and epibenthic investigations (Cory, 1967; Nair, 1962; Skerman, 1958). These relatively greater surface areas allow growth of larger species and individuals. Therefore assemblages which are more typical of those found on natural substrata may develop. Small settling surfaces such as glass microscope slides, oyster shells, commercial tiles, etc., probably do not offer sufficient surface for normal community development. Also, mechanical disturbance or grazing by animals may eradicate an assemblage on a small surface whereas on a larger surface only a portion might be disturbed, thus leaving sufficient material for analysis.

By using discs of different substances it has been possible to make preliminary evaluation concerning the effectiveness of various materials as artificial substrata or habitats. The data suggest that the communities developing on rubber and concrete are similar (Fig. 1) and that initially these materials are superior to untreated steel as a substratum for most sessile invertebrates. It has been reported that when used as artificial habitats in the marine environment 'plated' steel is more attractive than concrete in attracting fish (anonymous, 1967). Considering the current interest in developing artificial reefs (Carlisle, Turner & Ebert, 1964; Unger, 1967) these observations are important considerations in recommendations made for reef construction and habitat improvement.

SUMMARY

Since June 1967, larval settling, colonization, growth, competition, mortality and community stability have been studied on a variety of hard substrata at stations in Gloucester Harbour, Mass., Long Island Sound, N.Y., and off Sandy Hook, New Jersey, Charleston, South Carolina and St Thomas, V.I.

Standard techniques include the use of divers to recover fouling discs 0.05 m² in surface area which were made from concrete, rubber, steel, glass, aluminum and wood. This methodology has enabled us to compare community development, biomass and productivity at widely separate stations.

Diving observations and study of finfish digestive tracts have enabled us to detail the role that reef-dwelling finfish play in community development and stability. Echinoderms are extremely important as predators in the boreal, Virginian and Carolinian zoogeographic provinces whereas reef fishes are the dominant predators and control community stability in the tropical environments of the Virgin Islands.

60   J. B. PEARCE AND J. R. CHESS

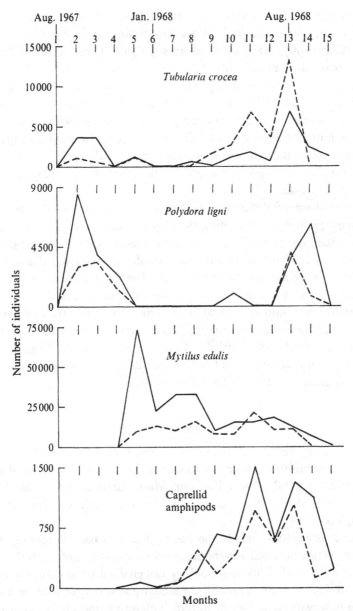

Fig. 1. The abundance of four dominant organisms on concrete (solid lines) and rubber discs (broken lines) at the Sea Bright, New Jersey site during the period of August 1967 to October 1968. The hydroid *Tubularia crocea*, competes directly with the polychaete *Polydora ligni*, for available surface area. These two forms decline in number during the winter months and are replaced by the mussel *Mytilus edulis*. Caprellid amphipods are closely associated with the *Tubularia* and occur in greatest number when the hydroid is abundant. The abundance of an aeolid nudibranch is also closely correlated with the abundance of *Tubularia*. All four of the dominant organisms are preyed upon by both finfish and the seastar, *Asterias forbesii*.

REFERENCES

ANONYMOUS (1967). Sea winds – reefs from ruins. *Sea Secrets* **11** (8), 4.

CARLISLE, J. G., TURNER, C. H. & EBERT, E. E. (1964). Artificial habitat in the marine environment. *Bull. Dep. Fish Game St. Calif.* 124.

CORY, R. L. (1967). Epifauna of the Patuxent River estuary, Maryland, for 1963 and 1964. *Chesapeake Sci.* **8** (2), 71–89.

FAGER, E. W., FLECHSIG, A. O., FORD, R. F., CLUTTER, R. I. & GHELARDI, R. J. (1966). Equipment for use in ecological studies using SCUBA. *Limnol. Oceanogr.* **11** (4), 503–9.

HEDGPETH, J. H. (1957). Obtaining ecological data in the sea. In *Treatise on Marine Ecology and Paleoecology*, Vol. 1, *Ecology*, pp. 53–86 (ed. J. H. Hedgpeth). Geol. Soc. America, Mem. 67.

HOLME, N. A. (1964). Methods of sampling the benthos. In *Adv. Mar. Biol.*, Vol. 2, pp. 171–260 (ed. F. S. Russell). London: Academic Press.

HOPKINS, T. L. (1965). A survey of marine bottom samplers. In *Progress in Oceanography*, Vol. 2, pp. 213–56. Oxford: Pergamon Press.

NAIR, N. B. (1962). Ecology of marine fouling and wood-boring organisms of western Norway. *Sarsia* **8**, 1–88.

SKERMAN, T. M. (1958). Marine fouling at the Port of Lyttelton. *N.Z. Jl Sci.* **1**, 224–57.

THORSON, G. (1957). Bottom communities. In *Treatise on Marine Ecology and Paleoecology*, Vol. 1, *Ecology*, pp. 461–534 (ed. J. H. Hedgpeth). Geol. Soc. America, Mem. 67.

UNGER, I. (1967). Artificial reefs – a review. *American Littoral Society, Highlands, N.J.* Spec. Publ. No. 4, 74 pp.

SETTLEMENT OF MUSSEL LARVAE
MYTILUS EDULIS ON SUSPENDED
COLLECTORS IN NORWEGIAN WATERS

B. BØHLE

Institute of Marine Research, Directorate of Fisheries, Bergen, Norway

INTRODUCTION

In order to determine the period of the main settlement of larvae of the mussel (*Mytilus edulis* L.) and find out to what extent primary and secondary settlement (Bayne, 1964) occurred on collectors suspended from rafts, experiments were performed between 20 May and 22 August 1967 at Snarøya in the inner Oslo Fjord.

MATERIAL AND METHOD

Plates measuring 15 × 15 × 2 cm made from foam plastic (polystyrene) with slightly rough surfaces were used as spat collectors. They were attached to two metal frames in positions chosen at random and suspended 1 m below the sea surface from a raft anchored at 15 m from the shore.

On 20 May, 18 plates (series A) were suspended. Every third day a plate was removed and replaced by a new one (series B) which was left until 13 July. In series C (23 May–10 July), single plates were suspended for successive 3 day periods. Two additional plates were suspended on 13 July and 1 August respectively and left until 22 August.

Algae with adhering spat were removed from the plates taken for investigation. The spat were then separated by shaking the algae in sea water and 4% formalin and subdivided in a plankton divider (Wiborg, 1951). When the spat were numerous, only one-fourth or one-eighth of the sample was examined. When the spat were less numerous, all the spat were counted.

Planktonic mussel larvae were collected with a plankton net (mesh size 125 μm), towed for 5 min in the upper 4 m, at a distance of not more than 50 m from the raft. Sea temperature was recorded with a continuous recorder (Negretti and Zambra, T 350 Model A) placed at 2 m depth at Huk, 3 km from Snarøya. It is assumed that the temperature was almost the same at the two localities.

The spat were measured in a stereoscopic microscope with ocular micrometer, one division corresponding to 12.5 μm. Spat less than 533 μm were arranged in 50 μm groups, larger spat in 1000 μm groups.

RESULTS AND DISCUSSION

Larvae in the plankton

In the Oslo Fjord, the mussels start spawning in early May at a sea temperature of 8 °C (Bøhle, 1965). In 1967, this temperature was reached on 10 May (Fig. 1).

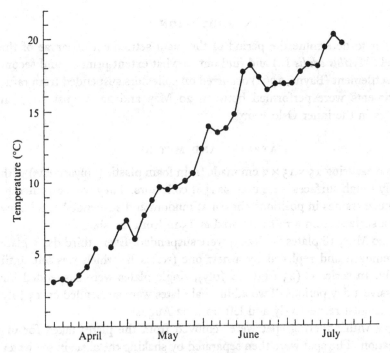

Fig. 1. Sea temperature at Huk, recorded daily at 09.00 a.m. at 2 m depth, and plotted as 3 day means.

Mussel larvae reached a maximum size of 250 μm on 29 May (Table 1), and were abundant until 22 June. On 25 June, very few were caught but on 28 June small larvae (100 μm) were abundant. These larvae probably derived from a second spawning in mid-June when the sea temperature reached a temporary peak (Fig. 1). Only very few larvae larger than 300 μm were caught in net hauls.

Table 1. *Length of planktonic larvae taken with net of mesh size 125 μm.*

Date	Mean length (μm)	Maximum length (μm)	Number measured
26 May	159	225	102
29 May	189	250	137
1 June	193	262	164
4 June	180	275	259
7 June	184	287	251
10 June	171	275	322
13 June	174	287	499
16 June	200	287	226
19 June	181	275	267
22 June	176	262	315
25 June	—	—	—
28 June	135	375	129
1 July	165	375	106
4 July	146	325	149
7 July	149	325	109
10 July	160	250	269
13 July	157	350	371

Settlement on plates

On the plates, mussel spat were mainly attached to algal filaments. The following genera of algae were found attached to the plates:

CHLOROPHYCEAE (Green algae)
Enteromorpha
Urospora
Ulothrix

BACILLARIOPHYCEAE (Diatoms)
Nitzschia
Licmophora
Fragilaria
Synedra

Diatoms were most abundant with *Nitzschia* as the dominating genus. The algae initially represented a typical presummer association (Grenager, 1957), and were replaced by *Enteromorpha* in late June.

On the plates of series A, which were allowed to accumulate algae and spat from 20 May onwards, the first spat were observed on 4 June at a density of 790 per plate (Table 2). On 10 June, 20740 spat per plate were recorded. There was a rise to 23480 on 19 June with a further steep rise to 33796 on 25 June (Fig. 2). Accordingly, the main spatfall occurred on 4–10 and 19–25 June. The density of spat on plates of series A decreased from 33796 to 12576 in the period 25 June–13 July, the slight rise on 10 July being presumably a sampling error. There was agreement between the length of the largest planktonic larvae encountered and the smallest spat (250 μm group) settled on the plates between 4 and 19 June (Table 1, Fig. 2). After 25 June, spat 1 mm in length were most abundant and on 13 July some spat measured nearly 4 mm.

On the plates of series B, which were exposed at progressively later dates and left in the sea until the end of the experiment, the spat were less numerous, with a maximum number of 11 568 on the plate suspended on

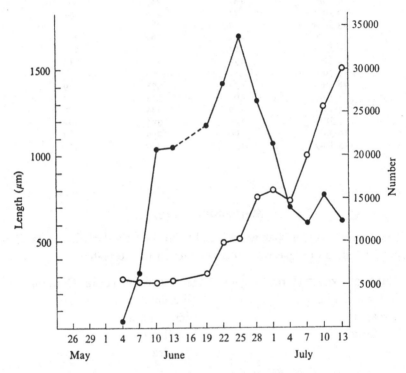

Fig. 2. Number (solid circles) and mean length (open circles) of mussel spat in series A at date of removal.

23 May which was left in the sea for the longest time (Table 3). On the plates suspended after 16 June, very few spat were recorded, though in series A, spatfall occurred between 19 and 25 June. The reason that spatfall did not occur on the series B during this period is ascribed to the scarcity of algae to which the larvae might become attached.

In series C, in which the plates were immersed for equal intervals of 3 days duration, spat were recorded from 4 to 22 June, with a maximum of 423 on 16 June (Table 4). These spat were mostly 250 μm in length. However, from late June onwards, occasional specimens of larger size up to

1000 μm were observed. These had probably crawled from neighbouring plates fastened to the same frame. The plates in series C had been suspended for too short a time to allow algal filaments to grow and the settlement of planktonic larvae on these plates was negligible compared with the spatfall in series A (14485 spat 7–10 June).

Table 2. *Density and length of spat settled on plates suspended 20 May (series A) and removed on successive dates*

Removal	Number of spat per plate	Mean length (μm)	Maximum length (μm)	Number measured
23 May	0	—	—	—
26 May	0	—	—	—
29 May	0	—	—	—
1 June	0	—	—	—
4 June	790	233	320	219
7 June	6255	267	312	136
10 June	20740	266	400	180
13 June	20870	276	375	147
16 June	*—	—	—	—
19 June	23480	315	550	121
22 June	28520	498	750	212
25 June	33796	516	1050	270
28 June	26360	754	1460	201
1 July	21360	800	1400	392
4 July	14020	739	2060	406
7 July	12176	1007	3050	498
10 July	15504	1287	3500	296
13 July	12576	1502	3950	418

* The plate was lost.

Table 3. *Density and length of spat settled on plates suspended on successive dates and removed 13 July (series B)*

Suspended	Number of spat per plate	Mean length (μm)	Maximum length (μm)	Number measured
23 May	11568	1346	3800	446
26 May	7208	1373	4015	454
29 May	5312	1483	4000	473
1 June	4833	1130	4300	1267
7 June	3344	1314	4300	568
10 June	1059	1256	4010	536
13 June	550	794	4300	532
16 June	410	1103	4050	396
19 June	8	—	—	—
22 June	14	—	—	—
25 June	80	—	—	—
28 June	1	—	—	—
1 July	0	—	—	—
4 July	0	—	—	—
7 July	0	—	—	—
10 July	0	—	—	—

On the plates exposed between 13 July and 22 August (Table 4), a heavy settlement of mussel larvae occurred. From the growth rate of the spat in June–July (32 μm per day), it is calculated that this settlement occurred about 18 July. Small lamellibranch larvae (100 μm), probably mussel larvae, were commonly recorded in the plankton from 28 June to 13 July.

Table 4. *Density and length of spat settled on plates submerged in 3 day periods (series C), and on two plates submerged July–August*

Suspended	Removal	Number of spat per plate	Mean length (μm)	Maximum length (μm)	Number measured
23 May	26 May	0	—	—	—
26 May	29 May	0	—	—	—
29 May	1 June	0	—	—	—
1 June	4 June	0	—	—	—
4 June	7 June	185	249	312	164
7 June	10 June	353	257	300	125
10 June	13 June	416	258	375	72
13 June	16 June	423	256	375	157
16 June	19 June	206	227	450	138
19 June	22 June	179	292	575	144
22 June	25 June	56	—	—	—
25 June	28 June	31	—	—	—
28 June	1 July	25	—	—	—
1 July	4 July	22	—	—	—
4 July	7 July	0	—	—	—
7 July	10 July	0	—	—	—
(A) 13 July	22 August	18000	1351	4000	685
(B) 1 August	22 August	455	1807	5000	455

De Blok & Geelen (1958) found that mussel larvae settle preferentially on filamentous substrata. The larvae may detach themselves from the substratum if this is not suitable. In British waters, Bayne (1964) described the settlement taking place in two steps, a primary settlement (larvae of 250 μm) on filamentous algae and a secondary one, in which the same larvae (now juveniles of 900–1500 μm) having detached from the algal filaments migrate and attach to the mussel bed. These juvenile mussels were recorded in the plankton.

During the period 25 June–13 July, the number of spat (length 500–1500 μm) on the experimental plates of series A decreased from 33796 to 12576 (Fig. 2). Though the larvae settled almost entirely on the filamentous algae, the remaining spat on 13 July were mainly attached to the plates and very few to *Enteromorpha*. In the period 25 June–13 July, the filamentous algae vanished. It is assumed that part of the spat moved on to the plate itself before the algae loosened. The rest probably remained on the algae, sinking to the bottom. Very few mussel larvae larger than 300 μm were found in the plankton hauls. This fact is supported by previous investigations in the inner Oslo Fjord (T. Schram, personal communication). A planktonic stage preceding a secondary settlement seems therefore not to exist in Norwegian waters.

SUMMARY

In order to study the settlement of larvae of mussels (*Mytilus edulis* L.), plates of foam plastic fastened to metal frames were suspended from a raft for periods varying from 3 to 54 days in the period 20 May–22 August. Plankton hauls were taken every third day in the period 26 May–13 July.

The mussel larvae settled mainly 4–10 June and 19–25 June, attaching to filamentous algae. Part of the spat moved on to the plates before the algae disappeared. The remainder of the spat probably followed the algae sinking to the bottom. A planktonic stage preceding a secondary settlement seems not to exist.

I would like to express my gratitude to Cand. real. Thomas Schram, University of Oslo, for fruitful discussions and aid with identification of planktonic larvae. Sincere thanks are also due to the Institute of Marine Biology, Section A, University of Oslo, for laboratory facilities. Cand. real. Jan Rueness at the University Biological Station in Drøbak kindly identified the algae. Mr Hans Petter Olsen, Snarøya, kindly assisted with the arrangement of the experiment.

REFERENCES

BAYNE, B. L. (1964). Primary and secondary settlement in *Mytilus edulis* L. (Mollusca). *J. Anim. Ecol.* **33**, 513–23.

BØHLE, B. (1965). Undersøkelser av blåskjell (*Mytilus edulis* L.) i Oslofjorden. *Fiskets Gang* **51**, 388–94, *Fisken Hav.* 1965 (1), 19–25.

DE BLOK, J. W. & GEELEN, H. J. F. M. (1958). The substratum required for the settling of mussels (*Mytilus edulis* L.) *Archs néerl. Zool.* (Volume Jubilare) **13**, 446–60.

GRENAGER, B. (1957). Algological observations from the polluted area of the Oslofjord. *Nytt Mag. Bot.* **5**, 41–60.

WIBORG, K. F. (1951). The whirling vessel. An apparatus for the fractioning of plankton samples. *FiskDir. Skr.* serie Havundersøkelser. **9** (13), 1–16.

ON THE ECOLOGY OF YOUNG
IDOTEA IN THE BALTIC

ANN-MARI JANSSON

Department of Zoology and the Askö-Laboratory, University of Stockholm

AND ANN-SOFI MATTHIESEN

Institute of Statistics, University of Stockholm

INTRODUCTION

Three species of the genus *Idotea* penetrate into the Baltic Sea. Those are *I. chelipes* (Pallas). *I. baltica* (Pallas) and *I. granulosa* (Rathke). Their characteristic distribution in the field in relation to temperature, salinity gradient and degree of exposure to wave action has been investigated by Forsman (1956), Sywula (1964), v. Oertzen (1968) and Muus (1967). The oxygen consumption by adult *I. baltica* and *I. chelipes* has been tested by v. Oertzen (1965) and the diurnal activity of the same species by B-O. Jansson & Källander (1968). Life-cycles of the Baltic species are described by Muus (1967) from the Danish Belts. In the southern Baltic two generations per season are not uncommon but in the northern Baltic the populations have only one brood each year. The release of young takes place in June–August with a maximum in the first part of July (Jansson, 1966, 1969b). Some *Idotea* species are more or less bound to certain seaweeds. In the Baltic for instance *I. baltica* has about the same distribution as *Fucus vesiculosus* L. (Segerstråle, 1944) and in the northern part of the Baltic the juveniles of *I. baltica* and *I. chelipes* are most abundant in the vegetation of *Cladophora glomerata* (Kützing) (Jansson, 1966, 1969b). At a length of 6 mm *I. baltica* prefer *Ceramium* to *Cladophora* in choice experiments (Jansson, 1970). The significance of different algal species as food for *Idotea* was investigated by Ravanko (1969).

Muus (1967, p. 115) stated that 'it must be the mutual competition which decides the characteristic distribution of the *Idotea* species'. The present study aims to analyse whether there is any clear competition for food and (or) space particularly between young *I. baltica* and *I. chelipes* when developing together in the *Cladophora* belt.

THE INVESTIGATED AREA

The present investigation was carried out in the bay at which the Askö Laboratory is situated in the archipelago off the Swedish east coast, 60 km south of Stockholm. The bay is about 10 m deep and separated from the open sea by several islets and underwater reefs. The prevailing south-west winds cause an almost continuous agitation of the water mass, except for a period in winter when the bay is covered with ice. The salinity varies between 6 and 7‰ and the area is practically unaffected by water pollution (Jansson, 1969 a). A rich vegetation of *F. vesiculosus* grows from 0.5 to 7 m depth and during the summer *C. glomerata* forms characteristic belts on the rocks 0–0.5 m beneath the water surface.

METHODS

Sampling in the field

The field studies were concentrated on the *Cladophora* vegetation at 0–0.5 m depth. Quantitative samples were taken at short intervals during July to August in 1968 and 1969 at seven localities with different degrees of exposure to wave action. Water temperature, density and width of the algal belt were annotated at each sampling time. The algal material growing within 1 dm² was collected from two representative areas, the macroscopic animals sorted out and counted and the wet and dry weight of alga determined. The algal material was analysed as to coarseness of filaments, density and type of epiphytic growth.

Laboratory experiments

General

All laboratory experiments were carried out at 15 °C temperature and under natural light conditions. Only clearly active animals were accepted for the experiments. Glass bowls containing 10 ml and 25 ml filtered sea water were used for the 2–3 mm and 4 mm stages respectively. Every two or three days the experimental animals were moved to clean bowls containing fresh sea water.

Rearing experiments

In the rearing experiments the technique described by Jansson (1967) was used.

Competition experiments

In the typical *I. chelipes* habitat the *Cladophora* is finer in structure than it is on shores more exposed to wave action. Localities moderately exposed usually have the highest densities of epiphytic diatoms (Jansson, unpublished results). Our hypothesis when planning the competition experiments was that if any competition for food between *I. chelipes* and *I. baltica* existed it might be different on different types of *Cladophora*. *I. chelipes*, which is the more fragile species, would more likely be excluded from the outer localities, especially since it is unable to consume the coarse filaments of *Cladophora* growing there. *I. baltica* on the other hand might be favoured on a food rich in diatoms, as suggested by the rearing experiments on page 77 and in Jansson (1967). In order to investigate the importance of these specific differences in the ability to compete and survive, the following experiment (competition expt. I) was carried out. Forty small bowls (capacity of refrigerator) containing filtered water from the habitat were numbered and placed in order on a tray. A small amount of one of the four diets; fine *Cladophora*, coarse *Cladophora*, fine *Cladophora* plus diatoms, or coarse *Cladophora* plus diatoms was put in each bowl. The diets were allocated to the bowls by means of a table of random permutations. Newborn larvae were placed in pairs into the bowls containing each diet; two bowls with *I. chelipes*, two with *I. baltica* and six with one specimen of each species. The choice as to which species or combination of species should be put into which bowl was made by means of a table of random permutations. Through the three randomizations any systematic influence on survival due to intraspecific differences in fitness, differences in food supply, or differences in position in the refrigerator were eliminated. The remaining differences may be accounted for in a statistical analysis.

The second competitition experiment differed from the first in that the animals were older at the beginning of the experiment (10 days) and were given smaller quantities of fine *Cladophora* only.

Competition expts. I and II, were carried out with animals randomly chosen from the same two contemporary broods. These broods were released by females sampled in the *Fucus* vegetation a few days earlier and isolated in plastic boxes containing a *Cladophora* tuft.

In competition expt. III larvae about 4 mm long and approximately 1 month old, which had been sampled at locality 2, were used. Two diets were used – fine *Cladophora* and coarse *Cladophora*.

The experiments were read twice a day. Dead animals were removed, mounted in glycerine and determined to species.

Identification of the species

Descriptions of the first postmarsupial stages of *Idotea* are given by Matsakis (1956) and Naylor (1955). In the present investigation fixed specimens of 2–3 mm were determined to species according to Naylor (1955) using the numbers of aesthetascs present on the tip of the antennule – one in *I. chelipes* and two in *I. baltica* and *I. granulosa*: The aesthetascs are easily visible under a microscope in specimens mounted in glycerine. The determination was completed by measuring the width of the terminal segment of the antennule. *I. granulosa* has coarser appendages than the two other species at equal body-length.

At a length of about 4 mm the three species differ so much in external morphology (width/length, length of antenna, form of telson and so on) that an identification without a binocular microscope is possible, especially in living material. At this stage there are also specific differences in behaviour which will be dealt with later (see page 77).

In living material at the 2–3 mm stage it is almost impossible to distinguish between the species with certainty.

RESULTS

Field studies

Fig. 1 shows the total results of quantitative samples taken at seven localities on four different occasions during July and August 1969. The whole macroscopic fauna was counted but only the Arthropods are presented here. The Gammaridea, namely *Gammarus oceanicus* (Segerstråle) 1–5 mm dominate at the outer localities 5, 6 and 7 but at those within the laboratory bay the most frequent group consists of young *Idotea* except for location 3 which is dominated by chironomid larvae. *Jaera* sp., which plays but a small role in the sheltered bay, is commoner at locations 5, 6 and 7, evidently replacing much of the *Idotea* larvae. Fig. 1 also shows the distribution of the three different *Idotea* species in the area. At the most sheltered locality 1, *I. chelipes* is the dominating species but at locations 5, 6 and 7 it is completely absent. On locations 2, 3 and 4. *I. chelipes* and *I. baltica* occur together in about the same numbers.

In Fig. 2 the variations during July–August of the populations of *Idotea* at the different sampling localities is given. The histograms show how 2 mm long larvae were released in the *Cladophora* vegetation in great numbers in the middle of July. At location 3, *I. baltica* and *I. chelipes* were almost equal in numbers on 17 July but on the 23 July *I. chelipes* was the

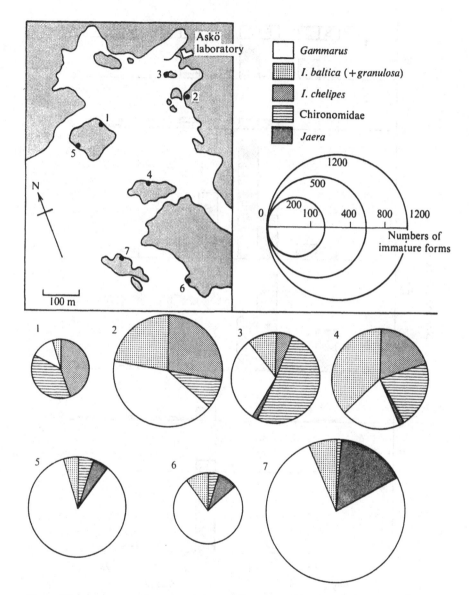

Fig. 1. The occurrence of young *Idotea* and *Gammarus*, Chironomid larvae and *Jaera* sp. amongst *Cladophora* in and around the laboratory bay at Askö. Salinity 6.3‰. Prevailing winds south-westerly. Areas of circles and sectors are directly proportional to the total number of animals from four samplings, each of 1 dm², taken between July and August 1969.

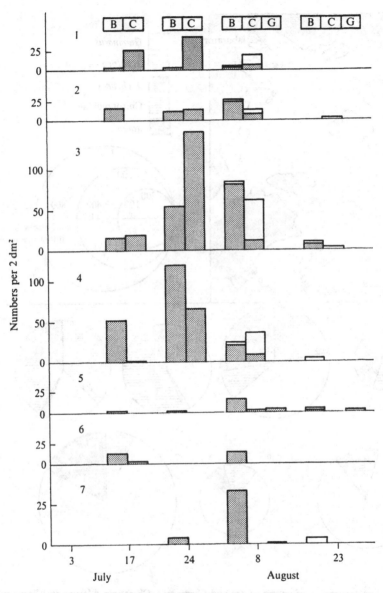

Fig. 2. Numbers per 2 dm² of young *I. chelipes* (C), *I. baltica* (B) and *I. granulosa* (G) at the stations 1–7 shown in Fig. 1. The localities were ranked in relation to degree of exposure to wave action, number 1 being the most sheltered shore. Shaded histograms 2 mm stages. Open histograms 4 mm stages.

more numerous. In August, *I. chelipes* dominated the 4 mm size groups at all inshore localities whereas *I. baltica* still had a great number of small individuals. At location 1 *I. chelipes* was the dominating species during the whole season whereas *I. baltica* played a greater role at locations 2 and 4. At locations 5, 6 and 7 a few specimens of *I. granulosa* were found but *I. baltica* was here the dominating species. Juveniles 4 mm in length were poorly represented at these localities.

Laboratory studies

Observations on the behaviour of I. baltica and I. chelipes, 4 mm juveniles

During the sampling and handling of the experimental animals two types of behaviour, differing in *I. baltica* and *I. chelipes*, were repeatedly observed.

Specimens of *I. baltica*, when sorted out into jars filled with water from the habitat and placed in daylight, actively swim around, especially at the bottom edges. Under corresponding conditions specimens of *I. chelipes* show a marked clinging behaviour. They aggregate in groups (see Fig. 3 a).

When a specimen of *I. baltica*, carefully held with a pair of forceps, is allowed to fall through the water column it usually rolls up into a ball (Fig. 3 b, B). *I. chelipes* handled in the same way becomes fully extended with its back arched and appendages fully outstretched (see Fig. 3 b, A).

Rearing experiments with I. baltica and I. chelipes 2 mm

Results from rearing experiments on *I. baltica* were published earlier by Jansson (1967). In Fig. 4 the results from two separate experiments are put together. The LD 50 curves show that *I. baltica* survives the first days of the postmarsupial development much better on a diet of diatoms than on pure *Cladophora*. *I. chelipes*, on the other hand, has a very poor survival as a whole during the first days of the experiment and there is hardly any difference in the LD 50 values if diatoms or *Cladophora* is given as food. During the experiments *I. chelipes* moulted more frequently than *I. baltica* and after 27 days it had reached a length of 5 mm. This length was not reached by *I. baltica* until the 36th day.

Competition experiments

a. Principles of survival diagrams. Fig. 5 expresses the results of the first experiment. The four diagrams show respectively the percentage survival of *I. baltica*, placed with *I. baltica*; *I. baltica*, placed with *I. chelipes*; *I. chelipes*, placed with *I. baltica*; and *I. chelipes*, placed with *I. chelipes*. The black parts of the diagrams represent animals still living as pairs, the

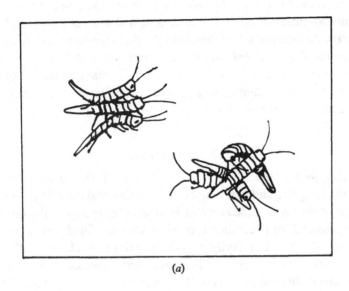

(a)

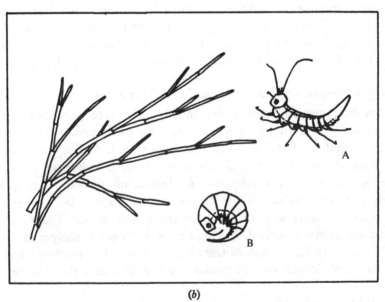

(b)

Fig. 3. (a) Clinging behaviour shown by *I. chelipes* (4 mm) when placed in a bowl without any material to settle upon. (b) The reactions of A, *I. chelipes*, and B, *I. baltica* when sinking through a column of water. Drawn from living material.

cross-hatched areas represent animals which have been left singly in a bowl.

If individuals of the same expected strength, for example animals of the same species, are kept in pairs and are not affecting each other's survival,

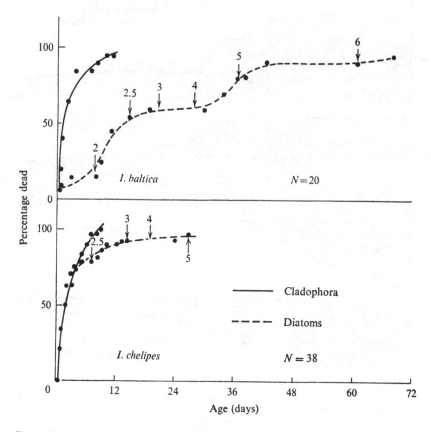

Fig. 4. Mortality and growth-rates of *I. baltica* and *I. chelipes* in rearing experiments at Askö in 1966 and 1967, wher eeither filaments of *Cladophora* or diatoms were given as nourishment. Natural lighting. Temperature 15 °C Salinity 6.3 ‰. The arrows indicate the time of ecdysis and the figures give the approximate length in millimetres of the larvae at successive stages.

one would, with sufficient material, expect that the maximum number of single individuals would be less than 25% of the original number of individuals and that this maximum would be reached at about the time when the total survival is 50%. We will now present the reasons for this statement.

Consider two individuals randomly chosen from identical populations and placed in a certain bowl – say bowl number i. Denote by $P_{i,t}$ the probability that an individual chosen from such a population and placed in bowl i will be alive at time t. Denote by $P'_{i,t}$ the probability of *one out of two*

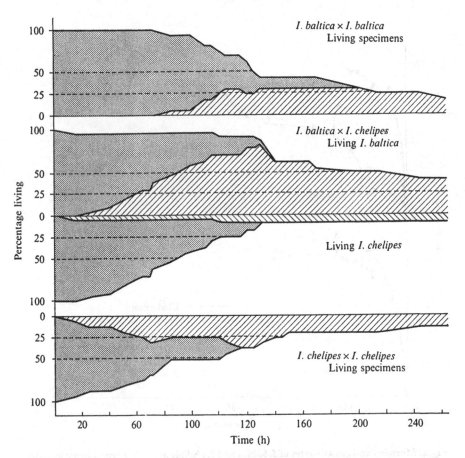

Fig. 5. Competition expt. I. Survival curves for newborn *Idotea*. Diagrams on top and bottom are each based on 16 specimens and diagrams in the middle on 24 specimens. Stippled areas represent percentage of animals surviving in pairs. Cross-hatching represents percentage of single animals.

animals surviving in the bowl at time t. Then if the two animals are not affecting each other's survival, we obtain the relation

$$P'_{i,t} = 2 P_{i,t} (1 - P_{i,t}),$$

It is easily seen that $P'_{i,t}$ reaches its maximum at $P_{i,t} = 0.5$; that is $\mathrm{Max}(P'_{i,t}) = 0.5$.

If we have n bowls and observe x_t bowls with single living animals at time t then the random variable x_t follows a generalized binomial distribution with mean, E, and variance, σ^2, defined by the relations

$$E(x_t) = \sum_{i=1}^{n} P'_{i,t},$$

$$\sigma^2(x_t) = \sum_{i=1}^{n} P'_{i,t}(1 - P'_{i,t}).$$

But since none of the terms $P'_{i,t}$ can exceed 0.5, $E(x_i)$, the number of surviving singletons, can never be greater than 0.5 n,

$$\text{Max}\,[E(x_t)] = 0.5\,n$$

and,

$$\text{Max}\,[\sigma^2(x_t)] = 0.25\,n$$

These maxima are reached only if there exists a time t when

$$P_{1t} = P_{2t} = \ldots = P_{nt} = 0.5 \quad \text{(the probability of survival in every bowl} = 0.5).$$

The total number of individuals involved in the experiments is $2n$. Thus the maximum expected number of single living individuals ($\text{Max}\,[E(x_t)] = 0.5n$), is 25% out of the total number ($2n$) *provided that the survival times are independent*. Looking at the diagram for *I. chelipes* $\times$ *I. chelipes* in Fig. 5 the maximum number of single individuals is 6 out of 16, i.e. 37%. The 25% point is reached at 65 h when the total survival is only 75%. This suggests that the partnership may have affected survival in some respect. Since this point was reached earlier than expected, a possible interaction would be negative for the survival of at least one of the two individuals. The same tendency can be observed on the *baltica* $\times$ *baltica* diagram. However, as our samples are small we should ask whether the observed high percentages could be explained by random variation.

To test whether these results are compatible with the hypothesis that the survival of each of the pair is independent of the other, it is convenient to consider the case when $P_{1,t} = P_{2,t}\ldots P_{n,t} = 0.5$. Then x_t follows a binomial distribution.

The assumption that equal survivorship (of 0.5 probability) applies to every bowl simultaneously, leads to the greatest possible value of the probability that the maximum number of singletons observed could arise from sampling errors about the expected value of 0.25. In the case of *I. chelipes* $\times$ *I. chelipes* and *I. baltica* $\times$ *I. baltica* we observed $6+5 = 11$ singletons. The probability of obtaining 11 or more singletons from $8+8$ dishes ($n = 16$) calculated from the binomial distribution will be

$$P(x_t \geqslant 11/\text{independence}) = \sum_{r=11}^{n} \begin{bmatrix} n \\ r \end{bmatrix} 0.5^n = 0.10.$$

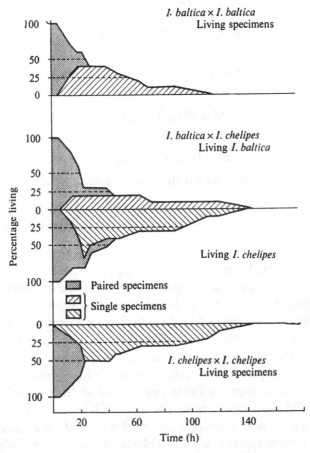

Fig. 6. Competition expt. II. Survival curves for 10-day-old juveniles. Each diagram is based on 10 specimens. Shading as in Fig. 5.

It is most likely that the $P_{i,t}$ values would not be identical in all the dishes, considering the different diets, and that the probability calculated above will be less than 0.10. Thus we conservatively reject the hypothesis of independence on the 10% level of significance. Comparing the total survival of *I. baltica* when placed with *I. baltica* and with *I. chelipes* (Fig. 5), one finds that *I. baltica* lives longer in company with *I. chelipes* than with its own species. *I. chelipes*, on the other hand, lives longer in company with a specimen of its own kind. If living with one of its own species can be regarded as reducing the survivorship of *I. chelipes* in comparison with an isolated specimen, the survivorship is apparently even further reduced when living in company with an individual of *I. baltica*. This indirect way of showing interaction between *I. baltica* and *I. chelipes* is used since the fact that *I. chelipes* is inferior to *I. baltica* when they are living together could be due to differences in fitness only and need not necessarily mean competitive interaction.

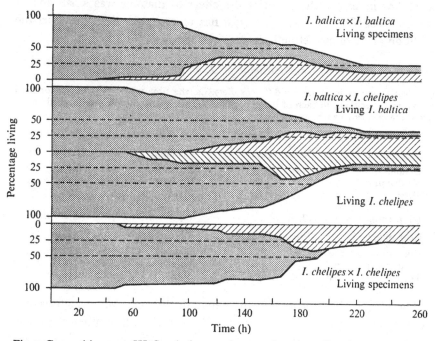

Fig. 7. Competition expt. III. Survival curves for 4 mm long juveniles, about 1 month old. Diagrams on top and bottom are each based on 20 specimens and diagrams in the middle on 18 specimens. Shading as in Fig. 5.

The results of competition expt. II using 10-day-old juveniles, summarized in Fig. 6, suggests intraspecific interaction in both species ($P[x_t \geqslant$ the observed value 9/independence] < 0.05, $n = 10$) but this time *I. chelipes* was the 'stronger' one, when the two species were living together. The difference in survivorship between *I. chelipes* × *I. chelipes* and *I. chelipes* × *I. baltica* seems to be small but *I. baltica* seems to die faster, at least in the beginning of the experiment, when in competition. Thus, older juveniles of *I. chelipes* surpass *I. baltica* in general 'strength' but the evidence of interaction between the species is weaker.

The results of competition expt. III, summarized in Fig. 7, suggest intraspecific interaction in both species ($P[x_t \geq$ observed value 14/independence] < 0.06, $n = 20$) but the difference in strength between *I. baltica* and *I. chelipes* placed together has disappeared. *I. baltica* appears to survive slightly longer in company with *I. chelipes* than with another individual of its own species, while for *I. chelipes* there is little difference.

Effect of diets. Table 1 summarizes the results of survival in competition expt. I. Diatoms improved the survival of *I. chelipes* significantly. For *I. baltica* in pair with *I. baltica* the effect of diatoms was small and of dubious benefit, but when *I. baltica* has to compete with *I. chelipes* the addition of diatoms distinctly lowered its survival.

Table 1. *Competition expt. I, effect of different diets*

(Median longevity in hours for different diets, i.e. the time when 50 % of the original number of specimens in each of the groups (a)–(d) were first observed to be dead. After 490 h the experiment was ended.)

(a) *I. chelipes* in pair with *I. chelipes*

Cladophora	Fine	Coarse	Total
Diatoms	135	125	135
No diatoms	70	85	85
Total	85	125	85

(b) *I. baltica* in pair with *I. chelipes*

Cladophora	Fine	Coarse	Total
Diatoms	160	160	160
No diatoms	336	>490	>490
Total	220	190	220

(c) *I. chelipes* in pair with *I. baltica*

Cladophora	Fine	Coarse	Total
Diatoms	140	85	110
No diatoms	72	90	72
Total	85	90	90

(d) *I. baltica* in pair with *I. baltica*

Cladophora	Fine	Coarse	Total
Diatoms	150	125	150
No diatoms	220	140	145
Total	220	140	145

The effects of the structural difference in the *Cladophora* material are more complex and difficult to interpret from the small number of experimental units. However, *I. baltica* in pair with *I. baltica* seems to live longer on fine *Cladophora*. There are no differences in the total median longevity of animals fed on fine *Cladophora* between pairs of the same or different species. The corresponding totals for coarse *Cladophora* show distinct differences – both species live longer in companion with *I. chelipes* than with *I. baltica*. In competition expt. III, however, no clear effects of different *Cladophora* material could be seen from the results.

DISCUSSION

Within the investigated area young *I. baltica* and *I. chelipes* occur together in great numbers at the four localities in the laboratory bay. At the end of August the groups of 4 mm long juveniles almost exclusively consist of

I. chelipes. It is difficult to judge to what extent the absence from the upper algal belts of *I. baltica* of larger size is caused by competition with *I. chelipes*, since several factors interact. *I. chelipes* grows faster than *I. baltica*, hence it reaches a length of 4 mm earlier. Even those localities from which *I. chelipes* is completely absent have very few large *I. baltica*.

The experiments lead us to believe that competition occurs under laboratory conditions. When the animals are given the opportunity to move away from each other competition leading to mortality is certainly rare. A more biologically adequate method of measuring the effect of competition would be to determine the physiological state of the experimental animals after a certain period of time together, rather than to construct survival curves. The technique used does not exclude predation and cannibalism which are extreme 'results of competition' (Milne, 1961). Such behaviour was in fact observed on a few occasions under special stress, as when the experimental animals had just been moved into new bowls.

Mortality was not caused by a lack of food, since the surviving specimen was able to live on the remaining nourishment in the bowl for a very long period of time. Nevertheless, starvation might be the cause of death if there were competition for space; space in terms of algal filaments, which are at the same time food for the young *Idotea*. The structure of the algal material seems, according to the results from competition expt. I, primarily to affect the competition for space. Fine *Cladophora* offers more filaments to which to cling and a greater possibility of avoiding contact with the competitor. The different survival times of *I. chelipes* and *I. baltica* on fine *Cladophora* probably correspond to a greater natural mortality on the part of *I. chelipes* during the first postmarsupial period. When diatoms are added, the competitive potential of *I. chelipes* increases significantly, as a result of which it has a deleterious effect upon the survival of *I. baltica*. As a whole it is difficult to interpret the effects of different diets in the competition experiments, as survival might simply depend upon the fact that different types of diets offer different amounts of space. What constitutes sufficient space, we have not been able to determine as yet, but our results from competition expt. II, where very little Cladophora was added, gave some information. At the start of this experiment the two animals were apparently already 'within each other's range of perception' which, according to Schein & Fohrman (1955), could mean 'visual, olfactory, tactile or other' contact. One of the specimens died very soon after. In the other two experiments, this perception range was probably not reached during the first 2 days.

The existence of intraspecific competition is clearly seen from almost all the diagrams (Figs. 5–7). The effects of interspecific competition changed during postmarsupial development. In the newly released brood *I. baltica* was superior to *I. chelipes*. At that stage interspecific competition had a more serious effect on *I. chelipes* than intraspecific competition. At an age of 10 days, the survival ratio of the larvae was reversed. When the larvae reached a length of 4 mm, interspecific competition had almost disappeared but again *I. baltica* survived somewhat better in company with *I. chelipes* than with its own kind.

The question whether competition as studied in the laboratory has any importance in the field is difficult to answer satisfactorily. The distribution of the two *Idotea* populations in the investigated area is schematically

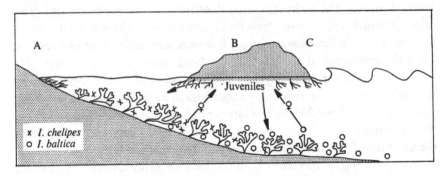

Fig. 8. The proposed dynamics of populations of *I. baltica* and *I. chelipes* in the northern baltic area. Distribution according to different depth and degree of exposure to wave action is shown. (A, sheltered; B, moderately exposed; and C, very exposed shore.) The arrows show the migration of pregnant females from *Fucus* to *Cladophora* and the transport of juveniles from *Cladophora* to *Fucus*. *I. chelipes* larvae are transported near the water surface into sheltered bays and *I. baltica* larvae fall and swim downwards to deeper zones.

illustrated in Fig. 8. Adults of *I. chelipes* are most common among the fronds of *F. vesiculosus* on sheltered and shallow localities, while the adults of *I. baltica* dominate the more turbulent areas. In between there is a zone which both species cohabit. The following is an attempt to analyse the mechanisms behind the observed facts. In July, when the females are about to release their broods, they seek out approximately the same degree of exposure to wave action. For *I. baltica* the upper layers are too turbulent at locality C, and for *I. chelipes* there is too much detritus at locality A. Thus a concentration of juveniles forms on the moderately exposed shores B. During the first postmarsupial development when the animals show little mobility due to their immaturity, they are restricted to their existing habitats. Competition, even leading to predation, may play a role in survival at that stage, and at first *I. baltica* is favoured by the coexistence with *I.*

chelipes. During later development *I. baltica* becomes very mobile. It actively swims and is thus apparently able to avoid the lack of space and food caused by the competitive potential of the growing *I. chelipes* population. *I. chelipes* is at all times the less active swimmer. In adults this can clearly be seen from the figures showing diurnal activity in *I. chelipes* compared with *I. baltica*, published by Jansson & Källander (1968).

The intense clinging tendency shown by *I. chelipes* assures that the animals remain in the *Cladophora* belt even in August when the vegetation is very thin (Jansson, 1970). At that period the larvae sit even more closely in the habitat than in the experimental bowls, and intraspecific competition certainly decides the number of surviving *I. chelipes*.

When water agitation increases, the strong clinging behaviour becomes insufficient to protect the larvae from being swept away. The stretching behaviour of *I. chelipes* juveniles (see Fig. 3) in such a situation will keep them floating in the surface water and they will tend to be captured again by the upper algal vegetation. Since the species is more fragile, many specimens certainly get lost in the surf. *I. baltica* on the other hand, when exposed to turbulent water, actively loses its hold on the upper algal vegetation. It rolls into a ball (see Fig. 3), a feature which will protect it and very likely assist it in sinking to a depth where water movement is less strong. Here the animal is able to swim and seek out a place to settle. In our opinion the observed differences in the behaviour pattern of *I. chelipes* and *I. baltica* greatly contribute in separating the two species; thereafter physical and chemical factors of the environment may play a part in determining the success of each species.

SUMMARY

1. Quantitative samples indicate a concentration of *I. baltica* and *I. chelipes* juveniles on moderately exposed shores.
2. For the two *Idotea* species studied there are differences in behaviour pattern.
3. In rearing experiments *I. chelipes* grows faster in length than *I. baltica* on a diet of diatoms.
4. To investigate intra- and interspecific competition in *Idotea* larvae at different ages, three randomized experiments were designed. Specimens were kept in pairs and their survival studied.
5. A statistically significant intraspecific competition was found in all experiments. (Level of significance $P < 0.10$.) An indirect analysis points to interspecific competition in the youngest stages.

6. Juvenile animals of both species are thought to congregate at moderately wave-exposed localities, and to find their way to the characteristic habitat mainly as a result of differences in behaviour.

REFERENCES

FORSMAN, B. (1956). Notes on the invertebrate fauna of the Baltic. *Ark. Zool.* **9**, 389–419.

JANSSON, A-M. (1966). Diatoms and microfauna – producers and consumers in the *Cladophora* belt. *Veröff. Inst. Meeresforsch. Bremerh.* (Sonderbd) **2**, 281–8.

JANSSON, A-M. (1967). The food-web of the *Cladophora*-belt fauna. *Helgoländer wiss. Meeresunters.* **15**, 574–88.

JANSSON, B-O. (1969a). Asköarkipelagen. *Zool. Revy.* **31** (1–2), 7–8.

JANSSON, A-M. (1969b). Competition within the *Cladophora* belt. *Limnologica* **7**, 113–17.

JANSSON, A-M. (1970). Production studies in the *Cladophora* belt. *Thalassia jugosl.* **6**, 143–55.

JANSSON, B-O. & KÄLLANDER, C. (1968). On the diurnal activity of some littoral peracarid crustaceans in the Baltic Sea. *J. exp. mar. biol. Ecol.* **2**, 24–36.

MATSAKIS, J. (1956). Développement postembryonaire d'*Idothea viridis* (Slabber) provenant de l'Etang de Leucate. *Vie Milieu* **7**, 287–330.

MILNE, A. (1961). Definition of competition among animals. In *Mechanisms in Biological Competition. Symp. Soc. exp. Biol.* XV, pp. 40–61.

MUUS, B. J. (1967). The fauna of Danish estuaries and lagoons. *Meddr Danm. Fisk.-og Havunders.* **5**, 316.

NAYLOR, E. (1955). The comparative external morphology and revised taxonomy of the British species of *Idotea. J. mar. biol. Ass. U.K.* **34**, 467–93.

V. OERTZEN, J-A. (1965). Stoffwechselaktivitätsmessungen (Sauerstoffverbrauch) an Invertebraten der Fucuscoenose aus der mittleren Ostsee. *Zool. Anz.* **175**, 166–73.

V. OERTZEN, J-A. (1968). Untersuchungen über die Besiedlung der Fucusvegetation der Gewässer um Hiddensee. I. Ökologischer Teil. II. Produktionsbiologischer Teil. *Z. Fisch.* NF **16**, 252–77.

RAVANKO, O. (1969). Benthic algae as food for some evertebrates in the inner part of the Baltic. *Limnologica* **7**, 203–5.

SEGERSTRÅLE, S. G. (1944). Über die Verbreitung der *Idothea*-arten im baltischen Meeresgebiet Finnlands. Soc. Sci. Fennica, *Commentat. biol.* **9** (15), 1–15.

SCHEIN, M. W. & FOHRMAN, M. H. (1955). Social dominance relationships in a herd of dairy cattle. *Br. J. Anim. Behav.* **3**, 45–55.

SYWULA, T. (1964). A study on the taxonomy, ecology and the geographical distribution of species of genus *Idothea* Fabricius (Isopoda, Crustacea) in Polish Baltic. I. Taxonomical part. II. Ecological and zoogeographical part. *Bull. Soc. Amis. Sci. Lett. Poznan,* Ser. D **4**, 141–200.

CHOICE OF ALGAE BY
LARVAE OF *SPIRORBIS*, PARTICULARLY
OF *SPIRORBIS SPIRORBIS*

E. W. KNIGHT-JONES, JULIE H. BAILEY
AND M. J. ISAAC

Department of Zoology, University College of Swansea

INTRODUCTION

The little tubicolous worms, which are found so characteristically on the brown alga *Fucus serratus* L., were called *Serpula spirorbis* (serpulidae), by Linnaeus (1758). Daudin (1800) called them *Spirorbis borealis* and many biologists have followed him, whilst others (Bush, 1904; Rioja, 1942) have used the name *Spirorbis spirorbis*, for Linnaeus's description was at least as good as Daudin's. The older name has never lapsed and we are inclined to use it now, not only because it has priority, but because the more popular name *borealis* was applied for more than a century in a rather loose and confusing way, to a complex of closely related forms, which are best regarded as separate species (de Silva & Knight-Jones, 1962; Gee & Knight-Jones, 1962; L'Hardy & Quiévreux, 1962). These are found attached to specific algae and their larvae choose these algae for settlement. The larvae will very rarely settle upon the algae favoured by another species (de Silva, 1962).

S. spirorbis is less specific than some of its close relatives, in its choice of alga for settlement. In laboratory experiments at Menai Bridge its larvae settled on *Laminaria saccharina* L., *Dictyota dichotoma* (Huds.), *Fucus spiralis* L. and *Fucus vesiculosus* L. just as readily as on *Fucus serratus*, or even more so (Gross & Knight-Jones, 1957). In those experiments *F. vesiculosus* was more favoured than *F. serratus*, but in other situations that order of preference seems to be reversed (see for instance Pyefinch, 1943). The reverse preference is shown clearly in a population from Bracelet Bay, Mumbles Point, near Swansea (grid ref. 629872, 1 in O.S. sheet 153). That population, which was studied by de Silva (1962), is attached exclusively to *F. serratus*. A very different situation is found at Abereiddy, Pembrokeshire (grid ref. 795315, 1 in O.S. sheet 151), where there is an old quarry flooded by the sea. Spirorbinae are abundant there and the depth distribution of five sublittoral species has already been noted (Bailey, Nelson-Smith & Knight-Jones, 1966). *F. serratus* is scarce or absent in the quarry, however, which harbours a dense population of *S. spirorbis* attached exclusively to *F. vesiculosus*. We thought it interesting to compare the larval behaviour of this population with that at Mumbles Point and, whilst we had abundant larvae, to attempt some preliminary experiments bearing on the question of what it is, in *Fucus*, that promotes settlement of *S. spirorbis*.

METHODS

The embryos of *S. spirorbis* are incubated for about a fortnight (Garbarini, 1936) and the larvae can be obtained from the adults in great numbers, particularly during the peak periods of liberation, which occur at about the moon's quarters (Garbarini, 1933; de Silva, 1967). Given favourable conditions, the larvae swim for no more than a few hours (Knight-Jones, 1953). Larvae used in the following experiments were obtained from adults that had been collected less than 10 days previously. The adult stocks were kept in subdued light in large glass tanks of sea water, which was stirred by vigorous bubbling with air and changed daily. In the evening and overnight the tanks were covered with a black cloth.

In the morning this cover was removed and the adults were transferred, in batches of a few thousand, to fresh sea water in smaller bowls, which were placed in a north-facing window and left unstirred. Larvae were soon liberated and swam to the side nearest the light, from which they were pipetted or siphoned away into clean beakers. The bowls and beakers had been swabbed vigorously with cotton wool, to discourage settlement of larvae on the glass (Knight-Jones, 1951).

The settlement experiments were carried out in flat-bottomed glass dishes, which were generally placed on white paper. Since the larvae are photosensitive the light-absorbing properties of experimental substrata were matched as closely as possible and the experimental pairs were equally illuminated. In some experiments even illumination was ensured by rotating the dishes on turntables at a slow constant speed (Crisp & Ryland, 1960). The algae were collected from Mumbles Point no more than 3 days before the experiments and were kept before use in shallow dishes with relatively large volumes of sea water. The pieces offered as substrata for settlement were several centimetres long, having been cut with scissors to exclude air bladders and matched for equal surface area and intensity of pigmentation.

In analysing the results statistically, it was clearly unjustifiable to treat individual larvae as units, because the larvae are gregarious during settlement. Each experiment was therefore treated as a unit and the result subjected to an angular transformation (number settled on control/total number settled $= \sin^2\theta$), which renders the variance independent of the mean and so promotes homogeneity in treatment. In deriving a mean value of θ for each series, however, each result was weighted according to the number of larvae involved, to minimize the effect of batches with large errors due to small numbers of larvae settling. P was then derived from $45 - \theta/\Delta\theta$, with the help of the appropriate table (Fisher & Yates, 1948).

CHOICE BETWEEN *FUCUS* SPECIES

In a series of 16 experiments, larvae from Abereiddy quarry stock of *S. spirorbis* were allowed to choose between *F. serratus* and *F. vesiculosus* (Table 1). The results indicated a distinct preference for *F. vesiculosus* ($P < 0.02$). In a further series of 24 experiments larvae from the Mumbles Point stock were given a similar choice (Table 2). They showed a significant preference for *F. serratus* ($P < 0.05$), supporting the idea that there was a difference between the two stocks.

Table 1. *Choice of S. spirorbis larvae between F. serratus and F. vesiculosus*

(Larvae from adults on *F. vesiculosus* at Abereiddy.)

F.s.	F.s.	F.v.	F.v.
7	2	117	170
20	13	167	173
0	36	24	0
3	7	56	27
1	80	57	0
0	80	124	0
1	62	121	9
0	38	121	12
32	318	787	391
350		1178	

Since the larvae were tending to choose the algal species to which their parents had been attached, it seemed possible that they might have been conditioned to it temporarily, as a result of having been incubated in close proximity. The two stocks were attached exclusively to the preferred species of alga, so samples were obtained of a third stock of *S. spirorbis*, from rocks about 100 m north-east of Menai Bridge pier, Anglesey (grid ref. 560721, 1 in O.S. sheet 106). From that situation, adults were obtained attached quite abundantly to both species of *Fucus*. These were separated into different tanks and thus yielded two batches of larvae, one from parents on *F. serratus*, the other from parents on *F. vesiculosus*. These larvae, allowed to choose as before, settled equally readily on both species of algae (Table 3). This indicates that the immediate situation of individual parents is of no direct importance. It supports the idea, however, that there may be, within a single species of *Spirorbis*, genetically determined differences in settlement behaviour between stocks from different localities.

Table 2. *Choice of S. spirorbis larvae between*
F. serratus and F. vesiculosus

(Larvae from adults on *F. serratus* at Mumbles Point.)

F.s.	F.s.	F.v.	F.v.
39	28	0	10
7	311	0	0
3	21	67	0
44	22	2	0
2	11	14	0
10	6	17	0
62	11	8	2
58	8	8	0
56	4	2	67
430	1	17	23
10	6	190	5
173	0	125	13
894	429	450	120
	1323		570

Table 3. *Choice of S. spirorbis larvae between*
F. serratus and F. vesiculosus

(Using larvae from Menai Bridge.)

Adults on *F. serratus*		Adults on *F. vesiculosus*	
F.s.	F.v.	F.s.	F.v.
610	495	201	255
126	302	18	15
1266	950	626	484
19	113	175	304
35	123	0	8
96	259	28	64
311	229	413	163
2463	2471	1461	1293

EFFECT OF REDUCED ILLUMINATION

It has been noticed that, in darkness, settlement occurs less readily on *Fucus* (Knight-Jones, 1951), so experiments were carried out to see whether reduced illumination would differentially affect choice between algae. Ten series of experiments were carried out with *S. spirorbis* from Abereiddy quarry, at each of four levels of illumination (Table 4). In each series the larvae showed their usual predilection for *F. vesiculosus* and the overall results which demonstrate this are highly significant ($P < 0.02$ for Table 4; $P < 0.001$ for Tables 1 and 4 combined). The numbers attaching themselves to the upper and lower surfaces of the algae were recorded separately, but are approximately equal, and this relationship too, showed no very marked change at different illuminations. The total numbers settling, however, were considerably reduced when the light was dimmer.

Table 4. *Choice of S. spirorbis larvae between F. serratus and F. vesiculosus in different illuminations*

(Larvae from adults on *F. vesiculosus* at Abereiddy point.
Upper surface and Lower surface of *F. serratus* and *F. vesiculosus*.)

Light background				Black background			
5000 lx		1500 lx		500 lx		50 lx	
F.s.	F.v.	F.s.	F.v.	F.s.	F.v.	F.s.	F.v.
58	408	21	3	0	8	0	9
144	410	6	15	0	28	4	19
8	100	94	0	86	0	23	40
129	88	235	11	111	1	31	93
83	87	2	39	0	28	3	0
63	160	0	88	1	8	1	1
24	56	18	0	0	79	2	2
45	180	47	13	0	58	0	0
0	98	13	7	0	88	0	16
11	121	26	4	0	22	0	66
0	54	1	66	1	50	1	3
0	69	0	88	3	71	0	0
56	6	65	0	0	60	0	3
51	8	65	0	6	45	0	4
21	136	18	51	8	63	1	114
55	125	0	55	3	42	0	39
26	33	0	58	2	19	10	1
15	32	0	54	0	46	3	2
0	17	2	78	12	87	0	0
3	99	1	47	11	92	0	0
276	995	234	302	109	482	40	188
516	1292	380	375	135	413	39	164
792	2287	614	677	244	895	79	352
3079		1291		1139		431	

Totals: 1729 *F. serratus*, 4211 *F. vesiculosus*.

A similar experiment, but at three levels of illumination, was carried out with larvae from Mumbles Point (Table 5), which continued to prefer *F. serratus* ($P < 0.001$).

Table 5. *Choice of S. spirorbis larvae between F. serratus and F. vesiculosus in different illuminations*

(Larvae from adults on *F. serratus* at Mumbles Point.
Upper surface and Lower surface of *F. serratus* and *F. vesiculosus*.)

5000 lx		1000 lx		50 lx	
F.s.	*F.v.*	*F.s.*	*F.v.*	*F.s.*	*F.v.*
2	21	11	3	23	10
4	30	65	2	19	1
44	2	24	9	34	54
56	2	57	18	49	5
7	80	71	3	5	42
5	65	32	50	1	4
36	9	18	13	23	0
18	14	34	17	58	0
10	9	17	10	32	0
12	16	4	24	15	2
14	0	22	7	24	8
34	0	72	10	53	3
78	4	87	5	15	44
60	0	63	10	24	45
18	23	10	2	33	2
22	32	41	12	63	3
22	3	10	0	10	20
15	0	18	5	19	10
10	0	6	1	3	2
22	0	25	3	10	5
241	151	276	53	202	182
248	159	411	151	311	78
489	310	687	204	513	260
799		891		773	

Totals: *F. serratus* 1689, *F. vesiculosus* 774.

It was considered that the decline in numbers settling on *Fucus* at reduced illumination, which is recorded in Table 4, might have resulted directly from the larvae having difficulty in finding the *Fucus* visually. In other words, with a dark background, in dim light, the photonegative larvae might not be directed so strongly to the light-absorbing *Fucus*, because of the lack of contrast. By way of testing this possibility, nine pieces of dark slate were matched with pieces of *F. serratus* and left in sea water overnight, to accumulate the bacterial film which is necessary for the settlement of *S. spirorbis* on such surfaces. Larvae from Abereiddy quarry were then allowed to choose between these pieces of slate and *Fucus*, at three levels of illumination. The larvae were distributed carefully between the dishes, so that experiments at different illuminations involved larvae

Table 6. *Choice of S. spirorbis larvae between F. serratus and slate*

(Slate in pieces of similar size in three separate dishes at each of three different illuminations.)

5000 lx		1000 lx		50 lx	
Fucus	slate	Fucus	slate	Fucus	slate
48	169	67	26	1	216
47	178	6	49	8	26
62	120	0	90	0	153
157	467	73	165	9	395

Table 7. *Choice of S. spirorbis larvae between F. serratus and slate*

(Filmed slate at two different illuminations.)

5000 lx		50 lx	
F. serratus	Slate	F. serratus	Slate
81	61	0	9
57	11	61	132
173	97	3	15
73	5	40	96
38	66	8	22
2	99	5	60
28	3	1	0
18	6	2	44
12	16	10	9
17	15	0	4
497	380	130	391

that were similar in numbers and parental origin. Once more, the numbers settling on *Fucus* declined markedly at reduced illuminations, but the numbers settling on the slates remained high (Table 6). These results were confirmed by a similar series of experiments at two levels of illumination (Table 7). In bright north light the pieces of *F. serratus* were almost or quite as favourable for settlement as the filmed slates, but in deep shadow they were much less so ($P < 0.001$). The effect was not due to the larvae failing to find the *Fucus*. They were observed to find and crawl upon the *Fucus* in great numbers, but they usually swam away again without much delay. A preliminary experiment indicated that the *Fucus* became less unfavourable in reduced illumination, if the water was stirred (Table 8).

Table 8. *Effect of water stirring on the choice of S. spirorbis larvae between F. serratus and slate*

(In stagnant conditions, reduction in illumination resulted in the *Fucus* becoming relatively unfavourable for settlement, but this effect was less pronounced if the water was stirred.)

No rotation or stirring				Dishes rotated and stirred			
5000 lx		50 lx		5000 lx		50 lx	
F. serratus	Slate	F. serratus	Slate	F. serratus	Slate	F. serratus	Slate
97	14	12	58	900	4	157	9
83	2	1	18	330	126	29	255
332	130	1	237	320	150	1	223
219			0	119		44	
731	146	14	313	1669	280	231	487

VARIOUS PHYSICAL AND CHEMICAL TREATMENTS OF FUCUS

Brief immersion of *F. serratus* in sea water heated to 45 °C or higher rendered it unfavourable for settlement. Table 9 shows the result of experiments in which pieces of *Fucus* were thus immersed for 1 min in sea water at five different temperatures. They were then arranged round the peripheries of two large, circular, slowly rotating glass troughs of sea water, in such a way that *Fucus* raised to each temperature was represented twice in each trough. Abundant larvae were added, from Abereiddy stock, and settlement occured on *Fucus* which had been exposed to temperatures of up to 40 °C, but not 45 °C. Very many larvae were trapped in mucus strands on the bottom of the trough and were thus unable to complete the settlement behaviour pattern which is a prerequisite for metamorphosis. Evidently heating to 45 °C caused the *Fucus* to produce this mucus in great

Table 9. *Effect of the immersion of F. serratus in heated sea water on S. spirorbis settlement*

(Immersion at each temperature for 1 min.)

25 °C	30 °C	35 °C	40 °C	45 °C
1	181	4	7	0
0	0	2	8	0
0	3	3	9	0
6	7	4	4	0
7	191	13	28	0

quantities and it then became unsuitable for settlement. Immersion in sea water at 40 °C for a longer period (1 h) also made *Fucus* unsuitable for settlement (Table 10).

Table 10. *Effect of the immersion of F. serratus in heated sea water and formaldehyde on S. spirorbis settlement*

(In 40 °C sea water for 1 h, in 40 % formaldehyde for 5 min.)

Sea water		Formaldehyde	
Treated	Untreated	Treated	Untreated
0	41	5	277
0	36	0	265
0	51	1	394
0	35	0	352
0	41	2	328
0	46	2	235
0	55	3	264
0	34	2	226

For further experiments, pieces of *Fucus* were immersed in solutions of formaldehyde followed by thorough rinsing for about half an hour in running tap water. Larvae would not settle on this *Fucus*, but did so readily on controls (Table 10), which had been similarly rinsed in tap water. Immersion for 10 s in 0.1 N hydrochloric acid or sodium hydroxide made the *Fucus* unsuitable for settlement, but similar brief immersion in 0.01 N solutions had no appreciable effect (Table 11). The effects of acid treatment were most drastic. Larvae tended to stick to the *Fucus* so treated, but were unable to crawl upon it and died without metamorphosing. Settlement on the control *Fucus* in the same dish was thus reduced. Similarly, many larvae were found lying, inactive and unmetamorphosed, around the *Fucus* that had been treated with sodium hydroxide, but some others succeeded in settling and metamorphosing, even on *Fucus* which had been treated with N sodium hydroxide and was markedly swollen in places.

Evidently the firm but somewhat mucilaginous surface of the *Fucus* can be made unsuitable for settlement by various simple treatments. Too much surface mucus is unfavourable to the crawling larvae, trapping them in great numbers. Probably the *Fucus* reacts by mucus-production to various irritant treatments and certainly too much surface mucus will prevent settlement.

Table 11. *Effect of acid and alkali treatment of F. serratus on the settlement of S. spirorbis*

Treatment time	2 min		1 min		10 s		10 s	
Rinse time	30 min		15 min		20 min		20 min	
	N		0.1 N		0.1 N		0.01 N	
	Treated	Untreated	Treated	Untreated	Treated	Untreated	Treated	Untreated
Sodium hydroxide	0	118	0	5	3	1	12	188
	2	166	8	6	4	12	119	25
	89	5	2	70	1	157	83	134
	1	100	59	6	0	80	67	152
Hydrochloric acid	0	1	0	0	1	3	47	165
	0	1	0	12	0	0	64	139
	0	9	0	0	0	29	205	20
	0	9	0	0	0	0	198	33

DISCUSSION

Spirorbis spirorbis is particularly characteristic of sheltered marine environments such as the Menai Straits, where it forms abundant populations. The
bulk of such a population is attached to *F. serratus* and the vertical zona-
tions of the polychaete and alga coincide closely, lying between low water
marks of neap and spring tides. *F. vesiculosus* occupies a higher and wider
range, extending throughout the mid-littoral zone (see for instance, Moyse
& Nelson-Smith, 1963). It can survive a considerable amount of desiccation and may become fairly dry and brittle in hot sunshine, absorbing
water and reviving when the tide returns. Recently settled *Spirorbis* have
thin mucous tubes and could not survive such conditions. It is only on the
lowest plants of *F. vesiculosus*, which occur in or adjoining the *F. serratus*
zone, that *Spirorbis* is found. To settle on the higher *F. vesiculosus* plants
would be suicidal and it may at first seem rather remarkable that natural
selection has not discouraged the *Spirorbis* larvae from choosing this alga
for settlement.

It seems unlikely that the buoyant air bladders of *F. vesiculosus* are very
relevant to this problem. The fact that this alga floats up, whilst searching
Spirorbis larvae become photonegative and thus swim down, does not
prevent *Spirorbis* from becoming quite abundant on the lowest plants of
F. vesiculosus, in these sheltered localities. The old plants often become
very large, break loose from their holdfasts and may be carried by currents
for considerable distances before being cast ashore. It must help the distribution of the species to have small breeding populations rafted about in this
way. These are the pioneers, which would invade new areas, but they
would soon disappear from established stocks if there was strong selection
pressure against them. Evidently this pressure is not too strong, probably
for at least two reasons. First, it seems likely that gregariousness during
settlement takes very many of the pelagic larvae back to the zone occupied
by the adults. The mortality which no doubt occurs through settlement on
Fucus at high levels may thus be relatively small. Secondly, there is in
sheltered situations severe competition for space on *F. serratus* from various
encrusting Polyzoa (Ryland, 1962). These smother great numbers of
Spirorbis, but they are not found so abundantly on the lowest plants of
F. vesiculosus. Given gregariousness, to take the larvae back to low tide
levels, it may well be a positive advantage to choose *F. vesiculosus* as
a substratum.

On the exposed coasts of the Gower peninsula, just west of Swansea, the
situation is very different. Polyzoa are scarce on *F. serratus*, perhaps be-

cause of abrasion by wave-borne sand particles. *S. spirorbis* is scarce too and is indeed mostly confined to deep rock pools, remote from sandy areas. These pools are about or above half-tide mark and are cut off from the sea for 6 h or more, during low water, which is more than sufficient time for the larvae of *S. spirorbis* to complete their pelagic phase (Gee, 1963) and settle. *F. vesiculosus* is comparatively scarce in these pools. It occurs round them and may hang into them, but its fronds are obviously more likely to be washed about and abraded than those of *F. serratus*, which lie permanently immersed in the pools. It is not surprising that a marked preference for *F. serratus* has been selected for, in the small populations of *S. spirorbis* which manage to survive on these exposed coasts. They seem to be endangered more by the turbulent conditions than by competition from Polyzoa.

Conditions in the Abereiddy quarry are extremely sheltered and it seems that *F. serratus* scarcely survives there. The connection with the sea is said to be less than 100 years old, but it does not follow that selection has brought about a marked change within that time. The adjoining coast is exposed to wave action, but it is less so than the Gower and there are other quite sheltered areas not far away, from which *F. vesiculosus* plants bearing *S. spirorbis* may have drifted, to introduce into the quarry a population that was already inclined to favour that algal species. The area is rather remote from Swansea and we have scarcely begun to survey it as yet. It seems quite likely, however, that these *Spirorbis* may have developed their special predilection for *F. vesiculosus* since they were introduced into the quarry. This follows not only from a consideration of how quickly artificial selection may bring about striking changes in the behaviour of other animals, but from the distribution in Wales of certain strains of *S. spirillum* (L.) which favour *Laminaria saccharina*.

Spirorbis (*Circeis*) *spirillum* is a sublittoral species, which favours organic substrata. It is found on certain hydroids, particularly *Abietinaria abietina* (L.), some Polyzoa such as *Cellaria*, and the insides of *Buccinum* shells inhabited by large *Pagurus bernhardus* (L.). One very closely related form, which is found characteristically attached to lobsters and crawfish, is sometimes regarded as a distinct species, *Spirorbis armoricanus* Saint-Joseph (see Gee, 1964), or alternatively as a mere variety.

In Norwegian fjords *S. spirillum* is recorded from *Laminaria* (Bergan, 1953). Similarly, in Loch Sween in Scotland and Lough Ine in Ireland, it occurs abundantly on *L. saccharina*. We have never observed it, however, on that alga in strictly natural situations on the Welsh coasts, not even in Milford Haven, which is in many respects like a sea-loch, though with stronger currents and more turbid water. The only species we have seen abundantly there on *Laminaria* are the dextral *Spirorbis* (*Janua*) *pagenstecheri* Quatrefages and sinistral *Spirorbis* (*Spirorbis*) *inornatus* L'Hardy & Quiévreux (1964). There are abundant populations of *S. spirillum* on *Laminaria*, however, in the large and essentially artificial harbours of Holyhead and Fishguard, where there is deep, quiet, clear water, as in a sea-loch. These seem to be special strains since this quite abundant species does not settle on *Laminaria* elsewhere in Wales. It is possible that these strains, which favour *Laminaria*, have been brought from Scottish or Irish harbours attached to coastal shipping, for *S. spirillum* is one of the species commonly recorded from ships' fouling exposure-plates in northern Britain (Meadows, 1969). It seems quite likely, however, that they have arisen independently in the two Welsh harbours, in response to the same selection pressures that operate in sea-lochs. Fishguard Harbour was completed about 1906 (Gilpin, 1960), but Holyhead Harbour is much older.

It may easily be imagined how selection of such strains, leading to gradually increasing skill in the choice of suitable and specific substrata, may have contributed to speciation in the Spirorbinae (de Silva, 1962; Crisp, 1965). The capacity for self-fertilization in these forms would have assisted such divergence (Gee & Williams, 1965), which may in any case continue under divergent selection pressures despite gene flow of the same amount as is given by random mating (Millicent & Thoday, 1960). We may expect to find examples of such speciation in other parts of the world, when the taxonomy of Spirorbinae is better known. It is interesting to note that Harris (1969) recently described from Tristan da Cunha two new sinistral species, *S. inventis* and *S. scoresbyi*, which had simple collar setae and therefore seem quite closely related to *S. perrieri*, previously recorded from that island by Day (1954). All three species could be called *Romanchella* Caullery & Mesnil (see Bailey, 1969), a subgenus which is otherwise represented throughout the world very sparsely (it includes only one other well-established species). To have these three species of *Romanchella* around this small island, all attached to *Macrocystis* or other algae, seems a significant situation which deserves further study.

Our finding that *Fucus* becomes less favourable for settlement in dim light, does not indicate a phenomenon of much significance to *S. spirorbis*, which liberates larvae only when the light is fairly bright. It suggests something of interest in the physiology of *Fucus*, which would be worth a special investigation. For further studies on the question of what qualities of algae promote settlement, the methods used by Williams (1964) and Gee (1965), involving the adsorption of algal extracts on to surfaces filmed by bacteria, seem very promising. Our limited experience of such methods, however, with larvae of *S. spirorbis*, suggested that it would be better to use species such as *S. corallinae* de Silva & Knight-Jones, or *S. rupestris* Gee & Knight-Jones, which are more highly specific to their favoured algae. It is, however, comparatively difficult to obtain larvae in abundance from these species, particularly because their larvae do not readily swim (Gee, 1963).

SUMMARY

Larvae from Swansea populations of *S. spirorbis* preferred to settle on *F. serratus* rather than *F. vesiculosus*, but this preference was reversed in a population from a flooded quarry at Abereiddy, Pembrokeshire. Both species of *Fucus* were about equally favourable to larvae from Menai Bridge, irrespective of whether their parents had been attached to *F. serratus* or to *F. vesiculosus*.

In reduced illumination *Fucus* became less favourable for settlement, especially in stagnant water, compared with granite and slate. *F. serratus* was made unsuitable for settlement by heating to 40 °C for 1 h, or to 45 °C for 1 min, or by brief immersion in acid, alkaline or formaldehyde solutions.

Some speciation amongst Spirorbinae seems to have been related to the choice of certain algae as specific substrata.

REFERENCES

BAILEY, J. H. (1969). Methods of brood protection as a basis for reclassification of the Spirorbinae. *Zool. J. Linn. Soc.* **48**, 387–407.

BAILEY, J. H., NELSON-SMITH, A. & KNIGHT-JONES, E. W. (1966). Some methods for transects across steep rocks and channels. *Rep. Underwat. Ass.* 1966–67, 107–11.

BERGAN, P. (1953). The Norwegian species of *Spirorbis*. *Nytt Mag. Zool.* **1**, 27–48.

BUSH, K. J. (1904). Tubicolous annelids of the tribes Sabellides and Serpulides from the Pacific Ocean. *Harriman Alaska Expedition* **12**, 169–355.

CRISP, D. J. (1965). Surface chemistry, a factor in the settlement of marine invertebrate larvae. *Bot. Gothoburg.* **3**, 51–65.

CRISP, D. J. & RYLAND, J. S. (1960). Influence of filming and of surface texture on the settlement of marine organisms. *Nature, Lond.* **185**, 119.

DAUDIN, F. M. (1800). *Recueil de mémoires et de notes sur les espèces peu connues de Mollusques, Vers et Zoophytes*. Paris.

DAY, J. H. (1954). The polychaeta of Tristan da Cunha. *Results Norw. scient. Exped. Tristan da Cunha* **29**, 1–35.

de SILVA, P. H. D. H. (1962). Experiments on choice of substrata by *Spirorbis* larvae. *J. exp. Biol.* **39**, 483–90.

de SILVA, P. H. D. H. (1967). Studies on the biology of Spirorbinae. *J. zool. Lond.* **152**, 269–79.

de SILVA, P. H. D. H. & KNIGHT-JONES, E. W. (1962). *Spirorbis corallinae* n.sp. and some other Spirorbinae common on British shores. *J. mar. biol. Ass. U.K.* **42**, 601–8.

FISHER, R. A. & YATES, F. (1948). *Statistical Tables for Biological, Agricultural and Medical Research.* London.

GARBARINI, P. (1933). Rhythme d'émission des larves chez *Spirorbis borealis* Daudin. *C. r. Séanc. Soc. Biol.* **112**, 1204–5.

GARBARINI, P. (1936). Rhythme de croissance des oocytes et d'incubation des larves chez *Spirorbis borealis* Daudin. *C. r. Séanc. Soc. Biol.* **122**, 157–8.

GEE, J. M. (1963). Pelagic life of *Spirorbis* larvae. *Nature, Lond.* **198**, 1109–10.

GEE, J. M. (1964). The British Spirorbinae, with a description of *Spirorbis cuneatus* sp.n. and a review of the genus *Spirorbis*. *Proc. zool. Soc. Lond.* **143**, 405–41.

GEE, J. M. (1965). Chemical stimulation of settlement in larvae of *Spirorbis rupestris*. *Anim. Behav.* **13**, 181–6.

GEE, J. M. & KNIGHT-JONES, E. W. (1962). The morphology and larval behaviour of a new species of *Spirorbis*. *J. mar. biol. Ass. U.K.* **42**, 641–54.

GEE, J. M. & WILLIAMS, G. B. (1965). Self- and cross-fertilization in *Spirorbis borealis* and *S. pagenstecheri*. *J. mar. biol. Ass. U.K.* **45**, 275–85.

GILPIN, M. C. (1960). Population changes round the shores of Milford Haven from 1800 to the present day. *Fld Stud.* **1** (2), 23–36.

GROSS, J. & KNIGHT-JONES, E. W. (1957). The settlement of *Spirorbis borealis* on algae. *Rep. Challenger Soc.* **3** (9), 18.

HARRIS, T. (1969). *Spirorbis* species from the South Atlantic. *Discovery Rep.* **35**, 135–78.

KNIGHT-JONES, E. W. (1951). Gregariousness and some other aspects of the setting behaviour of *Spirorbis*. *J. mar. biol. Ass. U.K.* **30**, 201–22.

KNIGHT-JONES, E. W. (1953). Decreased discrimination during setting after prolonged planktonic life in larvae of *Spirorbis borealis*. *J. mar. biol. Ass. U.K.* **32**, 337–45.

L'HARDY, J.-P. & QUIÉVREUX, C. (1962). Remarques sur le polymorphisme de *Spirorbis borealis* Daudin. *C. r. hebd. Séanc. Acad. Sci., Paris* **255**, 2173–5.

L'HARDY, J.-P. & QUIÉVREUX, C. (1964). Observations sur *Spirorbis inornatus* et sur la systématique des Spirorbinae. *Cah. Biol. mar.* **5**, 287–94.

LINNAEUS, C. VON (1758). *Systema Naturae.* I. 10th edition, p. 787. Stockholm.

MEADOWS, P. S. (1969). Sublittoral fouling communities on northern coasts of Britain. *Hydrobiologia* **34**, 273–94.

MILLICENT, E. & THODAY, J. M. (1960). Gene flow and divergence under disruptive selection. *Science, N.Y.* **131**, 1311–12.

MOYSE, J. & NELSON-SMITH, A. (1963). Zonation of animals and plants on rocky shores around Dale, Pembrokeshire. *Fld Stud.* **1** (5), 1–31.

PYEFINCH, K. A. (1943). The intertidal ecology of Bardsey Island, North Wales. *J. Anim. Ecol.* **12**, 82–108.

RIOJA, E. (1942). Estudios anelidologicos. V. Observaciones acerca de algunas especies del genero *Spirorbis* Daudin de las Mexicanas del Pacifico. *An. Inst. Biol. Univ. Méx.* **13**, 137–53.

RYLAND, J. S. (1962). The association between Polyzoa and algal substrata. *J. Anim. Ecol.* **31**, 331–8.

WILLIAMS, G. B. (1964). The effect of extracts of *Fucus serratus* in promoting the settlement of larvae of *Spirorbis borealis*. *J. mar. biol. Ass. U.K.* **44**, 397–414.

SETTLEMENT AND ORIENTATED GROWTH IN EPIPHYTIC AND EPIZOIC BRYOZOANS

J. S. RYLAND AND A. R. D. STEBBING

Department of Zoology, University College of Swansea

INTRODUCTION

In a majority of bryozoans the colony forms a unilamellar adherent crust over the surface of stones, shells, algae or other substrata. There may be marked specificity with respect to the substratum utilized: thus *Membranipora membranacea* (L.) occurs almost exclusively on the fronds of *Laminaria* spp. and only rarely on other algae. *Electra pilosa* (L.), on the other hand, displays great catholicity in this respect, and is found on inert objects as well as on algae of many species (Ryland, 1962). *Scrupocellaria reptans* (L.) is another species found on a wide range of substrata, one of which is another bryozoan, the frondose species *Flustra foliacea* (L.). The purpose of this paper is to draw attention to some striking patterns of settlement and growth exhibited by *M. membranacea*, *E. pilosa* and *S. reptans* on certain organic substrata.

All the bryozoans discussed in this paper belong to the Cheilostomata. Two very different larval types occur in this order: first, the relatively long-lived planktotrophic larvae, known as cyphonautes, which are produced by species of *Electra* and *Membranipora*; second, the short-lived lecithotrophic larvae found in most other genera, including *Scrupocellaria*. That a cyphonautes (originally known as *Cyphonautes compressus* Ehrenberg) metamorphosed into the primary zooid of *E. pilosa* has long been known (full account in Kupelwieser, 1905); but it is only quite recently that Atkins (1955) demonstrated conclusively that *M. membranacea* also developed from a cyphonautes, although this had been suspected previously. It was shown later that two supposedly distinct larvae, known in the literature as *C. borealis* Lohmann and *C. schneideri* Lohmann, both belonged to *M. membranacea*. The two 'species' represented no more than variation in the shape of the shell (Ryland, 1964).

The fertile zooid of *E. pilosa* produces 10–20 ovoid eggs of about 60 μm in length (Marcus, 1926), which are shed through a tubular opening (the intertentacular organ) lying between the dorso-medial tentacles. Fertilization, by spermatozoa from another zooid, presumably occurs in the intertentacular organ during the discharge of the ova (Silén, 1966). Development proceeds in the sea until a young cyphonautes, which has far from reached its final form and organization, hatches from the egg. The larva continues to grow and develop, and is believed to be ready to settle about 2 months after hatching.

Fertilization in *M. membranacea* must be internal for, according to Lutaud (1961), oocytes and developing embryos are both present in the zooids during spring, and the young larvae are all liberated by mid-summer. Entry of sperm and discharge of larvae are presumably both effected through the intertentacular organ, which is present in *Membranipora* as in *Electra* (Hincks, 1880). The larvae continue to develop throughout their planktonic life and grow considerably (Atkins, 1955), although O'Donoghue (1926) thought that the larvae of the Pacific species *M. villosa* Hincks settled soon after their release.

Cyphonautes larvae are considerably modified from the basic trochophore from which they are presumably derived (Hyman, 1959). They are triangular in side view and strongly compressed between a bivalved shell, the two parts of which are held together by an adductor muscle. Fully developed *E. pilosa* larvae measure 400–500 μm across the base and nearly 400 μm in height; those of *M. membranacea* are 750–850 μm long and about 600 μm high. The apex or aboral pole, which is generally directed forwards as the larva swims, is surmounted by a sensory organ from which nerve fibres radiate; while the base, where the valves gape, is encircled by the ciliary corona. There is a tripartite alimentary canal, and the larva feeds on small organisms of the phytoplankton. Other important structures are the anteriorly situated pyriform gland and its associated tuft of long plume cilia (linked to the apical sensory organ by a neuromuscular tract) and the adhesive sac, which lies behind the adductor.

A cyphonautes about to settle glides in an erect position over the substratum, with the pyriform gland foremost and the plume cilia beating on, and presumably testing, the surface (Atkins, 1955). The larva adheres to the substratum with a secretion from the pyriform gland. The cilia of the corona gradually become quiescent. Immediately before settlement the cyphonautes moves around in circles until suddenly, as the adhesive sac everts, the valves separate and are pulled down on to the substratum. During this flattening the apical organ comes to lie just behind the pyriform gland. The larva is by now strongly attached; the shell valves fall away some time later.

The larval organization completely breaks down during metamorphosis and, in *Electra*, a new structure, the ancestrula or primary zooid of the colony, arises in its place. *Membranipora* is unusual among bryozoans in that metamorphosis gives rise to twin primary zooids (O'Donoghue, 1926; Atkins, 1955). The two zooids are joined proximally, but their distal or blastogenic ends diverge.

During metamorphosis there is a reversal of internal polarity: the distal end of the polypide and the blastogenic face of the ancestrula correspond to the posterior of the larva. This correlation is very important, for it reveals how the initial direction of colony formation depends on the orientation of the larva at the time of settlement. In *E. pilosa* the first two zooid buds arise distolaterally and the third between them (Pl. XV, figs. 5–8 in Barrois, 1877); in *Membranipora* the first daughter zooid arises medially, between the twin ancestrulae, with the second and third placed each side of the first (fig. 4 in Atkins, 1955; fig. 2 in Lutaud, 1961).

Young *Membranipora* colonies are roughly subcircular in outline, with budding taking place almost all around the periphery. Later growth, however, is restricted to one or a few broad, rapidly advancing lobes, the distal extremity of each being marked by a conspicuous blastogenic fringe (fig. 1 in Lutaud, 1961). Older colonies may form extensive patches.

The larva of *S. reptans* (Pl. X, fig. 11 in Barrois, 1877) differs in many ways from a cyphonautes. It is ovoid and shell-less, with the ciliary corona expanded so as to cover the whole surface. There is no alimentary canal, and the larva metamorphoses a few hours after its release. The apical sense organ, adhesive sac, pyriform gland and plume cilia are present and occupy the same relative positions as in a cyphonautes. The adhesive sac opens at the pole directly opposite the apical organ. The pyriform gland merges externally with a median groove which extends apically from the adhesive sac until it ends at the tuft of plume cilia midway between the poles. There is a bright red pigment spot on each side, some way behind the plume cilia. Although these are commonly termed 'eye-spots', their function as photo-receptors has not been proven by physiological methods. The larva and the ancestrula, but not metamorphosis, have been described by Barrois (1877), and an account of the early stages of astogeny (i.e. the post-metamorphic development of a colony) has been given by Lutaud (1953).

For behaviour and settlement it is necessary to refer to work on larvae of the closely related genus *Bugula*. In *B. simplex* Hincks the larva and its metamorphosis have been comprehensively described by Calvet (1900: as *B. sabatieri* Calvet), and its settlement behaviour by Grave (1930) as *B. flabellata* (Thompson). Settlement and metamorphosis in another species, *B. neritina* (L.), have been studied by Lynch (1947). These larvae commonly swim with their long axis tilted forwards, so that both the apical organ and the plume cilia are in advance and the median groove is directed downwards.

Prospecting *B. neritina* larvae explore the surface with their plume cilia, while continuously rotating counterclockwise about an axis perpendicular to the opening of the pyriform gland: the apical organ is therefore facing horizontally. As with cyphonautes, secretion from the pyriform gland appears to hold the larva to the selected surface before it settles. At the moment of fixation the median groove grasps the substratum; then the adhesive sac is suddenly everted, spreading as a disc beneath the larva. The median groove releases its hold, and the apical organ returns to a more or less dorsal position. The larva then rotates about its perpendicular axis until the eye-spots are located on the lighted side. It is of interest that orientation is accomplished after eversion of the internal sac, not before. As metamorphosis proceeds in both *Bugula* and *Scrupocellaria* the larval mass starts to elongate in the apical direction, so that the ancestrula stands upright. As in *Electra* and *Membranipora* the distal, blastogenic end of the ancestrula corresponds with the final position of the apical organ.

It appears from Lynch's (1947) description and figures (especially Pl. II, fig. 11) that the disposition of frontal (corresponding to upper in *Electra*) and basal surfaces is determined by the orientation of the meta- morphosing larva to light. The transverse plane of the first zooids is clearly at right angles to the direction of illumination; but whether it is the basal or the frontal surface that faces the light is not stated.

The *Scrupocellaria* colony develops by the extension and bifurcation first of its main axis and then of the constituent rami (Fig. 4b), all of which are composed of a double line of zooids. Numerous rhizoids passing to the substratum anchor the colony and convert its plane of growth from vertical to horizontal.

MATERIAL

Laminaria hyperborea (Gunn.) Fosl., the alga on which *M. membranacea* is generally most plentiful, was collected from 1 to 2 m by diving in Brandy Cove, Gower Peninsula, South Wales. On *Fucus serratus* L., *Membrani- pora* has been found abundantly at two localities only: Church Island in the Menai Strait, North Wales and Spiddal, on the north shore of Galway Bay, Ireland. *E. pilosa*, on the other hand, is readily found on *F. serratus*, and collections were made at Pwll du Bay, Gower Peninsula; Broad Haven, south Pembrokeshire, Wales; and Solva, a sheltered shore near St Davids, west Pembrokeshire.

S. reptans has been studied on *Flustra foliacea* and on the fronds of *Laminaria*. *Flustra*, which occurs sublittorally in localities subjected to strong tidal currents, was dredged from 15 to 20 m off Oxwich Point, Gower

Peninsula. *Laminaria digitata* (Huds.) Lamour. and *L. saccharina* (L.) Lamour. bear abundant *Scrupocellaria* and other epiphytic bryozoans only in habitats where shelter from wave action is combined with extensive water movement. These conditions obtain in a rapids system at Keeraunagark, Cashla Bay, in Connemara, Ireland, where our material was collected.

METHODS

The observations on the orientation of ancestrulae or growth of the bryozoans were made using a sheet of transparent perspex inscribed with a circle and divided for convenience into eight equal segments of 45° each (Fig. 1a). A line bisecting the most distal segment was extended beyond the circle to mark 0 or 360° and was used to orientate the perspex on the substrate organism, so that it was parallel to the axis of the alga or to the lines of zooids of the *Flustra*. The centre of the circle was placed over the ancestrula and that segment of the circle in which the growing edge of the bryozoan was furthest from the centre was counted as the direction of growth. Thus, if a bryozoan colony was growing proximally, parallel to the axis of the substrate organism or within 22.5° either side of this line, it would fall within the 180° segment. The method of assessment in *E. pilosa* had to be slightly different. Here it was necessary to score each actively growing arm of the star-shaped colonies. When assessing the orientation of ancestrulae a stereomicroscope was used.

Colonies on which the observations were made were picked at random, with the proviso that they fell within a specified size range (see below). Certain of these were, however, excluded from consideration: first, colonies whose growth had been obstructed or interfered with in any way by another sessile organism or by irregularities in the substratum; second, colonies which had grown so that they had reached or overlapped the edge of the substrate organism; the third, colonies damaged in any way.

RESULTS

Membranipora membranacea

Orientation of the twin ancestrulae of *M. membranacea* (and therefore of the larva at the time of settlement) on the fronds of *L. hyperborea* is random (Table 1, Fig. 1b). However, early on in the astogeny of the colony, when it consists of only 6 to 15 zooids, orientated growth develops (Fig. 1c). Even the first or second rows of daughter zooids may sometimes show a turning of the direction of growth towards that which characterizes older colonies.

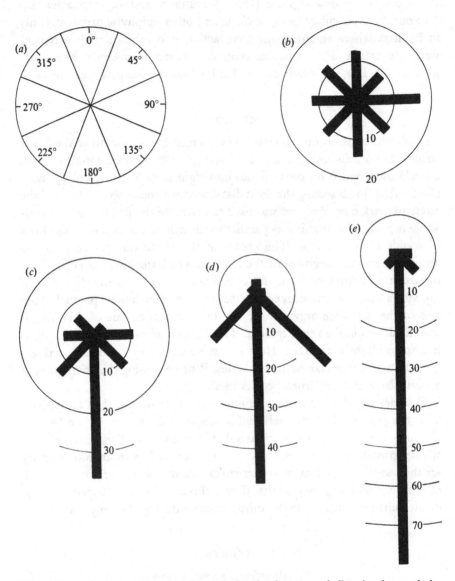

Fig. 1. The orientation of ancestrulae and the development of directional growth in *Membranipora membranacea* on *Laminaria hyperborea*. (*a*) Diagram to show the arrangement of octants used for recording. (*b*) Orientation of ancestrulae. (*c*) Orientation of growth in colonies < 35 mm diameter. (*d*) Orientation of growth in colonies 15–35 mm diameter. (*e*) Orientation of growth using all colonies. The results are expressed as the percentage falling within the range of each octant.

Lutaud (1961) noticed the orientated growth of *Membranipora*, but did not comment upon its significance. One of her drawings (fig. 2 in Lutaud, 1961) shows a very young colony in which a direction of growth, different from the orientation of the ancestrula, is developing.

By the time colonies have reached a diameter of 1.5–3.5 cm most have acquired a roughly proximal growth orientation (Fig. 1 *d*), although the proportion of colonies that are growing straight downwards (180°) is not as great as it is subsequently. When colonies of *Membranipora* of all sizes are considered together (Fig. 1 *e*) there is a very marked preponderance (80%) growing straight downwards. Observations of similarly orientated growth have been made when *L. digitata* is the substrate alga.

M. membranacea may sometimes be locally common on *F. serratus*. On this alga the orientation of the ancestrula (Table 1) is not random (χ^2 for 7 degrees of freedom is 24.00; $P < 0.005$). Although the sample (79 colonies) is not very large, there is a clear tendency for the ancestrulae to face proximally. Growth too is proximally directed and not, as in most of the other associations studied, towards the meristematic tissue or areas of frond extension.

Electra pilosa

E. pilosa is a widely distributed species. In a sheltered position it may be most plentiful on *F. serratus*; where wave action is more severe, *E. pilosa* tends to be commonest on the tufted red algae *Chondrus crispus* (L.) Stackh. and *Gigartina stellata* (Stackh.) Batt. Although much of the *F. serratus* is then devoid of *E. pilosa*, searching reveals a few plants heavily incrusted with this bryozoan. An impression was gained during the present study that young colonies of *E. pilosa* were most likely to be found in the vicinity of older colonies. This suggests that *E. pilosa* larvae respond to the presence of established colonies at the time of settling, a phenomenon already established for many unrelated marine species (see review by Knight-Jones & Moyse, 1961).

The ancestrulae are not randomly orientated on the *Fucus* fronds (Table 2). In a substantial majority (375/589) the median longitudinal axis of the ancestrula and the frond were parallel. A further 171 ancestrulae had their longitudinal axis oblique to that of the frond, and only 43 were at right angles.

It was also observed that many more ancestrulae faced the frond apices than faced the frond base (Fig. 3 *a*). Thus, considering orientation based on the 0 and 180° positions alone, the ratio was 257:118. Taking the combined settlement in the 315, 0 and 45° positions and that in the 135, 180 and 225° positions, the ratio was 361:176. The trend, however, is much more apparent on some shores than on others (Table 2).

Table 1. The orientation of Membranipora membranacea ancestrulae

(Orientation, and direction of growth of colonies expressed as the percentage falling within the range of each octant, o° being upwards (see text).)

	0°	45°	90°	135°	180°	225°	270°	315°	Number of observations
On *Laminaria hyperborea* (Brandy Cove; August 1969)									
Ancestrulae	12.6	8.8	17.0	11.3	14.5	11.9	11.9	11.9	159
All colonies < 3.5 cm diameter	9.5	3.4	10.6	11.3	36.9	13.5	8.2	6.6	379
Colonies 1.5–3.5 cm diameter	3.6	0.0	2.4	26.2	48.8	16.7	0.0	2.4	84
All colonies	1.2	0.0	3.6	6.7	80.6	4.8	2.4	0.6	165
On *Fucus serratus* (various localities; September 1969)									
Ancestrulae	11.4	5.1	19.0	8.9	27.8	8.9	8.9	10.1	79
All colonies	11.3	8.5	10.7	11.3	35.0	8.5	7.9	6.8	177

Table 2. The orientation of Electra pilosa ancestrulae

(Orientation, and direction of growth of colonies on *Fucus serratus*, expressed as the percentage falling within the range of each octant.)

	0°	45°	90°	135°	180°	225°	270°	315°	Number of observations
Ancestrulae (August 1969)									
Pwll du Bay (outer)	58.7	8.4	2.8	0.7	13.3	5.6	3.5	7.0	143
Broad Haven	44.8	15.2	1.0	6.7	15.2	5.7	4.8	6.7	105
Pwll du Bay (inner)	41.0	8.1	4.3	2.1	24.8	7.3	4.7	7.7	234
Solva	28.0	14.0	5.6	8.4	23.4	5.6	5.6	9.4	107
Colonies (August 1969)									
Solva	35.8	10.7	4.7	7.4	15.8	7.4	5.1	13.0	215

Old colonies of *E. pilosa* may cover much of the lower part of a *Fucus* plant, but the shape of small colonies can be observed on the younger parts of the thallus. Colonies are linear at first, stellate later (Fig. 4c). The initial stages of astogeny have been illustrated by Barrois (1877) and Waters (1924), while Marcus (1926) made a more extensive study. The first two zooid buds are symmetrically situated, arising distolaterally from the ancestrula; the third is distal and median, lying between the first two. Provided that there are no obstructions, these three zooids give rise to a straight, distally directed band-like lobe. There is no tendency for this lobe to turn early in astogeny, as was noted in *M. membranacea*. The first-formed pair of zooids then generally produce proximolateral buds, which develop into zooids lying alongside the ancestrula but facing proximally. From these arises a second straight band of zooids, in the line of the first but proximally directed. It follows from this rectilinear astogeny that the form of a young colony reflects precisely the orientation of the ancestrula. Up to four diagonal lobes may then arise (fig. 12 in Marcus, 1926). In a still older colony the original lobes may curve or be otherwise affected by the nature of the substratum, while the spaces between the lobes become filled as further zooids are produced. Although accurate estimation is often difficult, on *F. serratus* there seems to be a preponderance of apically directed growth even when, as at Solva (Fig. 3b), apical orientation of the ancestrulae is not strongly marked (Table 2).

Scrupocellaria reptans

Measurements of the size and position of *S. reptans* colonies on the bryozoan *Flustra foliacea* (Fig. 2) show that the youngest colonies are generally most common near the growing edge of the *Flustra* fronds, and their size increases proportionally to the distance between the point of initial attachment and the frond margin. When *Scrupocellaria* is growing on *Laminaria* spp. (Table 3), the highest numbers of small colonies (less than 1 cm across) are found near the junction of the stipe with the lamina, and densities decrease distally.

Ancestrulae of colonies of all sizes show a marked distal orientation (68% at 0°) on *Flustra* and later growth too is dominantly in the same direction (Table 4, Fig. 3c, d). However, the ancestrulae of *Scrupocellaria* initially point almost vertically, like those of *Bugula* (see p. 108), and are attached by an adhesive disc (formed from the internal sac of the larva) and a pair of rhizoids (Pl. X, fig. 16 in Barrois, 1877). Subsequent growth and the production of more rhizoids brings the young colony down parallel to the surface of the substratum. The flexibility of the three initial points of attachment would allow the ancestrula to be pulled over to point in any direction within a wide arc. Therefore, it is not clear if the dominant distal orientation of the ancestrulae of established colonies shown in Fig. 3c is the result of orientated settlement or of orientated growth pulling the

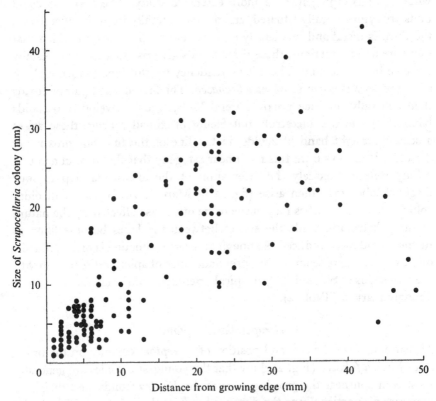

Fig. 2. Graph to show the relationship between the size of *Scrupocellaria reptans* colonies and the distance between their ancestrulae (i.e. the point of settlement) and the edge of *Flustra foliacea* fronds.

ancestrula over to point in that direction. Certainly growth of *Scrupocellaria* on *Flustra* has a very dominant distal orientation which is not apparent on *Laminaria digitata* (Table 4). It was not possible on the *Laminaria* fronds to distinguish the *Scrupocellaria* ancestrulae clearly enough to assess their orientation.

DISCUSSION

The demonstration that, in three epiphytic and epizoic bryozoans, both settlement and subsequent growth may be orientated in relation to the axis of the substrate organism poses two fundamental questions. First, how is the orientation achieved? Second, why should the orientation occur? In considering the first of these, it has to be appreciated that the factor(s) facilitating orientation of the metamorphosing larva and the growing colony need not be the same.

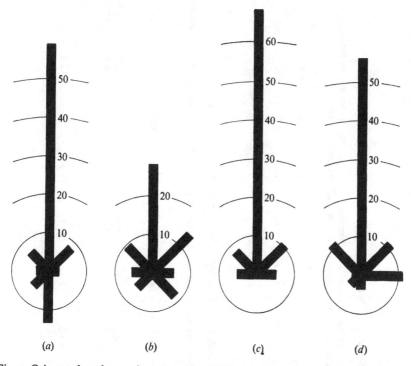

Fig. 3. Orientated settlement in two species of Bryozoa. (a) Orientation of *Electra pilosa* ancestrulae on *Fucus serratus* (outer Pwll du Bay; see Table 2). (b) Orientation of *E. pilosa* ancestrulae on *F. serratus* (Solva; see Table 2). (c) Orientation of *Scrupocellaria reptans* ancestrulae on *Flustra foliacea* (Table 4). (d) Direction of growth of *S. reptans* colonies on *Flustra foliacea* (Table 4). The results are expressed as the percentage falling within the range of each octant.

The orientation of the larva at metamorphosis

E. pilosa ancestrulae on *F. serratus* tend to face the frond tips; *M. membranacea* ancestrulae on the same alga are predominantly directed towards the frond base. (The larvae at metamorphosis were in each case facing the opposite way.) On *L. hyperborea*, however, *Membranipora* ancestrulae show no preferred orientation.

Orientation at settlement could be a response to any of the following factors: directional illumination, gravity, water movement and polarity in the surface of the substratum. Each of these must therefore be considered. Algae in the turbulent waters of the intertidal zone and just below will

Table 3. *The settlement of Scrupocellaria reptans on Laminaria*

(Settlement expressed as the mean number of colonies of diameter < 1 cm occurring per successive dm² along a line from the origin of the lamina to its free end, September 1969.)

	1	2	3	4	5	6	7	8	9	10
Laminaria digitata (10 fronds)	33.2	15.9	2.4	1.9	1.0	5.5	1.8	0.0	—	—
L. saccharina (15 fronds)	12.3	11.2	8.0	5.0	2.7	1.0	0.8	0.9	0.4	0.1

Table 4. *The orientation of Scrupocellaria reptans ancestrulae*

(Orientation, and direction of growth of colonies, expressed as the percentage falling within the range of each octant.)

	0°	45°	90°	135°	180°	225°	270°	315°	Number of observations
On *Flustra foliacea* (Oxwich Point; April 1969)									
Ancestrulae	68.4	10.5	5.3	1.3	0.0	1.3	5.3	7.9	76
Colonies	56.1	11.0	11.0	1.2	3.7	3.7	2.4	11.0	82
On *Laminaria digitata* (Cashla Bay; September 1969)									
Colonies	11.9	14.3	11.1	11.1	16.7	11.9	15.9	7.1	126

normally be in constant motion, so that the position of any given point on the frond relative to the direction of light and gravity will be continuously changing. It therefore seems unlikely that either of these factors could be responsible.

Orientation to water movement seems more probable. Algal fronds are swept to and fro by the waves and always stream in the direction of water flow, which is thus from the base towards the extremity of the frond. The almost sail-like shape of a cyphonautes suggests that an alignment of the larva in the direction of flow is likely to be adopted. If so, then the larvae of *E. pilosa* tend to face into the current. It may be significant that on a very sheltered shore, like that at Solva, a smaller proportion of larvae adopt this orientation compared with an exposed shore, like Pwll du Bay or Broad Haven, where the water movements are stronger (Table 2, Fig. 3 a, b). It seems curious, however, that the larvae of *E. pilosa* and *M. membranacea* settling on *F. serratus* should orientate in opposite senses, and that larvae of the latter should not orientate at all on *L. hyperborea*.

Finally, the possibility that the cyphonautes can detect a polarity in the surface of the alga cannot be excluded. Larvae will certainly settle along visible grooves or scratches in the surface, and they can also respond to the age of a frond (Ryland, 1959; present observations).

Regarding the distally directed orientation of *S. reptans* ancestrulae on *Flustra foliacea*, either water movement or surface polarity could again provide a mechanism. *Flustra* is confined to localities swept by strong currents, and observations made by divers confirm that the flexible fronds tend to lead downstream. The frond surface is ridged by the lateral walls of the zooids and clearly polarized on a submicroscopic scale (see Pl. XVI, fig. 1 a in Hincks, 1880). In addition, however, light must be considered. Since *Flustra* colonies are low, bushy and dense, and live sublittorally, they are not in continuous motion to the same extent that intertidal algae may be. Light will thus fall rather constantly on the fronds from above. Moreover, settling *Bugula* larvae are known to orientate with respect to the direction of light (see p. 108).

The orientation of growing colonies

Any of the four stimuli just considered (light, gravity, water current and polarity of the surface) might induce directional growth in the colonies of epiphytic bryozoans. In considering how these could effect *Membranipora*, it is important to know what is the normal posture of submerged *Laminaria* plants. Observations made by divers show that, in a gentle current, the stipes of *L. hyperborea* stand up vertically, while the laminae stream out horizontally. If there is a swell, the laminae swing from side to side in the wave orbits. Neither light nor gravity could provide a constantly uni-directional stimulus.

Since *Membranipora* colonies on *Laminaria* fronds grow towards the stipe, the growing edge moves simultaneously into the water current and towards the youngest part of the frond: thus either or both of these factors could be inducing the response. Basally directed growth on *Fucus*, on the other hand, extends into the water current but away from the youngest part of the frond. From this it might be concluded that orientated growth was a response to water movement. Moreover, Marcus (1926) has demon-strated that stolons of the ctenostomatous bryozoan *Farella repens* (Farre) grow into a water current. In the case of *Membranipora*, however, there is one reservation to be noted. On *Laminaria* the young colony rapidly turns towards the stipe and grows downwards as a relatively slender lobe; but on *Fucus* this rarely seems to happen. The colonies remain subcircular or oval and, although the greatest distance from the ancestrula to the periphery most often lies in the 180° octant, there is no rapidly advancing lobe.

Much of the apically directed growth apparent when *E. pilosa* grows on *F. serratus* originates through astogeny from the orientation of the ances-trula (p. 113). The production of longer or more numerous distally directed lobes, however, indicates growth which is rheonegative and towards the youngest part of the frond.

S. reptans colonies on *Flustra foliacea* display distally directed growth, but on *L. digitata* they do not (Table 4). Water movement cannot, then, here be the factor inducing orientation. Light, as explained earlier, would provide a directional stimulus on *Flustra* fronds, but not on *Laminaria*; so it could provide the orientating mechanism. Possibly significant in this respect is the work of Schneider (1963), who demonstrated positively photo-tropic growth in the branches of *Bugula*. Nevertheless, a response by the *Scrupocellaria* to an age gradient in the surface of the *Flustra* cannot be ruled out. In the Cashla Bay rapids, on the other hand, the water current physically opposes proximally directed growth, by flexing the more proxi-mal *Scrupocellaria* branches to point downstream. The current might in this way negate any response to age-dependent polarity in the surface of the *Laminaria*.

The biological significance of non-random settlement and orientated growth

The size and population density of epiphytes and epizoites are generally in proportion to the age of the substrate organism or its component parts. Thus, the older portion of *Fucus* thalli and *Laminaria* fronds may be

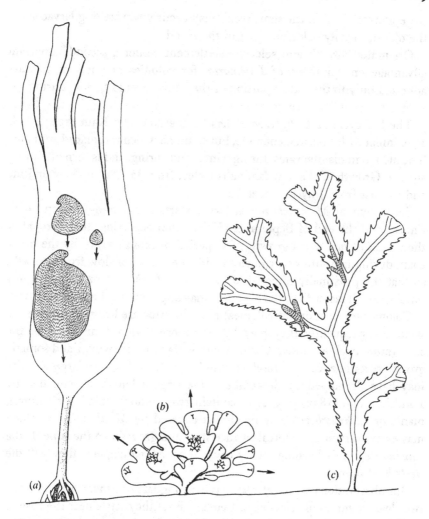

Fig. 4. Diagram summarizing the patterns of settlement and growth in three species of Bryozoa. (a) Proximally directed growth in colonies of *Membranipora membranacea* on a frond of *Laminaria hyperborea*. (b) Distally directed growth in colonies of *Scrupocellaria reptans* on *Flustra foliacea*. The distribution of young colonies shows that recent settlement has occurred only near the periphery of the fronds (see Fig. 2). (c) Distally directed growth in colonies of *Electra pilosa* on *Fucus serratus*.

densely covered with *Ectocarpus*, bryozoans, hydroids, *Spirorbis*, ascidians and so on, and an acute shortage of space may then be obvious. Where epiphytes, like *Membranipora*, ultimately become large, growth directed predominantly towards the youngest part(s) of the substrate organism will lead to a reduction in the intensity of competition for space (Fig. 4). When the epiphyte is small, the same result may occur when settling larvae avoid the oldest, heavily colonized part of the frond.

Orientated growth and selective settlement confer a second important advantage on epiphytes of *Laminaria*, for colonies or parts of colonies adherent only to the oldest portion of the lamina would be lost during the winter or following spring.

The life cycle of *L. hyperborea* has been studied by Kain (1963). It is a perennial with a permanent stipe, but a lamina which is replaced annually. The old frond disintegrates during winter and spring, and is torn off during storms. Growth of the new lamina is fastest from January to May or June and slowest from July to December.

Settlement of *M. membranacea* larvae starts about mid-June and continues until the end of September (Nair, 1962) occurring, therefore, after the period of fast frond growth. The period of rapid growth by the bryozoan, during late summer, coincides with the season of slow frond growth, so that the proximally directed lobes of the *Membranipora* colonies grow linearly much faster than the alga, and may approach the base of the frond.

Colonies (or portions of colonies) near the stipe are likely to survive the winter; consequently they may live for more than 1 year although the substratum itself is annual. Gametogenesis starts in late winter and somatic growth recommences in March (Lutaud, 1961), so that very large colonies may have developed by September. One large colony was estimated by Landsborough (1852, p. 352) to contain two million zooids. By autumn, plants of *L. hyperborea* may be almost covered by *Membranipora*, which may even overrun the intercalary meristem and grow down the stipe. If the *Laminaria* fronds become totally covered by *Membranipora*, they will die (*teste* K. Lüning).

Our observations on *S. reptans* larvae during late summer have provided two clear examples of selective settlement. First, they settle near the base of *L. digitata* and *L. saccharina* fronds (Table 3), where new growth takes place from the top of the stipe; and secondly, near the edge of *Flustra foliacea* fronds (Figs. 2, 4b), where the new growth is peripheral.

Even the largest colonies of *S. reptans* scarcely exceed 50 mm in length, and could never grow any significant distance towards the base of a *Laminaria* frond. However, by colonizing the youngest portion at settlement, they both attach themselves to that part of the lamina which will remain alive and intact longest, and select the area in which competition for space is least intense. In the case of settlement of *Scrupocellaria* on the periphery of *Flustra* fronds, reduction of competition seems to be the sole advantage, and this applies also to distally directed growth by *E. pilosa* on *F. serratus*.

Selective settlement behaviour of this kind appears to be comparable to that shown by *Alcyonidium polyoum* Hassall, investigated experimentally by Ryland (1959), in which the larvae were shown to settle preferentially on the youngest parts (the tips) of *F. serratus* thalli. Kato, Nakamura, Hirai & Kakinuma (1961) have described the distribution of *Clytia volubilis* and *Obelia dichotoma* on *Sargassum fulvellum*. In both species, the colonies develop from the tips of the leaflets. As *S. fulvellum* has apical meristems this implies that the planulae may settle preferentially on the youngest parts. Nishihira (1968) studied the pattern of distribution of three epiphytic hydroids on two species of eelgrass (*Clytia edwardsi* and *Tubularia mesembryanthemum* on *Zostera marina* and *Plumularia undulata* on *Phyllospadix iwatensis*). He suggested that settlement starts on the older parts of the older blades and, as the density of epiphytes increases, spreads to the younger parts. He also noted that growth is orientated towards the youngest parts of the blades.

SUMMARY

A short review is given of larval structure and settlement behaviour in bryozoans. Few observations have been published on the effect of environmental stimuli on larval orientation at metamorphosis, and none correlate these stimuli with the initial direction of colony growth.

Observations have been made on the orientation of ancestrulae in *Membranipora membranacea*, *Electra pilosa* and *Scrupocellaria reptans*. *M. membranacea* ancestrulae tend to face basally on *F. serratus*, but no orientation could be detected on *Laminaria hyperborea*. *E. pilosa* ancestrulae tend to face distally on *F. serratus*. Whatever the initial orientation, colony growth in *M. membranacea* is directed towards the base of the algal frond. In *E. pilosa* on *F. serratus* distally directed growth predominates. This is true also of *Scrupocellaria reptans* growing on *Flustra foliacea* but not on *Laminaria digitata* or *L. saccharina*, where growth is randomly orientated.

Our observations alone do not permit orientation at settlement and direction of growth to be explained in terms of a single environmental factor. Orientation at settlement in *Membranipora* and *Electra* is most likely a response to the direction of water flow; but in *Scrupocellaria* light could be the stimulus.

Directional growth in *Membranipora* may be a rheopositive response, although detection of an age dependent polarity in the surface of the substrate organism cannot be ruled out. A response to either substrate polarity or light would account for orientated growth of *Scrupocellaria* on

Flustra, but the factor does not operate when *Scrupocellaria* grows on *Laminaria.*

Scrupocellaria settles preferentially on the youngest part of a *Laminaria* frond. This appears to be a way in which small epiphytes increase their prospects of survival to the following season. As when *Scrupocellaria* settles on the edges of *Flustra* fronds, this behaviour also locates the young colonies on that part of the substrate organism where competition for space is least. Apically directed growth of *Electra* on *Fucus* probably serves the same purpose.

REFERENCES

ATKINS, D. (1955). The cyphonautes larvae of the Plymouth area and the meta-morphosis of *Membranipora membranacea* (L.). *J. mar. biol. Ass. U.K.* **34**, 441–9.

BARROIS, J. (1877). *Recherches sur l'embryologie des Bryozoaires.* Lille: Thèse de Paris.

CALVET, L. (1900). *Contribution à l'histoire naturelle des Bryozoaires ectoproctes marins,* 488 pp. Montpellier: Thèse de Paris.

GRAVE, B. H. (1930). The natural history of *Bugula flabellata* at Woods Hole, Massachusetts, including the behavior and attachment of the larva. *J. Morph.* **49**, 355–84.

HINCKS, T. (1880). *A History of the British Marine Polyzoa,* 2 vols. London: van Voorst.

HYMAN, L. H. (1959). *The Invertebrates, 5, Smaller Coelomate Groups.* New York: McGraw-Hill.

KAIN, J. M. (1963). Aspects of the biology of *Laminaria hyperborea.* II. Age, weight and length. *J. mar. biol. Ass. U.K.* **44**, 415–33.

KATO, M., NAKAMURA, K., HIRAI, E. & KAKINUMA, Y. (1961). The distribution pattern of Hydrozoa on seaweed with some notes on the so-called coaction among hydrozoan species. *Bull. biol. Stn Asamushi* **10**, 195–202.

KNIGHT-JONES, E. W. & MOYSE, J. (1961). Intraspecific competition in sedentary marine animals. *Symp. Soc. exp. Biol.* **15**, 72–95.

KUPELWIESER, H. (1905). Untersuchungen über den feineren Bau und die Metamorphose des Cyphonautes. *Zoologica, Stuttg.* **47**, 1–50.

LANDSBOROUGH, D. (1852). *A popular history of British zoophytes or corallines.* London: Reeve and Co.

LYNCH, W. F. (1947). The behavior and metamorphosis of the larvae of *Bugula neritina* (Linnaeus). *Biol. Bull. mar. biol. Lab., Woods Hole* **92**, 115–50.

LUTAUD, G. (1953). Premiers stades de la croissance zoariale chez le Bryozoaire cheilostome *Scrupocellaria reptans* Thompson. *Archs Zool. exp. gén.* **90**, 42–58.

LUTAUD, G. (1961). Contribution à l'étude du bourgeonnement et de la croissance des colonies chez *Membranipora membranacea* (L.), Bryozoaire Chilostome. *Annls Soc. r. zool. Belg.* **91**, 157–300.

MARCUS, E. (1926). Beobachtungen und Versuche an lebenden Meeresbryozoen. *Zool. Jb. (Syst.)* **52**, 1–102.

NAIR, N. B. (1962). Ecology of marine fouling and woodboring organisms of western Norway. *Sarsia* **8**, 1–88.

NISHIHIRA, M. (1968). Distribution pattern of Hydrozoa on the broad-leaved eelgrass and narrow-leaved eelgrass. *Bull. biol. Stn Asamushi* **13**, 125–38.

O'DONOGHUE, C. H. (1926). Observations on the early development of *Membranipora villosa* Hincks. *Contr. Can. Biol. Fish.* N.S. **34**, 249–63.

RYLAND, J. S. (1959). Experiments on the selection of algal substrates by polyzoan larvae. *J. exp. Biol.* **36**, 613–31.

RYLAND, J. S. (1962). The association between Polyzoa and algal substrata. *J. anim. Ecol.* **31**, 331–8.

RYLAND, J. S. (1964). The identity of some cyphonautes larva. *J. mar. biol. Ass. U.K.* **44**, 645–54.

SCHNEIDER, D. (1963). Normal and phototropic growth reactions in the marine bryozoan *Bugula avicularia*. In *The Lower Metazoa*, pp. 357–71 (ed. E. C. Dougherty). University of California Press.

SILÉN, L. (1966). On the fertilization problem in the gymnolaematous Bryozoa. *Ophelia* **3**, 113–40.

WATERS, A. W. (1924). The ancestrula of *Membranipora pilosa* L., and of other cheilostomatous Bryozoa. *Ann. Mag. nat. Hist.*, Ser. 9, **14**, 594–612.

SETTLEMENT AND GROWTH
PATTERN OF THE PARASITIC BARNACLE
PYRGOMA ANGLICUM

J. MOYSE

Department of Zoology, University College, Swansea, U.K.

INTRODUCTION

All the known species of the closely related balanid genera *Pyrgoma* and *Creusia* (Cirrepedia), live in association with corals or alcyonarians. The bases of these barnacles grow anchored in the host skeleton and the barnacle shell is covered by a thin layer of host coenosarc except for a small area where the cirral feeding net can protrude through the operculum. Recently, Ross & Newman (1969) have shown that one species, *Pyrgoma monticulariae*, has only vestigial cirral limbs and feeds by nibbling away at the coral coenosarc. Ross & Newman distinguish such fully parasitic forms which depend on their host for food as well as habitat from semiparasitic species whose dependence is not apparently nutritional. Since *P. anglicum* Sowerby (1820–25) is one of the latter, its behaviour at the time of settlement is of special interest. Adaptations associated with parasitism are sometimes simpler to unravel in partially than in fully parasitic forms. The phylogeny of the parasitic condition is being investigated by Ross & Newman. Brooks & Ross (1960), working on fossil material, suggested the name *Creusia* be suppressed – following Darwin (1854) who demoted it to subgeneric level and Withers (1929) who suggested that the *Pyrgoma*-form in which the shell 'compartments' are not clearly separated had arisen several times from the *Creusia* form in which the compartments are more obvious.

Most of the older taxonomic studies of extant species have good accounts of the range of coral hosts and growth form (e.g. Annandale, 1924; Broch, 1931; Hiro, 1931, 1934, 1935, 1938; Hoek, 1913).

Host specificity is a feature of all species but is not absolute. Although *P. anglicum* has been found on corals from at least eight genera (Rees, 1962), in Britain it is confined to the cup coral *Caryophyllia smithi* Stokes & Broderip (1829) even where this and the morphologically similar *Balanophyllia regia* Gosse occur intermingled on the same rock face (as was encountered during this study in Pembrokeshire).

P. anglicum, in common with many other warm water species, reaches its northern limit on the West coast of Britain despite the distribution of its host coral much further north in Shetland and Norway. Rees (1962) gives an exhaustive catalogue of British records of both coral and barnacle and says of the association 'Nothing is known of the relationships with the coral but it is reasonable to assume that heavy infestations must restrict the disk movements of the coral and possibly rob it of much food. It is noted that the barnacles appear gregarious, preferring to settle on each other rather than on the column of *Caryophyllia*.' This erroneous conclusion was based on a study of the preserved specimens available to him at the British Museum.

The only published account of settlement of any of these barnacles is that of Utinomi (1943) who described the behaviour of cyprids of *Creusia spinulosa* f. *angustiradiata* Broch when settling on the coral *Leptastrea purpurea* Dana.

The present study is based on laboratory experiments and analyses of random samples of adult *Caryophyllia* collected by aqualung diving from the large population at Martins Haven in Pembrokeshire, S. Wales.

HABITAT AND METHODS

In the area of the survey, the rocky intertidal region continues sublittorally as steep sided rocky crags down to a depth of 30 m below low tide level. For the first 10 m of sublittoral depth, the rock surfaces are dominated by *Laminaria* spp. and the *Caryophyllia* are small, stunted, not very numerous and bear few *Pyrgoma*. Below the *Laminaria* zone and especially below the red algal zone, the corals are larger and more numerous, often reaching a density of over 100/m². Their distribution is apparently random over the rocky ridges and in narrow clefts, as well as on flat surfaces.

The general rock surfaces are covered by a turf of sedentary organisms, dominance being shared by sponges and Bryozoa. This turf extends up the column of the larger corals. *Echinus esculentus* and other echinoderms move over even the steepest surfaces, grazing on the turf, and must frequently come into close contact with the corals.

Collection and analysis of samples

Generally each coral was severed at the narrowest point of its column just above the base, for which purpose a hammer and chisel were more effective than a geological hammer. Even so, smaller *Caryophyllia* shattered readily, so most of those collected were large specimens and some of these fractured radially when struck.

When it was intended to examine the positioning of barnacles on corals from vertical rock faces, shallow rectangular boxes about 3 cm deep were lined with very oily putty and taken down by the diver. The putty, even when cooled by the sea, was soft enough for the detached corals to be pressed into easily, and by careful co-operation between two divers the corals were orientated in the box in a similar way to their position on the rocks.

SETTLEMENT

Laboratory experiment on cyprid settlement

Some cyprids reared in the laboratory by the method described in a previous paper (Moyse, 1961) were pipetted into dishes containing live corals. After 2 days most had attached themselves to the living coral epithelium on the sides of the corals. After a further few days they had become firmly attached to the hard skeleton, the coral epithelium now covering the barnacle except for an area over the operculum. This agrees with the description of the settlement of *Creusia spinulosa* given by Utinomi (1943), who also observed that the cyprids become attached to the coral skeleton by sinking through rather than crawling under the coral epithelium.

Pyrgoma cyprids must be immune to the action of *Caryophyllia* nematocysts. The larvae of other sedentary groups such as bryozoa, other coelenterates, etc., are undoubtedly killed and removed from the coral flesh by the action of the nematocysts and cilia on its surface.

The use of the cyprid antennules to burrow through soft coenosarc tissue to attach to a deep solid surface seems to be achieved by a behavioural rather than structural adaptation from the condition in rock-living barnacles. Structurally the cyprid antennules appear little different from those of other balanids.

Settlement pattern in a field population

Adult *P. anglicum* grow around the rim of the *Caryophyllia smithi* 'cup'. Examination of corals collected just after the barnacle settlement season (September and October) indicated that *P. anglicum* cyprids settle at a mean distance of 2.5 mm below the coral rim (Fig. 1). Settlement at this level appears at first sight to be related to the grooving of the coral column. The radially placed sclerosepta are very prominent at the disc of the coral but more basally they gradually merge with the outside of the stem (Plate 1 b). The coral epithelium which extends from 5 to 10 mm down the side of the coral is sometimes undulating where it overlies these ridges; so perhaps the *Pyrgoma* cyprids come to rest in these undulations. Barnes (1955) has suggested that grooves may induce a slowing down of movements (low rugokinesis) of cyprids of barnacles.

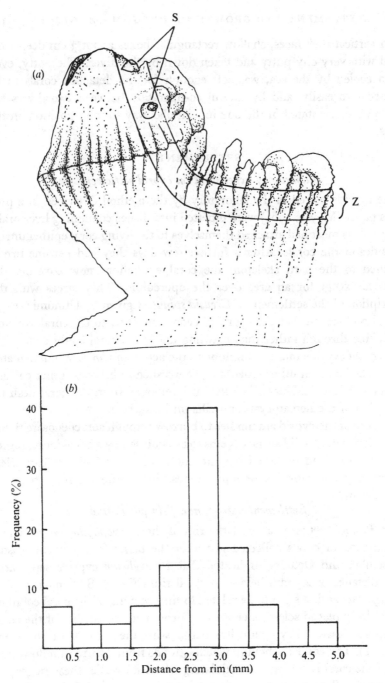

Fig. 1. Position of settlement of *Pyrgoma anglicum* spat. (*a*) Drawing from Plate 1*b* to show zone (Z) in which most settlement occurs, S, spat. (*b*) Histogram of settlement position of cyprids settling directly on *Caryophyllia*. The distance plotted is from the general upper level of the coral to the middle of the spat.

The deep groove formed where the basis of an adult *Pyrgoma* rises from the column of a *Caryophyllia* may be more attractive than a shallow sclero-septal groove and this position is indeed favoured for settlement. *Balanus balanoides* cyprids prefer depressions with a small radius of curvature (Crisp, 1961).

Settlement of *P. anglicum* cyprids sometimes occurs on adults of their own species and the position is then more usually related to the sclerosepta secreted by the coral on the adult barnacle than to the underlying grooves of the barnacle itself which usually run at a different angle. This suggests that some feature of the coral flesh in this region in addition to grooving, helps determine settlement position.

INCIDENCE OF *PYRGOMA* ON *CARYOPHYLLIA*

Mean numbers and aggregation on certain corals

The numbers of *Pyrgoma* on *Caryophyllia* were examined to determine the mean distribution and to seek evidence of gregariousness. In a random collection of one-hundred and seven large corals (all having a mean dia-meter of over 15 mm) made at a depth of 10–15 m below low tide level the distribution of *Pyrgoma* was variously plotted on probability paper against Poisson curves of the same mean (Fig. 2). The mean number of live barnacles per coral was 1.26 and 3.55 before and after a period of settlement respectively. All methods of plotting indicated a certain amount of aggrega-tion rather than strictly random distribution. The highest number per coral was 12 (see also Table 1). Higher numbers are recorded by Rees (1962).

Although aggregation on certain corals may be the result of gregarious settlement behaviour of the same sort as observed in shore barnacles (Knight-Jones, 1953) it could equally be the result of concentrated settle-ment on certain corals because of differences in their condition or siting. The figures do not provide conclusive evidence on this point, especially since they do not take into account coral recruitment and they are affected by the differing mortality rates of coral and barnacle.

Caryophyllia smithi in the *Laminaria* zone rarely bears *Pyrgoma* and abundant *Caryophyllia* at various depths off the south coast of the Pem-brokeshire Island of Skomer support no *Pyrgoma*. *Pyrgoma* densities, high enough to result in detrimental overcrowding, as described in littoral barnacles by Barnes & Powell (1950), were not seen.

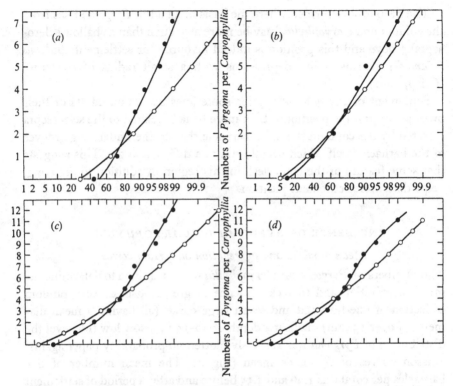

Fig. 2. Frequency distribution (●—●) of *Pyrgoma anglicum* on a random sample of 107 large *Caryophyllia*, plotted on arithmetic probability paper against Poisson distributions (○—○) corresponding to the same mean number of *Pyrgoma*. (*a*) Live adult *Pyrgoma* which were present before the annual spatfall. Mean 1.26. (*b*) Young *Pyrgoma* of the annual spatfall. Mean 2.29. (*c*) Total live *Pyrgoma* after spatfall. Mean 3.55. (*d*) Total *Pyrgoma* including dead individuals after the spatfall. Mean 3.95. Aggregation is apparent since the observed values indicate that more corals have high numbers of barnacles than they would were the settlement random (Poisson curve).

Table 1. *Preferential settlement of Pyrgoma anglicum spat on Caryophyllia smithi already bearing several adult Pyrgoma*

Number of live adult P. anglicum per coral	Number of cases	Mean number of spat per coral	Grouped mean numbers of spat per coral
0	47	1.935	
1	30	1.2	1.7
2	12	2.08	
3	5	3.2	
4	5	4.4	
5	5	3.0	3.33
6	1	6.0	
7	1	1.0	
8	1	3.0	

Distribution of Pyrgoma on individual Caryophyllia

It is a matter of simple observation that amongst corals bearing two or more *Pyrgoma* these tend to be aggregated at one or more points on the coral rim. The extent of this aggregation was measured as follows. If it is assumed that all corals are of the same diameter and exactly circular then two randomly placed points on the rim of each coral would be at any distance apart from $0 - \pi r$ (i.e. half the circumference) and in a large enough sample would be

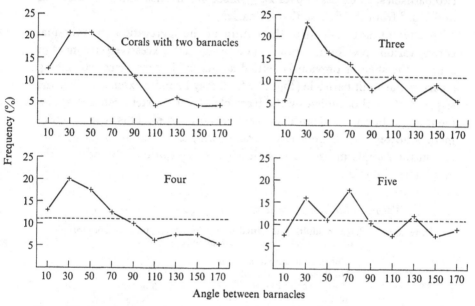

Fig. 3. Minor angles subtended at *Caryophyllia* centres by pairs of *Pyrgoma* on the rims of corals bearing respectively two, three, four and five *Pyrgoma*. Excessive occurrence at small angles apparent in all cases, is indicative of local aggregations on the coral rims. The dotted line in each case represents the expected distribution from random settlement.

at the same density throughout this distance. If, for these two points, barnacles are substituted, the same applies (except at small distances where the two barnacles would touch) unless some aggregating or dispersing influence occurs. However, as corals are neither circular nor of the same size, actual distances apart cannot be employed. A slightly less informative but more convenient datum is the smaller angle subtended by the two barnacles at the centre of the coral. If the barnacles settled randomly around coral rims, then these angles would be of even frequency between 0 and 180°. Information from corals with three or more barnacles can be used by plotting the smaller angles between all the possible pairs of barnacles.

Employing this method on corals bearing two to five barnacles, by the use of a transparent protractor (taking the angle to the nearest 5°), the distributions shown in Fig. 3 were produced. They reveal clear evidence of aggregation. Too many cases of small angles occur for distribution to be random. The aggregating influence extends as far as 90°, after which it levels off well above zero, suggesting random settlement at these larger angles. At angles of 5 and 10° the number is below the random level, the result of divergent growth of contiguous barnacles. In cases of a coral with two barnacles, an excess of over 25 % of second arrivals settled in angles of under 90° from the first settled barnacle.

In attempting to elucidate the nature of the aggregation on individual corals, various possibilities can first be eliminated. Excessive settlement of spat on the shell of previously settled adults might take place, colonizing a part of the adult barnacle (Fig. 1) which may be more extensive than the segment of coral occupied by the barnacle. To that extent random settlement might be confused with apparent gregariousness. This type of settlement is, however, quite unusual in corals bearing small numbers of barnacles and in the sample measured was clearly not the cause of the peak in the graph (see Table 2).

Table 2. *Position of settlement of Pyrgoma anglicum spat*

(In relation to *Pyrgoma* adults that settled in previous years, on a random sample of *Caryophyllia smithi*, autumn 1960 spatfall.)

Not in contact with an adult	79 %
On the cone of an adult	13 %
On the basis of an adult	3 %
In the groove between adult and the coral column	5 %

Frequently two or more *Pyrgoma* of the same year group, growing contiguously, are found to have settlement points separated by one or two sclerosepta. Similarly, examination of two contiguous adults of different year groups often reveals that the younger one settled a few scleroseptal grooves away from the first. Since two or more offset aggregation points sometimes occur, this type of aggregation might be related to the symmetry of the mesenteric directives of the coral. Corals were examined from this standpoint but no evidence was found of such an aggregation (see Fig. 4).

The distribution of *Pyrgoma* on corals from vertical rock faces, collected in putty boxes, shows clear indication of aggregation on the lower halves of polyps (Fig. 5). This type of aggregation probably arises as a result of the

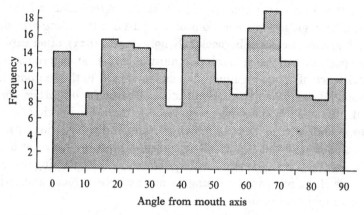

Fig. 4. Distribution of *Pyrgoma* around *Caryophyllia* plotted according to their angular distance from the nearest end of the mouth axis.

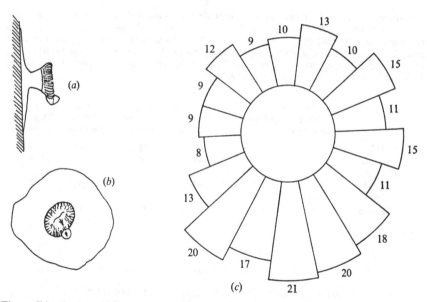

Fig. 5. Distribution of *Pyrgoma* around *Caryophyllia* collected from vertical rock faces. (*a*) Side view of *Caryophyllia* with one *Pyrgoma*. (*b*) View from over the coral disc showing how angle was measured. (*c*) Frequency distribution summarized to 20° angles, showing preponderance of *Pyrgoma* on lower surface.

photopositive swimming of the cyprids immediately before attaching, which would result in corals being approached mostly from below. Barnes, Crisp & Powell (1951) noted heavier settlement of *Elminius* cyprids on the lower surfaces of horizontally exposed plates.

Thus, although it seems quite possible that the aggregations of *Pyrgoma* result from truly gregarious behaviour on the part of the settling cyprid, as recorded for other species of barnacles (Knight-Jones, 1953), other explanations are possible and the final solution must be arrived at experimentally.

The existence of an aggregating influence is clearly to the reproductive advantage of *P. anglicum*. The concentration of barnacles on certain coral-cups will ultimately lead to the production of more broods than would result from random settlement with many isolated individuals since, as was demonstrated during the study, self-fertilization probably does not occur (see Table 3). By settling close to existing barnacles, the operating distance for the penis of the protandrous young barnacle will be reduced and earlier broods of nauplii may result.

Table 3. *Evidence for obligatory cross-fertilization and protandry in Pyrgoma anglicum*

Numbers of adult *Pyrgoma* (over 4.0 mm diameter) incubating embryos, categorized according to the presence or absence of other barnacles on the same coral, i.e. (*a*) corals bearing a single (adult) *Pyrgoma*, (*b*) corals having two or more adult *Pyrgoma* per coral and (*c*) corals with a single adult *Pyrgoma* accompanied by one or more young (settled the previous year). Those few with embryos in category (*a*) could have been fertilized by *Pyrgoma* individuals on neighbouring corals.)

	(a)		(b)		(c)	
	With embryos	Without embryos	With embryos	Without embryos	With embryos	Without embryos
12 July	3	3	28	15	2	0
18 Aug.	2	11	17	8	1	0
22 Aug.	3	6	3	1	3	1
2 Sept.	0	2	43	14	14	2
28 Sept.	1	13	11	27	1	1
Totals	9	35	102	65	21	4
% With embryos	20.4		61.1		84	

ORIENTATION AND GROWTH OF INDIVIDUAL *PYRGOMA*

Orientation of Pyrgoma anglicum *on coral rims*

The settled spat is orientated in a groove with the carinal end towards the rim of the coral. Thus at the time of metamorphosis the cyprid must be facing up the coral column. A few individuals settle the opposite way round. This orientation persists throughout the subsequent growth of the barnacles. After metamorphosis *P. anglicum* appears to be able to change its orientation only very slightly in the same way as *Balanus balanoides* does by spiralling growth (Moore, 1933), but not as that species does by gliding (Crisp, 1960).

Orientation of the cyprids along the length of the grooves may be partly rugotropic as in *B. balanoides* (Crisp & Barnes, 1954). The accuracy of alignment suggests that the cyprid, whilst in the act of settling, can probe through the thin layer of soft coral tissue with its antennules and distinguish the grooving of the hard coral skeleton.

The unidirectional aspect of the alignment is no doubt partly a response to the direction of the incident light, as in shore barnacles (Barnes *et al.* 1951). Even for corals on vertical rock faces the incident light comes mostly from a seaward direction.

Changes in the shape of the shell during growth

Recently settled *P. anglicum* closely resemble the spat of other thoracic barnacles, except that the shell is in the form of a single uninterrupted ring instead of separate parietal plates. As the spat grows, the sides become less steep and the opercular aperture relatively smaller and more elongated rostrally. Ultimately the operculum sinks deeper into this parietal cone and is not visible externally. The tergum and scutum comprising each opercular valve are fused in most species of *Pyrgoma* (Hiro, 1934), but in *P. anglicum* they retain their full articulation even when adult. The parietal shell, shaped like a truncated cone as in other barnacles, grows in size by addition all around its lower edge and by internal addition to its thickness.

The integument of the basis of a barnacle lying against the substratum is thin and cuticular in midlittoral and upper littoral barnacles but in lower littoral and sublittoral species it is calcified. Such a calcified basis may help to smother, kill and protect against other encrusting animals. The basis of *P. anglicum* is calcareous and starts to grow in the usual way as a flat plate which has radiating ridges and is not always symmetrically disposed around the settlement point. Growth of the basis spreads down the coral column if the spat has settled near the rim and, conversely, settlement lower on the column is adjusted upwards. The growth of two closely sited barnacles becomes similarly separated.

When a mean diameter of about 2–2.5 mm is reached, strikingly differential growth of the basis on the rostral side of the barnacle lifts it away from the coral column while on the carinal side it continues to spread up towards the rim until it reaches the limit of the sclerosepta. Thus the basis acquires a shallow cup-shape orientated almost at right angles to the coral axis. Further growth of the rostral and lateral sides of the basis, during the second and subsequent years of life, eventually brings the whole barnacle growth axis almost parallel to that of the coral. The junction between parietes and basis, which in most barnacles adjoins the substratum, is lifted

into full view in *P. anglicum*. But, in spite of growth irregularities, all parts
of this suture remain in one plane. Clearly there is no fusion between
parietal cone and basis since all growth in size of the barnacle shell takes
place at the suture. The ridges on the cone, which are accurately mirrored
by others on the basis, perhaps provide increased mechanical support to

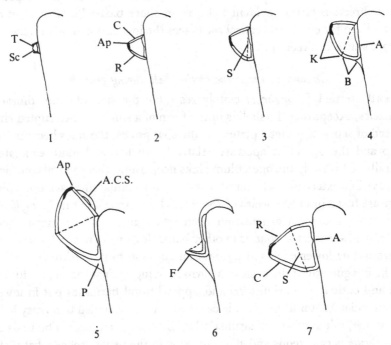

Fig. 6. Change in shape of the shell with growth of *Pyrgoma*. 1, Recently settled spat;
2, 3 and 4, conditions in spring, summer and autumn of the year following settlement;
5, adult; 6, broken remains of dead *Pyrgoma* being overgrown by coral; 7, growth form of
Pyrgoma settled in reversed position; A, point of settlement; A.C.S., adventitious coral
sclerosepta; Ap, apex of cone; B, basis; C, carinal side of cone; F, broken basis; P, coral
substances filling angle between barnacle and coral column; R, rostral region; S, suture;
Sc, scutum; T, tergum; K, parietal cone.

this line of potential weakness. In adult *Pyrgoma* growth of the basis keeps
accurately in step with growth of the coral, thus maintaining the suture
just above the coral rim. Scleroseptal plates may be secreted on the cone
and join those of the coral cup but the strong growth of the barnacle is
normally able to break any that bridge the gap.

A few cyprids settle the 'wrong way round' on the coral column and in
their subsequent growth the usual (and presumably inherent) outward
growth of the basis on the rostral side still occurs, sometimes resulting

in a *Pyrgoma* growing towards the base of a coral (Plate 2*a*). Excessive growth of the carinal side of the cone tending to cancel the effect of the basal growth is often seen and appears to be in response to the disadvantageous feeding position of the cirri, which strive to place themselves across water currents. Cyprids settling with an intermediate orientation, often exhibit slight torsion during growth, thus tending to lift the cone into the 'normal' position. These various effects are also seen in individuals settling on shells of their fellows.

The result of normal basal growth is to lift the apex of the cone into a position several millimetres above the rim of the cup (Fig. 6). Its orientation is such that the cirri beat away from the centre of the coral (Plate 1*a*), i.e. the cirral net will fish water currents approaching the coral rather than those that have just flowed over it. This type of orientation is advantageous in that it is less likely to result in entanglement of the cirral net in the coral tentacles. Similarly the fossil *P. prefloridanum*, growing on the compound coral *Manicina mayori*, has the carinal end of the shell situated on the side towards the dorsal central area of the corallum (Brooks & Ross, 1960).

INTERACTION BETWEEN CORAL AND BARNACLE

Effect on the coral

The corals appear defenceless against adult barnacles. If the cirri of a barnacle nudge a coral tentacle the latter withdraws and contracts but the cirri appear unaffected. A heavy infestation must indeed hamper the coral's food gathering as suggested by Rees (1962).

The corals may feed on some dead barnacles but it is not known if death itself is ever attributable to coral action. About 5 % mortality of recently settled *Pyrgoma* spat has been observed, the shells often penetrated by coral septal filaments and nematocysts. Since empty shells of dead adult *Pyrgoma* sometimes become completely embedded in new coral material they must have died whilst still covered by coenosarc, but for many, perhaps most barnacles, death occurs only after the coral flesh has been withdrawn (see below and Fig. 7). So, at most, only a minority of *Pyrgoma* become carrion for the *Caryophyllia* – an unimportant source of food.

Dependence of Pyrgoma anglicum on coral coenosarc

The cloak of *Caryophyllia* flesh keeps the *Pyrgoma* shell clean and free from epizoic growths. Young or vigorous corals keep their attached barnacles entirely covered except for the opercular opening, ageing ones give less complete protection. The extent of the coverage is obvious from the dirty

J. MOYSE

and encrusted surface beyond its protection. Withdrawal of the coenosarc
is progressive, starting at the more distant extremities. The differing stages
in this process were classified into eight patterns and notes made of their
frequency and whether the barnacles concerned were alive (Fig. 7). The

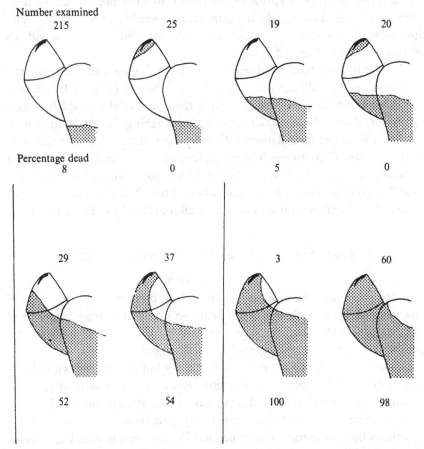

Fig. 7. Arbitrary stages used to classify progressive withdrawal of coral flesh from *Pyrgoma*
during senescence of *Caryophyllia*, their frequency of occurrence and incidence of dead
specimens at each stage. Shaded areas indicate regions from which the coenosarc has been
withdrawn.

analysis showed that the majority of barnacles well covered by flesh were
alive whilst those uncovered were nearly all dead. Most of those in an
intermediate state were also alive except those with the suture line badly
fouled. The exposed suture line may be a source of weakness, providing
access for infection, or its exposure may represent the curtailment of a necessary route of solute exchange between barnacle and coral.

DISCUSSION

Under the plankton-rich conditions of the coastal euphotic zone, the sub-littoral rock habitat is exploited by a host of sedentary planktotrophic forms amongst which competition is intense. Barnacles tend to be over-shadowed by the larger branching colonies of bryozoa, etc. Corals, represented in Europe by, for example, *Caryophyllia smithi*, often compete successfully for space. The barnacle *Verruca stroemia* may be found low down on the column of this coral but only *P. anglicum* has succeeded in adapting to settle on the living region of the coral. Barnacles of other genera have adapted to soft substrata and perhaps a chance preadaptive immunity to the coral nematocysts helped the first *Pyrgomas* to take the small behavioural step of settling on soft coral tissue.

The coral rim has a twofold advantage to *Pyrgoma*. It is free of inter-specific competing neighbours and epizoites and represents a vantage point for employment of the cirral feeding mechanism. Since there is no evidence of compensating advantage to the coral, the relationship of the *Pyrgoma* to the coral is therefore that of an obligate semiparasite. The precision of positioning and orientation of the cyprid at settlement – determining as it does the position of the adult – represents a significant behavioural adaptation. The inherent heterogonic growth of the carinal half of the basis must have been evolved later. It not only enabled the barnacle to grow in step with the coral but by elevating the cone it ensured its correct positioning for functioning of the shadow response.

Whilst the enveloping coral coenosarc serves an 'antifouling' function for the *Pyrgoma* shell, it is probably exploited physiologically too. Where it covers the ridged suture between basis and parietal cone, the intimate contact between host and parasite tissues, especially during active barnacle growth, must allow diffusion of solutes. It is noteworthy that in the much more fully parasitic *P. monticulariae* (Ross & Newman, 1969) the parietal plate is conspicuously irregular and folded in outline, thus greatly extending the length of the suture line, perhaps for diffusion. Such diffusion is much more likely to benefit *P. anglicum* than *C. smithii*, another reason for regarding it as semiparasitic rather than commensal. It is doubtful if there is any metabolic control of host coenosarc proliferation or of skeleton formation by *P. anglicum*. Such hypertrophy as is sometimes seen is similar to that occurring in a damaged coral. Coral growth tends to normalize the irregularity. Growth of the *Pyrgoma* shell merely fractures by mechanical stress any scleroseptal plates that start to form across the suture plane (see Plate 2 b). Ross & Newman (1969) are therefore wrong in saying that control over skeletal formation was the first important step in establishing the association of *Pyrgoma* with corals.

It is unlikely that such small corals as *Caryophyllia* could support as many *P. anglicum* as they do, if the latter as adults were totally parasitic. Nevertheless *P. anglicum* does represent a step towards the fuller parasitism seen in *P. monticulariae* and as such the early adaptations leading to parasitism are more easily distinguished.

I wish to thank Professor Knight-Jones for his suggestions and criticism and to acknowledge his assistance and that of Alan Osborn in the diving work involved in this study.

REFERENCES

ANNANDALE, N. (1924). Cirripedes associated with Indian corals of the families Astraeidae and Fungidae. *Mem. Indian Mus.* **8**, 61–8.

BARNES, H. (1955). Further observations on rugophilic behaviour in *Balanus balanoides*. (L). *Vidensk. Meddr. dansk naturh. Foren.*, bd. **117**, 341–8.

BARNES, H., CRISP, D. J. & POWELL, H. T. (1951). Observations on the orientation of some species of barnacles. *J. Anim. Ecol.* **20**, 227–41.

BARNES, H. & POWELL, H. T. (1950). The development, general morphology and subsequent elimination of barnacle populations, *Balanus crenatus* and *B. balanoides*, after a heavy initial settlement. *J. Anim. Ecol.* **19**, 175–9.

BROCH, H. (1931). Indomalayan Cirripedia. Th. Mortensen's Pacific Expedition, 1914–1916, *Vidensk. Meddr. dansk naturh. Foren.* **91**, 1–146.

BROOKS, H. K. & ROSS, A. (1960). *Pyrgoma prefloridanum*, a new species of cirriped from the Caloosahatchee Marl (Pleistocene) of Florida. *Crustaceana* **1**, 353–65.

CRISP, D. J. (1961). Territorial behaviour in barnacle settlement. *J. Exp. Biol.* **38**, 429–46.

CRISP, D. J. & BARNES, H. (1954). The orientation and distribution of barnacles at settlement with particular reference to surface contour. *J. Anim. Ecol.* **23**, 142–62.

DARWIN, C. (1854). A monograph of the sub-class Cirripedia. The Balanidae, the Verrucidae. *Ray Soc. Publs* 684 pp.

HIRO, F. (1931). Notes on some new Cirripedia from Japan. *Mem. Coll. Sci. Kyoto Univ. Ser.* B 7 (3), 143–59.

HIRO, F. (1934). A new coral-inhabiting barnacle *Pyrgoma orbicellae* n.sp. *Proc. imp. Acad. Japan* **10** (6), 367–9.

HIRO, F. (1935). A study of Cirripeds associated with corals occurring in Tanabe Bay. *Rec. oceanogr. Wks Japan* **7** (1), 45–72.

HIRO, F. (1938). Studies on the animals inhabiting reef corals. II. Cirripeds of the genera *Creusia* and *Pyrgoma*. *Palao trop. biol. Stn Stud.* **1** (3), 391–416.

HOEK, P. P. C. (1913). The Cirripedia of the Siboga Expedition, Cirripedia Sessilia. *Siboga Exped.* **31**, 129–275.

KNIGHT-JONES, E. W. (1953). Laboratory experiments on gregariousness during setting in *Balanus balanoides* and other barnacles. *J. exp. Biol.* **30**, 584–98.

MOORE, H. B. (1933). Change of orientation of a barnacle after metamorphosis. *Nature, Lond.* **132**, 969–70.

MOYSE, J. (1961). The larval stages of *Acasta spongites* and *Pyrgoma anglicum* (Cirripedia). *Proc. zool. Soc. Lond.* **137** (3), 371–92.

REES, W. J. (1962). The distribution of the coral, *Caryophyllia smithii* and the barnacle *Pyrgoma anglicum* in British waters. *Bull. Br. Mus. nat. Hist. Ser.* D **8**, no. 9.

ROSS, A. & NEWMAN, W. A. (1969). A coral-eating barnacle. *Pacif. Sci.* 23, 252–6.

SOWERBY, J. (1820–25). *The Genera of Recent and Fossil Shells*, 2 vols. London.

STOKES, C. & BRODERIP, W. J. (1829). Note (on *Caryophyllia*). *Zool. J.* **3**, 485–6.

UTINOMI, H. (1943). The larval stages of *Creusia*, the barnacle inhabiting coral reefs. *Annotnes. zool. jap.* **22**, 15–22.

WITHERS, T. H. (1929). The phylogeny of the cirripedes *Creusia* and *Pyrgoma*. *Ann. Mag. nat. Hist.* **10** (4), 559–66.

ELECTROPHORETIC EXAMINATION OF PARTIALLY PURIFIED EXTRACTS OF *BALANUS BALANOIDES* CONTAINING A SETTLEMENT INDUCING FACTOR

P. A. GABBOTT AND V. N. LARMAN

N.E.R.C. Marine Invertebrate Biology Unit, Marine Science Laboratories, Menai Bridge, Anglesey

INTRODUCTION

The gregarious behaviour displayed at settlement by barnacle cyprids was first described by Knight-Jones & Stevenson (1950) for *Elminius modestus*. Laboratory experiments with *Balanus balanoides* and other barnacles indicated that the gregarious response might be due to the recognition by the cyprids of the tanned cuticular surface of the adult barnacles (Knight-Jones, 1953). Later, Crisp & Meadows (1962, 1963) showed that aqueous extracts of whole barnacles, when applied to otherwise inert surfaces rendered them attractive at settlement thus simulating the gregarious response. The factor responsible for the chemical induction of settlement in cyprids of *B. balanoides* was shown to be non-dialysable, resistant to boiling in aqueous solution, and could be fractionated by precipitation with ammonium sulphate (Crisp & Meadows, 1962, 1963). The aim of our present investigations is to isolate and identify the chemical nature of the settlement factor.

In this paper we describe the partial purification and fractionation of an aqueous extract of *B. balanoides* containing the settlement-inducing factor. The heterogeneity of the partially purified fractions has been investigated by u.v.-spectrophotometry and acrylamide gel electrophoresis, and the activity of the fractions in promoting settlement of cyprids studied at various stages of purification.

EXPERIMENTAL PROCEDURES

The general scheme for the partial purification and fractionation of the barnacle extract is summarized in Fig. 1. Other methods were as follows.

Group separation and desalting by chromatography on Sephadex G-50 (Determan, 1968)

The E2, E3 and E3′ extracts (Fig. 1) were applied to a column (1 × 22.5 cm) of Sephadex G-50, equilibrated with 0.02 M phosphate buffer, pH 7.0 and

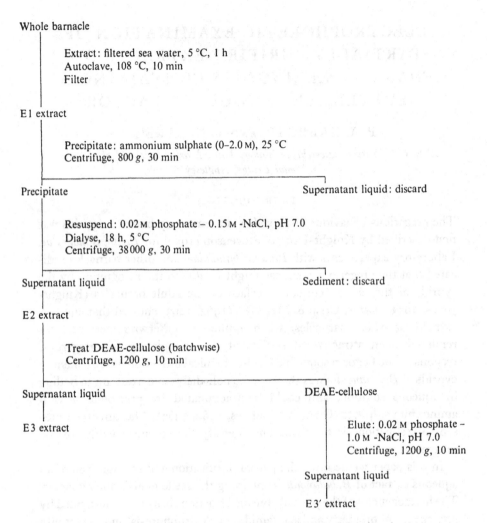

Fig. 1. General scheme for the partial purification and fractionation
of the barnacle extract.

eluted with the same buffer. The eluent from the column was monitored
for absorption at 280 nm.

Concentration of protein from dilute extracts

The pooled fractions containing diluted extracts from the Sephadex G-50
column were concentrated by the addition of dry Sephadex G-25 (coarse)
followed by centrifugal filtration in a basket centrifuge (Determan, 1968).

Determination of protein concentration

Extracts were analysed for total protein content, without nucleic acid interference, by the u.v.-spectrophotometric method of Groves, Davis & Sells (1968) using bovine serum albumin as standard.

Analytical gel electrophoresis

Gel electrophoresis was carried out in glass tubes 8.5 × 0.75 cm using a continuous tris-EDTA-boric acid (0.1 M), pH 9.2 buffer system, and a current of 6 mA/tube. The gels (5 % acrylamide-Cyanogum 41*) were prerun for 20 min to remove persulphate. The extracts, containing c. 250 μg total protein, were mixed with an equal volume (100 μl) of 20 % sucrose Sephadex G-200 (Broome, 1963) and layered under the buffer on top of the gel surface.

Staining

Proteins were stained with 1 % amido black 10 B in 7 % acetic acid and the gels destained electrolytically in 7 % acetic acid. Protein-bound carbohydrates were stained with a periodic acid-Schiffs (PAS) reagent as described by Keyser (1964). Nucleic acids were stained with acridine orange as described by Richards, Coll & Gratzer (1965), and the gels destained electrolytically in 7 % acetic acid.

Preparative gel electrophoresis

Preparative gel electrophoresis was carried out in the apparatus described by Hauschild-Rogat & Smith (1968)†, modified to allow continuous mixing of the buffer between the electrode compartments by means of a microcentrifugal pump. Separations were carried out using a continuous tris-EDTA-boric acid (0.1 M), pH 9.2 buffer system in a 5 % acrylamide (Cyanogum 41) gel column 5.0 cm in height. The gel was prerun for 1 h at 45 mA to remove persulphate and u.v.-absorbing substances from the gel (Gordon & Louis, 1967). The sample (E3 extract) containing c. 25 mg total protein was mixed with an equal volume (2 ml) of 20 % sucrose–Sephadex G-200 and layered under the buffer on top of the gel surface. A current of 50 mA was passed for 1 h until all of the components had entered the gel and then the buffer mixing pump was started. Electrophoresis was continued for 18 h at 15 °C and the elution buffer monitored for absorption at 280 nm.

* Cyanogum 41 is a product of the American Cyanamid Company distributed in the UK by B.D.H. Ltd.
† Available commercially from Shandon Scientific Company Ltd, London.

Assay of settlement activity

The method of assaying settlement activity was essentially the same as that described by Crisp & Meadows (1963).

Cyprids of *B. balanoides* were collected from the plankton and kept in the laboratory at 2 °C for 24–48 h prior to each assay. Slate panels, each with ten small pits 2 mm diameter drilled on the upper surface were used as the test surface. The slates were treated for 1 h at room temperature, either by soaking the upper surface in the appropriate extract (surface adsorption) or by placing 5 μl of the extract in each pit and allowing the slates to air-dry; control slates were treated in the same manner with 0.02 M phosphate buffer, pH 7.0. All extracts were compared at equal protein concentrations *c.* 1.0 mg/ml. After treatment the panels were rinsed in filtered sea water before being placed in the experimental dish with the cyprids.

The treated panels were arranged in order at the periphery of a large cylindrical glass trough containing the barnacle cyprids. The trough was rotated slowly on a turntable, and illuminated from the side at an intensity of *c.* 150 lumens/sq ft. An air jet was played on the water surface diametrically opposite to the light source; in this way cyprids not actively exploring the test surfaces were encouraged to swim off. At the end of each assay (4 h) the panels were removed, rinsed in fresh water and the number of cyprids settled on each panel counted. In each assay all treatments were replicated at least six, and usually 12 times on separate slate panels and the results expressed as the number of cyprids settled per slate.

RESULTS AND DISCUSSION

Partial purification and fractionation of the barnacle extract

A standard extract was prepared by crushing whole barnacles in filtered sea water, and partially purified by boiling (E1 extract) and precipitation with ammonium sulphate (0 → 2.0 M). The ammonium sulphate precipitate was resuspended in 0.02 M phosphate buffer, pH 7.0, containing 0.15 M sodium chloride and dialysed for 18 h against the same buffer. The redissolved precipitate (E2 extract) was further separated into a protein fraction (E3 extract) and a nucleic acid containing fraction (E3' extract) by batchwise treatment with DEAE-cellulose. Fig. 2 shows the u.v.-absorbance spectra from 210 to 325 nm for the E2, E3 and E3' extracts. The absorbance ratio (280/260 nm = 1.00) for the E3 extract indicates a reduction in the nucleic acid content of this fraction from 10% or more of the total protein + nucleic acid in the E2 extract (280/260 nm = 0.70) to less than 4% (Warburg &

Christian, 1941). Subsequent electrophoretic examination of the E3 extract failed to detect any bands staining with acridine orange showing that removal of nucleic acid was essentially complete (see Fig. 4).

The E2, E3 and E3′ extracts were desalted by chromatography on Sephadex G-50 resulting in a 'group separation' of the high molecular

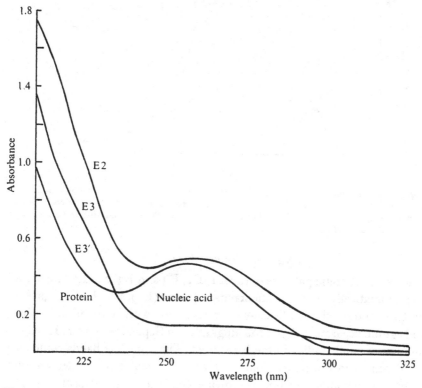

Fig. 2. U.V.-absorbance spectra for E2, E3 and E3′ extracts in 0.02 M phosphate buffer containing 0.15 M sodium chloride (total protein concentrations 88, 73 and 15 μg/ml respectively).

weight components from low molecular weight contaminants with a cut-off at *c.* 50 000 (Determan, 1968). The desalted extracts were concentrated and examined by analytical gel electrophoresis and assayed for settlement activity.

The E3 extract was further separated into two fractions E3(1) and E3(2) by preparative gel electrophoresis in tris-EDTA-boric acid (0.1 M), pH 9.2 buffer under similar conditions to those used for the analytical runs (Fig. 3). Fractions comprising the main peaks (8–11, 16–20) were concentrated and the tris-EDTA-boric acid buffer exchanged for 0.02 M phos-

phate pH 7.0 by batchwise treatment with Sephadex G-25 (fine) as described by Determan (1968). The pooled fractions E3(1) and E3(2) were
examined by analytical gel electrophoresis and assayed for settlement
activity.

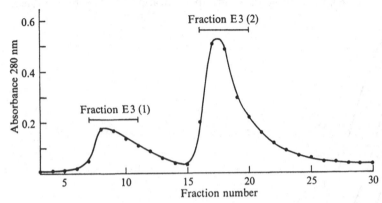

Fig. 3. Elution profile of E3 extract after preparative electrophoresis in 0.1 M tris-EDTA-
boric acid, pH 9.2 (5% acrylamide gel, height 5 cm). The elution buffer flow rate was
15 ml/h and fractions were collected at 10 min intervals. Tubes containing fractions E3(1)
and E3(2) were bulked as indicated.

Polyacrylamide gel electrophoresis

The electrophoretic patterns of the E1, E2, E3 and E3′ extracts are shown
diagrammatically in Fig. 4. The standard extract (E1), after boiling, gave six
protein bands as shown. Fractional precipitation with ammonium sulphate
and 'group separation' by chromatography on Sephadex G-50 reduced the
number of bands to five in the E2 extract. The same five bands could also
be identified in the pattern obtained when the standard extract was run
without the boiling treatment showing that the high molecular weight
components in the E2 extract are not denatured products resulting from
the boiling treatment.

The E2 extract gave three main bands; a fast moving protein component
(band 6) associated with a nucleic acid band and two protein components
(bands 2, 3) which were PAS-positive indicating protein-bound carbohydrate. In addition there was a slow-moving component (band 1) staining
for protein, nucleic acid and carbohydrate and a diffuse zone (band 4)
staining for protein and containing trace amounts of several protein components. None of the bands in the E2 extract stained for lipid with oil
red O.

Treatment of the E2 extract with DEAE-cellulose gave a protein fraction (E3 extract) and a nucleic acid containing fraction (E3′ extract). The

E3 extract contained the same five protein components present in the E2 extract but no nucleic acid bands; bands 2 and 3 again stained PAS-positive but not band 1. The nucleic acid fraction (E3') contained a fast-moving protein component associated with a nucleic acid band as in the E2 extract (band 6'), and trace amounts of bands 2, 3 and 4 as contaminating proteins. In addition E3' contained two slow-moving nucleic acid bands (1a', 1b') both free from protein; band 1a' also stained PAS-positive.

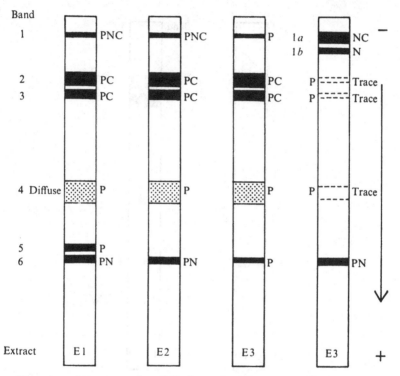

Fig. 4. Diagrammatic representation of the electrophoretic patterns of E1, E2, E3 and E3' extracts. Electrophoretic conditions: pH 9.2, 5% acrylamide gel, 6 mA per tube for 45–75 min; arrow indicates direction of migration. Staining: P = protein; C = carbohydrate; N = nucleic acid.

The E3 extract was further separated into two fractions E3(1) and E3(2) by preparative gel electrophoresis (Fig. 3) and the fractions examined by analytical gel electrophoresis. The electrophoretic patterns are shown in Fig. 5. Fraction E3(1) contained the fast-moving protein component of the E3 extract (band 6) and trace amounts of contaminating proteins from band 4. Fraction E3(2) contained bands 2 and 3 of the E3 extract and trace amounts of the fast-moving protein component (band 6) as contaminating protein. Presumably very small amounts of band 4 proteins were also present in the E3(2) fraction but these could not be detected by staining with amido black. Gels containing fraction E3(2) proteins were not tested with periodic acid-Schiff due to the limited material available for electrophoretic examination. It is a reasonable assumption, however, that bands 2 and 3 contained protein-bound carbohydrate as shown for the E2 and E3 extracts (Fig. 4).

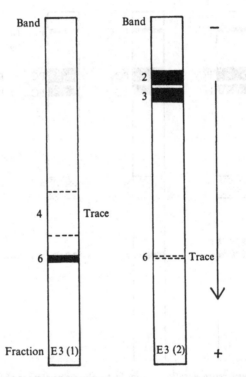

Fig. 5. Diagrammatic representation of the electrophoretic patterns of fractions E3(1) and E3(2) obtained by preparative electrophoresis of E3 extract. Electrophoretic conditions: pH 9.2, 5 % acrylamide gel, 6 mA/tube for 45 min; arrow indicates direction of migration. Staining: amido black 10 B.

Visual observation of the separated protein bands during the preparative electrophoretic run showed the fast-moving component (band 6) as a red–brown doublet but only one corresponding protein band was found by analytical gel electrophoresis and staining with amido black in both the E3 extract and the E3(1) fraction; the PAS-positive bands 2 and 3 appeared light brown and yellow–brown respectively.

Settlement activity

Table 1 shows the results of two experiments in which the E1, E2, E3 and E3' extracts were assayed for settlement activity. The total protein content of each extract was determined by the method of Groves *et al.* (1968) and adjusted to *c.* 1.0 mg/ml. In the case of the E1 extract the figure of 1.82 mg/ml (Table 1) was derived from the absorbance ratio at 280/260 nm (Warburg & Christian, 1941) since the u.v.-spectrophotometric method of Groves *et al.* (1968) is not strictly applicable to undialysed protein extracts

Table 1. *Settlement activity of E1, E2, E3 and E3' extracts*

Treatment	Protein concentration (mg/ml)	Number of cyprids settled per slate
Expt. 1. Surface adsorption (1 h), 2.5 ml/slate		
		surface + pits
E1 extract	0.93 (1.82)	45.0
E2 extract	0.99	16.8
E3 extract	1.01	22.7
Bovine albumin	1.0	1.5
Control–buffer	—	1.0
Expt. 2. Air-dried (1 h), 5.0 μl/pit		
		Pits only
E1 extract	0.93 (1.82)	34.8
E2 extract	1.16	27.3
E3 extract	1.15	25.8
E3' extract	1.06	34.5
Bovine albumin	1.00	2.0
Control–buffer	—	1.6

due to the possible interference by low molecular weight substances (loc. cit.). The results show that all the extracts were highly active in promoting the settlement of barnacle cyprids, the nucleic acid containing fraction E3' being at least as active as the nucleic acid free fraction E3. Since the concentration of the settlement factor cannot be determined independently of the total protein concentration the results cannot be used to determine specific activities but can only be ranked to indicate the presence or absence of a settlement-inducing factor. Slates treated with bovine serum albumen (1.0 mg/ml) were inactive as were control slates treated with phosphate buffer.

Fractions E3(1) and E3(2) obtained by preparative electrophoresis were assayed for settlement activity at total protein concentrations of *c.* 1.0 and 0.1 mg/ml. The results of this experiment are shown in Table 2. Both fractions were active in promoting settlement of cyprids at a total protein concentration of 1.0 mg/ml but were less active than the corresponding E3 extract at the same total protein concentration presumably due to loss of activity during electrophoresis. The activity of both fractions was considerably reduced by tenfold dilution with phosphate buffer. This last point is important since it is known from the electrophoretic patterns that fraction $E_3(2)$, containing mainly bands 2 and 3, also contained trace amounts of the $E_3(1)$ components, notably band 6, as contaminating proteins (Fig. 5). It might be argued therefore that the activity of the $E_3(2)$ fraction derived entirely from the contaminating $E_3(1)$ proteins. However, visual observation of the stained gels and examination of the elution profile

Table 2. *Settlement activity of fractions $E_3(1)$ and $E_3(2)$*

Treatment	Protein concentration (mg/ml)	Number of cyprids settled per slate
Expt. 3. Air-dried (1 h), 5 μl/pit		
		Pits only
E3 extract	1.27	28.3
Fraction E3(1)	1.22	18.2
Fraction E3(1) diluted × 10	0.12	7.0
Fraction E3(2)	0.95	17.1
Fraction E3(2) diluted × 10	0.09	5.5
Control–buffer	—	2.7

of the E_3 extract after preparative electrophoresis (Fig. 3) showed that the $E_3(1)$ proteins in the $E_3(2)$ fractions were present at a very low concentration and certainly less than 10% of the total protein. If they were responsible for the whole effect, then dilution of $E_3(1)$ to 10% of its original concentration would have given a solution as active or more active than $E_3(2)$. However the activity of $E_3(1)$ at 0.12 mg/ml was considerably less than that of $E_3(2)$ at 0.95 mg/ml. Hence it must be concluded that one or both of the band 2 and 3 proteins, the major components of fraction $E_3(2)$, are active in promoting the settlement of barnacle cyprids as well as band 6 protein of fraction $E_3(1)$. Comparison of the activity of fraction E_3', containing the nucleic acid associated band 6' protein and of fraction $E_3(1)$ which contained the nucleic acid free band 6 protein, suggests that the protein may be the active component, rather than the associated nucleic acid.

CONCLUSION

It is concluded that the fractionation of aqueous extracts of *B. balanoides* by ammonium sulphate precipitation yields a number of high molecular weight protein components, including some proteins associated with nucleic acid and others with carbohydrate. At least two of the components are active in promoting the settlement of barnacle cyprids under standard experimental conditions. One of the active components stained PAS-positive and is probably a glycoprotein or mucopolysaccharide-protein complex. The other active component was a protein, associated with the nucleic acid fraction.

REFERENCES

BROOME, J. (1963). A rapid method of disc electrophoresis. *Nature, Lond.* **199**, 79–80.

CRISP, D. J. & MEADOWS, P. S. (1962). The chemical basis of gregariousness in cirripedes. *Proc. Roy. Soc. Lond.* B **156**, 500–20.

CRISP, D. J. & MEADOWS, P. S. (1963). Adsorbed layers: the stimulus to settlement in barnacles. *Proc. Roy. Soc. Lond.* B **158**, 364–87.

DETERMAN, H. (1968). *Gel Chromatography*. New York: Springer Verlag New York Inc.

GORDON, A. H. & LOUIS, L. N. (1967). Preparative acrylamide gel electrophoresis: a single system. *Analyt. Biochem.* **21**, 190–200.

GROVES, W. E., DAVIS, F. C., Jr. & SELLS, B. H. (1968). Spectrophotometric determination of microgram quantities of protein without nucleic acid interference. *Analyt. Biochem.* **22**, 195–210.

HAUSCHILD-ROGAT, P. & SMITH, I. (1968). Preparative acrylamide gel disc electrophoresis. In *Chromatographic and Electrophoretic Techniques*. Vol. II. *Zone Electrophoresis*, 2nd ed., p. 475 (ed. I. Smith). London: Heineman.

KEYSER, J. W. (1964). Staining of serum glycoproteins after electrophoretic separation in acrylamide gels. *Analyt. Biochem.* **9**, 249–52.

KNIGHT-JONES, E. W. (1953). Laboratory experiments on gregariousness during setting in *Balanus balanoides* and other barnacles. *J. exp. Biol.* **30**, 584–98.

KNIGHT-JONES, E. W. & STEVENSON, J. P. (1950). Gregariousness during settlement in the barnacle *Elminius modestus* Darwin. *J. mar. biol. Ass. U.K.* **29**, 281–97.

RICHARDS, E. G., COLL, J. A. & GRATZER, W. B. (1965). Disc electrophoresis of ribonucleic acid in polyacrylamide gels. *Analyt. Biochem.* **12**, 452–71.

WARBURG, O. & CHRISTIAN, W. (1941). Isolierung und Kristiallisation des Gärungsferments Enolase. *Biochem. Z.* **310**, 384–421.

THE BIOLOGY OF LARVAE OF
OPHLITASPONGIA SERIATA FROM TWO
NORTH WALES POPULATIONS

W. G. FRY

Marine Science Laboratories, Menai Bridge, Anglesey

INTRODUCTION

The literature on sponges contains many papers concerned with morphogenesis and cell behaviour in sponge larvae, and yet there are but few detailed accounts of the behaviour of sponge larvae themselves. Studies on the behaviour of marine invertebrate larvae which possess complex sense organs, e.g. cirripedes (Crisp, 1965), molluscs (Carriker, 1961; Bayne, 1964, 1969; D'Asaro, 1967), polyzoa (Ryland, 1959; Crisp & Williams, 1960), annelida (Wisely, 1960; Williams, 1964), tunicates (Grave & Nicoll, 1940), have demonstrated complex behaviour patterns which may change as the larvae develop, or in response to external stimuli, and which result in settlement in regions most suitable for growth and reproduction.

The apparent absence of sensory organs, and of nervous integration in sponges (see Jones, 1962, cf. Pavans de Ceccatty, Thiney & Garrone, 1970) suggests that sponge larvae are either extremely unselective in their settlement preferences, or that they exhibit a pattern of behaviour which undergoes change independently of external stimuli, or that sensory and motor 'nervous' systems affecting the behaviour of the whole animal must be sought in the ultrastructure of the individual cells.

Warburton (1966), Bergquist & Sinclair (1968) and Bergquist, Sinclair & Hogg (1970) have discussed recently the behaviour of larvae of several species of Demospongiae from eastern North American and New Zealand waters, and the last authors have reviewed the literature on sponge larval behaviour. These researches show that the larvae of some species do respond to certain obvious physical features of the environment, such as gravity and light, and that the responses to such stimuli may change during larval life.

Bergquist *et al.* (1970) came to the general conclusion that the settlement of larvae on the shore is not haphazard, and thus the final distribution of sponges is not determined solely by differential selection in different microhabitats. These authors, and Warburton (1966), worked with freeswimming larvae from littoral or shallow-water specimens, but Borojević's

(1967) description of a flattened, creeping larva from a deeper water species (*Polymastia robusta*) supports the general conclusion that sponge larvae are diverse in both structure and behaviour, according to the needs of the post-larval sponge.

However, neither Warburton (1966) nor Bergquist *et al.* (1970) were able to show in every case a clear relationship between larval behaviour patterns and the distribution of post-larval sponges on the shore. One such puzzling larval behaviour pattern, described by Bergquist *et al.* was that of *Ophlitaspongia seriata* (Grant, 1826; Demospongiae) in New Zealand waters. This species is very common on the north Welsh coast, and personal observations on larvae from this region give a conflicting picture of the larval life-cycle and behaviour.

LIBERATION OF LARVAE

Materials and methods

Whole specimens of *O. seriata* were collected, still attached to their rock substratum over the whole of the basopinacoderm, and were maintained in the laboratory individually in jars kept in a constant environment system (Gruffydd & Baker, 1969) (see Fig. 1). The 'food' material supplied to the sponge consisted of a 100 ml of flagellate culture *Micromonas squamata* (mean diameter 3.24 μm), *Isochrysis galbana* and *Monochrysis lutheri* (mean diameters 3.85 μm), and *Tetraselmis suecica* (mean diameter 6.27 μm), mixed in varying proportions according to which algal cultures were ready for harvesting on a particular day. To the 100 ml of algal suspension were added 100 ml of a dead bacterial culture. The bacterial cultures were made by adding 50 g of bacteriological peptone to 5 l of Menai Straits sea water. Air was bubbled through the culture gently for 4 or 5 days at room temperature, after which time the culture was autoclaved for 15 min at 120 p.s.i. To the cooled, dead culture, was then added 50 ml of a stock antibiotic solution made up of 3 g of penicillin and 5 g of streptomycin sulphate in 1 l of filtered sea water. New cultures were made up as proved necessary and, although it was not practicable to assay the cultures, each culture probably contained a preponderance of *Pseudomonas* and *Arthrobacter* species (G. D. Floodgate, personal communication). The treatment of the bacteria produced quantities of disrupted bacterial cells, and hence of minute organic particles as well as whole dead bacteria. Thus the range of size of potential food offered to the sponge encompassed the ranges of sizes of particles considered by Rasmont (1968) to be suitable for sponge nutrition.

Sponges were collected in June and July of 1967 and 1968 from the well-protected shore between Church Island and Ynys Welltog in the Menai Strait and from the fully exposed, irregular coast at Bodorgan, Anglesey (Fig. 2). In both localities, as in the rest of its geographical range, the species occupies a narrow horizontal band from the approximate level of mean high water of neap tides to 1 or 2 m below extreme low water of

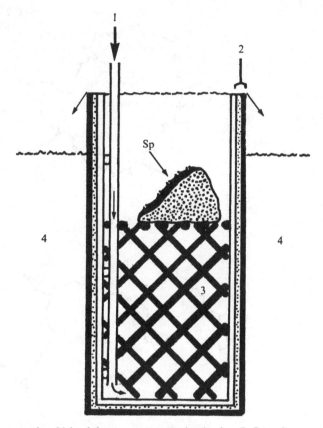

Fig. 1. Apparatus in which adult sponges were maintained. 1, Inflow of sea water; 2, layer of black paint over white paint on glass, 3, PVC mesh supporting specimen; Sp, sponge attached to rock substratum; 4, controlled temperature recirculating water.

spring tides. Specimens were collected occasionally from slightly greater depths as encrustations on shells of living *Chlamys opercularis*, but investigations by Mr R. Smith while diving do not indicate that the species is common below the lowest tide levels. From the literature, as well as from personal observations, it appears that, in the entire range of the species on the eastern North Atlantic coasts, specimens are found only where there is considerable water movement, generated either by tidal streams or by waves.

Unlike the situation described by Simpson (1968*b*) for *Microciona pro-lifera*, specimens of *O. seriata* on the north Welsh coasts do not appear to undergo any histological regression during the winter months. Thin sections of tissue taken at various periods of the year show no signs

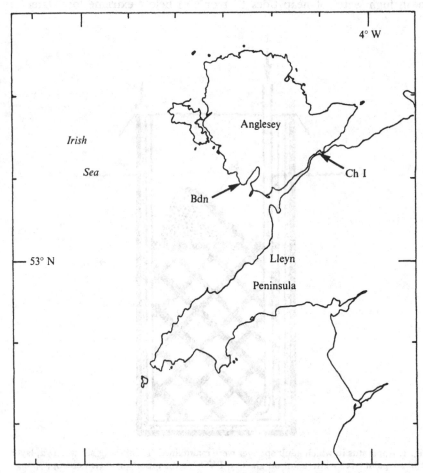

Fig. 2. Map of part of the North Wales coast, showing collecting sites. Ch. 1, Church Island in the Menai Strait; Bdn, Bodorgan.

of deterioration of the aquiferous system, nor changes in the numbers and proportions of archaeocytes, archaeocyte-derivatives, collencytes or pinacocytes.

For the 1967 studies, six specimens were collected from each locality during the last week of July. This period was chosen because repeated monthly samples over the two previous years had shown that embryos were recognizable in the tissues only during July and August. In 1968, the specimens were collected on the 12th and 13th of June. Subsequent treatment was the same in both years. They were placed in an apparatus held initially at 15 °C, and the temperature raised by a small part of a degree each day. Ten days after collection, when the temperature of the apparatus had reached 18 °C, the first larvae were liberated. Thereafter, the temperature was maintained at 18 °C.

Throughout the period of maintenance of the sponge specimens in the laboratory, sea water was kept flowing through all the apparatus for 18 out of each 24 h. During the night, and the latter part of each day, every jar was covered with a black plastic sheet. Each morning, the plastic sheets were removed, the sea water flow stopped for 6 h, and 200 ml of 'food' material added to each jar.

Control of liberation

Once larval liberation had begun, it was complete within 2 h of removing the covers and stopping the flow of water in the apparatus. It was observed during the routine daytime liberation of larvae that, after the water flow had stopped, the oscular papillae of the sponge became greatly extended and the force of the exhalant currents increased. Shortly afterwards, larvae began to be ejected with considerable force from the oscula. It is possible, therefore, that increase in the velocity of the exhalant current, or increase in the rate of transport through the aquiferous system, accompanied by dilation of the peripheral exhalant canals, is the immediate cause of larval liberation.

In the course of this operation, three changes took place, any one of which might have involved the liberation of larvae; the cessation of water flow, a considerable increase in illumination, and the addition of food material. However, on two occasions the sea water flow stopped accidentally during the night. Though the sponges were in almost complete darkness, some larvae were liberated. Also, if the sea water was kept flowing through the apparatus, larval liberation occurred and was completed within 2 h of uncovering and illuminating the specimens, whether or not food material had been added. On the other hand, the addition of food material to covered jars through which sea water was flowing was not followed by any liberation of larvae.

Hence, the presence of additional food material can be discounted as a stimulus for larval liberation, and either a marked increase in illumination or diminution of water flow are factors which could induce larval liberation.

Because of the wide variation in numbers of larvae liberated by individual sponges from day to day and by different specimens in one day, it was not possible to make statistically significant comparisons of the numbers of larvae liberated following cessation of water flow alone, increased illumination alone, or cessation of water flow together with increased illumination. However, O. *seriata*, as an intertidal organism, is subject to regular wide changes in the degree of illumination and speed of water flow, and these two factors can therefore be considered relevant to larval liberation in the field.

Although the specimens maintained in the laboratory in 1968 were collected in mid-June and their environmental temperature raised to 18 °C within 10 days of collection, larvae were not liberated until 29 July. This was only 4 days before the date on which larval liberation had begun during the previous year, when the specimens had been collected during the latter part of July. It appears that, although the laboratory conditions may well have affected the productivity of individual specimens, they did not affect markedly the duration of the period of larval liberation. In 1967, larval liberation occurred between 2 August and 8 September. In 1968, liberation occurred between 29 July and 30 August – a period shorter by only 5 days than the 1967 season.

In 1967, the Bodorgan specimens, although equal in number and of the same range of size (12–40 oscula) as the Church Island specimens, produced substantially fewer larvae than did the latter. The average rate of larval liberation was 6.85 larvae/specimen/day for the Bodorgan material, and 15.01 larvae/specimen/day for the Church Island material. Because of this difference, work in 1968 was concentrated upon Church Island specimens and only four Bodorgan specimens as against eight Church Island specimens were kept to provide larvae. In 1968, the figures were 0.87 larvae/specimen/day for the Bodorgan material and 7.94 larvae/specimen/day for the Church Island material.

Differences between populations of Ophlitaspongia seriata

As well as differences in larval output, some evidence of skeletal differences between the two populations of O. *seriata* exist, and have been presented elsewhere (Fry, 1970). It was observed during maintenance of live specimens that the membranous contractile oscular papillae, which are visible only when the sponge is active, are much more strongly developed in the Church Island specimens than in those from Bodorgan. This is probably because the latter inhabit a hydrodynamically more violent regime. Possibly related to this difference in environment is the fact that Bodorgan specimens are characterized by their possession of a cuticle, which appears from its staining reactions to be composed of spongin. A similar cuticle could not be found in Church Island specimens.

Bodorgan specimens are more difficult to maintain in the laboratory, which receives Menai Straits water, than are Church Island specimens. Intact specimens collected from Bodorgan during summer and winter months die back completely within 8–10 weeks, although kept in a constant slow flow of water, liberally supplied with bacterial and algal food, and kept at a variety of temperatures within the ranges which they normally encounter. Church Island specimens, on the other hand, can be maintained apparently indefinitely under the same conditions.

Whereas samples of mechanically macerated Church Island specimens can traverse all the stages of reconstitution in Menai Strait and Bodorgan sea water to produce new and perfectly functional sponges, the reconstitution process in Bodorgan cell macerates, whether in Menai Strait or in Bodorgan coastal water, proceeds little further than the flattening of the diamorph. (For terminology, see Borojević *et al.* 1968). At most, an aquiferous system begins to differentiate, but it remains disorganized. Few, if any, oscules are formed, and the reconstituting Bodorgan sponges rapidly die back and break up.

To determine whether the difference in ability to reconstitute arose from differences in the ability of nucleolate cells from the two populations to aggregate, an assessment was made of aggregation ability of cells from a Bodorgan and from a Church Island specimen. Approximately equal volumes of tissue were cut from each, broken up into small fragments, and the cells dissociated by squeezing through fine bolting-silk into crystallizing dishes containing 200 ml of filtered fresh sea water from the collecting site of the sponge. Ten ml of cell suspension were added to 25 ml of sea water, either from the Menai Strait or from Bodorgan, in a Petri dish. Changes with time in density of aggregates per unit area were measured by counting under a microscope field the number of groups of four or more non-flagellated cells hourly in the Petri dishes. Each test was replicated five times, thus every point in Fig. 5 represents the mean of 25 counts of aggregates. The experiment was performed during March when the sea surface temperatures in the Menai Strait and Bodorgan vicinity are approximately 6 and 7 °C respectively (Fig. 3). However, the specimens whose tissues were used in the experiment had been kept in running sea water at 18 °C in the laboratory, and liberally supplied with food suspension for 1 week before the experiment was performed.

The most significant result of the experiment is that the origin of the sea water used (see Figs. 3, 4), has little effect on reconstitution ability. The Church Island cell suspensions produced the same average densities of aggregates in Bodorgan water as they did in Menai Strait water, and the

6

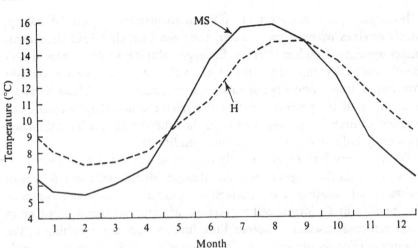

Fig. 3. Mean monthly sea surface temperatures in the Menai Strait and off Holyhead. MS, temperatures off Menai Bridge pier 1961–8; H, Irish Sea temperatures off Holyhead, data from Hughes (1966).

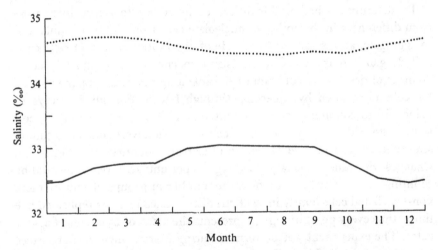

Fig. 4. Mean monthly sea surface salinities in the Menai Strait and off Holyhead. Solid line, data from Menai Bridge pier 1955–68; dotted line, data from Holyhead rise 1954–64.

final density of aggregates of Bodorgan cell suspensions was the same in Menai Strait and Bodorgan sea water (Fig. 5). The density of aggregates from the Menai Straits specimens, however, was less than that of the Bodorgan specimens, a difference which may result from differences between the densities of the cells in the initial suspensions. The initial cell densities were estimated by weighing on Oxoid membranes, the fixed (sea water Bouin) and dried (50 °C) remains of the cell suspension from each piece of sponge. The Church Island cell suspension produced an aggregation density of 0.00251 g/cm^2, while the Bodorgan cell suspension produced an aggregation density of 0.0129 g/cm^2. A higher density of cells settling on the substratum would result in fewer, but more massive, diamorphs, because of the shorter distance which will have to be traversed by single cells before they encounter others. The difference in aggregation rate is not necessarily indicative of differences in cell mobility nor of the power of cells to adhere to one another.

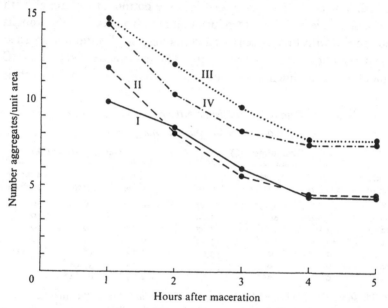

Fig. 5. Reaggregation behaviour of cells of Church Island and Bodorgan specimens of *O. seriata* in Menai Strait and Bodorgan coast water. I, Church Island cells in Menai Strait water; II, Church Island cells in Bodorgan water; III, Bodorgan cells in Bodorgan water; IV, Bodorgan cells in Menai Strait water.

BEHAVIOUR OF LARVAE

Larval activity

After expulsion from the parent sponge, the larvae swim rapidly upwards to the surface, with their more pointed end forwards, rotating about their long axis and following a looping path within which they gyrate in the same direction as they rotate about their longitudinal axis. Warburton (1966), Bergquist & Sinclair (1968) and Bergquist *et al.* (1970) have recorded this pattern of swimming in other species of sponge, and the last mentioned authors have termed it 'corkscrew' swimming. The lateral and vertical extents of this looping swimming vary at different times in the same larva. When the larvae strike the surface of the water, they either turn and perform more looping excursions, which bring them back to the air/water interface, or else they remain with their pointed end at the interface, rotating but not moving laterally.

Larvae liberated from a single Church Island specimen were distributed equally in five waxed troughs of internal dimensions 72 × 40 × 15 mm. Into the bottom of each trough was placed a 72 × 40 mm coverglass, which had been washed in fuming nitric acid, thoroughly rinsed in distilled water and oven-dried. The coverglass was sealed into the bottom of the trough with a hot needle. The larvae were introduced into the troughs, and the troughs filled to the rim with filtered sea water containing stock antibiotic solution in the concentration given above (p. 156). The troughs were kept at 18 °C throughout the experiments. Periodic rapid surveys were made of their activities.

Table 1. *Changes in activity patterns of larvae after release from the parent sponge*

Hours after liberation	Swimming (%)			Metamorphosing (%)		
	In midwater	At water surface	Along bottom	On sides	On bottom	At water surface
8	14.08	83.35	2.57	0	0	0
10	10.90	86.30	2.80	0	0	0
22	14.56	81.47	3.97	0	0	0
24	18.57	77.88	3.55	0	0	0
26	0	0	53.65	2.39	28.09	15.87
34	0	0	54.60	2.39	37.44*	5.57*
45	0	0	0	2.39	97.61*	0

* These values are affected by the sinking of post-larvae from the interface. Such post-larvae did not necessarily survive.

Table 1 indicates the percentages of the time spent swimming freely at the interface in still water, and the changes in these activities with time. The results revealed that under these conditions the larvae spent the major part of their time in the first 24 h hanging at the air/water interface. The percentage of time spent in looping swimming in midwater was surprisingly small and constant. Between 24 and 26 h after liberation, looping swimming in midwater and rotating hanging at the water surface ceased. Approximately half the larvae had by then begun metamorphosis while the remainder were creeping over the surface of the coverglass. Between 34 and 45 h after liberation, all the larvae had begun metamorphosis. Warburton (1966), Simpson (1968b) and Bergquist et al. (1970) referred to larvae attached by one end – usually the pole which is anterior in looping swimming. All those larvae of *O. seriata* whose settlement and onset of metamorphosis were observed came to rest lying on one side. The first sign of metamorphosis was the emergence of a stream of cells from the 'posterior' pole.

Considering first the pattern of activities represented by Table 1, the most remarkable phenomenon is the abrupt change from moving in the open water to creeping over the substratum. Warburton (1966) described a similar change in activity in *Microciona prolifera* and *Cliona celata*, and concluded that it was brought about by a period of active downwards swimming. Downward swimming was observed in *O. seriata*, but persistent downward swimming can be induced by water turbulence at any stage in the larval life prior to settlement. In fact, the power of swimming of the larvae appeared to diminish between 24 and 26 h after liberation, and thereafter the speed of rotation and the lateral and vertical extent of the looping swimming became progressively less. Table 1 also gives the sites of settlement and metamorphosis of the larvae. A very small percentage of the larvae metamorphosed on the vertical walls of the trough, and surprisingly, more than 15 % of the larvae metamorphosed at the water surface to produce perfect and functional small sponges. The high proportion settled on the bottom of the trough is also remarkable since on the shores around North Wales *O. seriata* cannot be found growing on upward facing horizontal surfaces; it always occurs on overhanging walls or ledges.

It is possible that the waxed walls of the troughs used in the experiment were inimical to settlement. Consequently, pieces of glass were sealed on to the vertical walls of the trough and the experiment repeated using five troughs each containing 30 larvae. In this experiment, the larvae were presented with equal areas of vertical and horizontal glass surface and the vertical surfaces contained equal areas of wax and glass. In every case, less than 5 % of the larvae metamorphosed on vertical surfaces, and some of these larvae attached themselves in crevices between the vertical glass plates and the adjoining wax walls. In all the experiments, several larvae metamorphosed at the water surface.

Reactions to physical stimuli

The larvae of *O. seriata* show a pattern of reactions to physical stimuli similar to that recorded by Warburton for *M. prolifera* and *C. celata*. During the first 24 h of larval life they show strong negative geotaxis, but during the ensuing 2 h this reaction disappears and is never regained. Negative geotaxis and its subsequent loss are not modified by changes in the intensity of illumination, by changes in direction of a source of parallel light rays, by changes in temperature between 5 and 20 °C, by the degree of oxygenation of the water, nor by the time of day during which the other conditions are varied.

If, however, during the first 24 h of free larval life the surface of the water is disturbed, all larvae resting at the interface begin looping swimming immediately. If the agitation is stopped the larvae again come into contact with the interface and the activity pattern shown in Table 1 is resumed. However, if the surface of the water is agitated continuously by a gentle jet of air, larvae never come to rest against the water surface while the turbulence persists. Their looping swimming carries them downwards and then up again until they reach the strongly turbulent water, when they turn and repeat the looping swimming. If a strong jet of air is applied for some time, all the larvae will soon be found swimming actively along the bottom of the container, or resting while swimming with their pointed ends ('anterior') against the bottom of the container.

Warburton (1966) has reported that, between 20 and 30 h after liberation, larvae of M. prolifera and C. celata develop a strongly positive geotactic patter of behaviour. As has been stated above, in O. seriata general mobility diminishes after 24 h of swimming, and if left undisturbed the larvae do not swim downwards strongly but gradually sink while performing looping swimming of greatly diminished extent. On the other hand, if disturbed by touching or by water movement, a larva of more than 26 h of free life will perform strong looping swimming for a short period.

The influence of a solid oversurface

Evans (1899) first reported metamorphosis of freshwater sponge larvae at the air/water interface, but since that date the phenomenon does not appear to have been recorded or discussed. As Evans observed for Spongilla lacustris, the larvae of O. seriata metamorphosing at this site do so with their basopinacoderm apposed to the interface, and their primary osculum facing downwards. Histological examination suggests that they are functionally and structurally normal sponges.

The propensity for larvae of O. seriata to metamorphose at the water surface and the large proportion of time which they spend hanging at the water surface suggest that contact with overhanging surfaces might play an important role in determining the eventual sites of settlement of the larvae.

Experiments using a smooth glass oversurface were first attempted on the assumption that, in the field, accidental contact with an overhanging surface might be followed by settlement on such a surface or by movement along the surface until a suitable steep face was encountered. Experiments were conducted using waxed troughs with a 72 × 40 mm coverglass fixed to the bottom and a similar coverglass laid across the middle of the top of the trough, to offer an 'oversurface' of 1600 mm² area which made contact at the sides of the trough with vertical glass surfaces of 40 × 15 mm. Larvae were placed in the trough, which was then filled with sea water containing antibiotics until the underneath of the coverglass was just wetted. Care was taken to ensure that no air bubbles were trapped beneath the coverglass. The troughs were maintained at 18 °C. The experiment was repeated five times with 30 larvae in the trough.

Contrary to expectation, no larvae attached and metamorphosed on the vertical glass walls. However, all the larvae came to rest within 6–8 h with their more pointed ends against the glass oversurface, and nearly all settled and metamorphosed on the glass bottom of the trough directly below the glass oversurface. In all five experiments, the final distribution of the metamorphosing larvae on the bottom of the trough was found to be significantly contagious according to Fisher's coefficient of dispersion. In contrast, in experiments not involving glass oversurfaces the settlements were not found to be significantly different from random.

Of the five groups of 30 larvae used in the above experiment, only four attached and metamorphosed at the glass oversurface itself. Several others settled in crevices where the coverglass was sealed to the bottom of the trough. A few even moved through small gaps left by imperfect sealing and attached under the glass; metamorphosis was never completed at such sites.

As the presence of a solid oversurface did not appear to induce settlement on contiguous vertical surfaces, but the position of such an oversurface did influence the site of attachment and distribution of the larvae on the bottom of the trough, further experiments were performed. Oversurfaces of square coverglasses were suspended rigidly in the middle of the surface of the water contained in a glass dish. Four factors were varied: (1) the number of larvae (2) the area of solid oversurface (3) the area of the total water surface (4) depth of the water. In all experiments care was taken to ensure that the water surface was disturbed as little as possible, and that no air bubbles formed under the solid oversurface.

The results of the experiments are presented in Tables 2 and 3.

These results can be summarized as follows.

(1) In still water, only when the area of the oversurface was at least 22.6% of the total water surface did all the larvae come to rest under the oversurface before sinking (Table 2).

(2) If metamorphosing sponges made contact after attachment, they always fused completely, so that their aquiferous systems were continuous.

(3) The greater the density of larvae concentrated at the oversurface before sinking, and the smaller the distance between the oversurface and the substratum, the greater was the clustering of the metamorphosed larvae and hence the higher the percentage of larval fusion (Table 3).

Table 2. *Relationships between total water surface areas and solid oversurface areas and their effect upon aggregation of larvae*

(Each cell gives the percentage of the total water surface area made up of solid–water interface.)

Areas of solid oversurface (mm^2)	Area of total water surface (mm^2)		
	2880	4101	7084
462	16.04	11.26	6.52
950	32.98*	23.16*	13.41
1600	55.55*	39.01*	22.59*

* Entries where aggregation occurred.

Table 3. *Effect of larval density and depth of water upon fusion (F) of post-larvae settled under conditions inducing aggregation (Table 2)*

(F indicates the occurrence of at least one fusion mass of two or more post-larvae in three experiments.)

Depth (mm)	Square millimetres of solid oversurface per larva									
	9.24	15.40	19.00	31.67	32.00	46.20	53.33	77.00	95.00	158.00
15	F	F	F	—	—	—	—	—	—	—
80	F	F	F	—	—	—	—	—	—	—

Thus, when the oversurface of 950 mm^2 was suspended in the middle of a free water surface of 2880 mm^2 the aggregated larvae sank together. In a depth of 15 mm of water approximately 80 % of the settled larvae subsequently fused with adjacent post-larvae; in a depth of 80 mm of water, the aggregates sometimes split into two while sinking, but at least 50 % of the settled larvae subsequently fused with their neighbours.

FUSION AND POST-LARVAL MORTALITY

In an attempt to raise successive generations of sponges, and to discover successful techniques for rearing specimens in the laboratory, some 2000 larvae liberated in 1967, the majority from Church Island, were allowed to settle and metamorphose. As the possible significance of aggregation and fusion had not been detected, the larvae were settled on coverglasses in the absence of any solid oversurface. Although many of the glass sheets bore large numbers of post-larvae, no cases of fusion were observed.

The coverglasses bearing the young sponges were held vertically in racks and immersed in large tanks whose temperature could be controlled to ± 0.25 °C. The initial fall of 3.2 °C from the temperature of larval liberation to the mean monthly surface temperature of the Menai Straits and Irish Sea for September (Fig. 3) was accomplished in steps of approximately 0.25 °C performed over 14 days. Thereafter the tank temperature was diminished daily so that it coincided at the middle of the month with the monthly mean for Menai Strait water. The young sponges were maintained in slowly flowing sea water for 18 out of the 24 h and during the remaining 6 h were fed as described above.

Mortality of post-larvae was assessed by counting the sponges at weekly intervals and calculating the cumulative percentage of mortality. In nearly all cases of death, the first signs were the disappearance of choanocyte chambers and retraction of the cells into a diamorph. No sponge which showed this change later showed signs of redifferentiation, and the appearance of a diamorph was recorded, therefore, as the death of the sponge.

The results are shown graphically in Fig. 6. At the end of 14 days, there was 88.6% mortality among the Church Island larvae and 86.1% among the Bodorgan larvae. By the end of the 10th week after settlement, all post-larvae had disappeared or were in diamorph stage.

The experiment was repeated in 1968 and the effect of fusion on post-larval mortality was tested. A hundred and forty-nine larvae from Church Island were allowed to settle and metamorphose in isolation. In addition nine batches of Church Island larvae were settled in confined areas where fusion could take place. The fusion masses consisted mainly of two, three or four past-larvae.

Fig. 6 shows that among the isolated post-larvae the greatest mortality occurred during the first 14 days, as in the previous year. Moreover, the initial mortality was 45.8% for the isolated post-larvae, whereas only two of the seven two-cell fusion masses died in the first 14 days, amounting to 14% mortality, since the other fusion masses survived. There was no further mortality of the fusion masses over a period of 40 weeks.

The curves shown in Fig. 6 for isolated post-larvae fit only approximately the survivorship relation derived from a constant mortality rate β:

$$n/n_0 = e^{-\beta t}.$$

On isolated Church Island post-larvae, the best fit gives values of β of approximately 0.37 (weeks^{-1}) for the 1967 observations and of 0.12 (up to 10 weeks) falling to 0.09 (after 20 weeks) for the 1968 observations.

It is possible to calculate the survivorship of fusion masses on two assumptions: (1) that the survivorship of component post-larvae equals that of isolated post-larvae (2) that the longevity of the fusion masses is equal to that of the longest surviving post-larva. The survivorship (N/N_0) of fusion masses each containing x post-larvae will then be given as

$$N/N_0 = 1 - (1 - e^{-\beta t})^x.$$

From this formula, the number of surviving fusion masses can be compared with the observations on individual post-larvae shown in Fig. 6. The

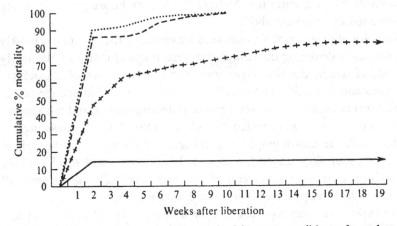

Fig. 6. Cumulative percentage mortalities under laboratory conditions of post-larvae of *O. seriata*. ..., isolated Church Island post-larvae 1967; ---, isolated Bodorgan post-larvae 1967; +++, isolated Church Island post-larvae 1968; solid line, fusion masses 1968.

Table 4. *Calculated survivorship of isolated and fused post-larvae*

Calculated survivorship

x	10 weeks $\beta = 0.12, t = 10$	20 weeks $\beta = 0.09, t = 20$	40 weeks $\beta = 0.09, t = 40$
1 (isolated larvae)	0.3	0.165	0.027
2	0.51	0.30	0.055
3	0.66	0.42	0.08
4	0.76	0.52	0.105
5	0.83	0.60	0.13

survivorship at 10, 20 and 40 weeks has been calculated from the above formula for $x = 1$ to $x = 5$ post larvae per fusion mass. The values for β were chosen to predict the observed mortalities of isolated post-larvae. Table 4 records the predicted values.

Entering the table at $x = 1$, the values of survivorship (1 − mortality) for $t = 10$ and 20 weeks, agree as required with the 1968 Church Island curve for isolated individuals, while the value predicted at $t = 40$ weeks (97.3% mortality) is obtained by extrapolation from $t = 20$. The observed survivorship of the fusion masses containing 2, 3 and 4 post-larvae each was 0.86 (only 14% mortality), and it remained unchanged at 10, 20 and 40 weeks. Thus, the survival of fusion masses was considerably greater than the calculated values in the table. The fact that the instantaneous mortality constant β for isolated post-larvae was not constant must reduce the force of the comparison since the theory is based on a constant mortality rate. Nevertheless, the evidence points strongly to an increase in the chances of survival as a result of fusion.

Neither the isolated post-larvae nor the fusion masses produced any larvae before the conclusion of the experiment, hence the effect of fusion upon reproduction cannot be assessed.

From some other observations, it can be concluded that larvae from Bodorgan and Church Island specimens will fuse. Single larvae from each region allowed to settle in a drop of water on a coverglass kept in a water-saturated atmosphere fused after metamorphosis on six occasions. Larger scale fusions were attempted but difficulties in identifying the larvae hampered the experiments. The larvae of *O. seriata* are dark blood-red, and vital staining, in order to be recognized, had to be so intense that it affected the behaviour of the larvae – in some cases completely inhibiting movement. However, on three occasions 30 unstained larvae, derived in equal numbers from Church Island and Bodorgan sponges, gave rise to single fusion masses when settled in the restricted environment of an inverted tube, the bottom end of which was 63.5 mm^2 in area.

LARVAL BEHAVIOUR AND SETTLEMENT ON THE SHORE

If the picture of the larval behaviour pattern of *O. seriata* and its change with time which can be obtained from laboratory observations and experiments is a true representation of events in the field, then it can be concluded that the larvae of this species are highly specialized for colonizing the littoral niches exploited by the adults. In some parts of the shore populations of this sponge are dense; for example, a survey of 420 m^2 exposed by spring tides at Church Island revealed 478 specimens of *O. seriata*. Even if the average productivity of these specimens were no higher than that achieved by specimens kept in the laboratory in 1967, it would lead to the production of 18 larvae/m^2/day during the liberation season.

It appears that maximum liberation of larvae will occur during slack water when the adults are submerged. The avoidance of turbulent water by looping swimming, and the positive thigmotaxis, will cause the larvae to concentrate well below the water surface or in crevices or under over-hanging rocks. This must result in reduced dispersal of the larvae from their sites of liberation; they will not inevitably be transported passively to sites of obligatory settlement.

DISCUSSION

Earlier reports (e.g. Ankel & Eigenbrodt, 1950; Rasmont, 1967) indicate that fusion of young sponges is probably a commonplace occurrence in fresh water. Although Wilson (1907) reported that *Microciona prolifera* larvae could be induced to fuse, and *Polymastia robusta* post-larvae (Boro-jević & Lévi, 1967) and *O. seriata* post-larvae from the Brittany coast (R. Borojević, personal communication) will fuse when metamorphosing in small volumes of water, natural fusions have never been demonstrated unequivocally, despite the fact that their occurrence has long been suspected (see Burton, 1949). If fusion is a commonplace event in the sea, its signifi-cance to the reproductive biology, the speciation mechanisms and the taxonomy of marine sponges must be considered.

In an earlier paper (Fry, 1970) I have shown that there are readily detect-able, if small, differences in skeletal composition between the Church Island and Bodorgan populations of *O. seriata*, and that both these populations differ widely from two Breton populations. In this paper evidence for addi-tional structural and physiological differences has been given. Yet, despite such differences, the larvae from the two north Wales populations show identical behaviour patterns in the laboratory.

The pulsating water movements in the Menai Strait and along the south-western coast of Anglesey (Harvey, 1968) are such that an interchange of larvae between Church Island and Bodorgan appears just possible, but the direction of the residual flow from the south-western exit of the Menai Strait implies a greater invasion of Bodorgan by Church Island larvae than vice versa. The isolation of the Bodorgan and Church Island regions by several miles of sandy and muddy shores suggests further that interchange of larvae in any one breeding season will be small. Nonetheless, if there can be intermixture of larvae with behaviour patterns inducing aggregation and fusion, and such fusion confers increased chance of survival, it does not seem unreasonable to suppose that among both Bodorgan and Church Island populations are specimens which – at least initially – were composed of post-larval cells originating from both regions.

Rasmont (1970) has discussed the reproductive advantages conferred upon the freshwater sponge *Ephydatia fluviatilis* by fusion of young asexually produced sponges. In this species, fusion of gemmule contents leads, under certain conditions, to an increase in productivity of new gemmules, and to an earlier onset of gemmulation. From this phenomenon and other data Rasmont (1970) has suggested that fusion of gemmule contents provides the young sponges with an excess of metabolite-rich, undifferentiated cells, which become reconcentrated in a new generation of gemmules. It is possible that fusion masses of O. *seriata* similarly contain undifferentiated, nutritive-rich cells in excess of their basic architectural needs, and that this excess provides metabolite reserves as well as reserves of cells necessary for accomplishing metamorphosis and early stages of growth.

In addition, there are genetic advantages which might arise from aggregation and fusion. The possession of a free-swimming larval stage in the life cycle permits an otherwise sedentary organism to attempt to colonize a large area and wide range of environments. If all O. *seriata* specimens are not genetically identical and homozygous, the initial result of larval liberation will be the production of a randomly dispersed, and well-mixed, population of various genotypes.

An initial random mixing of organisms followed by their aggregation during settlement is very clearly seen in gregarious barnacles such as *Balanus balanoides*, in which cyprids settle preferentially among members of their own species (Crisp, 1961; Crisp & Meadows, 1963). It is highly significant that such barnacles are hermaphrodite and indulge in cross-fertilization (Clegg, 1957; Crisp & Patel, 1961). Such cross-fertilizations involve relatively elaborate reproductive organs and behaviour patterns in B. *balanoides* and *Elminius modestus*. Complex organs and behaviour patterns are not present in hermaphrodite sponges, such as O. *seriata*. Cross-fertilization is traditionally considered to be achieved by the broadcasting of spermatozoa, which are trapped in the choanocyte chambers of other sponges.

However, the broadcasting of spermatozoa into the open water is normally associated with marked hypertrophy of gonadial tissues, as in many molluscs, echinoderms, tubicolous annelids. Comparable large-scale production of gametes has been recorded very rarely in sponges, and indeed it is only in a very few sponge species that gametogenesis and fertilization have been detected (e.g. Dendy, 1914; Gatenby, 1920; Duboscq & Tuzet, 1937; Tuzet & Pavans de Ceccatty, 1958; Simpson, 1968*a*, *b*). It is probably because of the scarcity of such data that the controversy over the significance of sexual reproduction in sponges has been revived recently (Sivaramakrishnan, 1951; Simpson, 1968*b*; Bergquist *et al.* 1970).

There seems to be no doubt, despite this controversy, that gametogenesis does occur widely in demosponges, and equally that special asexual reproductive and dispersal stages are of common occurrence. It is possible that post-larval fusion, by bringing together cell-masses of different genotypes, permits fusion of gametes of different genotypes within a common envelope. Such gametes need not differentiate to show the conventional form of free-swimming spermatozoa.

Complete fusion of post-larvae, followed by 'cross-fertilization' within a common envelope would appear to be a highly efficient method of maintaining a large heterogeneous gene-pool within which there is the greatest chance of the maximum possible number of genetic recombinations occurring after every period of gametogenesis with minimum wastage of nuclear material. Such a hypothesis is not at variance with observations of synchronous genesis of normal spermatozoa and ova (e.g. Simpson, 1968b), nor with the observation by Bergquist et al. (1970) of apparently synchronous sexual and asexual production of larvae, while it would account satisfactorily for the scarcity of observations of 'conventional' spermatogenesis in sponges.

On the other hand, the hypothesis does not account for some data from freshwater sponges. Rasmont and his coworkers (Rasmont, 1970) have isolated strains of Ephydatia fluviatilis which differ in some skeletal physiological attributes. Although the gemmules in this species are apparently asexually produced, the contents of gemmules of different strains will not fuse after hatching.

Clearly, many researches remain to be performed before we shall even be in a position to catalogue the types of reproductive processes occurring in sponges. Such widely differing reproductive methods as those described for freshwater sponges, marine Ceractinomorpha and Tetractinomorpha (Lévi, 1957) must permit widely different methods of genetic exchange within species. The remarkable geographical ranges, the morphological plasticity and the vague taxonomic boundaries amongst sponges, which have for so long perplexed taxonomists, are not necessarily so incomprehensible once so wide a range of reproductive methods is allowed for.

SUMMARY

Experiments extending over a period of 2 years have shown that the larvae of Ophlitaspongia seriata, though apparently insensitive to variations of other physical factors within the range used, showed a consistent pattern of behaviour in the presence of turbulence and of an overlying solid surface. Larval behaviour in this species leads to the formation of aggregations whose members fuse on contact during metamorphosis. This behaviour pattern is relevant to the distribution of adults on the shore.

Under laboratory conditions, fused post-larvae show a reduced mortality compared with isolated post-larvae. Metamorphosing larvae from two neighbouring populations which differ morphologically and physiologically will fuse under laboratory conditions. It is suggested that the fusion of post-larvae from neighbouring environments is not only analogous to the gregarious settlement of larvae of other marine organisms, but also provides an economical method of mixing genotypes. It seems highly likely that many sponge 'specimens' are genetically heterogeneous populations of cells. Such a hypothesis is not incompatible with the morphological plasticity and the wide geographical range of sponge 'species'.

The researches described in this paper were carried out during the tenure of the Red Hand Compositions Research Fellowship at Menai Bridge. Besides expressing my gratitude to the sponsors of this Fellowship, I wish to thank Professor D. J. Crisp, Dr Ll. Gruffydd, Dr J. Rees and Dr J. H. Simpson for advice and criticism at various stages of the work, and for their critical readings of the manuscript. In addition, I would like to express my gratitude to Sir George Meyrick and Mr O. J. Williams for allowing me access to the unspoiled shores bordering their lands.

REFERENCES

ANKEL, W. E. VON & EIGENBRODT, H. (1950). Uber die Wuchsform von *Spongilla* in sehr flachen Raümen. *Zool. Anz.* **145**, 195–204.

BAYNE, B. L. (1964). The responses of the larvae of *Mytilus edulis* to light and gravity. *Oikos* **15**, 162–74.

BAYNE, B. L. (1969). The gregarious behaviour of the larvae of *Ostrea edulis* L. at settlement. *J. mar. biol. Ass. U.K.* **49**, 327–56.

BERGQUIST, P. R. & SINCLAIR, M. E. (1968). The morphology and behaviour of larvae of some intertidal sponges. *N.Z. J. Mar. Freshw. Res.* **2** (3), 426–37.

BERGQUIST, P. R., SINCLAIR, M. & HOGG, J. J. (1970). Adaptations to intertidal existence: reproductive cycles and larval behaviour in Demospongiae. In *The Biology of the Porifera* (ed. W. G. Fry). *Symp. zool. Soc. Lond.* **25**, 246–71.

BOROJEVIĆ, R. (1966). Étude expérimentale de la différenciation des cellules de l'éponge au cours de son développement. *Devl Biol.* **14** (1), 130–53.

BOROJEVIĆ, R. (1967). La ponte et le développement de *Polymastia robusta* (Démosponges). *Cah. Biol. mar.* **8**, 1–6.

BOROJEVIĆ, R. (1970). Différenciation cellulaire dans l'embryogenèse et la morphogenèse chez les Spongiaires. In *The Biology of the Porifera* (ed. W. G. Fry). *Symp. zool. Soc. Lond.* **25**, 467–90.

BOROJEVIĆ, R. & LÉVI, C. (1965). Morphogenèse experimentale d'une éponge à partir de cellules de la larve nageante dissociée. *Z. Zellforsch. mikrosko. Anat.* **68**, 57–69.

BOROJEVIĆ, R. & LÉVI, P. (1967). Le basopinacoderme de l'éponge *Mycale contarenii* (Martens). Technique d'étude des fibres extracellulaires basales. *Jl Microsc.* **6**, 857–62.

BOROJEVIĆ, R., FRY, W. G., JONES, W. C., LÉVI, C., RASMONT, R., SARÀ, M. & VACELET, J. (1968). Mise au point actuelle de la terminologie des éponges. *Bull. Mus. Nat. Inst. Nat.* (2) **39** (6), 1224–35.

BUCHAN, S., FLOODGATE, G. D. & CRISP, D. J. (1967). Studies on the seasonal variation of the suspended matter in the Menai Straits. I. The inorganic fraction. *Limnol. Oceanogr.* **12** (3), 419–31.

BURTON, M. (1949). Observations on littoral sponges, including the supposed swarming of larvae, movement and coalescence in mature individuals, longevity and death. *Proc. zool. Soc. Lond.* **118**, 893–915.

CARRIKER, M. R. (1961). Interrelationships of functional morphology, behaviour and autecology in early stages of the bivalve *Mercenaria mercenaria*. *J. Elisha Mitchell scient. Soc.* **77**, 168–241.

CLEGG, D. J. (1957). Some observations on pairing in *Balanus balanoides*. *Ann. Rep. Challenger Soc.* **3** (9), 18–19.

CRISP, D. J. (1965). The ecology of marine fouling. In *Ecology and the Industrial Society*, *5th Symp. Brit. Ecol. Soc.* pp. 99–117.

CRISP, D. J. (1961). Territorial behaviour in barnacle settlement. *J. exp. Biol.* **38**, 429–46.

CRISP, D. J. & MEADOWS, P. S. (1963). Adsorbed layers, the stimulus to settlement in barnacles. *Proc. Roy. Soc. Lond.* B **158**, 364–87.

CRISP, D. J. & PATEL, B. (1961). The interaction between breeding and growth rate in the barnacle *Elminius modestus* Darwin. *Limnol. Oceanogr.* **6** (2), 105–15.

CRISP, D. J. & WILLIAMS, G. B. (1960). Effect of extracts from fucoids in promoting settlement of epiphytic Polyzoa. *Nature, Lond.* **188**, 1206–7.

D'ASARO, C. N. (1965). Organogenesis, development and metamorphosis in the queen conch, *Strombus gigas*. *Bull. mar. Sci. U.S.* **15**, 358–416.

DENDY, A. (1914). Observations on the gametogenesis of *Grantia compressa*. *Q. Jl microsc. Sci.* **60**, 313–76.

DUBOSCQ, O. & TUZET, O. (1937). L'ovogenèse, la fécondation et les premiers stades du développement des éponges calcaires. *Archs zool. exp. gen.* **79**, 157–316.

EFREMOVA, S. M. (1970). Proliferation activity and synthesis of protein in the cells of fresh-water sponges during development after dissociation. In *The Biology of the Porifera* (ed. W. G. Fry), *Symp. zool. Soc. Lond.* **25**, 399–413.

EVANS, R. (1899). The structure and metamorphosis of the larva of *Spongilla lacustris*. *Q. Jl microsc. Sci.* **42** (4), 364–476.

EWINS, P. & SPENCER, C. P. (1967). The annual cycle of nutrients in the Menai Straits. *J. mar. biol. Ass. U.K.* **47**, 533–42.

FRY, W. G. (1970). The sponge as a population: a biometric approach. In *The Biology of the Porifera. Symp. zool. Soc. Lond.* **25**, 135–61.

GATENBY, J. B. (1920). The germ-cells, fertilisation and early development of *Grantia (Sycon) compressa. J. Linn. Soc. Lond. Zool.* 34, 261–97.

GRAVE, C. & NICOLL, P. A. (1940). Studies on larval life and metamorphosis in *Ascidia nigra* and species of *Polyandrocarpa. Pap. Tortugas Lab.* 32, 1–46.

GRUFFYDD, LL.D. & BAKER, W. F. (1969). An integrated multiple unit controlled temperature system for sea water aquaria. *Lab. Pract.* 18, 300–4.

HARVEY, J. G. (1968). The flow of water through the Menai Straits. *Geophys. J.* 15, 517–28.

HARVEY, J. G. & SPENCER, C. P. (1962). Abnormal hydrographical conditions in the Menai Straits. *Nature, Lond.* 195 (4843), 794–5.

HUGHES, P. (1966). The temperature and salinity of the surface waters of the Irish Sea for the period 1947–61. *Geophys. J. R. astr. Soc.* 10 (4), 421–35.

HUUS, J. (1939). The effect of light on the spawning in ascidians. *Arch. Norske Vidensk Akad. Nat. Hist. Kl. 1939* (4), 1–30.

JONES, W. C. (1962). Is there a nervous system in sponges? *Biol. Rev.* 37, 1–50.

LÉVI, C. (1957). Étude des *Haliscarca* de Roscoff. Embryologie et systématique des Démosponges. *Archs Zool. exp. gen.* 93 (1), 1–181.

LÉVI, C. (1964). Ultrastructure de la larve parenchymella de Démosponge. I. *Mycale contarenii* (Marten). *Cah. Biol. mar.* 5, 97–105.

LÉVI, C. (1970). Les cellules des éponges. In *Biology of the Porifera* (ed. W. G. Fry). *Symp. zool. Soc. Lond.* 25, 353–64.

LÉVI, C. & PORTE, A. (1962). Etude au microscope électronique de l'éponge *Oscarella lobularis* Schmidt et de sa larve Amphiblastula. *Cah. Biol. mar.* 3, 307–15.

MERGNER, H. (1970). Ergebnisse der Entwicklungsphysiologie der Spongilliden. In *The Biology of the Porifera* (ed. W. G. Fry). *Symp. zool. Soc. Lond.* 25, 365–97.

PAVANS DE CECCATTY, M., THINEY, Y. & GARRONE, R. (1970). Les bases ultra-structurales des communications intercellulaires dans les oscules de quelques éponges. In *The Biology of the Porifera* (ed. W. G. Fry). *Symp. zool. Soc. Lond.* 25, 449–66.

RASMONT, R. (1968). Nutrition and digestion. In *Chemical Zoology*, 2, Section 1, Porifera: 43–51 (ed. Florkin & Sheer). New York and London: Academic Press.

RASMONT, R. (1970). Some new aspects of the physiology of fresh-water sponges. In *The Biology of the Porifera* (ed. W. G. Fry). *Symp. zool. Soc. Lond.* 25, 415–22.

RYLAND, J. S. (1959). Experiments on the selection of algal substrates by poly-zoan larvae. *J. exp. Biol.* 36, 613–31.

SIMPSON, T. L. (1968a). The structure and function of sponge cells: New criteria for the taxonomy of Poecilosclerid sponges (Demospongiae). *Peabody Mus. Nat. Hist. Yale Univ. Bull.* 25, 141 pp.

SIMPSON, T. L. (1968b). The biology of the marine sponge *Microciona prolifera* (Ellis and Solander). II. The temperature-related, annual changes in func-tional and reproductive elements with a description of larval metamorphosis. *J. exp. mar. Biol. Ecol.* 2, 252–77.

SIVARAMAKRISHNAN, V. R. (1951). Early development and regeneration in Indian marine sponges. *Proc. Indian Acad. Sci.* **34**, 273–310.

TUZET, O. & PAVANS DE CECCATTY, M. (1958). La spermatogenèse, l'ovogenèse, la fécondation et les premiers stades du développement d'*Hippospongia communis. Bull. biol. Fr. Belg.* **92**, 333–48.

WARBURTON, F. E. (1966). The behaviour of sponge larvae. *Ecology* **47** (4), 672–4.

WILLIAMS, G. B. (1964). The effect of extracts of *Fucus serratus* in promoting settlement of larvae of *Spirorbis borealis* (Polychaeta). *J. mar. biol. Ass. U.K.* **44**, 379–415.

WILSON, H. V. (1907). On some phenomena of coalescence and regeneration in sponges. *J. exp. Zool.* **5**, 245–58.

WISELY, H. B. (1960). Observations on the settling behaviour of the tubeworm *Spirorbis borealis* Daudin (Polychaeta). *Aust. J. mar. Freshwat. Res.* **2**, 55–72.

HOST SELECTION AND ECOLOGY OF MARINE DIGENEAN LARVAE

B. L. JAMES

Department of Zoology, University College, Swansea

INTRODUCTION

Studies on the distribution of Digenea, which have prosobranch molluscs as primary hosts, in the intertidal zone on the rocky shores of mid and south Wales, as described by James (1968 a–d, 1969) and in this paper, show that those without free-swimming larvae occur in the supralittoral fringe on semi-exposed to extremely exposed shores. Those with two free-swimming larvae (the miracidium and cercaria) tend to occur nearer low tide level, on more sheltered shores. Digenea with one free-swimming larva (the cercaria) are more widely distributed.

These observations suggested that, as they cannot select their hosts, the species in the supralittoral fringe may be less specific than those with free-swimming larvae which often have elaborate sense organs (James, 1969) apparently for active selection of the host. A comparison of the host lists, however, does not confirm this suggestion.

The experiments described in this paper are an attempt to determine how specificity is maintained in some of the species with and in some without free-swimming larvae.

MATERIALS AND METHODS

Most host species (Table 1) were collected from Cardigan Bay, between Twr Gwlanod and the Dovey estuary, from Pembrokeshire, between St Govan's Head and Broad Haven, and from the Gower peninsular, between Mumbles Head and Worm's Head. A map and a list of the 18 stations from which the collections were made in Cardigan Bay is given by James (1968c) and maps showing most of the 25 stations in Pembrokeshire by Fischer-Piette, Gaillard & James (1964) and James (1963). A full list of the stations in this region is given by James (1968d). The collections on the Gower peninsula were made on 14 stations in the neighbourhood of Mumbles Head, Caswell Bay, Pwll Du Head, Hunts Bay, Heatherslade Bay, Pobbles Bay, Three Cliffs Bay, Oxwich Point, Port Eynon Point and Worm's Head. One host species, *Littorina saxatilis* (Olivi), was collected from 178 shores around the coast of Britain in the areas shown in the map given by James (1968e).

Table 1. *The number and location of the prosobranchs examined for Digenea*

The figures for *L. saxatilis* include adults and juveniles (both being infected with Digenea) but only adults for the other species.

Ballantine's (1961) exposure scale; 1 – extremely exposed, 2 – very exposed, 3 – exposed, 4 – semi-exposed, 5 – fairly sheltered, 6 – sheltered, 7 – very sheltered, 8 – extremely sheltered.

EHWS – level of high water on extreme spring tides; MHWS – level of high water on mean spring tides; MHWN – level of high water on mean neap tides; EHWN – level of high water on extreme neap tides; MTL – mean tide level; ELWN – level of low water on extreme neap tides; MLWN – level of low water on mean neap tides; MLWS – level of low water on mean spring tides; ELWS – level of low water on extreme spring tides.

		Location of collecting areas		
Primary hosts	Number examined	Region	Exposure grade (Ballantine's (1961) scale) of shore	Tide level
Patellidae				
1. *Patella aspera* Lamarck	1567	Cardigan Bay,	1–5	MLWN–MLWS
2. *P. intermedia* Jeffreys	570	Pembrokeshire	1–5	MTL–MLWN
3. *P. vulgata* L.	2607	and Gower	1–8	MHWS–MLWS
4. *Patina pellucida* (L.)	750	peninsula	2–7	MLWN–ELWS
Trochidae				
5. *Monodonta lineata* (da Costa)	1750		4–8	MHWS–MLWN
6. *Gibbula umbilicalis* (da Costa)	8127		3–8	MHWS–MLWS
7. *G. cineraria* (L.)	2730		4–8	MLWN–ELWS
Muricidae				
8. *Nucella lapillus* (L.)	3548		3–7	MHWN–ELWS
Littorinidae				
9. *Littorina littoralis* (L.)	10003		3–8	MHWS–ELWS
10. *L. littorea* (L.)	14953		3–8	MHWS–ELWS
11. *L. neritoides* (L.)	5200		1–8	EHWS + 50 ft.– MLWS
12. *L. saxatilis tenebrosa* (Montagu)	30697	Throughout Britain (see	1–4	EHWS + 100 ft.– MHWS
13. *L. saxatilis jugosa* (Montagu)	6712	James 1968 e)	4–8	EHWS–MHWN
14. *L. saxatilis rudis* (Maton)	14244		3–8	MHWS–MLWN
15. *L. saxatilis neglecta* (Bean)	3080		1–8	MHWS–MLWS

The degree of exposure to wave action of each shore was estimated by considering both physical and biological features as described by Ballantine (1961). His exposure scale (1–8) is used throughout this paper. Each shore was divided into the biologically defined zones of Stephenson & Stephenson (1949) and, wherever possible, the relationship of the zones to height above chart datum and tide levels calculated. A sample of each host species in each zone was collected, the size being determined partly by abundance and partly by the need for the sample to be sufficiently large to represent the entire zone. Later, the sample was dissected for Digenea. The techniques used to examine, identify and describe these parasites (Table 2) are given in detail by James (1964, 1965, 1968 b).

The species selected for experimental work are: *Podocotyle atomon* (Figs. 1, 2) with two free-swimming larvae and a three-host life cycle (James, 1969); *Cercaria lebouri* (Figs. 3, 4) with one free-swimming stage and a two-host life cycle (James, 1969); *Microphallus similis* (Figs. 5, 6) also with one free-swimming stage but a three-host life cycle (James, 1969) and *M. pygmaeus* form A (Figs. 7–10) with no free-swimming stages and a two-host life cycle (James, 1968 b).

It is hoped to test the specificity of these species to all their hosts. To date, the following experiments have been performed to test:

(1) Host selection by the cercariae of *P. atomon*.

(2) Substrate selection by the cercariae of *C. lebouri*.

(3) Host selection by the cercariae of *M. similis*.

In addition, the experiments listed below, which are in progress, are being used to investigate a number of problems but the preliminary findings reported here are relevant to this discussion on specificity.

(4) Excystment and development of the metacercariae of *M. similis*, in vertebrate laboratory hosts.

(5) Excystment of the metacercariae of *M. similis*, *in vitro*.

(6) Development of the metacercariae of *M. pygmaeus* form A, in vertebrate laboratory hosts.

(7) Development of the metacercariae of *M. pygmaeus* form A, *in vitro*.

(8) Development and hatching of the eggs of *M. pygmaeus* form A, *in vitro*.

The conventional techniques used in these experiments are described elsewhere (James, 1964; Bowers & James, 1967; James, 1969; Ractliffe, 1968; Jensen, Stirewalt & Walters, 1965; Richards, 1969; Pascoe, 1968).

THE DISTRIBUTION OF LARVAL DIGENEA IN THE INTERTIDAL ZONE ON ROCKY SHORES

Four species without free-swimming larvae, namely *Parvatrema homoeotecnum*, *Parapronocephalum symmetricum*, *C. brevicauda* and *M. pygmaeus* form B were found only in one subspecies of *L. saxatilis*, namely subsp. *tenebrosa* in the supralittoral fringe on semi-exposed (4) to very exposed (2) shores. *M. pygmaeus* form A also occurs almost exclusively in *L. saxatilis tenebrosa*, in the supralittoral fringe on exposed (3) to extremely exposed (1)

Table 2. *Digenean parasites of prosobranch molluscs from rocky shores*

Mainly in mid and south Wales, together with their hosts and the degree of exposure to wave action and tide level of the stations from which infected host specimens were obtained.

Parvatrema homeotecnum and *Microphallus pygmaeus* form B were found only in juvenile hosts; *Cercaria littorinae saxatilis* in juvenile *Littorina saxatilis* and adult *L. neritoides* but the remaining species only in spent adult hosts.

The hosts are identified by the number given to them in Table 1. Only the hosts recorded by me are listed, as it is well known that specificity varies with geographical distribution. In addition, the possibility of misidentification or misnomers of hosts and parasites cannot be excluded in the literature.

Occasional records outside the normal distribution are given in parenthesis.

For explanation of tide levels see Table 1.

Parasite	No. of free swimming stages in life cycle	Host	Exposure grade	Tide level
Gymnophallidae				
1. *Parvatrema homoeotecnum* James, 1964	0	12	3	EHWS + 20 ft.– MHWS
Echinostomatidae				
2. *Himasthla leptosoma* (Creplin, 1829)	2	10	(3) 5–8	MTL–ELWS
3. *H. littorinae* Stunkard, 1966	2	10	5	MLWN–ELWS
4. *Cercaria littorinae obtusatae* Lebour, 1911	2	9	5–6	MHWN–MLWN
Philophthalmidae				
5. *Parorchis acanthus* (Nicoll, 1906)	2	8	(3) 5–7	MHWN–ELWS
6. *Cercaria patellae* Lebour, 1907	2	1–3	3–8	MHWS–MLWS
Notocotylidae				
7. *Cercaria lebouri* Stunkard, 1932	1	9–12, 14	(3) 4–8	EHWS + 10 ft.– MLWS
Pronocephalidae				
8. *Parapronocephalum symmetricum* Belopolskaia, 1952	0	12	2–3	EHWS + 20 ft.– MHWS
Microphallidae				
9. *Microphallus similis* (Jagerskiold, 1900)	1	9, (10,) 12–14	4–6 (1–8)	MHWS–MTL (EHWS + 10 ft.– MLWN)
10. *M. pygmaeus* (Levinsen, 1881) form A James, 1968*b*	0	12 (13, 14)	1–3 (4–6)	EHWS + 40 ft.– MHWS (MHWS– MHWN)
11. *M. pygmaeus* (Levinsen, 1881) form B James, 1968*b*	0	12	3	EHWS + 7 ft.– MHWS
12. *Cercaria littorinae saxatilis* 1 James, 1969	1	11, 12, (14)	1–3 (7)	EHWS + 100 ft– MHWS (MTL)

Table 2 (*cont.*)

Parasite	No. of free swimming stages in life cycle	Host	Exposure grade	Tide level
Microphallidae (*cont.*)				
13. *C. littorinae saxatilis* 11 James, 1969	1	12	3	EHWS + 10 ft.- MHWS
14. *C. littorinae saxatilis* 111 James, 1969	1	14	6	MHWN
15. *C. littorinae saxatilis* IV James, 1969	1	12	1	EHWS + 40 ft.- EHWS + 30 ft.
Renicolidae				
16. *Renicola roscovita* (Stunkard, 1932)	1	11–14	3–7	EHWS–MHWN (EHWS + 10 ft.- MTL)
17. *C. emasculans* Pelseneer, 1906	1	10	3–5	MHWN–MLWS
18. *C. brevicauda* Pelseneer, 1906	0	12	3–4	EHWS + 10 ft.- MHWS
19. *C. parvicaudata* Stunkard and Shaw, 1931	1	10	4	MTL
Opecoelidae				
20. *C. linearis* Lesprés, 1857	2	5–7, 10	(3)5–7	MHWN–ELWS
21. *C. stunkardi* Palombi, 1938	2	6	(3)5–7	MHWN–ELWS
22. *C. littorinae* Rees, 1936	2	10	3–6	MTL–ELWS
23. *C. buccini* Lebour, 1911	2	10	5–8	MTL–ELWS
24. *Podocotyle atomon* (Rudolphi, 1809)	2	14	5–7	MHWN–MLWN
Heterophyidae				
25. *Cryptocotyle lingua* (Creplin, 1825)	1	10	4–7	MHWS–MLWS

shores. Very occasionally, however, this species is found also in *L. saxatilis jugosa* and *L. saxatilis rudis* in the upper midlittoral zone on fairly sheltered (5) and sheltered (6) shores. *M. pygmaeus* form A is a very common parasite, occurring throughout Britain (James, 1968*b*). The adult primary host is sometimes up to 85 % infected. *Parvatrema homoeotecnum* was found only in Cardigan Bay, in Pembrokeshire and on Plymouth Breakwater and *M. pygmaeus* form B only in Cardigan Bay but, when present, percentage infection of the juvenile host species may exceed 50 % and is usually very high (James, 1968*a, b*). *Parapronocephalum symmetricum* was found only on the Gower Peninsula and *C. brevicauda* in Pembrokeshire, both in adult hosts.

Most of the ten species with two free-swimming larval stages were found to be common in the regions where the hosts were collected but, unlike the species without free-swimming larvae, percentage infection, rarely exceeding 10%, was never very high.

Generally they occur exclusively on fairly sheltered (5) to extremely sheltered (8) shores (Table 2) in the midlittoral zone and infralittoral fringe, usually below mean high water of neap tides, but often only below mean tide level. The exceptions are *C. patellae* and *C. littorinae* which commonly occur on exposed (3) and semi-exposed (4) shores. *Himasthla leptosoma, Parorchis acanthus, C. linearis* and *C. stunkardi* also were found to occur occasionally on more exposed shores in Cardigan Bay.

Most were recorded from only one host species but *C. patellae* was recorded from three and *C. linearis* from four (Table 2).

Ten species found during this survey have only one free-swimming larval stage (the cercaria). *C. lebouri* was recorded from four-host species, *M. similis* from three, *C. littorinae saxatilis* I and *R. roscovita* from two and the remainder from one host species. Of the commonly occurring species, *C. littorina saxatilis* I was found mainly in the supralittoral fringe on exposed shores and *M. similis* in the midlittoral zone on semi-exposed (4) to sheltered (6) shores but the others occurred over a wider vertical and exposure range.

EXPERIMENTAL OBSERVATIONS

Host selection by the cercariae of Podocotyle atomon

In addition to *L. saxatalis rudis* (the primary host), the life cycle includes a variety of marine amphipods as intermediate hosts and marine fishes as vertebrate hosts (Hunninen & Cable, 1943; Srivastava, 1966; James, 1969). The cercariae (Fig. 1) escape from the sporocysts in the visceral haemocoel and from the primary host. They cannot swim but crawl along the substratum by alternately attaching the oral and ventral suckers. Eventually, they become attached to the substratum or vegetation, such as hydroid fronds, by means of the truncate tail (Fig. 1) which is modified into a third sucker. The upright cercariae then sway from side to side until a passing amphipod touches the anterior setae (James, 1969). They then grasp the amphipod with the oral sucker, detach the tail, penetrate the cuticle, using the stylet and penetration glands and later encyst in the haemocoel.

The experiments test the ability of the cercariae to recognize potential intermediate hosts. In the first experiment, about 1000 cercariae were placed in a dish of sea water at 15 °C and allowed to settle on hydroid fronds. A number of marine molluscs, Crustacea and fishes were then placed in the dish and allowed to swim freely. They were removed at regular intervals and their external surface examined for cercariae. In addition, a glass rod was moved gently through the hydroids at about 10 min intervals and then examined for attached cercariae. The potential hosts were dissected 2 days later to see if they contained encysted metacercariae. These procedures were repeated at 30 °C, in diluted sea water (17‰ at 15 °C) and in super saline sea water (50‰ at 15 °C).

Table 3. *The selection of intermediate hosts by the cercariae of* Podocotyle atomon *in various experimental conditions*

Percentage of cercariae attached to each host

Temperature ...	15 °C				15 °C Super saline sea water (50‰)		15 °C diluted sea water (17‰)		30 °C Normal sea water
Medium ...	Normal sea water								
Time after liberation from primary host (h) ...	1	3	6	12	1	3	1	3	1
1. *Gammarus locusta* (L.)	20	19	15	12	12	11	10	16	8
2. *Hyale nilssoni* (Rathke)	11	11	10	12	4	13	3	3	10
3. *Marinogammarus marinus* (Leach)	18	19	16	10	16	3	15	4	14
4. *Crangon vulgaris* Fab	8	10	9	9	7	0	19	8	12
5. *Carcinus maenas* (L.)	9	8	10	12	8	13	0	15	16
6. *Cancer pagurus* L.	8	9	9	9	19	16	9	0	18
7. *Onos mustellus* (L.)	7	5	12	10	9	14	13	9	2
8. *Blennius pholis* L.	8	6	6	10	6	15	9	14	16
9. *Littorina littorea* (L.)	5	6	7	8	7	15	11	14	0
10. Glass rod	6	7	6	8	12	0	11	17	4
Total number attached to all hosts (= 100%)	121	240	304	388	108	170	99	150	51

The results are summarized in the table (Table 3) which indicates that, within 3 h of leaving the primary host in normal sea water at 15 °C, the cercariae marginally prefer to settle on amphipods but later appear to become attached indiscriminately to any moving object, particularly in unfavourable conditions. Dissection revealed that only amphipods contained encysted metacecarirae. An examination of immobilized specimens with a binocular microscope showed that the cercariae fail to penetrate the molluscs and crabs but degenerate on entering the shrimps and fishes. The encysted metacercariae were encapsulated by an extremely hard fibrous substance (Fig. 2), presumably of host origin, in *Marinogammarus* and *Hyale*.

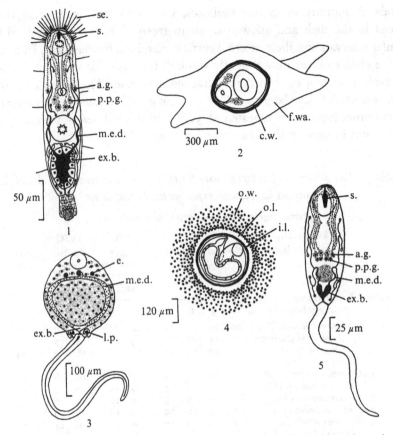

Figs. 1–10. Some digenean parasites which have intertidal prosobranchs as primary hosts.

Fig. 1. *Podocotyle atomon*, cercaria, dorsal view.

Fig. 2. *P. atomon*, encysted metacercaria encapsulated by fibrous wall presumably of host origin.

Fig. 3. *Cercaria lebouri*, cercaria, ventral view.

Fig. 4. *C. lebouri*, encysted metacercaria.

Fig. 5. *Microphallus similis*, cercaria, dorsal view.

Substrate selection by the cercariae of Cercaria lebouri

The cercariae (Fig. 3) escape from the rediae in the primary host and swim or crawl for a short time before settling on the substratum, vegetation or the external surface of marine animals. They then lose their tails and encyst. The vertebrate hosts, probably marine ducks or geese, become infected by eating the encysted metacercariae.

In experimental conditions (Table 4) in glass dishes containing sea water at 15 °C some cercariae began to encyst (Fig. 4) after swimming or crawling

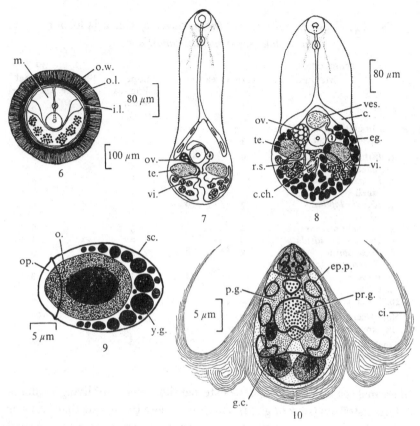

Fig. 6. *M. similis*, encysted metacercaria.

Fig. 7. *Microphallus pygmaeus* form A, metacercaria, ventral view.

Fig. 8. *M. pygmaeus* form A, adult, ventral view.

Fig. 9. *M. pygmaeus* form A, egg.

Fig. 10. *M. pygmaeus* form A, hatched miracidium, dorsal view.

(a.g., anterior penetration gland cell; c., caecum; c.ch., central chamber of Mehlis's gland; c.w., cyst wall secreted by metacercaria; ci., cilia; e., eye spot; eg., egg; ep.p., epidermal plate; ex.b., excretory bladder; f.wa., fibrous wall surrounding cyst, secreted by host; g.c., germinal cell; i.l., inner layer of inner cyst wall; l.p., locomotor pocket; m., metacercaria; m.e.d., main excretory duct; o., ovum; o.l., outer layer of inner cyst wall; o.w., outer cyst wall; op., operculum; ov., ovary; p.g., penetration gland cell; p.p.g., posterior penetration gland cell; pr.g., primitive gut; r.s., receptaculum seminis; s., stylet; se., setae; sc., sclerotin shell; te., testis; ves., vesicula seminalis; vi., vitelline gland; y.g., yolk globule.)

for 2 h. Those which had not encysted died after 8 h. In unfavourable conditions, such as an increase or decrease in salinity or an increase in temperature, encystment began almost immediately and those which had not encysted died after 1 h.

Table 4. *Substrate selection by the cercariae of* Cercaria lebouri *in various experimental conditions*

				15 °C		30 °C
Temperature ...				15 °C		30 °C
Medium ...	Normal sea water		Super saline sea water (50‰)	Diluted sea water (17‰)	Normal sea water
Time after liberation from primary host (h) ...	3	6	1	1	1
1. *Littorina littorea* shell	9	10	8	3	3
2. *L. saxatilis rudis* shell	12	13	6	5	2
3. *Nucella lapillus* shell	7	9	6	4	2
4. *Carcinus maenas*	4	6	3	0	0
5. *Pelvetia canaliculata*	0	2	2	1	1
6. *Fucus spiralis*	2	8	3	2	3
7. *F. serratus*	1	3	4	2	2
8. Rock	15	17	11	0	3
9. Glass	12	16	13	5	4
10. Small crustacea	0	0	0	0	0
No. of cercariae used in experiment	100	100	100	50	100
Total encysting on all substrata	62	84	56	22	20

In normal sea water at 15 °C, the external shell surface of living molluscs and the substratum (rock or glass) was chosen more frequently than crabs or algal fronds (Table 4) as a site for encystment by cercariae which settled within 3 h of release from the primary host. This preference was not so pronounced in cercariae which encysted later and was further reduced with an increase or decrease in salinity or with an increase in temperature (Table 4). Encystment never occurred on small rapidly moving Crustacea. In these experiments care was taken to ensure that each potential substratum offered about the same surface area for encystment.

Host selection by the cercariae of Microphallus similis

The cercariae (Fig. 5) escape from the primary host and swim for a short period before penetrating and encysting in the intermediate host. This species is more specific to the intermediate host than any other investigated, being found almost exclusively in the haemocoel of the hepatopancreas of *Carcinus maenas* (James, 1969). The adult worm is widely distributed, mainly in *Larus* species in the northern hemisphere (Deblock & Tran van Ky, 1966).

In the first experiment, 12 infected specimens of *L. saxatilis rudis*, from which cercariae were emerging, were placed for 7 days in a tank of aerated sea water with a number of Crustacea, including 10 *Carcinus maenas*, 8 *Cancer pagurus*, 20 *Crangon vulgaris*, 50 *Gammarus locusta* and 50 *Marinogammarus marinus*. The cercariae, which appear to have no sensory cells, swam freely in the tank, did not appear to seek out any of the Crustacea but were carried by respiratory currents into the gill chambers of the crabs and on to the limbs of the amphipods. On dissection, 10 *C. maenas* and 1 *C. pagurus* were infected with encysted metacercariae at various stages in development. Each *C. maenas* contained 25–250 metacercariae but only three were found in *C. pagurus*. None were found in any of the other species used in the experiment or in the control animals.

Encysted metacercariae (Fig. 6) were found mainly in the haemocoel of the hepatopancreas and cercariae in the gill chambers and the blood spaces under the gills. Thus, as suggested by Stunkard (1957), the cercariae probably enter the crab, using the stylet and penetration glands (Fig. 5), near the bases of the gills where the cuticle is thin and migrate to the hepatopancreas.

In the second experiment, the potential host species, which included *Blennius pholis* and *Onos mustellus* in addition to those listed in the previous experiment, were pinned down in a simple maze, in order that the mode of entry of the cercariae could be examined more carefully. The cercariae again swam aimlessly and were swept into the gill chambers of the crabs and fishes. Only *Carcinus maenas* was found to contain encysted metacercariae but *Cancer pagurus* and *O. mustellus* contained cercariae in the blood spaces beneath the gills.

In the third experiment, the crabs, shrimps, amphipods and fishes were allowed to feed on the digestive gland and gonad of *L. saxatilis* containing the sporocysts and cercariae of *M. similis*. Once again only *Carcinus maenas* was found to be infected with encysted metacercariae. This was the most efficient method of experimental infection, as every *C. maenas* became infected and the number of encysted metacercariae recovered in each was vastly in excess of the number recovered in the other experiments. This suggests that the cercariae may also enter the crab by penetration of the gut wall.

In the fourth experiment, cercariae were injected, with a hypodermic syringe, directly into the haemocoel of *C. maenas*, *Cancer pagurus* and *Crangon vulgaris*. Only a small portion (less than 1 in 100) of those injected into *C. maenas* survived to produce encysted metacercariae.

Excystment and development of the metacercariae of Microphallus
similis, *in vertebrate laboratory hosts*

The encysted metacercariae (Fig. 6) were removed from the hepatopancreas
of *C. maenas* and force fed to 12 *L. argentatus*, 2 *L. fuscus* and 2 *L. marinus*
chicks, to 12 ducklings, 4 pigeons, 4 guinea-pigs, 12 rats and 25 mice. All
the experimental animals, with the exception of 1 *L. argentatus* chick and
4 pigeons were found, on dissection, to be infected with adult *M. similis*.
Two of the 13 control *L. argentatus* were also found to be infected but none
of the other control species. About 150 encysted metacercariae were fed to
each animal and a mean of 3.3, 6.0, 12.5, 6.5, 6.2, 8.0 and 14.8 adults
recovered from *L. argentatus*, *L. fuscus* and *L. marinus*, the ducklings,
guinea-pigs, rats and mice respectively.

This suggests that excystment of the metacercariae, settlement and
development into adults is not dependent on specific stimuli, as the experimental mammals appear to be as heavily infected as the gulls which are the
natural hosts.

Excystment of the metacercariae of Microphallus similis, *in vitro*

In sterile aerated chick saline, at 37 °C, excystment occurs in 5 h if the cyst
is first treated with pepsin in hydrochloric acid at a pH of 3.5 for 2 h and
then with trypsin at pH 6.5 for 3 h. The pepsin digests the outer cyst wall
(Fig. 6) and trypsin the outer layer of the inner cyst wall. The contained
metacercaria, which is extremely contractile at this temperature, is released
when it ruptures the inner layer of the inner cyst wall which is not digested.
The rupture occurs in a flattened region (the escape aperture) of the
otherwise spherical cyst wall. In similar experimental conditions, but at
room temperature (20 °C), excystment takes 24 h. Excystment does not
occur if the cyst is treated only with pepsin in hydrochloric acid but will
occur in 24 h if treated only with trypsin at pH 6.5 at 37 °C.

Since similar conditions occur in almost any warm-blooded vertebrate,
this result confirms the previous experiments which suggested that excystment was not dependent on the presence of specific stimuli.

Development of the metacercaria of Microphallus pygmaeus
form A, in vertebrate laboratory hosts

The daughter sporocysts of *M. pygmaeus* form A, in the haemocoel of the
digestive gland and gonad of the primary host, contain metacercariae
(Fig. 7) which develop into adults (Fig. 8) when the primary host is eaten
by lariform, charadriiform or anseriform birds (James, 1968b).

The development into an adult (Fig. 8) involves very little increase in size but the body becomes wider, the caeca swollen (stage 1), the testes, ovary and vitelline glands increase in size (stage 2), sperm appears in the vesicular seminalis, receptaculum seminis or central chamber of Mehlis' gland or both (stage 3), phenolic egg shell precursors and phenolase develop in the vitelline glands (stage 5) and finally shelled eggs are formed in the central chamber and pass into the uterus (stage 6).

Adult winkles infected with sporocysts containing metacercariae were force fed to 1 herring gull, 12 ducklings and 50 mice. On dissection, 8 ducklings, the herring gull and all the mice were found to contain adult *M. pygmaeus*. The remaining ducklings and the control animals were not infected.

About 1 in 3000 metacercariae survived to develop into adults in the ducklings, 1 in 100 survived in the mice and 1 in 40 in the gull.

Dissection at regular intervals of samples of experimentally infected mice showed that the adults were situated in the duodenum and had developed to stage 1 in 12 h after infection. They developed to stage 3 after 18 h, to stage 5 after 24 h and stage 6, containing up to 45 eggs, after 48 h by which time they were situated in the anterior ileum. Later, the worms passed down into the posterior ileum (6 days) and rectum (9 days) and produced up to 70 eggs each. By 13 days, no worms remained in the intestine, all having passed out through the anus. Development was more rapid in the Herring gull but less rapid in the ducklings, than in the mice, stage 6 being attained within 24 h in the former and in 3 days in the latter.

Thus, although the Herring gull provides a more favourable environment for *M. pygmaeus* than the experimental hosts, the presence of specific stimuli or nutrients in the gut is not necessary for development and cannot account for the specificity observed in natural populations.

Development of the metacercariae of Microphallus pygmaeus *form A*, in vitro

The metacercariae (Fig. 7) develop into egg-bearing adults (stage 6) (Fig. 8) in the tissue culture medium 199 (Morgan, Morton & Parker, 1950) containing 10% bovine serum at 37 °C. In addition, it was necessary to use a gas phase of 5% carbon dioxide in 95% nitrogen, to maintain the pH at 6.8–7.2, to allow the metacercariae to settle on and become attached to tiny crevices in agar blocks and to agitate the incubation chamber. In these conditions 1 in 10 metacercariae survived to develop into egg-bearing adults. This is more than the survival attained *in vivo*. However, 5 days were required for each worm to produce the first egg and a maximum of only 20 eggs were obtained in 8 days, which is much slower than *in vivo*.

This result again suggests that the conditions necessary for development of the metacercariae of *M. pygmaeus* into adults probably occur in the gut of most warm blooded vertebrates.

It is interesting to note that the metacercariae died within a few minutes, *in vitro*, when subjected to a change of pH, from acid to alkaline, similar to that which occurs during the passage from the stomach to the duodenum of the vertebrate host. The reason for this is not known but it may account for the higher mortality, *in vivo*, than *in vitro*. The metacercariae of *M. pygmaeus* are unique among the Microphallidae in not being enclosed in a cyst. It is possible that the cyst may protect the metacercaria of *M. similis* from the effects of the change in pH, accounting for their lower mortality (1 in 10 survive) than *M. pygmaeus* (1 in 100 survive) in mice. The surrounding host tissue and sporocyst wall may afford some protection to the metacercariae of *M. pygmaeus* during infection.

Development and hatching of the eggs of Microphallus pygmaeus form A, in vitro

The operculated egg of *M. pygmaeus* (Fig. 9) contains a single ovum on leaving the vertebrate host. When the egg is eaten by the primary host, the ovum develops into a miracidium which hatches (Fig. 10), penetrates the gut wall and metamorphoses into a mother sporocyst (James, 1968 b).

A number of experiments are being carried out in a variety of media in order to discover the factors governing the rate of development and hatching of the miracidium.

Some preliminary results are summarized in the table (Table 5) which

Table 5. *Percentage of eggs of* Microphallus pygmaeus *containing fully formed miracidia or hatched after 20 days,* in vitro, *at 20 °C*

50–100 eggs counted in each experiment

Medium ...	Sea water				1 % sodium chloride			
Activator ... (0.001 M magnesium chloride)	Absent		Present		Absent		Present	
Reducing agent (0.02 M cysteine)	Absent	Present	Absent	Present	Absent	Present	Absent	Present
Air								
pH 5.8	13	5	6	7	17	5	5	10
pH 8.0	(34)	(5)	(43)	(29)	(36)	(13)	(30)	(8)
Gas phase 5 % carbon dioxide and 95 % nitrogen								
pH 5.8	15	14	8	14	9	12	12	14
pH 8.0	(58)	(62)	(36)	(7)	(68)	(76)	(62)	(89)

indicate that 1% sodium chloride solution is better than sea water, alkaline pH better than acid pH and a gas phase of 5% carbon dioxide and 95% nitrogen better than air in promoting development and hatching. The reducing agent and activator, which accelerates hatching in *Dicrocoelum lanceolatum* eggs (Ractliffe, 1968), appear to have little effect. Other experiments showed that a small proportion of the eggs hatched in media as diverse as the tissue culture medium 199 and distilled water.

These results indicate that the specificity of *M. pygmaeus* to the primary host is probably not a result of the occurrence of specific hatching factors in the molluscan intestine.

DISCUSSION

The absence of free-swimming stages in the life cycle may enable *Parvatrema homoeotecnum*, *Microphallus pygmaeus*, *Parapronocephalum symmetricum* and *Cercaria brevicauda* to occur above the level of the highest tides on exposed shores and their loss may be a response to the environmental conditions in this region. The susceptibility of miracidia to desiccation and their dependence on tiny cilia for locomotion may contribute to the restriction of most species with two free-swimming larvae to lower levels on more sheltered shores. In addition, the considerable variation in the dietary and environmental conditions between zones and between shores of different grades of exposure to wave action, may affect the susceptibility to infection and account for the fact that some primary host species have different parasites in different ecological niches. In other words, host and ecological specificity may be due, sometimes to the inability of the miracidium or cercaria to reach the host because of unfavourable environmental conditions and sometimes to failure to develop in a resistant host.

Host or site selection by free-swimming cercariae does not appear to contribute greatly to their specificity to the intermediate host, not, at least, in the species investigated in this paper. *M. similis*, with the least selective cercariae, is more specific than any other species to the intermediate host. Failure to penetrate the external surface of potential hosts, as in the cercariae of *Podocotyle atomon* with crabs and molluscs, and the destruction of the parasite in the 'wrong' host, as with the cercariae of *P. atomon* in fishes and shrimps and *M. similis* in *Cancer pagurus* and *Onos mustellus*, are probably more important factors in determining specificity in the intermediate host. Encapsulation by a fibrous wall in *Marinogammarus marinus* may also result in the destruction of the contained metacercaria of *P. atomon*.

Cheng (1967) in a review of the literature, quotes many examples of

7

rejection and destruction, including encapsulation of miracidia in the 'wrong' primary host. Heyneman (1966) found that miracidia and rediae surgically transplanted into the natural primary molluscan host develop normally but are rapidly destroyed on transplantation into the wrong hosts. This work, and Cheng's review, suggest that the failure of the free-swimming miracidia to penetrate or develop in the 'wrong' host are also probably more important factors in the determination of primary host specificity than host selection. The experiments described in this paper suggest that, in the absence of the necessity for specific hatching stimuli, the same is true for miracidia which hatch within the gut of the mollusc. It is interesting to note that there is some evidence, as reviewed by Wright (1966) and Cheng (1967), to suggest that innate and acquired humoral immune mechanisms to larval Digenea may occur in molluscs.

Thus, it may be concluded that the free-swimming miracidia and cercariae may be more important as means of dispersal than for primary or intermediate host selection.

For the species investigated in this paper, the conditions necessary for excystment or settlement or both of the metacercarial larvae and their development into adults are not specific and could be provided by almost any warm-blooded vertebrate. Nevertheless, guinea-pigs and mice are not infected with *M. pygmaeus* or *M. similis* in the field because they do not feed on marine winkles or crabs. Thus, in these species, vertebrate host specificity may be due primarily to the feeding behaviour of the vertebrate host. In other Digenea, however, the adults develop in fewer natural and experimental vertebrate hosts. They are more difficult to culture *in vitro* and other factors contribute to their specificity (Smyth, 1966; Rogers, 1962).

SUMMARY

The digenean parasites of intertidal prosobranchs on rocky shores in mid and south Wales, without free-swimming stages do not appear to select their hosts and occur in the supralittoral fringe of exposed shores. Those with two free-swimming larvae (the miracidium and cercaria) seem to actively select their hosts and tend to occur nearer low tide level on sheltered shores. Those with one free-swimming larva (the cercaria) are more widely distributed. The species which do not select the host, however, are as, if not more, host specific than those with active selection. Experimental observations on free-swimming cercariae indicate that host or site selection varies with species, age and environmental conditions but is never very precise and may be non-existent. Similarly, crude simulations of the environmental conditions in the molluscan or vertebrate intestine in chemically defined media, suggest that hatching, excystment, entry or settlement in the host, by non-swimming larvae is not dependent on responses to specific stimuli. In contrast, larvae are rapidly destroyed in the haemocoel of the 'wrong' primary or intermediate host but develop in the correct host. Thus, primary and intermediate host specificity may be due primarily to failure of the parasite to develop in 'wrong' hosts. The vertebrate host specificity of the parasites investigated in this paper however, appears to be due principally to the feeding behaviour of the vertebrate host.

I am grateful to my research students for help with some of the experiments, namely Dr D. Pascoe and Dr R. J. Richards (development of *M. pygmaeus* metacercariae, *in vitro*), Mr M. Zafar Hameed (hatching of *M. pygmaeus* eggs, *in vitro* and development of *M. pygmaeus* metacercariae, *in vivo*), Mr J. S. Thomas (excystment of *M. similis* metacercariae, *in vitro*) and Dr M. P. Harris and Mr A. Kalantan (development of *M. similis* metacercariae, *in vivo*). I am also grateful to Professor E. W. Knight-Jones for the provision of excellent working facilities.

REFERENCES

BALLANTINE, W. J. (1961). A biologically defined exposure scale for the comparative description of rocky shores. *Fld Stud.* **1** (3), 1–19.

BOWERS, E. A. & JAMES, B. L. (1967). Studies on the morphology, ecology and life cycle of *Meiogymnophallus minutus* (Cobbold, 1859) comb.nov. (Trematoda: Gymnophallidae). *Parasitology* **57**, 281–300.

CHENG, T. C. (1967). *Marine molluscs as hosts for symbiosis. Adv. mar. biol.*, Vol. 5. London and New York: Academic Press.

DEBLOCK, S. & TRAN VAN KY, P. (1966). Contribution à l'étude des Microphallidae Travassos, 1920 (Trematoda) XII. Espèces d'Europe occidentale. Création de *Sphairotrema* nov.gen. considérations diverses de systématique. *Annls Parasit. hum. comp.* **41**, 23–60.

FISCHER-PIETTE, E., GAILLARD, J-M. & JAMES, B. L. (1964). Études sur les variations de *Littorina saxatilis*. VI. Quelques cas qui posent de difficiles problèmes. *Cah. Biol. mar.* **5**, 125–71.

HEYNEMAN, D. (1966). Successful infection with larval Echinostomes, surgically implanted into the body cavity of the normal snail host. *Expl Parasit.* **18**, 220–3.

HUNNINEN, A. V. & CABLE, R. M. (1943). The life history of *Podocotyle atomon* (Rudolphi) (Trematoda: Opecoelidae). *Trans. Am. microsc. Soc.* **62**, 57–68.

JAMES, B. L. (1963). Subspecies of *Littorina saxatilis* around Dale. In Moyse, J. & Nelson-Smith, A. Zonation of animals and plants on rocky shores around Dale, Pembrokeshire. *Fld Stud.* **1**, 1–34.

196   B. L. JAMES

JAMES, B. L. (1964). The life cycle of *Parvatrema homoeotecnum* sp.nov. (Trematoda: Digenea) and a review of the family Gymnophallidae Morozov, 1955. *Parasitology* **54**, 1–41.

JAMES, B. L. (1965). The effects of parasitism by larval Digenea on the digestive gland of the intertidal prosobranch, *Littorina saxatilis* (Olivi) subsp. *tenebrosa* (Montagu). *Parasitology* **55**, 93–115.

JAMES, B. L. (1968a). The occurrence of *Parvatrema homoeotecnum* James, 1964 (Trematoda: Gymnophallidae) in a population of *Littorina saxatilis tenebrosa* (Mont.). *J. nat. Hist.* **2**, 21–37.

JAMES, B. L. (1968b). Studies on the life cycle of *Microphallus pygmaeus* (Levinsen, 1888) (Trematoda: Microphallidae). *J. nat. Hist.* **2**, 155–72.

JAMES, B. L. (1968c). The occurrence of larval Digenea in ten species of intertidal prosobranch molluscs in Cardigan Bay. *J. nat. Hist.* **2**, 329–43.

JAMES, B. L. (1968d). The distribution and keys of species in the family Littorinidae and of their digenean parasites, in the region of Dale, Pembrokeshire. *Fld Stud.* **2**, 615–50.

JAMES, B. L. (1968e). The characters and distribution of the subspecies and varieties of *Littorina saxatilis* (Olivi, 1792) in Britain. *Cah. Biol. mar.* **9**, 143–65.

JAMES, B. L. (1969). The Digenea of the intertidal prosobranch, *Littorina saxatilis* (Olivi). *Z. f. Zool. Systematik* **7**, 273–316.

JENSEN, D. V., STIREWALT, M. A. & WALTERS, M. (1965). Growth of *Schistosoma mansoni* cercariae under dialysis membranes in Rose multipurpose chambers. *Expl Parasit.* **17**, 15–23.

MORGAN, J. G., MORTON, H. J. & PARKER, R. C. (1950). Nutrition of animal cells in tissue culture. 1. Initial studies on a synthetic medium. *Proc. Soc. exp. Biol. Med.* **73**, 1–8.

PASCOE, D. (1968). *Studies on the metabolism of Digenea from sea birds and molluscs.* Ph.D. Thesis, University College, Swansea.

RACTLIFFE, L. H. (1968). Hatching of *Dicrocoelium lanceolatum* eggs. *Expl Parasit.* **23**, 67–78.

RICHARDS, R. J. (1969). *Nutrition and respiration of the germinal sacs of marine Digenea.* Ph.D. Thesis, University College, Swansea.

ROGERS, W. P. (1962). *The Nature of Parasitism.* New York and London: Academic Press.

SMYTH, J. D. (1966). *The Physiology of Trematodes.* Edinburgh and London: Oliver and Boyd.

SRIVASTAVA, L. P. (1966). The helminth parasites of the five-bearded rockling, *Onos mustellus* (L.) from the shore at Mumbles Head, Swansea. *Ann. Mag. nat. Hist.* **9**, 469–80.

STEPHENSON, T. A. & STEPHENSON, A. (1949). The universal features of zonation between tide marks on rocky shores. *J. Ecol.* **37**, 289–305.

STUNKARD, H. W. (1957). The morphology and life history of the digenetic trematode, *Microphallus similis* (Jäegerskioeld, 1900) Baer, 1943. *Biol. Bull. mar. biol. Lab., Woods Hole* **61**, 242–71.

WRIGHT, C. A. (1966). The pathogenesis of helminths in the Mollusca. *Helminth Abstr.*, **35**, 207–24.

EFFET DE LA TEMPERATURE SUR LE DEVELOPPEMENT ENDOTROPHE DES PLUTEUS

P. BOUGIS

Station Zoologique, Villefranche-sur-Mer, France

ENGLISH SUMMARY

After fertilization, the egg of the urchin develops into the pluteus using reserve nutrient. When the 'baguette somatique' reaches its maximum size, the lecithotrophic phase of growth may be considered complete. For *Paracentrotus lividus* the time in hours (T) to reach this stage is related to the temperature t °C by the linear equation

$$\log_e T = 9.05 - 1.59 \log_e t.$$

The maximum length of the 'baguette somatique', l, at a given temperature, t, varies in each experiment according to a parabolic equation:

$$l = l_{\max} - K(t_{\mathrm{opt}} - t)^2$$

where $l_{\max}$ is the theoretical maximum length at the optimal temperature t_{opt}. If the data of different experiments are replotted so that $l_{\max}$ and t_{opt} coincide, the points fit satisfactorily to a common parabola.

The influence of temperature appears to depend on two variables: $l_{\max}$, which may reflect the amount of material available in the egg, and t_{opt}, which may depend on proportions of nutrients and enzymes available. The value of t_{opt} is of theoretical interest only as it may exceed the upper lethal limit.

The percentage of eggs that develop normally in spring is high at 20–25 °C, and low at 15 °C. This confirms the critical temperature of 16 °C observed in summer by Hörstadius. The influence of temperature on the development of the pluteus obtanied by artificial fertilization is susceptible to variation throughout the year.

Après fécondation, l'oeuf d'oursin évolue en plutéus à quatre bras mettant en oeuvre les matériaux qu'il contient: c'est ce que nous avons appelé la phase endotrophe du développement (Bougis, 1967). La durée de cette phase est évidemment influencée par la température et, dans certaines limites, plus celle-ci sera élevée plus la formation du plutéus sera accélérée.

La baguette somatique constitue un bon index du développement du corps et, en prenant comme critère d'achèvement du plutéus endotrophe le moment où elle cesse de croître et devient constante, nous obtenons les valeurs suivantes pour l'oursin commun, *Paracentrotus lividus*: (°C)

$$10°:205 \text{ h} \pm 20; \ 20°:72 \text{ h} \pm 2; \ 25°:51 \text{ h} \pm 1.$$

En coordonnées logarithmiques les points figuratifs s'ordonnent sur une droite d'équation: $\log_e T = -1.59 \log_e t + 9.05,$

où T est le nombre d'heures pour atteindre la longueur maximale de la baguette somatique et t la température en degrés centigrades.

Mais la température, en dehors de cette action habituelle, intervient d'une autre façon sur la réalisation du plutéus endotrophe: la longueur maximale de la baguette somatique varie en effet notablement suivant la température à laquelle s'est déroulé le développement. Par exemple, dans l'expérience 145 du 4 Février 1969, nous obtenons les valeurs suivantes (longueur maximale exprimée en divisions du micromètre, égales à 10.1 μm):

(°C)   (145)   10°:19.9; 20°:24.3; 25°:24.8.

Les longueurs obtenues croissent donc avec la température sans qu'il y ait proportionnalité, l'allongement moyen par élévation d'un degré variant de 0.44, entre 10 et 20 °C, à 0.10 entre 20 et 25 °C.

Suivant les expériences, les résultats peuvent être notablement différents: ainsi dans l'expérience 141 la différence entre 20 et 25 °C est beaucoup plus importante que dans l'expérience précédente: (°C)

(141)   15°:19.5; 20°:22.5; 25°:24.6.

L'allongement moyen par degré entre 20 et 25 °C passe à 0.42.

Par contre dans l'expérience 148 il se produit une inversion: la longueur maximale est obtenue pour 20 °C et pour 25 °C existe une réduction de taille.   (148)   10°: 23.2; 20°: 26.5; 25°: 25.6.

Il apparaît donc, à première vue, que l'influence de la température sur la longueur maximale de la baguette somatique des plutéus est relativement anarchique. Cependant il est possible à partir des trois couples de données obtenues dans chaque expérience de déterminer l'équation d'une parabole passant par les point figuratifs, équation se formulant ainsi:

$$l = l_{ma} - k(t_{op} - t)^2,$$

où k est un coefficient; l la longueur maximale obtenue à la température t; l_{ma} la longueur maximale la plus élevée obtenue théoriquement à une température optimale t_{op}; l_{ma} et t_{op} déterminent le sommet de la parabole.

Nous obtenons ainsi les équations suivantes:

(138) Avril 1968: $l = 25.0 - 0.044\,(23.4 - t)^2$

(141) Octobre 1968: $l = 26.1 - 0.018\,(34.1 - t)^2$,

(145) Février 1969: $l = 24.8 - 0.023\,(24.7 - t)^2$,

(147) Avril 1969: $l = 24.7 - 0.023\,(22.1 - t)^2$,

(148) Mai 1969: $l = 26.5 - 0.034\,(19.8 - t)^2$,

(150) Juin 1969: $l = 26.2 - 0.039\,(23.0 - t)^2$.

Si nous reportons maintenant l'ensemble des données des expériences sur un même graphique en faisant coïncider les sommets des différentes paraboles (t_{op}, l_{ma}), nous obtenons la Fig. 1 où les points s'ordonnent de façon satisfaisante autour d'une parabole moyenne.

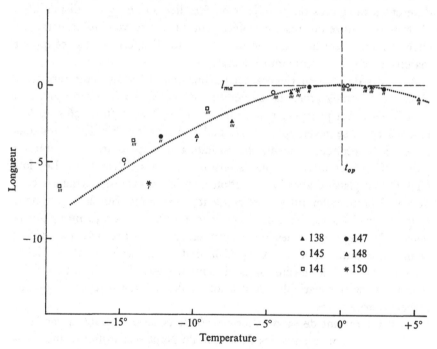

Fig. 1. Relation parabolique entre la longueur maximale de la baguette somatique des plutéus et la température de développement. Les points expérimentaux sont placés en fonction de la température optimale (t_{op}) et de la longueur maximale (l_{ma}) calculées pour chaque expérience. Longueur de la baguette somatique en division du micromètre (10.1 μm). Les traits près des points figuratifs correspondent au nombre d'élevages mesurés.

Nous pensons donc pouvoir conclure que la relation parabolique entre la température et la longueur maximale des baguettes somatiques n'est pas un simple artifice de calcul mais correspond à la réalité du phénomène. Cette relation permet d'expliquer aisément les variations importantes observées; dans l'expérience 145 la température de 25 °C était à peu de chose près la température optimale; dans l'expérience 148 la température de 25 °C se plaçait nettement à droite de la température optimale sur la partie descendante de la parabole; enfin dans l'expérience 141 les différents points se trouvaient sur la partie gauche ascendante, loin du sommet de la parabole, et l'allongement moyen par degré se trouvait très élevé entre 20 et 25 °C.

L'effet de la température sur le développement endotrophe des plutéus de *Paracentrotus lividus* nous apparaît ainsi dépendre de deux paramètres: la longueur maximale absolue l_{ma} et la température optimale t_{op}. La longueur maximale absolue reflète sans doute la quantité plus ou moins grande de l'ensemble des matériaux disponibles pour élaborer le plutéus; la température optimale pourrait traduire plutôt certains rapports entre les différentes catégories de matériaux ou être liée aux enzymes disponibles. Cette température optimale calculée, peut parfois n'avoir qu'un caractère effectivement théorique: elle est de 34 °C dans l'expérience 141 ce qui est manifestement une température léthale.

Au cours de ces expériences nous avons cherché à plusieurs reprises à obtenir des données pour la température de 15 °C: en Octobre 1968, cela fut effectivement possible. Par contre en Avril, Mai et Juin 1969, les lots placés à 15 °C se développèrent très mal et furent inutilisables. Dans tous ces essais la fécondation artificielle est faite à la température du laboratoire (20–23 °C) et les lots sont placés ensuite dans les thermostats. Les lots à 10 °C sont placés d'abord avec les lots à 15 °C puis transférés dans le bain de 10 °C pour éviter un refroidissement trop rapide. Sur la Fig. 2 nous avons reporté les résultats des expériences 147, 148 et 150 indiquant la densité des plutéus dans les lots expérimentés; ceux-ci reçoivent au départ la même quantité d'oeufs. A 25 °C la densité est légèrement inférieure à 20 °C. A 15 °C la densité est extrêmement basse. Par contre à 10 °C les densités sont très variables et des lots à densité normale sont obtenus à chaque expérience.

Il est intéressant de rapprocher ces résultats de ceux publiés par Hörstadius (1925) concernant des *P. lividus* de Naples et utilisant une technique semblable. En Février et Mars le développement normal des plutéus ne se produit qu'entre 8 et 23 °C. En Aout il n'est obtenu qu'entre 16 et 29 °C. Cette dernière température critique de 16 °C s'accorde parfaitement avec nos mauvais résultats à 15 °C. Nous nous expliquons mal, par contre, comment, à 10 °C, certains lots peuvent présenter un excellent développement.

En définitive si nous considérons les résultats différents de Hörstadius en hiver et en été et les variations que nous avons notées de la température optimale, nous pouvons conclure que l'effet de la température, sur le développement des plutéus obtenus par fécondation artificielle, est susceptible de présenter des différences importantes au cours de l'année, qui méritent d'être étudiées de façon approfondie, et devraient nous conduire à une meilleure connaissance de l'écologie de la reproduction chez les oursins.

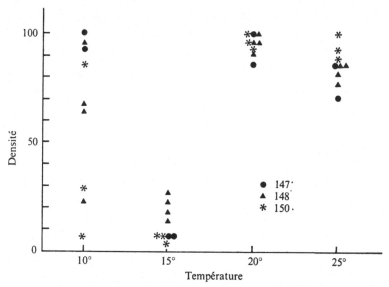

Fig. 2. Densité des plutéus dans les élevages à différentes températures. La plus forte densité à 20 °C est considérée comm égale à 100.

Ce travail a été réalisé avec la collaboration technique de M. C. Corre et B. Hirel (C.N.R.S.); nous sommes heureux de les remercier.

RÉFÉRENCES

BOUGIS, P. (1967). Utilisation des plutéus en écologie expérimentale. *Helgoländer wiss. Meeresunters* **15**, 59–68.

HÖRSTADIUS, S. (1925). Temperaturanpassung bei den Eiern von *Paracentrotus lividus* Lk. *Biologia gen.* **1**, 522–36.

LA FORME ELLIPTIQUE DE LA RELATION TEMPERATURE–DUREE DE DEVELOPPEMENT EMBRYONNAIRE CHEZ LES COPEPODES PELAGIQUES ET SES PROPRIETES

M. BERNARD

Institut Océanographique d'Alger

ENGLISH SUMMARY

When accurately measured, the period of embryonic development of two warm water copepods, *Temora stylifera* and *Centropages chierchiae*, appears to be a second-order function of temperature, with a minimum. Several curvilinear functions can be fitted to the data with only slightly different residual sums of squares, but the ellipse seems the most convenient for three reasons. (1) In warm waters embryonic development often occupies a minimum period. This has ecological significance and should be accounted for by the mathematical model. (2) There are high and low temperature boundaries outside which development does not occur (e.g. between 6 and 28 °C); instead of development gradually slowing down it stops and mortality increases sharply. An asymptotic function such as that based on Q_{10} theory therefore is not suitable. (3) The ellipse does not offer any theoretical background at the cellular level such as might be derived from enzyme reaction kinetics or changes in physical properties such as viscosity. Such assumptions are inappropriate to a comprehensive model extending over the whole of embryonic development.

Some of the mathematical properties of the ellipse are convenient for predictions and ecological comparisons, such as the angular coefficient K, and the shape $(a-b)/a$. The intercepts of the two vertical tangents on to the temperature axis represent the extreme limits within which development can be completed.

Des femelles mûres de *Temora stylifera* et *Centropages chierchiae*, copépodes pélagiques de mer chaude, sont isolées dans des cristallisoirs de 20 ml contenant de l'eau de mer sans antibiotiques de salinité comprise entre 36.60 et 37.20‰, et placées dans un thermostat. Quand les femelles sont vraiment mûres, la ponte se produit quelques minutes après l'isolement; on

peut en déterminer l'instant avec une précision de l'ordre de 15 min, et le
choisir tel que l'éclosion ait lieu à un moment convenable pour l'obser-
vateur. Les valeurs expérimentales obtenues sont figurées dans la Tableau 1.

Tableau 1.

T. stylifera		C. chierchiae	
Température (°C)	Durée en heures	Température (°C)	Durée en heures
8.0	136	6.5	180
8.7	120	7.6	138
11.1	84	7.7	122
12.4	72	9.7	90
16.2	38	17.2	38
17.4	27	20.0	23
20.0	18	23.0	16
21.6	16	23.5	14
22.0	17	23.8	17
24.0	12	24.0	16
24.5	13	24.8	18
27.0	15	26.0	21
27.9	23	26.7	25
28.0	23	27.7	40

Les points s'alignent sur une courbe qui présente un minimum net ainsi
que deux branches ascendantes asymétriques. Cette forme se retrouve pour
les oeufs de *Calanus helgolandicus* en Méditerranée (données inédites) ainsi
que pour divers oeufs de poissons (par exemple Kinne et Kinne, 1962) et
d'insectes (Birch, 1944).

Nous avons d'abord pensé à ajuster à cette courbe la fonction précédem-
ment employée pour les oeufs de copépodes par McLaren (1963, 1965,
1966), et Corkett (1969) qui est la fonction de Bělehrádek (1957):
$D = a(T-\alpha)^b$ où D est la durée du développement embryonnaire, T la
température et a, b et α des coefficients spécifiques. Mais les auteurs
précédents ne se plaçaient pas dans des conditions telles que le minimum et
le ralentissement ultérieur puissent être décelés. Dans le cas des mers
chaudes, ou si l'on pousse l'expérimentation jusqu'au voisinage des températures léthales, une fonction monotone ne suffit plus.

Nous avons essayé d'ajuster une ellipse pour les raisons suivantes: nous
préférons une fonction dépourvue d'asymptotes par suite de considérations
théoriques qui seront développées ultérieurement (cf. Bernard 1970); certaines propriétés de l'ellipse peuvent recevoir une interprétation écologique,
ce qui n'est pas le cas des autres fonctions du second ordre.

Le programme d'ajustement par la méthode des moindres carrés, du
calcul des paramètres et du tracé de la courbe théorique a été écrit par
Mr Saoud, du Centre de Calcul de l'Institut d'Etudes Nucléaires d'Alger.

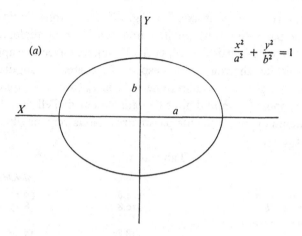

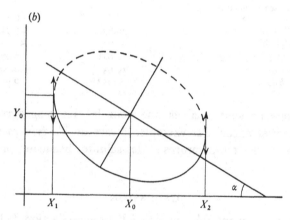

Fig. 1 (a) forme canonique de l'ellipse rapportée à ses axes. Le grand axe est a, le petit, b. L'équation est

$$\frac{x^2}{a^2} + \frac{y^2}{b^2} = 1.$$

(b) Ellipse correspondant aux données expérimentales. Seule la portion en trait plein a une signification biologique. Voir le texte.

Il est reproduit dans un article plus détaillé (Bernard, 1970). Les deux équations canoniques obtenues sont les suivantes :

T. *stylifera:* $8.5 \times 10^{-3} (x - 18.82)^2 + 6.4 \times 10^{-5} (y - 133)^2 = 1,$

C. *chierchiae:* $0.0148 (x - 17.06)^2 + 0.123 \times 10^{-3} (y - 226.6)^2 = 1,$

dans lesquelles x est la température en degrés Centigrade et y la durée du développement en heures. La Fig. 1 a montre une ellipse rapportée à ses axes avec son expression canonique où les paramètres a et b sont respectivement le grand axe et le petit axe. En Fig. 1 *b* nous avons l'ellipse réelle, dont seule la partie en trait plein a une signification biologique. Les abcisses des tangentes verticales, X_1 et X_2, définissent avec une approximation de 2^0 les limites thermiques de l'espèce. Le coefficient angulaire K est la tangente de l'angle α. Les coordonnées X_0 et Y_0 sont celles du centre de l'ellipse. Le rapport $(a-b)/a$ définit l'aplatissement de l'ellipse. Tableau 2 donne les valeurs de ces différents paramètres. Nous verrons plus loin leur sens et leur utilité.

Tableau 2

	T. stylifera	*C. chierchiae*
Grand axe *a*	124.8	90.1
Petit axe *b*	10.8	8.23
Coordonnées centre		
X^0 (°C)	18.82	17.06
Y^0 (en heures)	132.90	226.60
Abcisses tg vertic		
X_1 (en °C)	6.82	6.41
X_2 (en °C)	30.83	27.71
$X_2 - X_1$	24.01	21.30
Coefficient angulaire K	0.041	0.075
Résidu	19.95	20.00
$(a-b)/a$	0.912	0.909

Ces résultats ne sont valables que pour la race biogéographique du bassin W-méditerranéen. Par exemple, la race de *T. stylifera* en Mer du Levant supporte des températures et des salinités beaucoup plus élevées.

DISCUSSION

Beaucoup de travaux ont été consacrés à l'étude de l'action de la température sur la durée de croissance des organismes. On s'accordait autrefois à interpréter les résultats comme une suite de droites d'Arrhenius formant une série de points d'inflexion, et par la théorie du Q_{10}. Mais de nombreuses critiques ont été élevées par les auteurs récents, dont les mesures sont plus précises, contre cette interprétation.

On les trouvera résumées dans les publications de McLaren (1965) et Watt (1968). L'ajustement aux données de fonctions rectilinéaires paraît désormais incorrect. Par contre un grand choix de courbes nous est proposé par les auteurs modernes, mais ce sont généralement des exponentielles plus ou moins compliquées ou des fonctions puissances. En fait, plusieurs courbes différentes peuvent être ajustées aux données expérimentales avec des résidus très voisins; il s'agit de choisir celle à qui l'on peut donner une signification générale soutenue par une théorie. Watt (1968) ajuste trois de ces fonctions: celle de Pradhan, d'Eyring et une logistique, à quatre séries de résultats expérimentaux, et trouve des résidus assez variables d'une série et d'un modèle à l'autre. L'ajustement d'une ellipse à nos propres résultats donne des résidus du même ordre de grandeur que ceux obtenus par Watt avec les exponentielles, qui vont de 1.71 à 101.35. La Fig. 2 montre les deux courbes théoriques calculées à l'ordinateur.

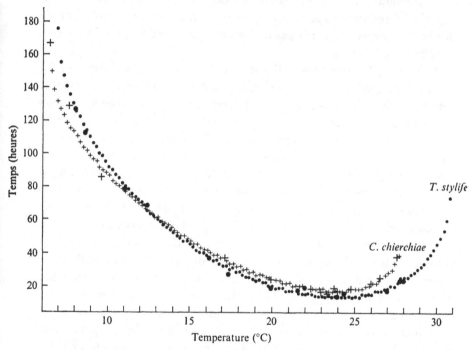

Fig. 2. Courbes théoriques calculées à l'ordinateur d'après les fonctions elliptiques obtenues *T. stylifera* et *C. chierchiae* (voir texte). Les grands cercles noirs et les grandes croix sont les valeurs expérimentales.

Si la ponte se produit dans un milieu qui est déjà à la température fixée pour l'expérience, on n'observe pas de larves anormales à l'éclosion aux extrémités 'froide' et 'chaude' de la courbe. Le voisinage des températures léthales n'entraîne qu'une augmentation rapide de la mortalité. Nous n'avons pas observé non plus de latence aux températures inférieures à la limite de développement comme en observent Costlow & Bookhout (comm. pers.) pour les oeufs de Décapodes: si les oeufs de *Temora* et *Centropages* sont maintenus trop longtemps (3 à 4 jours) à des températures basses (4° à 6°), puis amenés progressivement à une température normale, le développement ne commence pas et le cytoplasme est déjà lysé partiellement. Il en est de même au-dessus de 28° : les oeufs se lysent et ne peuvent reprendre leur développement s'ils sont remis à des températures convenables.

Il existe donc, au moins pour les espèces de mer chaude, des limites thermiques assez nettement marquées en-dehors desquelles on ne peut admettre l'existence de réactions biochimiques infiniment ralenties, ni un état de latence, car le cytoplasme se désintègre. C'est la raison pour laquelle la présence d'asymptotes, comme en possède la fonction de Bělehrádek par exemple, n'a pas de justification théorique pour les espèces sténothermes 'chaudes'.

On peut d'ailleurs contester l'ensemble des justifications théoriques sur lesquelles se fondent les modèles mentionnés ci-dessus parce qu'elles se réduisent à une thermo-dynamique de système fermé, celle de la réaction enzymatique. Or l'oeuf cesse en quelques minutes d'être une *cellule* au métabolisme relativement simple pour devenir un *organisme* dans lequel les inter-actions régulatrices prennent de plus en plus d'importance. Nos observations pas plus que celles de nos prédécesseurs ne tiennent compte du fait que le taux de croissance ainsi que les synthèses cytoplasmiques se modifient profondément à partir de la gastrulation. Avant ce stade, certaines synthèses, telle celle d'ADN, se produisent avec une intensité variable selon le stade de segmentation atteint (Stich, 1962). Il est donc illusoire de tenter d'atteindre la complexité du système multicellulaire au moyen d'une seule fonction. A l'échelle d'observation que nous avons adoptée, nous ne pouvons rendre compte que des réactions intégrées de l'organisme entier en relation avec son milieu. Nous ne demandons par conséquent à notre fonction théorique, l'ellipse, rien d'autre qu'une signification écologique et la possibilité d'être employée dans un but prédictif.

L'avantage d'une fonction elliptique est de donner une expression mathématique à la mort de l'organisme, les températures extrêmes correspondant, aux erreurs expérimentales près, aux abcisses des tangentes verticales X_1 et X_2. Comme il suffit de 5 points expérimentaux pour déterminer entièrement une ellipse, on peut donc calculer la portion de la courbe qui a une réalité biologique ainsi que l'intervalle thermique maximum dans lequel l'espèce peut vivre. Les conditions optimales sont comprises entre l'abcisse du centre et celle de la deuxième tangente verticale, et il est facile de préciser l'abcisse de la tangente horizontale, qui est celle du minimum de durée de développement.

Un paramètre important est celui qui définit la courbure de la fonction: b dans le cas de la fonction de Bělehrádek. Pour l'ellipse, c'est l'aplatissement: $(a-b)/a$, qui varie peu d'une espèce à l'autre.

Le coefficient angulaire K définit la *sensibilité thermique* de l'espèce, c'est à dire l'influence plus ou moins grande que la température exerce sur la durée de développement. L'ordonnée du centre Y_0 caractérise, pour un même coefficient angulaire K, le type de développement de l'oeuf. L'oeuf a un développement rapide ou lent selon les propriétés et l'abondance de son vitellus (McLaren, 1966 et nos propres observations).

L'ellipse étant une figure riche en propriétés analytiques, l'extension de ce genre de recherche à des types de métabolisme varié conduira sans doute à la découverte de paramètres intéressants du point de vue écologique. Mais il faudra attendre encore longtemps avant d'être en mesure d'établir un modèle unique rendant compte à la fois des phénomènes à l'échelle cellulaire et des relations de l'organisme avec le milieu extérieur.

RÉFÉRENCES

BĚLEHRÁDEK, J. (1957). Physiological aspects of heat and cold. *A. Rev. Physiol.* **19**, 59–82.

BERNARD, M. (1970). Quelques aspects de la biologie de *Temora stylifera* (Copépode pélagique) en Méditerranée. Essai d'écologie expérimentale. *Pelagos, bull. Inst. Océanogr. Alger* **11**, 1–96.

BIRCH, L. C. (1944). An improved method for determining the influence of temperature on the rate of development of insect eggs. *Aust. J. expt. Biol. med. Sci.* **22**, 277–83.

CORKETT, C. J. (1969). Techniques for breeding and rearing marine calanoid copepods. 'Cultivation of marine organisms and its importance for marine biology', *Int. Symp., Biol. Anst., Helgoland 1969.*

KINNE, O. & KINNE, E. M. (1962). Rates of development in embryos of a cyprinodont fish exposed to different temperature-salinity-oxygen combinations. *Can. J. Zool.* **40**, 231–53.

MCLAREN, I. A. (1963). Effects of temperature on growth of zooplankton, and the adaptive value of vertical migration. *J. Fish. Res. Bd Can.* **20**, 685–727.

MCLAREN, I. A. (1965). Some relationships between temperature and egg size, body size, development rate and fecundity of the copepod *Pseudocalanus.* *Limnol. Oceanogr.* **10** (4), 528–38.

MCLAREN, I. A. (1966). Predicting development rate of copepod eggs. *Biol. Bull. mar. biol. Lab., Woods Hole* **131** (3), 457–69.

STICH, H. F. (1962). Variations on the DNA content in embryonal cells of *Cyclops strenuus. Expl Cell Res.* **26**, 136–43.

WATT, K. E. F. (1968). *Ecology and Resource Management.* New York: McGraw-Hill.

THE EFFECT OF
CYCLIC TEMPERATURES ON LARVAL
DEVELOPMENT IN THE MUD-CRAB
RHITHROPANOPEUS HARRISII

J. D. COSTLOW, Jr. and C. G. BOOKHOUT
Duke University Marine Laboratory, Beaufort, North Carolina,
and Department of Zoology, Duke University, Durham, North Carolina

INTRODUCTION

It is generally recognized that temperature, acting either independently or simultaneously with other environmental factors, is one of the major physi cal factors in relation to survival, duration of stages, and distribution of marine larvae. In spite of the fact that virtually nothing in the estuarine environment is constant, most investigators, including ourselves, have gone to considerable effort and expense to insure that temperatures used in the laboratory culture of marine larvae were as constant as possible. Having determined the effect of a variety of constant temperatures and salinities on larval development of several species of Brachyura (Costlow, 1967; Cost- low, Bookhout & Monroe, 1960, 1962, 1966; Ong & Costlow, 1969) we have become interested in how regular, daily, cycles of temperature might affect survival, duration of individual larval stages, and the overall time required for development and in what way larval development under cyclic changes in temperature might differ from that observed previously for constant temperatures.

METHODS

Larvae of the small xanthid mud-crab, *Rhithropanopeus harrisii* (Gould), were selected for study because of the available data on the effect of constant temperatures, the high survival rates which we have consistently realized with previous experiments, and because in the natural environ- ment the larvae of *R. harrisii*, ranging from the Mirimichi Estuary in Canada to Lake Maricibo, Venezuela, would encounter daily cycles of temperature comparable to some of those simulated in the laboratory environment.

Ovigerous females were obtained from the vicinity of the Duke Univer- sity Marine Laboratory, Beaufort, North Carolina, placed in finger bowls containing filtered sea water of 25‰, and maintained in culture cabinets

[211]

at the experimental cycles of temperature in which the larvae would be cultured. Control series of ovigerous crabs were also maintained at constant temperatures, except in the case of 15 and 35 °C. In these two constant temperatures the females were maintained at 20 or 30 °C and, at the time of hatching, the first stage zoeae were moved to 15 or 35 °C respectively. Larvae hatched at 30 °C were also used in the 35–40 °C experimental cycles.

The constant temperatures, 5 and 10 deg.C cycles of temperatures, the number of replications for each temperature, and the total number of larvae used in each temperature series are shown in Table 1. The 5 deg.C cycles of temperature were controlled by a Honeywell temperature programmer unit in which the controlling disc was cut to produce equal periods (6 h) of maximum temperature, minimum temperature, and increase and decrease in temperature during each 24 h period. Thus, at the 20–25 °C cycle, the temperature began to rise from 20 °C at 04.00 h, reached the maximum of 25 °C at 10.00 h, maintained this temperature until 16.00 h, and decreased to 20 °C again by 22.00 h, remaining at this minimum until 04.00 h the next morning. The 10 deg.C cycles were designed and controlled in the same way but obviously the rate of increase to the maximum or decrease to the minimum was twice that observed for the 5 deg.C cycles.

At the time of hatching the first-stage zoeae (Fig. 1) were removed into freshly filtered sea water of the same temperature and salinity, separated into smaller glass bowls of ten larvae per bowl, fed recently hatched *Artemia* nauplii, and returned to the appropriate culture cabinet. During each day of larval life the larvae were changed to freshly filtered sea water, the *Artemia* nauplii replaced, and the stage of development and survival recorded. When the megalops stage was reached the larvae were maintained individually in plastic compartmented boxes to avoid cannibalism or fighting. In view of the observations on effect of *Artemia* from different sources on development of Brachyura larvae (Bookhout & Costlow, 1970) it should be noted that all *Artemia* used in these experiments were obtained from the San Francisco area and no abnormalities in development were observed.

All larvae, whether in constant or cyclic temperatures, were subjected to a 14 h day in cabinets illuminated by MacBeth Examolites.

Table 1. *Survival and duration of larval stages of R. harrisii cultured at 25‰*

| No. of series | No. of zoeae | Tempera-ture (°C) | Survival | | | | | | Duration in days | | | | | | | | |
| | | | Hatch to megalopa | | Megalopa to crab | | Hatch to crab | | Hatch to megalopa | | | Megalopa to crab | | | Hatch to crab | | |
			No.	%	No.	%	No.	%	Min.	Max.	Av.	Min.	Max.	Av.	Min.	Max.	Av.
3	150	15	87	58.0	32	36.8	32	21.3	54	83	64.8	15	40	21.0	72	100	86.8
4	421	20	380	90.3	351	92.3	351	83.4	16	23	18.2	6	21	12.5	26	39	32.9
4	442	25	389	88.0	359	92.2	359	81.2	10	14	11.1	4	11	6.9	15	22	18.7
4	434	30	288	66.4	252	87.5	252	58.1	7	13	8.7	2	7	3.6	11	14	12.7
1	100	35	5	5.0	2	40	2	2.0	8	9	8.6	3	4	3.5	11	12	11.5
8	400	15–20	253	63.2	230	90.9	230	57.5	21	37	23.5	8	26	13.2	30	50	36.6
4	400	20–25	337	84.2	306	90.8	306	76.5	11	18	12.5	5	18	8.2	17	30	20.8
6	400	25–30	329	82.2	292	88.7	292	73.0	7	9	7.8	3	9	4.1	10	17	11.9
5	400	30–35	362	90.5	334	92.2	334	83.5	7	11	7.7	3	8	3.8	10	15	11.5
2	100	35–40	0	0	0	0	0	0									
13	650	15–25	503	77.4	437	86.9	437	67.2	15	20	16.8	6	22	9.0	21	39	25.8
20	1000	20–30	826	82.6	738	89.3	738	73.8	9	13	10.3	3	11	5.3	13	25	15.6
22	1150	25–35	733	63.7	525	71.6	525	45.7	5	11	7.7	2	10	3.4	9	18	11.0

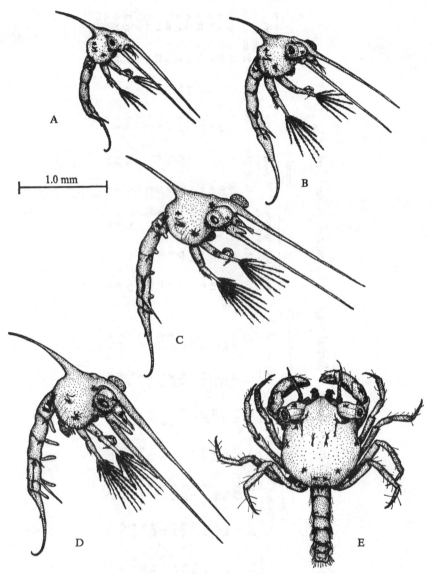

Fig. 1. Zoeal stages one through four (A, B, C, D) and megalops (E) of the *R. harrisii*.

RESULTS

Survival of larvae was observed at all experimental temperatures, both cyclic and constant, other than 35–40 °C. At this temperature none of the zoeae lived more than 24 h.

A comparison of survival of larvae maintained at 10 deg.C cycles and at comparable constant temperatures (Fig. 2) indicates that in general, survival of the four zoeal stages and the one megalops stage is similar. At 15–25 °C survival was lower than that observed at 20 or 25 °C constant but higher than that resulting at 15 °C. At 20–30 °C, survival was slightly lower than at 20 or 25 °C but higher than at 30 °C. At 25–35 °C survival was lower than at 25 or 30 °C but considerably higher than that observed at 35 °C. These results show no apparent trend in the survival of the larvae

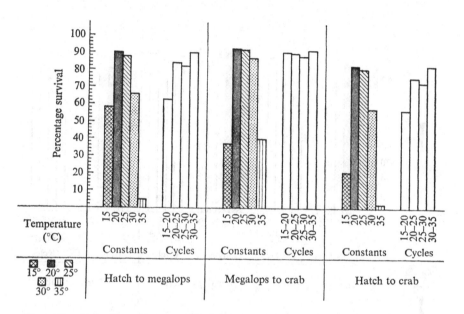

Fig. 2. Survival of larval stages of *R. harrisii* reared in the laboratory at constant and 10 deg.C daily cycles of temperature.

when maintained in cycles of 10 deg.C. Although there was some survival at the extreme low temperature (15 °C) and the extremely high temperature (35 °C), it was unusually low (21.3 and 2.0% respectively).

Total survival, i.e. survival to the first crab stage, tended to be lower in the 20–25 and 25–30 °C cycles of temperature than at comparable constant temperatures (Fig. 3). However, a larger percentage of larvae survived to metamorphosis at 15–20 °C than at 15 °C, but survival at a constant temperature of 20 °C was higher than at the cycle of temperature. Survival at 25–30 °C was lower than at 25 °C but considerably higher than at 30 °C.

Survival at 30–35 °C was higher than at either 30 or 35 °C and represents the maximum survival to metamorphosis for any of the experimental series of larvae.

In most of the experimental series, survival within the megalops stage was higher than that observed over the four zoeal stages. It would appear that having successfully completed zoeal development, the megalops stage is less affected by the conditions of temperature than the earlier stages. Although some increase in megalops survival contributes to the overall figure for total survival of larvae at 30–35 °C, most of the increased survival was in the four zoeal stages.

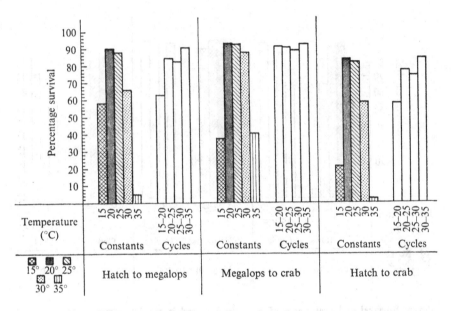

Fig. 3. Survival of larval stages of *R. harrisii* reared in the laboratory at constant and 5 deg.C daily cycles of temperature.

As was expected, the longest period required for complete development to the first crab was at 15 °C (Fig. 4). As the temperature was increased, whether through cycles of temperature or through constant temperatures, there was a corresponding decrease in the time required for development. Above 25 °C a temperature change produced relatively little difference in the time to metamorphosis, suggesting that the maximum rate of development had been reached at this point. There was, generally, a greater range in the time required for total development in 10 deg.C cycles than at either 5 deg.C cycles of temperature or constant temperatures.

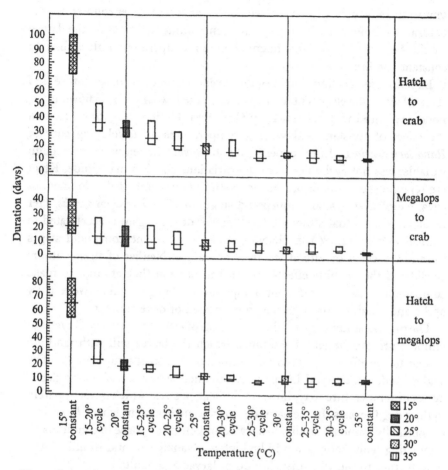

Fig. 4. Comparison of time required for development of larval stages of *R. harrisii* in the laboratory at constant temperatures and 5 and 10 deg.C daily cycles of temperature.

DISCUSSION

A number of earlier authors have noted the wide distribution of adults and larvae of the small xanthid mud-crab, *R. harrisii*. Bousfield (1955) reported this species in the Miramichi Estuary, Canada, Rodriguez (1963) described it from Lake Maracaibo, Venezuela, and others have listed *R. harrisii* as part of the fauna from a number of intermediate areas (Wurtz & Roback, 1955; Chamberlain, 1962; Pinschmidt, 1963). It has also been shown from laboratory experiments that the larvae will successfully metamorphose to the juvenile crab at temperatures as low as 15 °C (Chamberlain, 1962) and as high as 30 °C (Costlow *et al.* 1966), a range of temperature which the larvae would be expected to encounter in the natural environment. In the natural environment, however, the larvae would be subject to daily fluctuations in temperature rather than to constant conditions.

Experiments dealing with regular cyclic changes in temperature and their effect on development are rare and, of the few which are known to us, none deal with the development of Crustacea. Grainger (1959) considered the effect of constant and varying temperatures on developing eggs of *Rana temporaria* L. In his experiments, however, the temperature, though variable, was not cyclic nor were the variations equal in magnitude. Khan (1965) followed a similar procedure in studies on length of development of eggs, naupliar stages, and copepodid stages of *Acanthocyclops viridis*. Her results indicated that although development of the eggs was accelerated to a small extent (12.2%) in changing temperatures, the copepodid stages were not affected by alterations in temperature. Neither of these authors considered the possible effects on survival nor did they attempt to make comparisons between different temperature changes which were truly cyclic and persisted through the entire period of development.

The present study suggests that the effects of either 5 or 10 deg.C cycles of temperature over much of the range which the larvae will withstand are similar to the effects observed for comparable constant temperatures. The slightly reduced survival in larvae maintained at certain of the 10 deg.C cycles of temperature may, however, indicate a stress which is not apparent in the 5 deg.C cycles. In the natural estuarine environment where *R. harrisii* are normally found the ability to survive and metamorphose in conditions of fluctuating temperatures would be highly advantageous and would contribute further to the wide distribution of larvae and adults.

Specific zoeal stages of some species, *Sesarma cinereum* (Costlow *et al.* 1960) and *Hepatus ephiliticus* (Costlow & Bookhout, 1962) have been shown to be particularly sensitive to particular salinities or temperatures. Although this has not been observed for zoeae of *R. harrisii*, the relatively consistent high survival of the megalops stage further suggests that it is the zoeal stages which are more susceptible to environmental stresses than the megalops stage.

The rather startling survival rates at a cycle of 30–35 °C (83.5%), especially in light of survival at a constant 30 °C (58.1%) and 35 °C (2.0%), suggest a number of possible explanations, all of which are purely speculative at the present time, and also raise a number of questions. It is possible that the high survival at 30–35 °C is largely due to the elimination of some deleterious organisms in the laboratory environment, perhaps protozoans, and is not directly associated with the temperature tolerance of the larval stages. One may also postulate the presence of distinct physiological types within the larval stages which are, through the cyclic temperatures, provided with more favourable conditions for survival than through constant temperatures. Directly related to this could be the existence within the larvae of a sequence of enzyme systems, alternate pathways perhaps, with different optimum temperatures. At constant temperatures these alternate pathways either are not operative or operate at minimum efficiency while within the regularly recurring cycles of temperature, the alternate enzyme systems are utilized and each is exposed to an optimum temperature during some portion of the 24 h daily cycle.

The results of the present study raise a question concerning the validity of certain established procedures, such as LD 50s, and their use and applicability to physiological studies when only constant temperatures are employed. The inadequacy of available statistical methods for comparing data from experiments involving constant factors and cyclic factors is also emphasized.

It is to be hoped that some of the postulates presented can be more closely examined within the near future and thus provide us with a better understanding of environmental factors and how they contribute to larval survival, development, and distribution in the constantly fluctuating environment of the estuaries.

SUMMARY

Larvae of the mud-crab, *Rhithropanopeus harrisii* (Gould), were cultured, from hatching to the first-crab stage, at a constant salinity (25‰) combined with eight 24 h cycles of temperature to determine the effect of cyclic temperatures on survival and length of larval life and to compare the effects of constant and cyclic temperatures.

Length of larval life was not affected by either 5 or 10 deg.C cycles of temperature. The longest period required for complete development to the first crab was at 15 °C and as the temperature was increased, either through cycles or through constant temperatures, there was a decrease in the length of larval life.

Survival of larvae within the 10 deg.C cycles of temperature was similar to that observed for the comparable constant temperatures representing the mean temperature of the cycle. At most of the 5 deg.C cycles of temperature there was no apparent difference in survival from that observed in comparable constant temperatures. In spite of reduced survival at both 30 and 35 °C the highest survival was observed at a cycle of 30–35 °C. Several possible explanations are given for the increased survival at such a relatively high cycle of temperature.

These studies were partially supported by a grant (GB-5711) from the National Science Foundation. The authors are indebted to Mrs J. Herring and Mrs Berta Willis for their technical assistance throughout the study.

REFERENCES

BOOKHOUT, C. G. & COSTLOW, J. D. (1970). Nutritional effects of *Artemia* from different locations on larval development of crabs. *Helgoländer wiss. Meeresunters* 20, 435–42.

BOUSFIELD, E. L. (1955). Ecological control of the occurrence of barnacles in the Miramichi Estuary. *Bull. natn. Mus. Can.* 137, 1–69.

CHAMBERLAIN, N. A. (1962). Ecological studies of the larval development of *Rhithropanopeus harrisii* (Xanthidae, Brachyura). *Tech. Rep. Chesapeake Bay Inst.* 28, 1–47.

COSTLOW, J. D., Jr. (1967). The effect of salinity and temperature on survival and metamorphosis of megalops of the blue crab *Callinectes sapidus. Helgoländer wiss. Meeresunters* 15, 84–97.

COSTLOW, J. D. & BOOKHOUT, C. G. (1962). The larval development of *Hepatus epheliticus* (L.) under laboratory conditions. *J. Elisha Mitchell scient. Soc.* 77, 33–42.

COSTLOW, J. D., BOOKHOUT, C. G. & MONROE, J. (1960). The effect of salinity and temperature on larval development of *Sesarma cinereum* (Bosc) reared in the laboratory. *Biol. Bull. mar. biol. Lab., Woods Hole* 118, 183–202.

COSTLOW, J. D., BOOKHOUT, C. G. & MONROE, J. (1962). Salinity-temperature effects on the larval development of the crab, *Panopeus herbstii* Milne-Edwards, reared in the laboratory. *Physiol. Zoöl.* 35, 79–93.

COSTLOW, J. D., BOOKHOUT, C. G. & MONROE, J. (1966). Studies on the larval development of the crab, *Rhithropanopeus harrisii* (Gould). I. The effect of salinity and temperature on larval development. *Physiol. Zoöl.* 39, 81–100.

GRAINGER, J. N. R. (1959). The effect of constant and varying temperatures on the developing eggs of *Rana temporaria* L. *Zool. Anz.* 163, 267–77.

KHAN, M. F. (1965). The effect of constant and varying temperatures on the development of *Acanthocyclops viridis* (Jurine). *Proc. R. Ir. Acad.* 64, 117–30.

ONG, KAH SIN & COSTLOW, J. D. (1969). The effect of salinity and temperature on the larval development of the stone crab, *Menippe mercenaria*, reared in the laboratory. *Chesapeake Sci.* (In Press).

PINSCHMIDT, W. C., Jr. (1963). Distribution of crab larvae in relation to some environmental conditions in the Newport River Estuary, North Carolina. Manuscr. Duke Univ. (unpubl.).

RODRIGUEZ, G. (1963). The intertidal estuarine communities of Lake Maracaibo, Venezuela. *Bull. mar. Sci. Gulf Carib.* 13, 197–218.

WURTZ, C. B. & ROBACK, S. S. (1955). The invertebrate fauna of some Gulf Coast rivers. *Proc. Acad. nat. Sci. Philad.* 107, 167–206.

LES VELIGERES PLANCTONIQUES
DE PROSOBRANCHES DE LA REGION DE
BANYULS-SUR-MER (MEDITERRANEE
OCCIDENTALE): PHYLOGENIE
ET METAMORPHOSE

CATHERINE THIRIOT-QUIEVREUX

Laboratoire Arago, 66, Banyuls-sur-Mer, France

ENGLISH SUMMARY

The structural characteristics of the shell and velum of veligers near meta-morphosis allows them to be classified into two series which correspond to the systematic arrangement of the families of Prosobranchs (after the views of Taylor & Sohl, 1962, for the adults).

In the first series, the vela are all bilobed and the shells of small size $(300-600\mu)$; in the second series the vela are all tetralobed or hexalobed and the shells of larger size (500μ to more than 1 mm).

The metamorphosis observed in relation to the adult alimentary system demonstrates equally the classification of families in two series. In fact, the process of metamorphosis does not cause any important morphological change between larva and juvenile in series I (e.g. *L. neritoides*); by contrast, in the families of series II, metamorphosis coincides with a dramatic change in structure (e.g. *Simnia spelta*; *Atlanta lesueuri, Firoloida desmaresti*.

All veligers of series I collected at Banyuls are meroplanktonic, and meta-morphosis is accompanied by a change in the mode of life from planktonic to benthic. Furthermore, the adults are usually herbivorous. Planktonic veligers of molluscs are microphagic, feeding on phytoplankton.

The species of series II are either meroplanktonic or holoplanktonic but the adults (with the exception of the Aporrhaidae) are always carnivorous. At metamorphosis there is major reorganization of the alimentary tract.

L'étude des variations saisonnières qualitatives et quantitatives des popula-tions planctoniques a été entreprise pendant plusieurs années consécutives dans la région de Banyuls-sur-Mer (Méditerranée occidentale) dans le cadre d'un travail d'ensemble sur le plancton de la région.

Trois points situés sur une ligne perpendiculaire à la côte, correspondant à des profondeurs différentes du fond (55, 95 et 850 m) ont été suivis régulièrement et ont permis d'établir les cycles saisonniers de présence dans le plancton des Gastéropodes (31 espèces ou genres méroplanctoniques et 16 espèces holoplanctoniques).

Les principales caractéristiques morphologiques des véligères de Gastéropodes récoltées à Banyuls-sur-Mer ont été observées, et dans plusieurs cas, la métamorphose et le début de la croissance ont pu être précisés.

Tous ces résultats ont été exposés dans une thèse de Doctorat d'Etat (Thiriot-Quiévreux, 1969).

L'objet de cette communication sera de considérer l'ensemble de ces véligères (au stade proche de la métamorphose) sous un angle systématique et biologique, qui nous permettra de distinguer deux grandes séries de véligères correspondant à la classification des adultes.

Dans le tableau suivant, j'ai noté la liste des familles de Prosobranches (d'après l'ordre systématique de Taylor & Sohl, 1962), ainsi que les caractères morphologiques du vélum et de la coquille des véligères planctoniques correspondantes, et le régime alimentaire des adultes (Fretter & Graham, 1962).

Aux deux séries de familles délimitées correspondent deux séries de véligères.

Dans la série I (Littorinidae aux Eulimidae), nous choisirons un exemple: la véligère de *Littorina neritoides* possède une coquille brune, de trois tours de spire avant la métamorphose et mesure 350 μm de hauteur; le vélum est bilobé et incolore, le pied pigmenté de violet noir. La métamorphose est 'simple' c'est-à-dire ne provoque pas de bouleversement morphologique important: les individus larvaires et juvéniles sont très semblables, seuls la chute du vélum et le développement du pied, marquent le changement de mode de vie (pélagique à benthique). Le régime alimentaire de la larve et de l'adulte est herbivore.

Cette série est très homogène, particulièrement en ce qui concerne les caractéristiques morphologiques: les véligères ont toutes des vélums bilobés et des coquilles de petite taille (300 à 600 μm). Les métamorphoses sont de type 'simple' dans la plupart des cas. De plus, tous les adultes de cette série se nourrissent d'algues ou de particules détritiques, à l'exception des Eulimidae (dernière famille de la série), qui sont parasites d'Echinodermes. Il n'y a donc pas de changement important dû au mode alimentaire, le seul changement correspond au passage de la vie planctonique à la vie benthique et se traduit par la chute du vélum et l'allongement du pied.

Dans la série II (Aporrhaidae aux Turridae), les véligères ont toutes des vélums tétralobés ou hexalobés, et les coquilles sont de taille plus grande

($500\ \mu$m à 1 mm). La grand surface natatoire du vélum traduit une adaptation à la vie pélagique supérieure à celle des véligères de la série I. Certaines espèces d'ailleurs se récoltent en plus grande abondance aux stations éloignées de la côte (*Lamellaria, Atlanta, Carinaria*).

Tableau 1

Prosobranches

		Veligeres (stade proche de la métamorphose)		
Familles	Espèces exemple:	Velum	Taille coquille	Adultes régime alimentaire
Serie I				
Littorinidae	*Littorina neritoides*	Bilobé	$350\ \mu$m	Diatomés, particules détritiques, algues
Rissoidae	*Rissoa lineolata*	Bilobé	$350\ \mu$m	Particules détritiques, algues
Turritellidae	*Turritella communis*	Bilobé	$300\ \mu$m	Phytoplancton, particules détritiques
Caecidae	*Caecum* sp.	Bilobé	$300\ \mu$m	Diatomés
Cerithiidae	*Bittium reticulatum*	Bilobé	$300\ \mu$m	Particules détritiques
Cerithiopsidae	*Cerithiopsis tubercularis*	Bilobé	$500\ \mu$m	Particules détritiques, éponges
Triphoridae	*Triphora perversa*	Bilobé	500–$600\ \mu$m	Particules détritiques, éponges
Aclididae	*Aclis* sp.	Bilobé	$500\ \mu$m	?
Eulimidae	*Eulima* sp.	Bilobé	500–$600\ \mu$m	Parasite d'Echinoderme
Serie II				
Aporrhaidae	*Aporrhais pespelicani*	Hexalobé	1 mm	Microphage, algues
Lamellariidae	*Lamellaria perspicua*	Hexalobé	1 mm	Tuniciers
Eratoidae	*Erato laevis*	Tétralobé	$800\ \mu$m	Botrylles (Ascidies)
Cypraeidae	*Trivia* sp.	Tétralobé	$800\ \mu$m	Botrylles (Ascidies)
Ovulidae	*Simnia spelta*	Tétralobé	$500\ \mu$m	Gorgones
Atlantidae	*Atlanta lesueuri*	Hexalobé	$600\ \mu$m	Carnivore
Carinariidae	*Carinaria lamarcki*	Hexalobé	1 mm	Carnivore
Pterotracheidae	*Pterotrachea coronata*	Tétralobé	$700\ \mu$m	Carnivore
Naticidae	*Natica alderi*	Tétralobé	$750\ \mu$m	Bivalves
Nassariidae	*Nassarius incrassatus*	Tétralobé	$760\ \mu$m	Coprophage
Turridae	*Philbertia gracilis*	Tétralobé	1 mm	Carnivore

La métamorphose provoque un changement morphologique plus ou moins brutal selon les familles de cette série.

Chez *Simnia spelta*, avant la métamorphose, la coquille de la véligère est brune et ornée de cordons longitudinaux recoupés de côtes transversales formant un test treillissé; le manteau est incolore, le pied pigmenté de brun et le vélum tétralobé. Après la métamorphose, la coquille devient de couleur blanche et l'ornementation est beaucoup plus fine. Le manteau se développe tout autour du labre et est orné de nombreuses taches pigmentées brun rouge.

Chez *Atlanta lesueuri*, c'est à la métamorphose que commence l'édification de la carène et que la tête et le pied de l'animal acquièrent la forme définitive de l'adulte.

Les Pterotracheidae montrent les métamorphoses les plus brutales: les individus larvaire et juvénile sont totalement différents quant à leur morphologie externe. Chez *Firoloida desmaresti*, à la métamorphose (et en quelques heures), l'animal perd sa coquille et s'allonge considérablement; la nageoire et le mufle se développent très rapidement.

Ainsi, chez toutes ces espèces, la métamorphose marque des transformations morphologiques importantes, traduisant non seulement un mode de vie différent, pélagique à benthique pour *Simnia*, mais aussi un changement de régime alimentaire. Les véligères se nourrissent essentiellement de phytoplancton, par contre, les adultes sont carnivores: *Simnia* vit et se nourrit sur les Gorgones, et *Atlanta* avale des véligères ou d'autres Mollusques. Quant aux Pterotracheidae, qui ingèrent de grosses proies, la métamorphose correspond à une meilleure adaptation à la vie pélagique et montre une évolution phylogénique nette de cette famille.

Toutes les familles de cette série sont carnivores, à l'exception des Aporrhaidae, première famille de cette série. Les métamorphoses, plus complexes, traduisent le plus souvent un changement de mode de vie et de régime alimentaire.

L'étude de l'organogenèse, effectuée chez *Atlanta lesueuri*, a permis de mettre en évidence les transformations histologiques du tube digestif à la métamorphose.

Le tube digestif de la véligère au stade proche de la métamorphose comprend une bouche, un oesophage cilié avec quelques cellules glandulaires dans sa partie orale, un estomac bordé d'un bouclier muqueux et contenant une plaque dentée dans la lumière stomacale, un intestin composé de trois parties: intestin antérieur, cilié et sans sécrétion, intestin moyen à sécrétions de nature glucidique et intestin postérieur cilié. Les ébauches des glandes salivaires, de la radula et des cartilages radulaires sont visibles mais non fonctionnelles. La glande digestive (avec les différentes catégories cellulaires analogues à celles de l'adulte) débouche dans l'estomac.

A la métamorphose, il y a formation du mufle et mise en place de la radula, des cartilages radulaires et des glandes salivaires. L'oesophage postérieur perd ses cils et se transforme en jabot. L'estomac peu à peu perdra son bouclier muqueux et diminuera de volume ainsi que l'intestin antérieur. Toutes ces transformations permettront le fonctionnement du nouveau régime alimentaire carnivore.

Un autre caractère traduit une forme d'évolution très particulière chez le genre *Atlanta*: les ébauches de la gonade et de l'appareil copulateur chez le mâle apparaissent avant la métamorphose, indiquant là une précocité sexuelle tout à fait exceptionnelle chez les Prosobranches planctoniques.

La première série de véligères, correspondant aux familles primitives dans l'échelle de la classification montre des caractères morphologiques très homogènes et des métamorphoses simples.

La deuxième série, par contre, comprend des véligères de morphologie plus compliquée. Les métamorphoses marquent par leur complexité plus ou moins grande (aussi bien morphologique qu'histologique) le degré d'évolution adaptative de la position systématique des adultes.

Ainsi, pour les espèces de Prosobranches récoltées à Banyuls-sur-Mer, on retrouve chez les larves, le même ordre systématique que celui établi sur les adultes (Taylor & Sohl, 1962); de plus, il est possible de classer les véligères en deux séries, correspondant chez les adultes à deux types de mode alimentaire.

Il serait intéressant de comparer les véligères de toutes les familles de Prosobranches existantes, afin de vérifier si l'évolution larvaire et celle des adultes restent parallèles, à l'échelle de tout le groupe. En effet, dans le cas le plus général (Abeloos, 1956) la phase larvaire et la phase adulte varient chacune indépendamment selon ses propres modalités.

RÉFÉRENCES

ABELOOS, M. (1956). Les métamorphoses. *Coll. Armand Colin. Sec. Biologie.* No. 312.

FRETTER, V. & GRAHAM, A. (1962). *British Prosobranch Molluscs.* Ray Soc. Lond.

TAYLOR, D. W. & SOHL, N. F. (1962). An outline of Gastropod classification. *Malacologia* **1** (1), 7–32.

THIRIOT-QUIEVREUX, C. (1969). Contribution à l'étude écologique et biologique des Mollusques du plancton de la région de Banyuls-sur-Mer. *Thès. Doc. Etat, Fac. Sci., Paris.*

CONCERNING THE FOURTH ANTENNULAR SEGMENT OF THE CYPRIS LARVA OF BALANUS BALANOIDES

P. H. GIBSON* AND J. A. NOTT

N.E.R.C. Unit of Marine Invertebrate Biology, Marine Science Laboratories, Menai Bridge, Anglesey, North Wales, UK

INTRODUCTION

The cyprid is the final larval stage of the barnacle, which can settle permanently on any suitable substratum encountered and metamorphose into the adult form. Work on settlement behaviour, particularly of *Balanus balanoides*, has shown that the larvae can detect certain properties of the substratum. They have been found to orientate with respect to the direction of water current (Crisp & Stubbings, 1957) and surface contour (Crisp & Barnes, 1954) and to respond specifically to extracts of the same species adsorbed on to the attachment surface (Crisp & Meadows, 1962, 1963). This work, together with that of Knight-Jones (1953, 1955) on gregariousness and Crisp (1961) on territorial behaviour of settling cyprids, is *a priori* evidence that the larva is well equipped with mechanosensory and chemosensory structures.

When exploring, the cyprid 'walks' on the substratum with its antennules, using the attachment disc on these appendages to provide efficient but temporary adhesion. Furthermore, the antennules are the only structures on the cyprid in contact with the substratum during preliminary exploration, when the cyprid tends to hold the longitudinal axis of the body perpendicular to the attachment surface. If the larva does not swim off after 'walking', the subsequent exploration behaviour involves movement, with frequent turning, over a restricted area and the ventral region of the carapace is held against the substratum. At this stage the cirri and particularly the caudal setae (Crisp & Barnes, 1954) could also touch the substratum.

When the cyprid is walking on the substratum the fourth segment of the antennule, which bears setae, has been observed to move vigorously and independently after the attachment disc has temporarily adhered to the surface; some of the setae can be seen to come into regular contact with the attachment surface. In view of this behaviour and the identification of

* Present address: Department of Zoology, The University, Newcastle upon Tyne, UK.

a number of sensory structures on the attachment disc (Nott & Foster, 1969; Nott, 1969) a similar investigation was made on the structure of the fourth antennular segment of the cyprid of *B. balanoides*.

MATERIALS AND METHODS

Cyprids were taken from the Menai Strait and narcotized in 0.05 % Nembutal solution in sea water at 4 °C for 24 h. Specimens used for observations on the fine structure of the fourth segment were fixed at room temperature with 3 % glutaraldehyde in aqueous 0.01 M phosphate buffer at pH 7.3–7.4. Sodium chloride (2 g) and five drops of 2 % calcium chloride solution were added to every 100 ml of buffer. The cyprids were washed for 3 h in three changes of buffer and then postfixed for 3 h in buffered 1 % osmium tetroxide solution. Sodium chloride (3.45 g) was added to every 100 ml of buffer solution used for washing and postfixation. All procedures after narcotization were carried out at a reduced pressure of 400 mm of mercury.

The larvae were dehydrated in a graded series of alcohols and embedded in Araldite epoxy resin at 400 mm mercury. The specimens were sectioned with glass knives on a Cambridge Huxley ultramicrotome, stained with a saturated, aqueous solution of uranyl acetate and lead citrate (Reynolds, 1963) and examined in an A.E.I. EM 6 B electron microscope.

Specimens prepared for the scanning electron microscope were fixed and dehydrated similarly but not embedded. They were transferred from absolute alcohol to trichlorotrifluoroethane (Fluorisol) which was allowed to evaporate off slowly over a period of 24 h. The dry larvae were vacuum coated with a layer of carbon and a second layer of gold and then examined in the Cambridge Stereoscan.

EXTERNAL MORPHOLOGY OF THE FOURTH SEGMENT

The fourth segment arises midway along the lateral side of the third segment, which has extended distally to form the attachment organ (Plate 1 a). The width of the fourth segment is about 8.5 μm proximally but increases distally to about 14 μm. The length of the segment is about 54 μm on the abaxial side and 34 μm on the adaxial side, as it is inclined towards the attachment disc at an angle of about 45° to the third segment.

At its base the segment has cuticular ridges; the largest projects about 11 μm on the abaxial side and is continuous round the base except at the midpoint on the adaxial side. Distally the segment bears five terminal setae and four subterminal setae (Fig. 1). Details of the terminal setae A–E and the subterminal setae 1–4 are summarized in Table 1.

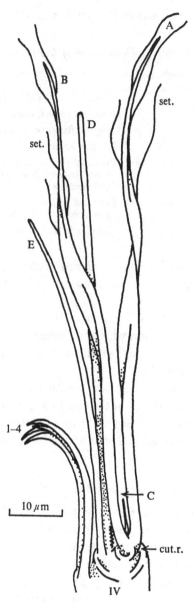

Fig. 1. Terminal region of the fourth segment shown diagrammatically; terminal setae A–E and subterminal setae 1–4.

Key to abbreviations for Figures and Plates:

a.d., attachment disc; a.m., amorphous material; b., basal body of cilium; bd., band of thickening; c., cilium; c.r., ciliary root; cut., cuticle; cut.r., cuticular ridge; d., dendrite; ex., extracellular space; g., gap; extracellular space extends proximally between dendrites; h., haemocoel; m., mitochondrion; mb., membrane complex where muscle is attached to cuticle; n., nucleus; p.p., postciliary process; r., scolopale rod; s., seta; s.c., sheath cell process; set., setule; t., tube containing postciliary processes; v., vesicle; A, B, C, D and E, terminal setae and associated sensory tissues; 1, 2, 3 and 4, subterminal setae and associated sensory tissues; III and IV, third and fourth segment.

Table 1. *Dimensions of the setae*

Seta	Length (μm)	Diameter at base (μm)	Other features
A⎱B⎰	115	2.5	⎰Proximal region, circular in T.S., distal ⎱ region, flattened in T.S. and setulate
C	10	0.7	Minute, setulate
D	110	2.5	Proximal region; c. 16 annular ridges. Distal region; ridges are irregularly arranged and branched
E	70	1.0	Proximal region; circular in T.S. Distal region; not observed in T.S.
1–4	40	2.8	Circular in T.S. Curved

The peripheral edge of the terminal end of the segment forms a ridge which extends distally and is continuous except at the midpoint on the preaxial side. There is a second, less prominent ridge which partially encircles the base of seta B and separates it from seta D. The cuticle at the distal end of the segment, from which the terminal setae arise, is thin and contrasts with the thickened wall of the segment at the base of the sub-terminal setae 1–4 (Plate 1 b).

All four subterminal setae (1–4) have similar morphology. They all curve away from the longitudinal axis of the fourth segment towards the attachment disc and each tapers distally to form a pointed tip (Plate 1 a). They arise from a ledge on the adaxial side which has a bevelled edge without ridges.

INTERNAL FINE STRUCTURE AND MORPHOLOGY

The arrangement of some of the structures within the fourth segment is shown in Fig. 2. There is a muscle connected to the abaxial wall of the segment, originating from the opposite wall of the third segment (Nott & Foster, 1969). Two cell bodies, one in the proximal and the other in the distal region of the segment, probably give rise to the sheath cell processes and the amorphous material, which occurs mostly at the extremities of the segment and supports the sensory structures. The haemocoel extends into the segment.

The dominant structural features within the segment are the scolopidia which form the sense organs (Fig. 3 and see reviews, Howse, 1968; Laver-ack, 1968). Dendrites from sensory cells situated proximally to the third segment, enter the base of the fourth segment and in the proximal region these are irregularly shaped in transverse section and contain vacuoles and

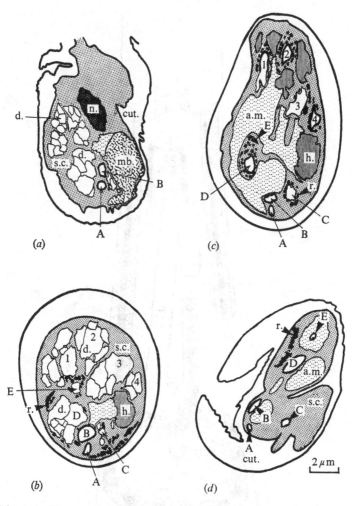

Fig. 2. Schematic diagrams of transverse sections along the fourth segment. A–E and 1–4 are the sensory tissues associated with the setae of the same classification (Fig. 1). (a) Proximal region. The cuticle is contiguous with that of the third segment. Sense organs associated with terminal setae A and B are shown as electron dense tubes. (b) Intermediate region with the dendrites (d) forming into groups comprising the sense organs associated with each of the setae. (c) Intermediate region distal to (b) where most of the sense organs have electron dense tubes. (d) Extreme distal end of the segment showing the arrangement of the scolopidia A–E. Scolopidia 1–4 have entered the subterminal setae (not shown).

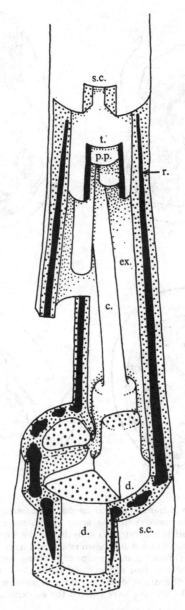

Fig. 3. Diagram of a generalized scolopidium of the fourth segment. Distally, the sheath cell process (s.c.), tube (t.) and postciliary processes (p.p.) extend into the lumen of the seta (not shown). The sheath cell process with its enclosed scolopale rods (r.) have been shown cut away to reveal one of the cilia (c.) arising from a dentrite and entering the tube as a post ciliary process.

mitochondria (Plate 1 c). The dendrites are separated into a number of groups each enclosed in processes of the sheath cell. Within the group they are in intimate contact with each other although the sheath cell processes can sometimes penetrate between some of the dendrites within a group (Plate 1 c). In the midregion of the segment some of the groups are further subdivided.

In the distal half of the segment each dendrite constricts abruptly to form a cilium, which continues distally in an extracellular space bounded by the sheath (Plate 1 d). The cilia have at least two or three roots which penetrate proximally along the dendrite (Plate 1 d, Plate 2 a) and the extracellular space appears to extend proximally between the dendrites (Plate 1 d) for a distance of about 250 nm. Scolopale rods arise in the sheath cell processes in this region (Plate 1 d) and extend distally as shown in Table 2.

The cilia converge distally to form a compact group of postciliary processes enclosed in a tube, which consists of an electron-dense layer between the membrane of the sheath cell and the membranes of the postciliary processes (Plate 1 e). The postciliary processes contain randomly arranged microtubules which usually do not exceed 20 in number and which decrease in number distally. Mitochondria and vacuoles do not occur in the sensory processes distal to the ciliary region. A tube, with associated sheath and postciliary processes enters the base of each seta and extends distally within the lumen. The fine structure of the distal region of each seta, including the termination of the tube, requires further investigation.

The sheath, tube and postciliary processes associated with setae A and B traverse the length of the fourth segment because, unlike the other sense organs, the ciliary regions occur proximally, within the junction of the second and third segments. The tube of A and B, together with its sheath and sensory processes, penetrates into the proximal region of the setae (Plate 2 d, h) but the sensory processes penetrate further into the flattened distal region where they appear to occur free within the lumen of the seta without being packed within a tube (Plate 2 i). In the short seta C, the tube fills the lumen (Plate 2 c) which is apparently open to the exterior at the distal end (Plate 2 b). A tube and sheath enter setae D and E; D has about 20 distal sensory processes in the tube and the relationship between these and the thin, ridged wall of the seta (Plate 2 e) requires further investigation. At the proximal end of D the cuticular wall is about 550 nm thick, but distally it is only 130 nm thick. The reduction in thickness occurs about 10 μm from the base of the seta where the external annular ridge conformation of the cuticle changes to irregularly branched ridges. At least three cilia have been observed in the scolopidium of D, but whether the distal sensory processes arise from these or directly from the dendrites has not been determined. In the other sense organs the frequency of the dendrites, cilia and postciliary processes remains constant within each scolopidium.

The sheath and tube extend at least 20 μm along the length of the subterminal setae; beneath the thin (0.1 μm) cuticular wall of these setae there are circular bands of thickening as shown in Plate 2*f*, *g*. Further structural features of the sensory setae are summarized in Table 2.

Table 2. *Frequency of sensory processes associated with each seta*

Seta	Number of dendrites	Number of cilia	Number of postciliary processes	Extent of scolopale rods
A	3 ?	3 ?	3 ⎫	Ciliary region only
B	5 ?	5 ?	5 ⎭	
C	4	4	4	Extend distally from ciliary region
D	6	3+	c. 20	Ciliary region to base of seta
E	1	1	1	Ciliary region only
1–3	6	6	6 ⎫	Ciliary region to cuticle at base of seta
4	3	3	3 ⎭	

DISCUSSION

When the cyprid explores the substratum, movements of the fourth segment of the antennule can be observed. When viewed from a direction perpendicular to the substratum the setae appear to be moved in a series of sweeping strokes over the substratum. When viewed parallel to the surface of attachment the subterminal setae are seen to be brought into regular contact with the substratum. However, since there is only one muscle attached asymmetrically to the abaxial wall, the two actions observed must be different views of a single movement. Thus, the fourth segment appears to be actively employed during exploration and, moreover, research on the structure shows that all the setae are sensory.

Tentative suggestions on the function of some of the setae can be made directly from the structures observed. The bases of the subterminal setae are rigidly supported by the thick cuticle of the ledge (Plate 1*b*) but the setae have thin cuticular walls supported by annular bands of thickening (Plate 2*f*, *g*). Therefore, each subterminal seta is probably a flexible structure and the scolopidium may detect the degree of distortion of the seta as the pointed tip is brushed over the substratum by the action of the fourth segment. An analogy can be made with the function of a gramophone stylus, which converts the frequency and amplitude of surface irregularities into electrical impulses. This may be the mechanism which enables the larva to select rough in preference to smooth surfaces for exploration (Crisp & Barnes, 1954).

The terminal setae arise from the relatively thin cuticle at the distal end of the segment (Plate 1b). The proximal regions of A and B have thick cuticular walls (Plate 2d, h) and each seta could be deflected, with some of the resulting movement occurring in the thin cuticle of the segment around the bases. The distal regions of A and B are flattened and bear setules, and deflection could be effected by the resistance offered to water movement relative to the larva. Cyprids prefer to explore sites subjected to water movement and can orient with respect to the direction of the current (Crisp & Stubbings, 1957) before cementing to the substratum. Alternatively, or in addition, these two long setae may enable the larva to detect some of the physical properties of the attachment surface. When the setae come into contact with the substratum they could limit the extent to which contractions of the muscle could move the fourth segment; with the antennule attached to a convex structure the distal end of the fourth segment could be drawn closer to the attachment disc than on a concave surface. The cilia of the scolopidia of the setae A and B are within the third segment proximal to the base of the fourth segment where they could act as receptors detecting the position of the fourth segment. Crisp & Barnes (1954) found that cyprids settled in pits and grooves in preference to convexities on the substratum.

Seta C is relatively short, thin-walled and possibly open-ended and could therefore be chemosensory. Seta D could be a mechanoreceptor with the cuticular ridges providing increased resistance to water movement or it could be a chemoreceptor, with the large surface area of thin cuticle allowing diffusion of chemical substances through to the numerous postciliary processes. This chemosensory mechanism is thought to occur in the aesthetascs of some arthropods (Ghiradella, Case & Cronshaw, 1968). The structure of seta E has not been observed in detail.

At this stage the function of the sense organs is based on observations on larval behaviour and structure, and analogies with the receptors of other arthropods. Further work on the fourth segment for a more precise understanding of its functions would require more fine structure observations on individual sense organs together with physiological experiments.

We thank Mr D. C. Williams for all the photographic work and Dr D. A. Dorsett for criticism of the manuscript. We are grateful to Professor D. J. Crisp, F.R.S. for supporting the investigation.

REFERENCES

CRISP, D. J. (1961). Territorial behaviour in barnacle settlement. *J. exp. Biol.* **38**, 429–46.

CRISP, D. J. & BARNES, H. (1954). The orientation and distribution of barnacles at settlement with particular reference to surface contour. *J. Anim. Ecol.* **23**, 142–62.

CRISP, D. J. & MEADOWS, P. S. (1962). The chemical basis of gregariousness in cirripedes. *Proc. Roy. Soc. Lond.* B **156**, 500–20.

CRISP, D. J. & MEADOWS, P. S. (1963). Adsorbed layers: the stimulus to settlement in barnacles. *Proc. Roy. Soc. Lond.* B **158**, 364–87.

CRISP, D. J. & STUBBINGS, H. G. (1957). The orientation of barnacles to water currents. *J. Anim. Ecol.* **26**, 179–96.

GHIRADELLA, H. T., CASE, J. F. & CRONSHAW, J. (1968). Structure of aesthetascs in selected marine and terrestrial decapods: chemoreceptor morphology and environment. *Am. Zool.* **8**, 603–22.

HOWSE, P. E. (1968). The fine structure and functional organisation of chordotonal organs. *Symp. zool. Soc. Lond.* **23**, 167–98.

KNIGHT-JONES, E. W. (1953). Laboratory experiments on gregariousness during setting in *Balanus balanoides* and other barnacles. *J. exp. Biol.* **30**, 584–98.

KNIGHT-JONES, E. W. (1955). The gregarious setting reaction of barnacles as a measure of systematic affinity. *Nature, Lond.* **174**, 266.

LAVERACK, M. S. (1968). On the receptors of marine invertebrates. *Oceanogr. Mar. Biol. Ann. Rev.* **6**, 249–324.

NOTT, J. A. (1969). Settlement of barnacle larvae: surface structure of the antennular attachment disc by scanning electron microscopy. *Mar. Biol.* **2**, 248–51.

NOTT, J. A. & FOSTER, B. A. (1969). On the structure of the antennular attachment organ of the cypris larva of *Balanus balanoides* (L.). *Phil. Trans. Roy. Soc. Lond.* B **256**, 115–34.

REYNOLDS, E. S. (1963). The use of lead citrate at high pH as an electron-opaque stain in electron microscopy. *J. biophys. biochem. Cytol.* **17**, 208–13.

THE FINE STRUCTURE OF THE
TROCHOPHORE OF *HARMOTHOE IMBRICATA*

P. L. HOLBOROW

Gatty Marine Laboratory, University of St Andrews, Scotland

INTRODUCTION

The annelid trochophore has been studied with the light microscope since the middle of the last century and the overall anatomy is reasonably well known. Most workers described external features and development and Thorson (1946) reviews much of this work. E. B. Wilson (1892) gave an account of the cell lineage of *Nereis* and several other workers made similar studies on other species. D. P. Wilson (1932) sectioned the mitraria larva of *Owenia fusiformis* for the light microscope and Segrove (1940) cut the *Pomatoceros* trochophore, but certain of the internal organ systems were not seen in detail.

Electron microscopical study has been sadly neglected, the only work so far reported being a note on the eye of the trochophore of *Neanthes succinea* by Eakin & Westfall (1964).

This investigation is an overall examination with the electron microscope of the trochophore of *Harmothoë imbricata* (Polynoid polychaete) and some comments will be made on comparative features and apparent functions of parts.

METHOD

The trochophores were collected as they were released from under the elytra of adult worms brought into the laboratory. The larvae were fixed in a mixture of one volume of sea water containing the trochophores and an equal volume of osmium tetroxide at 4 °C for half to one hour. The embedding medium was Araldite.

Two separate methods were used for preparing material for the scanning electron microscope. Dr Vernon Barber of Bristol prepared several samples of specimens by freeze drying, followed by carbon and gold coating (see Holborow, Laverack & Barber, 1969, for details of this method and further references).

A second set of specimens was prepared in St Andrews by critical point drying. The animals were fixed in the usual way, washed, and placed in nylon mesh bags. They were passed through graded alcohols to absolute

[237]

alcohol and transferred to amyl acetate which is miscible with absolute alcohol and with liquid carbon dioxide. The bag of specimens was then quickly placed in a chamber and the amyl acetate was washed off with a flow of liquid carbon dioxide. When all the amyl acetate was removed, the chamber was sealed and the pressure was raised by heating the chamber until the critical pressure was reached at which the liquid and gas phases of carbon dioxide occur simultaneously. This avoids surface tension effects which may cause breaking or sticking together of the cilia. This 'critical point drying' procedure was first described by Anderson (1951). The dry specimens were tapped from the bag on to stubs painted with sellotape solution and coated under vacuum with gold palladium. They were viewed in a Cambridge Stereoscan scanning electron microscope.

RESULTS AND DISCUSSION

The features of the living animal seen under the light microscope consist of apical 'tufts', a 'tuft' between the apex and mouth, prototroch, neurotroch and two somewhat kidney-shaped eyes, set slightly back, with some cilia on the surface nearby. The shape of this trochophore is characteristic of polynoid trochophores. The apex is more conical than the posterior pole, with a slight apical dip, and there is a prominent upper lip. In the living animal ciliary movement in the gut can be observed but the internal organs cannot be seen with any clarity.

The overall scanning electron microscope view (Plate 1 a, b) shows with greater clarity the same external features seen with the light microscope. In addition, the protuberances of gland openings may be seen. There are two pairs of large openings on each side, two smaller single ones more posteriorly and nearer the prototroch. A set of four small projections at the right-hand end of the akrotroch, just above the first cilia, may also be gland pores. These are the only gland pores which are asymmetrically placed.

Further asymmetry occurs in the arrangement of apical cilia (Plate 2 a). The cilia lie in five lines around the apex in a formation which more resembles a trapezium than a ring. Three longer lines of cilia lie towards the right. These are approximately 23, 22 and 25 μm long. There is usually a distinct right angle on the dorsal side between the 22 and 25 μm lines. To the left are two 13 μm long lines of cilia with an angle of about 60° between them. The dorsal lines of cilia are close together with gaps of about 2 μm, but to the left and the ventral side there are gaps of 5, 8 and 10 μm between the lines of cilia. All but one of the lines are reasonably straight. It is not known whether this consistently irregular formation is characteristic of this species or whether it occurs among other trochophores with apical rings of cilia. Gravely (1909) shows a neat apical ring on his diagram of a polynoid trochophore but Fuchs (1911) shows a gap in the apical ring of the trochophores of *Nepthys* and *Glycera*. This suggests that some irregularity may be present, but a scanning microscope study is needed before the precise arrangement of these cilia can be known.

With both the light and scanning microscope the cilia are very often found pointing away from the mouth. The reasons for this and their organization are problematical.

The cilia are 8 μm long. A few cilia of the same length occur centrally. The transmission electron micrograph (Plate 2b) shows that the cilia are 5–8 in a group and have oriented basal bodies with a lateral rootlet opposite the basal body, that is, parallel to the body surface, and a long root penetrating centrally. The cells have long thin extensions towards the blastocoel, but it is not known whether these connect directly with prototroch cells. The nerve bundle at the apex contains over 100 axons and some of these may derive from apical cells, although the extensions seen often bypass the bundle.

The akrotroch consists of four lines, 8–10 μm long, of 10 μm long cilia, separated by spaces of 2–7 μm. It runs about one-quarter of the way around the upper ventral side of the animal.

The organization of the prototroch, also, is not as straightforward as previously thought. Four or five rows of cells make up the prototroch band, all the cells being joined distally by desmosomes. The upper row bears short, widely spaced single cilia, the middle two rows bear the long cilia, the fourth and fifth rows bear the shorter cilia. Transmission electron micrographs of longitudinal sections show the two central cells bearing a continuous line of cilia. The basal bodies in both cells are uniformly oriented but the rootlets project into the cells in opposite directions. In certain animals the main, 20 μm long cilia are found to be uniformly oriented and are grouped into bundles of some 40–50 cilia. Each bundle is shared between two of the long, narrow main cells of the prototroch. Each of these cells bears up to 16 half bundles, making *c.* 350 in all. The smaller cilia (14 and 6 μm long) may also form bundles of 5–15 cilia grouped together. In none of these groups is there any connection between the membranes of the cilia. As the formation of cilia into groups occurs only occasionally and simultaneously with uniform orientation of direction of all the prototroch cilia around the animal, it is possible that this is a reaction mediated by nerves. For figures and further discussion of this, see Holborrow *et al.* (1969).

The neurotroch runs from the mouth to the anus and is a broad, tapering

band of 5 μm long cilia with lateral and longitudinal rootlets like those of the
apical cilia. The basal bodies are uniformly oriented pointing away from
the mouth and the neurotroch cilia therefore beat towards the anus. A tuft
of 3–5 cilia lies some 16 μm beyond the end of neurotroch slightly to the
right of the anus (Plate 1a). The mouth is relatively large, being about
40 μm long and 20 μm wide. A group of long cilia occurs on the left side
only of the mouth (Plate 1a, b). Gravely (1909) describes two tongues of
cilia, to either side of the mouth in the Polynoidae, but may be assuming
a symmetry which is certainly not the rule with these trochophores.

In a panorama of about one-third of the animal (Plate 3) the prototroch,
neurotroch, gullet and blastocoel are the dominating features. The prototroch cells are densely packed with mitochondria and ribosomes, particularly at the periphery, and have a large, distinctive nucleus and nucleolus.
To either side of the neurotroch there are cells packed with unstained,
membrane-bound vesicles 0.6–0.9 μm long and 0.5–0.8 μm broad. These
may be lipid. Similar cells are found just beneath the cuticle in various
other regions. Beneath these are a pair of active glandular cells with widely
spaced cisternae of granular endoplasmic reticulum, large golgis, some
mitochondria and islands of secretion droplets. These do not open to the
exterior and could be the site of hormone secretion to control growth and
metamorphosis. The neurotroch cells are roughly oblong with tapering
proximal extensions going under the glandular cells towards two or three
small axons. No synapses have yet been found in this region.

The blastocoel contains some recently divided cells and in section the
area of the cytoplasm is small relative to the nucleus. One in Plate 3 appears
to arise from the wall of the gullet and is one of the larval mesoderm cells
which produces musculature of the gullet. At least one of the other cells
floating in the blastocoel gives rise to a solenocyte, part of the larval
protonephridium.

The gullet (Plate 4a) is densely lined with cilia with oriented basal bodies
indicating that the direction of beat of all the cilia is towards the stomach.
There are two rootlets for each cilium, one parallel with the cuticle, the
other projecting centrally into the cells. The stomach is lined with cilia and
spaced, 1 μm long microvilli, some of which are clavate. There are two or
three populations of cells making up the stomach wall. The majority bear
only a few cilia and a moderately dense carpet of microvilli, some bear
tufts of long cilia (Plate 4a) and some are glandular and have a dense array
of microvilli at the surface with perhaps one cilium. There is a small area of
stomach wall where the tissue is highly vacuolated. This lies between the
stomach and gullet, near the opening of the gullet to the stomach. A cluster

of unusual cilia in the gullet overlie this tissue and project into the opening of the gullet to the stomach.

In structure these cilia are unique, each having an expanded membrane and extra filaments (Plate 4b). The membranes are fused by a uniform electron-dense substance similar to that between ciliary membranes in the flame cell of the fish tapeworm *Diphyllobothrium latum* (von Bonsdorff & Telkkä, 1966). The central filaments normally have the same orientation, as do the basal bodies. This indicates that the cilia beat in the same direction (Gibbons, 1961). As the membranes are fused it can be assumed that the cilia beat in unison. The function would appear to be a valve action, with the thin tissue of the stomach transmitting stomach pressure to the unusual cilia and the cells bearing them. The group of enlarged cilia could then move into and close the opening between stomach and gullet.

The secreting cells of the stomach are typical of exocrine secreting tissue, with close-packed cisternae of granular endoplasmic reticulum (Plate 5a) (Fawcett, 1966). The droplets of secretion are, however, unusual and distinctive, having an internally layered structure (Plate 5a). The only similar droplets to these are mineralized granules in the Malpighian tubules of *Rhodnius* (Wigglesworth & Perry, 1967) but the trochophore granules are only 0.3 μm across whereas the *Rhodnius* granules are between 0.8 and 2 μm across.

The glands opening on to the surface are quite different from stomach gland cells in internal structure and in the type of secretion produced (Plates 5b, 6a). There appears to be more than one type of externally opening gland but the basic type of secretion is probably mucus. All the glands open on to the surface by a projecting pore supported by one ring of 1 μm long microvilli in some cases or by three rings in others (Plate 6b). The cuticle is also pushed upwards to the mouth of the pore and projects beyond the microvilli to further raise the height of the opening. The opening is 0.6 μm in diameter and the whole projection is 1.7 μm across, 2–3 μm in height. Nørrevang (1965) describes microvilli-surrounded gland openings in an enteropneust and suggests that the function of the microvilli is to ensure a free release of mucus by raising the outlet above the body surface.

The larval protonephridium consists of one or two solenocytes which open into an intracellular duct, the cells of which add cilia to the duct. Such protonephridia are well known in the literature (Goodrich, 1945). The solenocyte arises from a cell floating free in the blastocoel (Plate 7a) and consists of a cilium contained in a tube, the lumen of which is 0.7–0.8 μm in diameter. The tube is supported by 15 interconnecting rods (Plate 7b). The rods appear to be composed of clusters of microtubules, unlike those in the flame cell of a rotifer, which bear striations like ciliary roots (Mattern & Daniel, 1966). Kümmel & Brandenburg (1961) studied with the electron microscope and compared various choanocytes and solenocytes, but were more interested in a gross comparison than the details of fine structure. Their diagram of the solenocyte of *Glycera* shows 17 rods in the wall, but even the gross structure of the rods appears to have been modified for the diagram.

Certain features of the fine structure of choanocytes do correspond with features of the solenocyte. The choanocyte of a sponge described by Fjerdingstad (1961) and the choanocyte of an enteropneust described by Nørrevang (1964) both consist of a flagellum surrounded by a collar of rods which Nørrevang calls microvilli. The membranes of the microvilli are separate in both types of choanocyte, whereas in the solenocyte the membrane of the rods is often found to be continuous (Plate 7*b*). In the choanocytes and trochophore solenocyte the rods are linked by fine fibrils, some 40 Å across. The enteropneust choanocyte and the trochophore solenocyte both have projections of these microfibrils into the lumen of the tube. Nørrevang suggested that the substance of the microfibrils is mucus and that the inward projection indicates an inward sweep of water between the rods. As the trochophore rods are on the whole interlinked, flow between them would not be great. Also, the filaments often link the central cilium with the rods. It is therefore unlikely that these filaments are mucus arranged under the influence of flow. They bear some resemblance to the mucopolysaccharide filaments on the surface of intestinal villi (Fawcett, 1966).

Around the outside of the tube of the solenocyte is a thin layer of cytoplasm separated from the rods by large vacuoles. Some of these vacuoles appear to have been formed by thin folds of tissue which are occasionally found arched to enclose part of the blastocoel. This suggests that pinocytosis is taking place.

The tip of the solenocyte tapers as it enters the main protonephridial duct and up to eight of the rods enter and terminate in the duct lumen. The cells of the duct each contribute cilia until there are about 20. Plate 8*a* shows an early stage in the duct with three cilia and the basal bodies of two other cilia cut in transverse section. Plate 8*b* is a later stage in the duct showing 15 cilia and a basal body. Internal septate desmosomes are a feature of the duct cells (Plate 8*a*). Septate desmosomes have been reported in a planarian protonephridium (Pederson, 1961), and are thought to form a diffusion barrier.

The nervous system is quite complex. A group of some 100–150 axons occurs near the apex and radial nerves run on the ectoderm side of the blastocoel. Up to 50 axons run circularly under the prototroch and neuro-ciliary synapses have been found to the main prototroch cells (Plate 9a). A muscle patch occurs beneath these and there are neuromuscular junctions (Plate 9b).

The larval eye consists of a cup of pigment and a two-layered retina of microvilli, the first layer spaced by 700 Å, the second layer, nearest the pigment, spaced by 1000 Å (Plate 10a). It is difficult to ascertain whether the second layer is continuous with the first but it is possible that it is produced by branching of the microvilli. The eye is similar to that of the *Neanthes succinea* trochophore with which Eakin & Westfall (1964) experienced the same difficulty.

One further structure warrants comment. This is a problematical body of whorled membranes apparently derived from cilia (Plate 10b). Apart from the presence of a central pair of filaments in the cilia, this structure bears some resemblance to the receptor region of the eye of the Mollusc, *Cardium* (Barber & Wright, 1969). The cilia branch and coil, rapidly losing their filaments, and the resulting array of membranes are seen longitudinally and transversely cut in different areas in the same section. This body is thus distinctly different in formation from the ciliary-derived lamellate bodies in the eye of a chiton (Boyle, 1969) and the eye of a ctenophore (Horridge, 1964). The neatly whorled membranes in the *Pecten* eye (Barber, Evans & Land, 1967), and the stacks of membranes in *Branchioma* (Krasne & Lawrence, 1966) also appear to have originated in a different manner. In a comparative chart of the derivation of membranes from cilia in eyes by Eakin (1965), there is no indication of branching cilia. This structure represents a further category in ciliary-derived membranes, but whether it becomes the eye has yet to be determined.

Although there are no cilia in the fully developed adult eye, this does not rule out the possibility that it may be derived from this structure. The cilia appear in only a small part of the mass of membranes and diminish in frequency as the size of the mass increases.

Eakin (1963) places annelids in a group of animals with rhabdomeric-type eyes derived from microvilli, by contrast with animals with eyes derived from cilia. Dorsett & Hyde (1968) support this view in a study of the eye of *Nereis virens*. An exception is a report of cilia in the eyes of *Branchioma* (Lawrence & Krasne, 1965; Krasne & Lawrence, 1966). In *Harmothoë*, no cilia have yet been found in the adult eye, but the present work reports that an eye-like structure of ciliary origin is in the process of development in the larva.

SUMMARY

Light microscopy of living and sectioned trochophores gives a general but limited view of the anatomy. The scanning electron microscope reveals the arrangements of cilia and gland openings and a certain amount of asymmetry is found. The apical region has five lines of cilia arranged in a roughly trapezoid form. Four short lines of cilia make an akrotroch running one-quarter of the way around the ventral side of the animal. The prototroch cilia are in three rows around the girth of the trochophore. The neurotroch is a broad band of short cilia running from the mouth and terminating in front of the anus. A small patch of cilia lies to the right on the other side of the anus and a tongue of long cilia is found at the left of the mouth.

The transmission electron microscope study completes the interpretation of external organization and the fine structure of all the internal organ systems is described. The functional and comparative significance of some of the findings is discussed.

Grateful acknowledgement is made to Professor M. S. Laverack who initiated and aided this study. Part of the work was carried out under a Science Research Council (UK) grant (B/SR/1871) for a Research Assistantship to Professor M. S. Laverack.

Dr J. L. S. Cobb helped in the early stages of the work and is to be thanked for Plates 6a and 8a. Dr V. Barber is thanked for scanning electron microscope studies and for Plate 2a.

Members of Edinburgh University are thanked for help with the second scanning microscope study. The gold–palladium coating was carried out by courtesy of Dr D. Bradley of the Zoology Department. Dr A. R. Dinnis of Electrical Engineering authorized the use of their scanning microscope under Mr J. Goodall.

Thanks are offered also to Mr J. Stevenson for preparation of photographs and to Mrs D. Hunter for typing the manuscript.

REFERENCES

ANDERSON, T. F. (1951). Techniques for the preservation of three-dimensional structure in preparing specimens for the electron microscope. *Trans. N.Y. Acad. Sci.* **13**, 130–4.

BARBER, V. C., EVANS, E. M. & LAND, M. F. (1967). The fine structure of the eye of the mollusc *Pecten maximus*. *Z. Zellforsch. mikrosk. Anat.* **76**, 295–312.

BARBER, V. C. & WRIGHT, O. E. (1969). The fine structure of the eye and optic tentacle of the mollusc *Cardium edule*. *J. Ultrastruct. Res.* **26**, 515–28.

BONSDORFF, C.-H. V. & TELKKÄ, A. (1966). The flagellar structure of the flame cell in fish tapeworm (*Diphyllobothrium latum*). *Z. Zellforsch. mikrosk. Anat.* **70**, 169–79.

BOYLE, P. R. (1969). Fine structure of the eyes of *Onithochiton neglectus* (Mollusca: Polyplacophora). *Z. Zellforsch. mikrosk. Anat.* **102**, 313–32.

EAKIN, R. M. (1963). Lines of evolution of photoreceptors. In *General Physiology of Cell Specialization*, pp. 393–425 (ed. D. Mazia & A. Tyler). New York: McGraw-Hill.

EAKIN, R. M. (1965). Evolution of photoreceptors. *Cold Spring Harb. Symp. quant. Biol.* **30**, 363–70.

EAKIN, R. M. & WESTFALL, J. A. (1964). Further observations on the fine structure of some invertebrate eyes. *Z. Zellforsch. mikrosk. Anat.* **62**, 310–32.

DORSETT, D. A. & HYDE, R. (1968). The fine structure of the lens and photoreceptors of *Nereis virens*. *Z. Zellforsch. mikrosk. Anat.* **85**, 243–55.

FAWCETT, D. W. (1966). *An Atlas of Fine Structure. The Cell.* Philadelphia and London: W. B. Saunders Company.

FJERDINGSTAD, E. J. (1961). Choanocyte collars in *Spongilla lacustris* (L.). *Z. Zellforsch. mikrosk. Anat.* **53**, 645–57.

FUCHS, H. M. (1911). Note on the early larvae of *Nepthys* and *Glycera*. *J. mar. biol. Ass. U.K.* **9**, 164–70.

GIBBONS, I. R. (1961). The relationship between the fine structure and the direction of beat in gill cilia of a lamellibranch mollusc. *J. biophys. biochem. Cytol.* **11**, 179–205.

GOODRICH, E. S. (1945). The study of Nephridia and genital ducts since 1895. *Q. Jl microsc. Sci.* **86**, 113–392.

GRAVELY, F. H. (1909). Polychaete larvae. *L.M.B.C. Mem. typ. Br. mar. Pl. Anim.* **19**, 1–79.

HOLBOROW, P. L., LAVERACK, M. S. L. & BARBER, V. C. (1969). Cilia and other surface structures of the trochophore of *Harmothoë imbricata*. *Z. Zellforsch. mikrosk. Anat.* **98**, 246–61.

HORRIDGE, G. A. (1964). Presumed photoreceptor cilia in ctenophores. *Q. Jl microsc. Sci.* **105**, 311–17.

KRASNE, F. B. & LAWRENCE, P. A. (1966). Structure of the photoreceptors in the compound eyespots of *Branchiomma vesiculosum*. *J. Cell Sci.* **1**, 239–48.

KUMMEL, V. G. & BRANDENBURG, J. (1961). Die Reusengeibelzellen (Crytocyten). *Z. Naturf.* **166**, 692–7.

LAWRENCE, P. A. & KRASNE, F. B. (1965). Annelid photoreceptors. *Science, N.Y.* **148**, 965–6.

MATTERN, C. F. T. & DANIEL, W. A. (1966). The flame cell of a rotifer. *J. Cell. Biol.* **29**, 552–4.

NØRREVANG, A. (1964). Choanocytes in the skin of *Harrimania kupfferi* (Enteropneusta). *Nature, Lond.* **204**, 398–9.

NØRREVANG, A. (1965). On the mucous secretion from the proboscis in *Harrimania kupfferi* (Enteropneusta). *Ann. N.Y. Acad. Sci.* **118**, 1052–69.

PEDERSEN, K. J. (1961). Some observations on the fine structure of planarian protonephridia and gastrodermal phagocytes. *Z. Zellforsch. mikrosk. Acad.* **53**, 609–28.

SEGROVE, F. (1940). The development of the Serpulid *Pomatoceros triqueter*, L. *Q. Jl microsc. Sci.* **82**, 467–540.

THORSON, G. (1946). Reproduction and larval development of Danish marine bottom invertebrates. *Meddr. Kommn Danm. Fisk. og Havunders.* Ser. Plankton. **4**, 1–523.

WIGGLESWORTH, V. B. & PERRY, M. (1967). In *The Ultrastructure of the Animal Cell*, p. 159 (ed. L. T. Threadgold). London and Oxford: Pergamon Press.

WILSON, D. P. (1932). On the Mitraria larva of *Owenia fusiformis. Phil. Trans. R. Soc. Ser.* B **211**, 231–334.

WILSON, E. B. (1892). The cell lineage of *Nereis. J. Morph.* **6**, 361–480.

DEVELOPPEMENT LARVAIRE DE
MICROSPIO MECZNIKOWIANUS

C. CAZAUX

Institut de Biologie Marine de l'Université de Bordeaux, Arcachon

ENGLISH SUMMARY

In the Bassin d'Arcachon the polychaete *Microspio mecznikowianus* (Claparède, 1869) reproduces throughout the year. The pelagic larval stages can be collected in great numbers especially in summer.

The male produces spermatophores; the female deposits the egg-mass in a flat elongated sac compressed against the wall of the tube.

In the present work the development of the worm has been followed through metamorphosis up to reproductive maturity. Each characteristic stage is described and commented upon. The trochophore develops and begins to swim within the egg sac; the ciliary apparatus is feeble and produces slow movement only. Metamerization precedes eclosion and consists of the simultaneous appearance of two segments whose setae are visible by transmitted light.

The liberated larva generally possesses three setigerous segments. The now powerful ciliary apparatus produces the rapid swimming of the pelagic phase.

During the entire pelagic life the metatrochophore has a characteristic silhouette with a narrow elongated body, clearly marked 'epaulettes', very long iridescent dorsal setae and typical white–gold pigmentation. Metamorphosis takes place when the larva has 17 segments, 1 month after eclosion. The metatrochophore ceases swimming, loses its long larval setae and the larva takes up its benthic existence. The first hooded-tridentate crochets appear on the eleventh segment – two or three to each neuropodium.

In conjunction with the similar work of Hannerz (1956) on *M. atlantica* this study of the larval development of *M. mecznikowianus* shows that these two spionids are distinct and should have the status of separate species.

INTRODUCTION

L'objet de ce travail est l'étude du développement larvaire de *Microspio mecznikowianus*, annélide polychète sédentaire appartenant à la famille des Spionidae. Cette étude permet de différencier *M. mecznikowianus* d'une espèce voisine *M. atlantica* avec laquelle elle est souvent confondue.

[247]

C'est Claparède qui décrivit l'espèce sous le nom de *Spio mecznikowianus* (Claparède, 1869). Langerhans (1881) décrivit une autre espèce à laquelle il donna le nom de *S. atlanticus*, le caractère distinctif principal étant l'emplacement des premières soies encapuchonnées ventrales: au neuvième segment chez *S. atlanticus*, au onzième chez *S. mecznikowianus*. Le nom de genre de ces deux Spionidae fut modifié par Mesnil (1896) qui créa le nom de *Microspio*.

Söderström reprenant l'étude de ces deux Spionidae estime en 1920 qu'ils appartiennent à une seule et même espèce: *M. mecznikowianus*. Fauvel (1927) et Hartman (1959) se rallient à cette assertion. Hartman catalogue l'espèce sous le nom de genre *Paraspio* créé en 1881 par Czerniavsky.

Dans son étude du développement larvaire des Spionidae, Hannerz (1956) montre que *M. atlantica* doit être définitivement considérée comme une espèce indépendante car, aux stades larvaires, les soies encapuchonnées ventrales apparaissent dès le neuvième segment, ce caractère étant conservé chez tous les adultes du même endroit.

L'étude du développement larvaire exposée dans ce travail confirme la thèse de Hannerz.

M. mecznikowianus se rencontre dans le sable vaseux des plages semi-abritées du Bassin d'Arcachon (Eyrac, la Vigne, le Phare) où elle colonise le niveau des basses mers. On la rencontre aussi en dragages sur fonds vaseux ou sablo-vaseux.

MATERIEL ET METHODES

Deux séries d'observations ont été effectuées: d'une part l'analyse du développement larvaire à partir de la ponte au laboratoire, d'autre part la recherche et l'étude des stades pélagiques recueillis dans le plancton du Bassin d'Arcachon et des stades benthiques récoltés dans le sable vaseux des plages semi-abritées. Cette méthode permet de comparer les différentes phases du développement dans leur déroulement au laboratoire avec celles réalisées dans la nature.

La fécondation expérimentale ne put aboutir. Les pontes récoltées sur les grèves sous forme de sacs ovigères permirent une étude continue en élevage, depuis l'oeuf jusqu'à la métamorphose et même dans certains cas, jusqu'à la maturité génitale des individus.

Les élevages effectués pour ces observations ont été menés dans des coupelles de verre à fond plat, de 6 cm de diamètre et 40 cm³ de contenance totale, emplies d'eau de mer filtrée et renouvelée chaque jour, couvertes de plaques de verre et baignant jusqu'à mi-hauteur dans une eau de mer courante à la même température que celle du Bassin (20 à 21 °C durant ces observations). Les larves étaient nourries à l'aide d'une culture de *Phaeodactylum tricornutum*. Les individus en métamorphose, les stades post-larvaires et les adultes ont été maintenus dans des boîtes de Pétri dont le fond était couvert d'une mince couche de sable propre.

Tous les dessins ont été effectués à la chambre claire; les larves étaient préalablement anesthésiées au chlorure de magnésium à 7% puis placées sur lame dans une goutte de mélange eau de mer–chlorure de magnésium. L'introduction dans cette goutte d'un grain de sable de diamètre très légèrement inférieur à celui des larves étudiées les protégeait de l'écrasement par la lamelle, qui, par sa pression légère, interdisait tout déplacement pouvant être provoqué par l'action de la ciliature.

DEVELOPPEMENT

Le mâle produit des spermatophores qui sont expulsés par la néphridie. Claparède & Mecznikow (1869) dans leur étude du développement larvaire de quelques polychètes montrèrent que les formations observées pour la première fois par Kölliker en 1848 (il les avait prises pour des grégarines), différenciées dans l'appareil néphridien des mâles mûrs, sont en réalité des spermatophores. Ces auteurs ont étudié et illustré la différenciation, la morphologie et le trajet des spermatophores. L'étude du spermatophore de *M. mecznikowianus* fut reprise par Cerruti (1908).

Des spermatophores furent également observés chez *Pygospio elegans* par Söderström (1920) plus simples que ceux de l'espèce étudiée ici. Ces structures semblent caractéristiques de la sous-famille des Spioninae (Hannerz, 1956) et comme le fait remarquer cet auteur, le développement des spermatophores est lié au fait que les espèces de cette sous-famille déposent leurs oeufs sous forme de pontes agglomérées en sacs ovigères. D'après Söderström il y a fécondation interne, les spermatozoïdes abandonnent les spermatophores déposés dans des réceptacles dorsaux de la femelle pénétrant à l'intérieur de l'organisme maternel. La fécondation n'a pu être observée à Arcachon.

Les larves de *M. mecznikowianus* sont présentes toute l'année dans le plancton d'Arcachon avec une abondance particulière de juin à août. La plupart des stades s'élèvent avec facilité et peuvent être étudiés en même temps que ceux nés au laboratoire de pontes récoltées dans le sédiment.

Le sac ovigère est allongé, de 1 à 2 mm de long, déposé à l'intérieur de la galerie habitée par la mère et plaqué contre la paroi. Son enveloppe incolore et transparente renferme une centaine d'oeufs blancs subsphériques. Il n'est pas divisé en plusieurs compartiments par des cloisons transversales comme celles qui peuvent être observées dans les pontes de *Polydora* par exemple. La croissance de la trochophore et la différenciation de la métatrochophore s'effectuent dans le sac aboutissant à l'apparition d'une larve qui se libère pour mener une vie pélagique de quelques semaines comme la larve de *M. atlantica* étudiée par Hannerz à la différence de ce que laissaient prévoir les dessins de Claparède et Mecznikow représentant cette larve volumineuse au début de son développement.

Phase d'incubation

Trochophore

Longueur 250 μm (Fig. 1 *a*, *b*). Elle est épaisse, plate ventralement et renflée dorsalement, de contour elliptique en vue dorsale, une fois et demie plus longue que large. Le pharynx et le rectum sont déjà visibles par transparence, la bouche est ouverte au centre d'une zone ciliaire buccale triangulaire large dont les angles latéraux, visibles en vue dorsale, sont contigus à un prototroque aux cils ténus et courts disposés en deux arcs latéraux; chaque arc se compose de deux cellules ciliaires transversales, insérées sur deux protubérances latérales qui limitent en arrière le prostomium, ce sont les ébauches des épaulettes caractéristiques des larves de Spionidae que Haecker (1898) et Gravely (1909) nomment 'ombrelle'. Ce terme est repris par Dean & Hatfield (1963). Hannerz (1956) désigne cette région par l'expression 'lateral parts of the peristomium', Mesnil (1896) et Casanova (1952) utilisent le terme 'épaulettes' retenu ici. Le télotroque ne comporte que deux cellules ciliaires latérales. Sur la ligne du prototroque se situe dorsalement deux îlots ciliaires larges, ébauches des organes nucaux. Cette trochophore ne possède pas d'yeux. Elle glisse lentement à l'intérieur du sac ovigère sous l'action des deux troques encore peu puissants. La masse vitelline confère une couleur blanche à l'endoderme.

Métatrochophore

Longueur 300 μm (Fig. 1 *c*). Vingt quatre heures après le stade précédent la métamérisation a fait apparaître les reliefs de deux segments dont les sacs sétigères sont visibles par transparence. Cinq soies font saillie au premier segment. Une paire de faisceaux spiniformes s'est différenciée à chaque extrémité (les faisceaux spiniformes très développés aux stades larvaires de certaines familles comme celle des Syllidae, sont constitués par de fins faisceaux de quelques flagelles rigidies accolés). Quelques cils isolés couvrent la face antérieure arrondie et large du prostomium. Une paire d'yeux noirs latéraux se détache nettement sur le tégument blanchâtre. Ventralement, sur le milieu du deuxième segment, s'est différenciée la fossette ciliée typique des larves de Spionidae nommée 'ciliated pit' par Hannerz (1956), organe particulier marquant l'extrémité du neurotrochoïde et représentant d'après l'étude de ce dernier auteur le débouché de deux systèmes de glandes muqueuses. Le télotroque s'est complété d'une deuxième paire de cellules ciliaires latéro-ventrales.

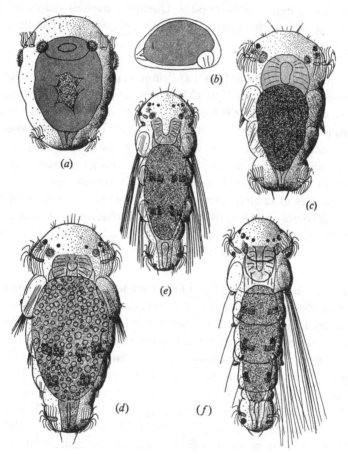

Fig. 1. (*a*) Trochophore de 250 μm, face dorsale. (*b*) Silhouette de la même larve vue de profil. (*c*) Métatrochophore de 300 μm, face dorsale. (*d*) Métatrochophore à l'éclosion, longueur 360 μm. (*e*) Métatrochophore 24 h après l'éclosion; longueur 480 μm. (*f*) Métatrochophore 6 jours après l'éclosion; longueur 560 μm. Les soies dorsales gauches ne sont pas représentées.

Métatrochophore — *Phase pélagique*

Longueur 360 μm, 3 sétigères (Fig. 1 *a*). Ce stade se situe deux jours après le stade précédent, au moment de l'éclosion. Le corps est renflé dans la région moyenne, le prostomium moins large. La masse endodermique contenant un abondant vitellus blanc est encore très volumineuse. Une deuxième paire d'yeux noirs est apparue. Le neurotrochoïde s'étend jusqu'au deuxième sétigère. Les soies encore courtes font saillie à chaque segment. Les segments 2 et 3 portent chacun un nototroque discontinu formé de cellules ciliaires transversales contiguës disposées symétriquement de part et d'autre de la ligne médio-dorsale. Quelques papilles ciliées sont disséminées à la surface du tégument, notamment près des yeux de la paire externe.

Longueur 480 μm, 3 sétigères (Fig. 1 e). Vingt quatre heures après l'éclosion, la larve possède trois paires d'yeux. Le corps est plus allongé, les segments bien dessinés. Les soies très irisées sont longues, celles du premier sétigère atteignent presque l'extrémité postérieure. Les cils plus puissants confèrent à la larve une nage rapide. Le vestibule s'est modelé, il est profond et large, très cilié. Le troisième sétigère possède un gastrotroque naissant. L'intestin de diamètre réduit contient des diatomées, les réserves sont presque épuisées et la larve commence à s'alimenter. Le tégument transparent et incolore porte trois paires de chromatophores jaune-soufre respectivement disposées à la base du prostomium, sur la face dorsale du deuxième sétigère et sur le pygidium.

Longueur 560 μm, 4 sétigères (Fig. 1 f). Ce stade est atteint six jours après l'éclosion. A chaque parapode se distinguent deux petits appendices courts, ébauches des lamelles dorsales et ventrales; les soies dorsales dépassent l'extrémité postérieure du corps. Les soies ventrales (deux par neuropodes) sont trois fois moins longues. Quand la larve est irritée, elle se hérisse en 'chétosphère'; cette attitude défensive caractéristique apparaît chez de nombreuses larves de Spionidae: le corps se roule en boule, les soies de chaque faisceau se dressant en éventail perpendiculairement à la surface du tégument. Le terme de chétosphère fut créé par Haecker (1898) qui, rencontrant des larves de *Laonice* fixées dans cette attitude, avait vu là une forme larvaire particulière et caractéristique d'un certain stade ontogénique.

Longueur 1150 μm, 12 sétigères, 9 jours après l'éclosion (Fig. 2 a). La région céphalique est plus large que le reste du corps, les épaulettes arrondies et peu saillantes sont en continuité avec les lèvres latérales du vestibule. Les palpes sont à peine ébauchés; la métatrochophore à huit sétigères de *M. atlantica* possède déjà des palpes très développés, celle à douze sétigères caractérisée par des parapodes bien marqués et des cirres allongés est prête à se métamorphoser (Hannerz, 1956). Autres caractères distinctifs entre les deux larves: la longueur des soies dorsales et la pigmentation. Chez *M. mecznikowianus* la pigmentation tégumentaire n'est plus jaune-soufre mais blanc brillant; les chromatophores blancs sont répartis par paires, ils se remarquent à l'avant des épaulettes, sur les lèvres latérales du vestibule, sur la face dorsale des segments 2 à 5 et sur la face latéro-ventrale du pygidium.

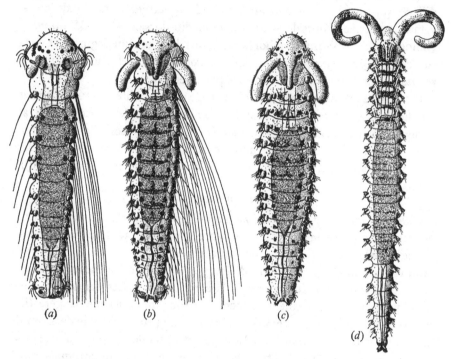

Fig. 2. (*a*) Métatrochophore 9 jours après l'éclosion; longueur 1150 μm. Les soies dorsales gauches ne sont pas représentées. (*b*) Métatrochophore 15 jours après l'éclosion; longueur 1530 μm. (*c*) Jeune ver à la métamorphose; longueur 1570 μm. (*d*) Jeune ver métamorphosé de 4.2 mm.

Longueur 1530 μm, 17 sétigères, 15 jours après l'éclosoin (Fig. 2*b*). Les épaulettes sont encore en relief malgré la croissance des palpes. Le prostomium s'étire postérieurement en une carène longitudinale bordée par les organes nucaux allongés, convergents vers l'arrière. Sur le pygidium et dorsalement par rapport à l'anus sont apparus deux cirres terminaux, c'est là un autre caractère de différenciation avec la larve de *M. atlantica* chez qui les cirres anaux n'apparaissent qu'après la métamorphose. La chute des longues soies larvaires provisoires est amorcée, elles sont remplacées par des soies capillaires de type adulte, courtes, légèrement limbées. On ne remarque encore aucune soie encapuchonnée. La ciliature s'est complétée; des gastrotroques à quatre cellules ciliaires sont portés par les segments 3, 5, 7, 9, 11, 13, 15 tandis que les nototroques composés de nombreuses cellules à cils courts sont visibles à chaque sétigère.

Outre les chromatophores blancs disposés maintenant en deux lignes longitudinales, sur la face dorsale est apparue une pigmentation noire plus diffuse qui forme à chaque segment (sauf les derniers) une bande transversale postérieure dorsale et une ligne médio-dorsale. Chez *M. atlantica* Hannerz décrit trois mélanophores sur le dos du deuxième segment et deux sur chacun des segments postérieurs. L'oesophage est encore court, le rectum s'étend sur les huit segments postérieurs.

Cette larve très attirée par la lumière, nage avec rapidité dans une attitude rectiligne, les palpes étant rabattus latéralement et leurs extrémités se rejoignant sous la face ventrale; pendant la nage les soies larvaires sont couchées le long du corps mais ne sont pas maintenues accrochées par les arcs ciliaires spéciaux ('Grasping cilia' de Wilson (1929) et Hannerz (1956)).

Phase benthique

Longueur 1570 μm, 17 sétigères, 1 mois après l'éclosion (Fig. 2c). Au moment de la métamorphose la larve cesse de nager et construit un tube parcheminé si elle est élevée en l'absence de toute particule, ou creuse le sable qui recouvre le fond des récipients d'élevages. Les dernières soies larvaires sont, après leur chute, incorporées à la paroi du tube. Les palpes, avec un sillon ventral finement cilié atteignent le début du troisième sétigère (ils sont beaucoup plus longs chez *M. atlantica*). Le prostomium reste arrondi (il est bilobé chez *M. atlantica*). Les épaulettes tendent à disparaître alors que le prototroque et le télotroque se désagrègent. Les nototroques subsistent, les gastrotroques disparaissent. Les lamelles dorsales acuminées sont doublées aux segments 2, 3 et 4 par un bourgeon dorsal, ébauche des trois premières paires de branchies (Hannerz signale que chez *M. atlantica* la première paire de branchies apparaît seule, sur le troisième sétigère). A partir du onzième sétigère (le neuvième chez *M. atlantica*) se distinguent des soies encapuchonnées tridentées au nombre de deux à trois par neuropodes. L'oesophage est long et étroit. Le corps est devenu fusiforme, de section légèrement aplatie dorso-ventralement. Les chromatophores blancs conservent la même disposition et la même intensité; selon l'état d'activité de la larve ils peuvent devenir étoilés ou même s'étendre en un fin réseau qui couvre toute la face dorsale. La face ventrale est incolore.

Jeune ver métamorphosé

Longueur 4.2 mm, 23 sétigères (Fig. 2*d*). Le prostomium est arrondi à l'avant. Les palpes sont gros et maintenus légèrement enroulés; étendus vers l'arrière ils atteignent à peine le début du huitième sétigère (ils dépassent l'extrémité postérieure chez *M. atlantica*). Il y a encore trois paires d'yeux. Les organes nucaux allongés atteignent le troisième segment. Il existe une rame dorsale provisoire au premier segment. Les branchies existent maintenant sur les segments 2 à 8. Les soies dorsales capillaires sont semblables à celles de l'adulte; à partir du onzième segment les neuropodes présentent une ou deux soies encapuchonnées tridentées accompagnant les soies capillaires ventrales. Le jeune ver possède quatre cirres anaux. Les palpes brunâtres portent trois manchons pigmentés blancs. Les cinq premiers segments sont dorsalement marqués par des ellipses de pigment brun; à partir du sixième le pigment blanc-brillant dorsal des stades précédents a diffusé en bandes transversales visibles jusqu'à l'extrémité postérieure du corps. Le pygidium et les cirres anaux ventraux sont bruns.

CYCLE LARVAIRE ANNUEL

La répartition quantitative des larves de *M. mecznikowianus* dans le Bassin d'Arcachon a été étudiée au cours d'un cycle annuel dans une station du Bassin stable par son hydrologie (Bouchet, 1968).

Le volume de plancton recueilli varie suivant la saison; il est notamment important au moment de l'explosion phytoplanctonique de printemps, cependant chaque prélèvement étant effectué dans les mêmes conditions (heure, durée, profondeur, vitesse) on peut considérer que le volume d'eau filtrée est pratiquement constant et que la quantité de zooplancton recueillie donne une image exacte de sa densité et de sa qualité réelle. Les nombres cités représentent les nombres moyens mensuels de larves de *M. mecznikowianus* pour des traicts de 10 min.

J.	F.	M.	A.	M.	J.	J.	A.	S.	O.	N.	D.
5	14	13	16	110	731	407	576	1074	223	8	1

L'examen de ces données révèle que la période optimale de reproduction de *M. mecznikowianus* se situe en été dans la station étudiée. D'autre part cette espèce apparaît comme ayant une reproduction pérannuelle puisque les larves se rencontrent tout au long de l'année à des stades variés dans chaque pêche; c'est l'une des Polychètes les mieux représentées à l'état larvaire dans le plancton d'Arcachon par son pourcentage élevé durant la majeure partie de l'année.

Tableau 1. *Comparaison entre les deux espèces:* M. mecznikowianus *et* M. atlantica

Ce tableau rassemble les caractères morphologiques larvaires et post-larvaires principaux de *M. mecznikowianus* et *M. atlantica* et fait apparaître la différence existant entre ces deux espèces.

	M. mecznikowianus	*M. atlantica*
Métatrochophore de 12 sétigères	Longueur 1150 μm. Soies du premier segment dépassant le pygidium	Longueur 900 μm. Soies du premier segment moitié plus courtes
Larve à la métamorphose	17 sétigères, 1570 μm. Palpes courts. Cirres anaux dorsaux ébauchés. 2 mélanophores dorsaux au segment 2. Soies dorsales du segment 1 dépassant le pygidium	12 sétigères, 900 μm. Palpes longs. Pas de cirres anaux. 3 mélanophores dorsaux en triangle au segment 2. Soies dorsales du 1er sétigère moitié plus courtes
Jeune stade benthique	Avec 23 sétigères: Palpes atteignant le 8e seg. Soies encapuchonnées à partir du 11e segment	Avec 17 sétigères: Palpes dépassant le pygidium. Soies encapuchonnées à partir du 9e segment

REFERENCES

BOUCHET, J. M. (1968). Etude océanographique des chenaux du Bassin d'Arcachon. *Thèse. Fac. Sci. Bordeaux.*

CASANOVA, L. (1952). Sur le développement de *Polydora antennata* (Claparède). *Archs. Zool. exp. gén.* **89** (3), 95–101.

CERRUTI, A. (1908). Ricerche sulla anatomia e sulla biologica de *Microspio mecznikowianus* Claparède, con speciale reguardo di nefridi. *Rc. Accad. Sci. fis. mat., Napoli, sér.* 2, **13** (12), 1–35.

CLAPAREDE, E. (1869). Les Annélides Chétopodes du Golfe de Naples. Seconde partie. *Mém. Soc. Phys. Hist. nat. Genève* **20** (1), 1–225.

CLAPAREDE, E. & MECZNIKOW, E. (1869). Beiträge zur Kenntniss der Entwickelungsgeschichte der Chaetopoden. *Z. wiss. Zool.* **19**, 163–205.

CZERNIAVSKY, V. (1881). Materialia ad zoographiam Ponticam Comparatam. Fasc. III. Vermes. *Bull. Soc. imp. nat. Moscow* **55**, 213–363.

DEAN, D. & HATFIELD, D. A. (1963). Pelagic larvae of *Nerinides agilis* (Verrill). *Biol. Bull. mar. biol. Lab., Woods Hole* **124** (2), 163–9.

FAUVEL, P. (1927). Polychètes sédentaires. *Faune Fr.* **16**, 1–494.

GRAVELY, F. H. (1909). Studies on polychaete larvae. *Q. Jl microsc. Sci.* **53**, 597–627.

HAECKER, V. (1898). Die pelagische Polychaeten und Achaeten-Larven der Plankton Expedition der Humboldt-Stiftung. *Ergebn. Atlant. Ozean Planktonexped.* Humboldt.–Stift. **2**, 1–50.

HANNERZ, L. (1956). Larval development of the Polychaete families Spionidae Sars, Disomidae Mesnil and Poecilochaetidae N. Fam., in the Gullmar Fjord (Sweden). *Zool. Bidr. Upps.* **31**, 1–204.

HARTMAN, O. (1959). Catalogue of the Polychaetous Annelids of the world. Part II. *Al. Hanck. Found. Publ.* (Occasional papers), **23**, 355–628.
KÖLLIKER, A. (1848). Beiträge zur Kenntniss niederer Thiere. *Z. wiss. Zool.* **I**, I.
LANGERHANS, P. (1881). Die wurmfauna von Madeira. *Z. wiss. Zool.* **34**, 87–143.
MESNIL, F. (1896). Etudes de morphologie externe chez les Annélides. I. Les Spionidiens des côtes de la Manche. *Bull. scient. Fr. Belg.* **29**, 110–287.
SÖDERSTRÖM, A. (1920). Studien über die Polychaetenfamilie Spionidae. *Inaug. Dissertation Uppsala, Almquiert and Wicksells.*
WILSON, D. P. (1929). The larvae of the British Sabellarians. *J. mar. biol. Ass. U.K. (N.S.)* **16**, 221–69.

SOME MORPHOLOGICAL CHANGES THAT OCCUR AT THE METAMORPHOSIS OF THE LARVAE OF *MYTILUS EDULIS*

B. L. BAYNE*

Marinbiologisk Laboratorium, Grønnehave, Helsingør, Denmark

INTRODUCTION

The change from a pelagic larva to a benthic post-larva marks a critical stage in the life history of many marine invertebrates. At this time a site suitable for further development must be selected and the change from a larval to an adult organization accomplished. 'The fact that the sensory equipment and vigour of the larvae reaches its highest development immediately prior to settlement is no doubt correlated with the importance of (this selection) in ensuring survival' (Cole & Knight-Jones, 1949).

Early studies of this stage in marine lamellibranchs dealt descriptively with the Ostreidae (Ryder, 1884; Stafford, 1913; Prytherch, 1934; Erdmann, 1935) and Teredinidae (Sigerfoos, 1907) and the Protobranchia (Drew, 1899, 1901). More recent studies have emphasized the functional morphology of the larva and post-larva; these include the papers by Cole (1937, 1938) on *Ostrea edulis* L., Quayle (1952) on *Venerupis pullastra* Montagu, Creek (1960) on *Cardium edule* L., Allen (1961) on *Pandora inaequivalvis* L. and Ansell (1962) on *Venus striatula* Da Costa.

In 1922 Field published a comprehensive account of the anatomy of the common mussel, *Mytilus edulis* L. including a brief description of the larva based on the examination of whole mounts only. The present study was undertaken to re-examine the morphology of the larva and post-larva of *Mytilus* and to relate the findings to the general problems of metamorphosis in marine bivalves. Detailed descriptions of the morphology of *Mytilus* larvae will be given only where these were found to differ significantly from other species.

MATERIALS AND METHODS

Pediveligers were obtained from larvae reared in the laboratory (Bayne, 1965). Post-larvae were obtained by providing filaments of the red alga, *Polysiphonia lanosa* L. (Tandy), on which larvae in the cultures attached and metamorphosed. Only post-larvae (plantigrades) aged up to 2 days after metamorphosis will be considered in this publication.

* Present address: School of Biology, The University, Leicester, LE1 7RH, UK.

The material was fixed in the fluids of Bouin, Duboscq-Brasil or Zenker and embedded in paraffin wax, ester wax or polymerized methacrylate. Sections were cut at 10, 5 or 2 μm and stained in Toluidine blue, Heidenhain's Azan or Masson's trichrome stains.

THE MORPHOLOGY OF THE PEDIVELIGER LARVA

The pediveliger possesses a functional velum and foot (Carriker, 1961) and is the larval stage that immediately precedes settlement and metamorphosis. The lateral outline of the shell of the pediveliger of *M. edulis* is characteristic, with a bluntly pointed anterior margin, oval posterior margin and low umbones with wide bases. Jørgensen (1946), Sullivan (1948) and Rees (1950) have all described and illustrated the larvae of *M. edulis* for the purposes of identification.

The main morphological features of the pediveliger, as based on sectioned and living material, are shown in Fig. 1. Some aspects of the morphology will now be considered in more detail.

The velum

The velum occupies the anterior half of the mantle cavity. It is ciliated at its thickened margin with an inner band of long cilia and a peripheral band of shorter cilia (Fig. 2). The long cilia provide the main force for swimming; they also create the feeding currents that impinge on the peripheral cilia, which beat towards the mouth. The mouth (Fig. 1) is situated on the posterior rim of the velum.

A thin membrane connects the velum with the inner lobe of the mantle (Fig. 2). The most anterior point of connection is directly below the anterior adductor muscle; posteriorly the membrane passes around the mouth and upper oesophagus, adhering closely to the latter before fusing with the visceral mass at the base of the foot (Fig. 1). At the mouth this membrane is thickened and ciliated on its outer edge; this is the oral palp (Fig. 1). The cilia of the oral palp act in sorting and rejecting particulate matter collected in the feeding current.

The apical plate (Fig. 1) lies in the centre of the velum. It is a cone-shaped organ of small, darkly staining cells surrounding a shallow apical pit (Fig. 2). A few narrow, elongated cells at the base of the pit bear an apical tuft of short cilia. The entire apical plate region is closely associated with the cerebral ganglia (Figs. 1, 2).

There are three pairs of velar retractor muscles (Figs. 1, 2) with localized insertions on the shell and multiple insertions on the velum. Some fibres from the posterior retractors insert also on the walls of the oesophagus. Creek (1960) has described three velar retractors in the larva of *Cardium edule*, though other authors (Quayle, 1952; Ansell, 1962) have identified only two retractors in other bivalves.

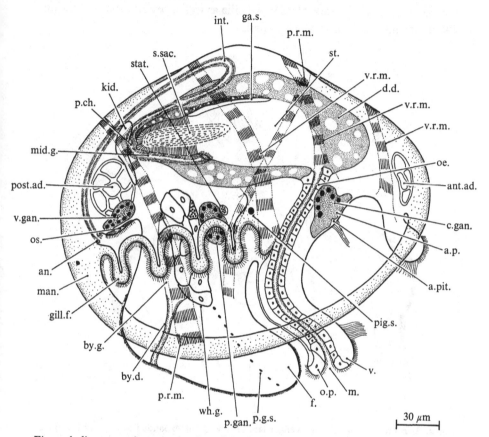

Fig. 1. A diagrammatic reconstruction of the pediveliger larva of *Mytilus edulis*. (Abbreviations: an., anus; ant.ad., anterior adductor muscle; a.p., apical plate; a.pit., apical pit; by.d., byssus duct; by.g., byssus gland; c.gan., cerebral ganglion; d.d., digestive diverticulum; f., foot; ga.s., gastric shield; gill.f., gill filaments; int., intestine; kid., kidney; m., mouth; man., mantle; mid.g., midgut; oe., oesophagus; o.p., oral palp; os., osphradium; p.ch., posterior chamber of mid-gut; p.gan., pedal ganglion; p.g.s., secretions of the phenol (or purple) gland; pig.s., pigment spot; post.ad., posterior adductor muscle; p.r.m., pedal retractor muscle; s.sac., style sac; st., stomach; stat., statocyst; v., velum; v.gan., visceral ganglion; v.r.m., velar retractor muscle; wh.g., white gland.)

The foot

The foot first appears as a knob of tissue between the posterior edge of the velum and the posterior adductor muscle at a shell length of 200–210 μm. It grows rapidly in length and becomes functional in crawling at a shell length of 260 μm. The foot has a distinct 'heel' and is ciliated along the entire ventral surface and dorsally along half its length. There is a ciliated pedal groove in the middle of the ventral surface of the foot from the tip to the heel; at the tip this groove enlarges slightly to form the pedal sucker. The larva crawls by means of its pedal cilia as well as by alternate extension, adherence and contraction of the foot.

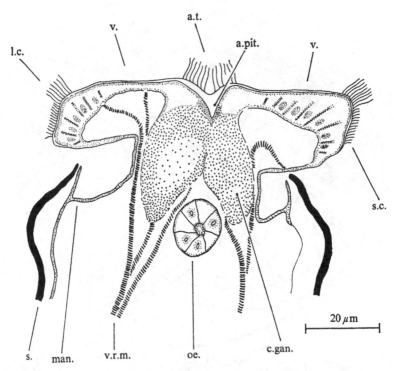

Fig. 2. A frontal section of a pediveliger larva showing the expanded velum and the apical plate. (a.pit., apical pit; a.t., apical tuft of cilia; c.gan., cerebral ganglion; l.c., long cilia of velum; man., mantle; oe., oesophagus; s., shell; s.c., short cilia of velum; v., velum; v.r.m., velar retractor muscles.)

The pedal musculature consists of two pairs of retractors (Fig. 1) and a complex of fibres within the foot. The posterior pair of retractors insert on the shell above the posterior adductor muscle from whence two straps of muscle pass dorsal and lateral to the visceral ganglia and insert posteriorly into the foot. The anterior pedal retractors are inserted anterior to the hinge; they pass peripherally around the visceral mass and enter the foot laterally to the pedal ganglia.

The byssus gland and associated structures are well developed in the pediveliger. In the description that follows (see Figs. 3 a, b), the larval system is related to the adult system as described by Williamson (1906), Brown (1952) and Pujol (1967).

Brown (1952) described three glands in the adult byssus system, the white gland, the purple gland and the byssus gland. These three glands are present also in the pediveliger. The white gland, called 'la glande du collagène' by Pujol (1967), is located at the base of the foot between the posterior adductor muscle and the pedal ganglia (Figs. 1, 3 b). It is a large, unpaired gland of large cells which stain blue in Azan and green in Massons stain. Brown (1952) stated that, with Azan, the white gland, stains 'in shades of blue, intensely blue towards the distal end of the foot and paler, mauvish blue towards the byssus gland'. In some pediveligers also this gland appears to consist of a posterior and an anterior region, the former staining more intensely than the latter.

Associated with the white gland are two or more cells that stain a bright red with Azan. Secretions from these cells pass through the tissues of the foot to the pedal sucker. These cells correspond, both in their staining properties and in the destination of their secretions, to the purple gland of the adult (the 'phenol' gland according to Smythe, 1954).

The byssus gland consists of a pair of pouches lying directly below the white gland (Figs. 3 a, b). A pair of ciliated ducts (Fig. 3 b) issue from these pouches and merge prior to passing into the pedal groove at the heel of the foot.

The byssus of a recently attached post-larva consists of a few single threads with terminal attachment discs; the stem and root of the adult byssus appear to be absent. The close association between the white gland and the pouches of the byssus gland suggests that secretions from the former pass through the pouches to form the byssus threads. The secretions of 'purple gland' cells, concentrated at the sucker, may contribute to the formation of the byssus threads as well as the important attachment discs. However, Pujol, Rolland, Lasry & Vinet (1970) have recently suggested that the larval byssus corresponds to the stem, and not to the threads, of the adult byssus in M. edulis. Further study of the post-larval stages is necessary to elucidate the relationships between the many components of the byssus complex.

The byssus system of the pediveliger is functional. Observations of living pediveligers (Bayne, 1965) have shown them to be capable of secreting a simple byssus within seconds of encountering a favourable substratum. It is this first secretion of a byssus that marks the end of the pelagic larval life.

The shell, the mantle, the ligament and the adductor muscles

During the growth of the larval shell the area of greatest marginal increment gradually shifts posteriorly from a midventral position. As the foot and the gill filaments (Fig. 1) are formed the posterior mantle cavity becomes enlarged to accommodate them.

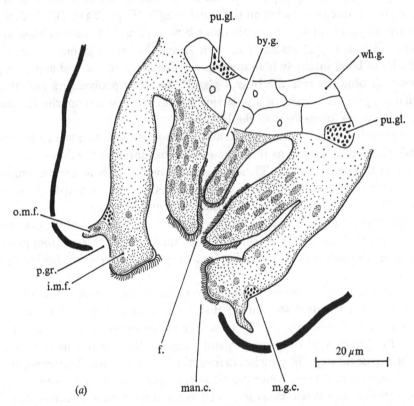

Fig. 3(a). A frontal section of a pediveliger larva showing the foot and byssus apparatus.

The mantle underlies the shell and is a single cell thick except at its margin, where it is thickened to form two folds (Fig. 3a). In the anterior half of the mantle cavity the marginal folds are thin but posteriorly they become very much thicker. A tract of cilia develops on the edge of the inner fold in the posterior half of the mantle cavity. These cilia may serve to control the flow of water through the mantle cavity; later they form a mantle rejection tract in the post-larva.

The edges of the mantle are separate ventrally between the two adductor muscles except for a region just below the posterior adductor where the two inner folds meet to form the siphonal septum. At first the fusion of these two folds is ciliary only; tissue fusion occurs later. The septum is ciliated on its outer edge. On each side of the septum the inner mantle fold is thickened to form two blocks of tissue, the presumptive gill tissue.

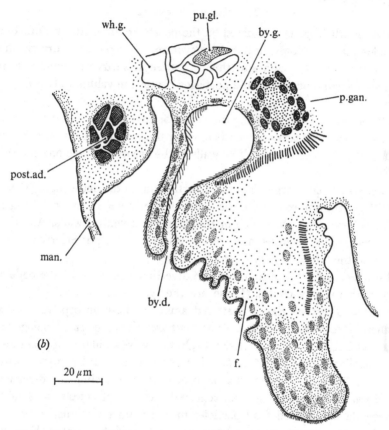

Fig. 3(b). A sagittal section of a pediveliger larva through the foot and byssus apparatus. (by.d., byssus duct; by.g., byssus gland; f., foot; i.m.f., inner mantle fold; m.g.c., mantle gland cells; man., mantle; man.c., mantle ciliary rejection tract; o.m.f., outer mantle fold; p.gan., pedal ganglion; p.gr., periostracal groove; post.ad., posterior adductor muscle; pu.gl., phenol (purple) gland cell; wh.g., white gland.)

Gland cells (Fig. 3a) are present throughout the length of the mantle margin in the outer fold dorsal to the periostracal groove between the outer and inner folds. These cells stain a bright red in Azan. They are 7–9 μm in length and they appear to discharge into the periostracal groove. Similar cells occur also in the thickened mantle margin beneath the hinge.

The larval hinge of *Mytilus* has been described by Rees (1950) and the larval and post-larval ligaments by Trueman (1950, 1951).

The anterior and posterior adductor muscles (Fig. 1) are both well developed in the pediveliger. They are situated on a line parallel to the hinge. At the time of metamorphosis the posterior adductor is slightly the larger of the two.

The alimentary system

The mouth (Fig. 1) is situated on the posterior rim of the velum. Food particles that are caught in the main swimming currents are driven on to the peripheral velar cilia and thence wafted towards the mouth. Some degree of sorting occurs on the posterior rim of the velum and on the oral palp.

The oesophagus passes along the posterior edge of the velum; as it enters the visceral mass it turns forwards to enter the stomach (Figs. 1, 4a) at the level of the cerebral ganglia. The walls of the oesophagus are one cell thick and lined with cilia.

There is a single, large digestive diverticulum in the pediveliger (Figs. 1, 4a). This communicates with the stomach over a wide area, including the entire anterior wall and much of the lateral and ventral walls. Anteriorly the diverticulum occupies most of the visceral mass; posteriorly it is displaced to the left of the style sac (Fig. 4b).

Two types of cell are found in the diverticulum. Most of the cells are large with many vacuoles. These are the typical absorptive cells of the molluscan digestive gland. Scattered amongst these absorptive cells are smaller, more darkly staining cells without vacuoles (Fig. 4a). No flagella or cilia have been observed on these cells, but they are similar in appearance to the 'flagellated cells' of other lamellibranch larvae (Millar, 1955; Ansell, 1962). Millar (1955) observed a band of muscle encircling the diverticula of *Ostrea edulis* larvae, and he reported rhythmical contractions of the diverticula that passed food particles to and from the stomach. No such muscles nor contractions have been observed in *Mytilus* larvae. However, the area of communication between the stomach and diverticula in *Ostrea* is rather less than in *Mytilus* and it may be that, in the latter, ciliary currents in the stomach and diverticulum are sufficient to bring about the circulation of the food.

The stomach is situated beneath the hinge. The oesophagus enters at the anterior end at the point where the tissues of the diverticulum and the stomach meet (Fig. 4a). An extensive part of the roof and sides of the stomach is covered by the gastric shield. The ventral and anterior edges of the shield on the left wall of the stomach are in-rolled into the lumen to

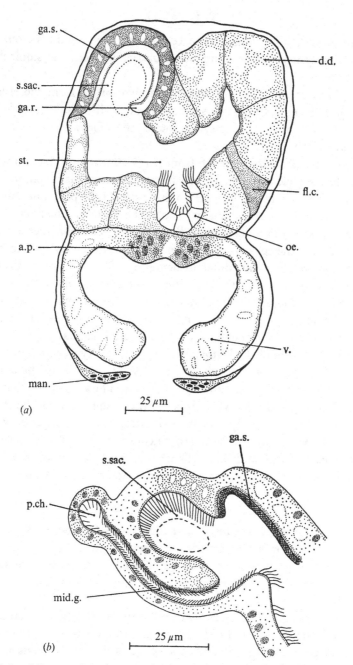

Fig. 4(a) An oblique section of a pediveliger larva through the stomach and the digestive gland, showing the point of entry of the oesophagus into the stomach. The velar cilia are excluded. (b) A sagittal section of a pediveliger larva showing the midgut and the posterior chamber. (a.p., apical plate; d.d., digestive diverticulum; fl.c., flagellated cell of the diverticulum; ga.r., gastric ridge; ga.s., gastric shield; man., mantle; mid.g., midgut; oe., oesophagus; p.ch., posterior chamber of the mid-gut; s.sac., style sac; st., stomach; v., velum.)

form the gastric ridge. This may serve to retain the food mass around the style; certainly it provides a functional barrier between the stomach and the diverticulum.

The posterior end of the stomach is drawn out to form the style sac (Fig. 4b). The cells of the style sac have large, well-defined vacuoles, and they are densely ciliated. The style is rotated in a clockwise direction when viewed anteriorly.

The midgut leaves the stomach from the ventral wall in the midline just anterior to the style sac (Fig. 4b). As it passes backwards it communicates dorsally with the style sac through a narrow slit. At the posterior end of the style sac the midgut turns dorsally to enter the posterior chamber (Figs. 1, 4b), which also communicates with the style sac. This chamber is a small, spherical organ, the cells of which resemble those of the style sac in having well-defined vacuoles, but differ in being less densely ciliated. From the dorsal wall of the chamber the intestine passes forwards above the style sac and stomach; it loops dorsally under the umbones and passes back on the right of the body to the anus (Fig. 1). The anus opens behind the posterior adductor muscle; it is encircled by a ciliated anal papilla.

A striking feature of the histology of the alimentary system is the abundance of large vacuoles. Vacuoles are present in the cells of the stomach, style sac, posterior chamber and diverticulum. These vacuoles contain no sectioned material. In the living larva, however, the region of the stomach and diverticulum is densely occupied with droplets of an apparently lipoid material. Bruce, Knight & Parke (1940) commented on the abundance of 'oil globules' in the epithelial cells of the alimentary tract of O. edulis larvae. These droplets may serve as a reserve food source (Millar & Scott, 1967).

The nervous system and sense organs

There are three pairs of ganglia and four associated sense organs in the pediveliger. No nervous connectives between ganglia have been identified.

The cerebral ganglia (Figs. 1, 2) lie dorsal to the apical plate and are closely associated with the apical plate and apical tuft of cilia throughout larval life.

The pedal ganglia (Fig. 1) are situated at the base of the foot anterior to the white gland. They are fused, as they are in the adult, but their paired origin is evident from the bilobed appearance. A pair of statocysts are fused to the pedal ganglia laterally.

The visceral ganglia (Fig. 1) lie ventrally to the posterior adductor muscle, linked by a long commisure. A strip of darkly staining, ciliated cells is situated across the roof of the mantle cavity ventral to the visceral ganglia and commissure. The appearance of these cells suggests a sensory function and the organ has been interpreted here as an osphradium. Its function remains obscure. However, unpublished observations indicate that pediveligers and young post-larvae are very sensitive to silt on the substratum and in the water. When crawling, the main water currents in the mantle cavity are produced by the pedal cilia. These currents impinge directly on the osphradium, which is therefore well situated to 'test' incoming water.

A pair of 'eye-spots' (Fig. 1) is present in the pediveliger. These are prominent pigment spots situated at the base of the first gill filament. Each spot consists of a number of small pigment granules deposited in the form of a cone in one cell of the mantle. These spots occur at the site of the greatest convexity of the shell, and their first appearance coincides with a change in the phototactic behaviour of the larva (Bayne, 1964).

The ctenidia

The first ctenidial filaments (Fig. 1) are formed at a shell length of 240–260 μm. Pediveliger larvae invariably possess 3–5 filaments situated on an oblique dorso-ventral line in the posterior mantle cavity, lateral to the foot. The most anterior pair is attached to the visceral mass, the posterior pair is attached to the presumptive gill tissue at the siphonal septum.

The first three gill filaments are formed almost simultaneously by the splitting of the ridge of presumptive gill tissue into three blocks which then become ciliated and grow into the mantle cavity. As they grow, these blocks of tissue split dorso-ventrally to form the cavities of the filaments, and the ventral tips become thickened into small knobs, the capitula. Additional filaments are formed by further growth and splitting of the presumptive tissue. Ctenidial development is therefore of the 'papillary' type (Rice, 1908; Raven, 1958).

The gill filaments of the pediveliger possess lateral and latero-frontal cilia. The ctenidia do not function in food-collection at this stage and movement of the cilia is sporadic and apparently uncoordinated.

The reno-pericardial system

The kidneys are present as a pair of simple tubes between the posterior adductor muscle and the end of the digestive diverticulum on the left and the style sac on the right. No openings into the mantle cavity have been demonstrated with certainty. The cells are cubical with clear cytoplasm and large nucleoli.

The pericardium consists of inner and outer membranes surrounding the intestine. The inner membrane lies close to the intestine; the heart has not yet been formed. The outer pericardial membrane extends to the kidneys on either side of the intestine; the outer membranes of each side are fused above and below the intestine. No reno-pericardial duct was identified.

MORPHOLOGICAL CHANGES THAT OCCUR
AT METAMORPHOSIS

The most dramatic of the morphological events that constitute metamorphosis in *Mytilus* are the loss of the velum, the reorientation of the mouth and the foot, and the growth of the labial palps and gill filaments. These are accompanied by the increasing complexity of the organ systems and by the secretion of the adult shell. All of these changes are preceded by the secretion of the first byssus.

The loss of the velum and the growth of the palps

The velum collapses and disintegrates within 48 h of the secretion of the first byssus. Disintegration proceeds from the posterior edge of the velum forwards, but does not include the cells of the oral palp, the oesophagus nor of the apical plate. As a result of the disappearance of the mass of velar tissue the mouth and the oesophagus are rotated forwards and come to lie close to the anterior adductor muscle (Fig. 5 *a*). Within 24 h of attachment a few ciliated cells of the velum remain in the velar cavity above the anterior adductor muscle. The cells of the apical pit disappear. The apical plate becomes incorporated with the oesophagus at the anterior border of the mouth. Also at this time a membrane grows out from the oesophagus to fuse with the mantle above the anterior adductor muscle.

Soon after the onset of metamorphosis the area surrounding the velum is invaded by phagocytic cells similar in appearance to those described by Cole (1938). Disintegrating velar tissue is not engulfed by the larva during metamorphosis nor is it shed externally; the disappearance of this tissue is apparently achieved entirely by phagocytosis.

The upper labial palp (Fig. 5 *b*) is formed coincidentally with the loss of the velum. The tissues of the upper lip of the mouth, including the apical plate, grow out laterally and ventrally into the mantle cavity. Within 2 days of attachment, the upper palp extends across the mantle cavity as a mobile, ciliated hood over the mouth. At its lateral extremities the palp is drawn

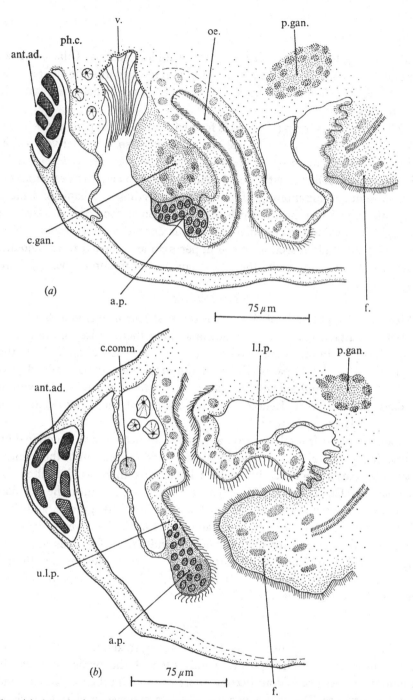

Fig. 5(a) A sagittal section of a pediveliger larva during metamorphosis, showing the degeneration of the velum and the incorporation of the apical plate into the mouth. (b) a later stage, showing the formation of the labial palps. (a.p., apical plate; ant.ad., anterior adductor muscle; c.comm., cerebral commissure; c.gan., cerebral ganglion; f., foot; l.l.p., lower labial palp; oe., oesophagus; p.gan., pedal ganglion; ph.c., phagocytic cells; u.l.p., upper labial palp; v., velum.)

out ventrally to a greater degree than its middle section. Cilia on these lateral extensions form a rejection tract directed towards the floor of the mantle cavity. Cilia on the middle section of the upper palp beat towards the mouth.

The lower labial palp (Fig. 5*b*) is formed from the oral palp of the larva. Within a few days of metamorphosis this palp also is a relatively large, mobile, ciliated fold of tissue under the mouth. At its lateral extremities it lies in close association with the first gill filaments, a condition essential for the subsequent functioning of the gill/palp feeding mechanism. Ciliary currents on this palp beat towards the mouth. Small food particles that are brought forward in the ctenidial feeding current are collected by the lower palp and the middle section of the upper palp and carried to the mouth. Larger particles drop on to the lateral portions of the upper palp and are rejected.

The ctenidia

With the disappearance of the velum the first pair of ctenidial filaments, which are attached to the visceral mass along their entire anterior face, move forwards in the mantle cavity and acquire a close association with the mouth and the labial palps (Fig. 6). Additional filaments are added posteriorly from the gill anlagen; usually 2–3 new filaments are added within 2 days of attachment. Reflection of these filaments does not occur until some time later (Rice, 1908).

Lateral, latero-frontal and frontal cilia are all present and functional on these filaments. However, the inhalent water current into the mantle cavity is created entirely by pedal cilia at this time, as described for *Macoma* by Caddy (1969). Food particles are swept against the gill filaments and carried either dorsally to the suprabranchial groove, and then forwards to pass over the first gill filament and on to the lower labial palp, or else to the ventral tips of the filaments and thence forwards on to the upper labial palp. Here, some particles are accepted into the mouth and other particles are discarded via the lateral edges of the palp and the marginal rejection tract of cilia.

The foot

As the velum disintegrates the foot grows in length and moves forwards in the mantle cavity. The 'purple gland' cells of the byssus increase in number and their secretions become more common in the foot and around the pedal sucker. The white gland and byssus pouches grow in size whilst maintaining their close association.

The mantle and the shell

The mantle remains bilobed for some days after metamorphosis, and in none of my sections has the third mantle lobe developed. However, in some sections a longitudinal split has appeared in the inner lobe of the mantle; this, together with the presence of a discrete and well-formed periostracal groove (Fig. 2) suggest that the middle mantle fold of the adult is formed from the inner larval fold (cf. Quayle, 1952).

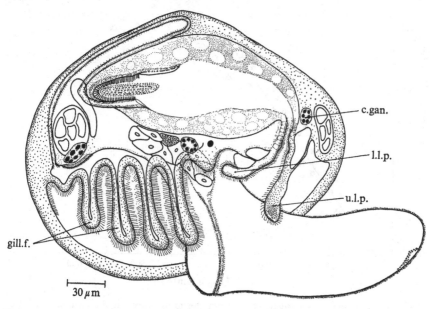

Fig. 6. A diagrammatic reconstruction of an early post-larva (plantigrade) of *Mytilus edulis* before secretion of the dissoconch shell. (c.gan., cerebral ganglion; gill.f., gill filaments; l.l.p., lower labial palp; u.l.p., upper labial palp.)

The tract of cilia on the mantle edge continues to develop after metamorphosis, extending forwards to the anterior adductor muscle. These cilia constitute the mantle rejection tract.

During metamorphosis there is no growth of the shell, but within 2 days of attachment, the secretion of the adult, dissoconch, shell has started. This shell is clearly distinguishable from the larval shell by its different pigmentation and more marked sculpturing. There is evidence of a further distinction between an interdissoconch and the dissoconch stage of shell growth (Ockelmann, 1965).

A striking feature of shell growth in *Mytilus*, as in other lamellibranchs, is the marked change of shape that occurs with secretion of the adult shell. This is illustrated in Fig. 7, where the lengths of larval and post-larval shells are plotted against an 'anterior dimension', UA, on the same shells (measured from the umbo through the anterior adductor muscle). These measurements were made on the right valves of larvae and post-larvae

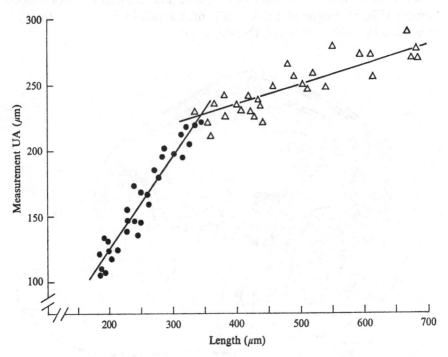

Fig. 7. An anterior measurement (UA) from the umbo through the anterior adductor muscle, plotted against length, for larvae (circles) and post-larvae (triangles) of *M. edulis* to illustrate the change of shape that takes place with secretion of the adult, dissoconch, shell.

taken from laboratory cultures. These data ('type D' data of Richards & Kavanagh, 1945) are not suitable for the application of an allometric relationship, but they do illustrate a correlation between the dimensions that reflects the change of shape that occurs at metamorphosis. This change of shape represents a posterior shift of the region of greatest marginal growth increment together with diminished growth at the anterior margin of the shell.

The alimentary system

With the forward movement of the mouth and oesophagus the anterior 'loop' of the latter disappears. The digestive diverticulum increases in size.

The nervous system and sense organs

The cerebral ganglia migrate forwards at metamorphosis to lie beside the anterior adductor muscle. The cells of the apical pit disintegrate and the apical plate is incorporated into the upper labial palp. The cerebro-visceral connective develops; it passes laterally over the posterior pedal retractor muscles close to the pedal ganglia. Cerebro-pedal connectives have not been observed. The statocysts and pigment spots are retained, and the osphradium grows in size.

DISCUSSION

There is a remarkable similarity in the general level of organization represented by the pediveliger larvae of different species of bivalves. A consideration of the literature (see citations in the Introduction) shows this organization to include: a large velum used in swimming and in feeding; a foot used in crawling; a ciliated palp on which the food-particles are sorted prior to being passed, via an oesophagus, into a stomach with style sac and large digestive gland; a simple intestine that is coiled in some species; a thin mantle with bilobed margin that forms a siphonal septum posteriorly; a nervous system that includes cerebral, pedal, and, in most species, visceral ganglia and a sensory system of statocysts, apical plate with or without a flagellum and in some species pigment spots; a few ($<$ 7) pairs of simple primary gill filaments; and a byssus system that, if not actually described as functional at this stage, is capable of secreting a byssus when the bottom habit is adopted.

The development of these planktotrophic pediveliger larvae (Thorson, 1950) follows a 'veliconcha' stage of development (Werner, 1939; Rees, 1950), the obligatory pelagic phase, during which there is considerable growth in size but little increase in morphological complexity (Raven, 1958; Creek, 1960). This veliconcha larva is equipped with the basic requirements of a pelagic, dispersive larval phase, namely a swimming organ, the velum, a functional gut and a shell.

The requirements of the pediveliger stage, however, are more complex, for it is during this time that the larva must select and colonize a suitable site on the bottom and must convert to the adult modes of feeding and locomotion. These requirements are reflected in the considerable sensory/ nervous complement of the pediveliger, and in the presence of the basic components of the adult gill/palp feeding mechanism. It is this latter that best illustrates the degree to which the pediveliger is prepared for the adult mode of life. The gill filaments and siphonal septum have no function in feeding during the larval phase; ciliary currents on the filaments are unable to act in feeding until a connection has been established between the filaments and the mouth, and such a connection is impossible until the velum has disappeared. Degeneration of the velum and formation of the labial palps occur rapidly at metamorphosis and, within 2–3 days in most species, the entire gill/palp feeding mechanism becomes functional, complete with ciliary sorting currents on the palps and division of the posterior mantle cavity into inhalent and exhalent regions. At the same time, the anterior migration of the mouth and foot permits the rapid growth of more gill filaments posteriorly and, in *Mytilus*, changing axes of shell growth also result in a much enlarged posterior mantle cavity. Furthermore, the ctenidial filaments of the larva are incapable of filtering the water until they effectively partition the mantle cavity; this they are unable to do until the foot moves anteriorly at metamorphosis, allowing the filaments to form marginal, interctenidial junctions.

In the few days following metamorphosis, the post-larvae of most species, being less than 2 mm high (Thorson, 1966) will be attached within the 'boundary layer' (Crisp, 1955) of reduced current flow. Under these conditions a rapid establishment of efficient feeding currents will be necessary in order to pass adequate volumes of water through the gills. Allen (1961) has discussed the metamorphosis of the larvae of *Pandora inaequivalvis*; in this species the larval life is short (< 4 days) and at metamorphosis the gill filaments are very rudimentary. Allen suggests that until these filaments have increased sufficiently in size after metamorphosis it is the ciliated foot, held close to the mouth and labial palps, that creates the feeding currents. Such a mechanism may also operate in other species (Caddy, 1969).

The obvious morphological similarities that exist between different species of marine lamellibranchs may be referred to the similar problems posed by a planktotrophic larval development. There are important differences between species, of course. Studies of the larval shell and hinge indicate characteristic specific differences (Rees, 1950) and differences of internal morphology, such as the greater complexity of the byssus in *Mytilus* as compared with other larvae (e.g. Quayle, 1952) are apparent. Nevertheless, the basic requirements of substrate selection and change from larval to adult feeding mechanism must be common to most species. The similarities between species in morphology and behaviour at the time of metamorphosis are to be expected. It is the changes that occur after metamorphosis, the reorientation, growth and increasing complexity of the organ systems, and the changing axes of shell growth, that establish recognizable specific differences. Yonge (1962) has suggested an inverse relationship between the degree of morphological specialization of the adult and the rapidity with which the adult morphology is established after metamorphosis.

The pediveliger larva of *Mytilus edulis* illustrates well the co-existence in a terminal larval stage of both larval and adult systems. It may be postulated that the adult systems are inhibited from developing fully at this time, and that at metamorphosis disruptive changes occur in the larval tissues so removing the inhibition and allowing constructive growth of the adult tissues to occur. Two considerations are implied in this view; first that the larval tissue, which may be senile (Berrill, 1947) responds by degeneration and involution to an altered, and perhaps an unfavourable, environment, namely the benthic, as opposed to the pelagic habitat; and secondly that the inhibition which is responsible for preventing further development of the adult system resides in the larval tissues and is removed as these tissues degenerate.

The larval tissues of the pediveliger of *Mytilus* do react to sublethal, but unfavourable situations by involution, more readily than the adult tissues (Bayne, 1965). It is true also that the larval velum may be responsible for 50% of the total larval metabolic rate (Zeuthen, 1949). During the delay of metamorphosis degenerative changes occur in the larval tissues (Bayne, 1965). But whereas in true metamorphosis the loss of the velum is accompanied by development of the adult system, these constructive changes do not occur during the delay of metamorphosis. As a consequence, in a pediveliger at stage three of the delay of metamorphosis the velum has been lost, but the palps and ctenidia have not developed. Such a larva cannot feed and does not grow but, immediately on encountering a suitable substrate, attachment can occur, followed by all the constructive changes of normal metamorphosis. If an inhibition of the adult system is present in the pediveliger, this does not reside solely in the larval tissues nor is it dependent on the high metabolic rate of the larval tissues. Coordination between the disruptive and constructive changes of metamorphosis requires a more positive stimulus that may be associated in some way with the act of attachment by the byssus. Attachment is in turn dependent on the reception of a suitable hierarchy of stimuli, so illustrating the integrated nature of behavioural, morphological and physiological change that constitute metamorphosis.

SUMMARY

The pediveliger larva of *Mytilus edulis* is similar in its general morphology to the larvae of many other species of marine bivalves. These similarities include a large velum and foot with retractor muscles, 3–5 paired gill filaments which are not yet functional, a simple mantle, an alimentary system of oral palp, oesophagus, stomach with crystalline style, digestive glands and intestine, a sensory system of statocysts, pigment spot and apical plate with cerebral, pedal and visceral nerve ganglia, and a byssus system. *Mytilus* differs from other species in the complexity of the byssus system (distinct white gland, purple gland cells and byssus gland with paired ducts emptying to the pedal groove) and in details of the alimentary system (open connection between midgut and style sac, presence of a posterior chamber in the midgut) and sensory system (well-developed osphradium). The gross morphological changes that occur at metamorphosis are also common to *Mytilus* and other bivalves. These changes include the collapse and disappearance of the velum, the incorporation of the apical plate into the labial palps of the post-larva, the reorientation of the organs in the mantle cavity, and the functioning of the adult gill/palp feeding mechanisms within 2 days of metamorphosis. Metamorphosis is followed by a marked change in shell shape. It is suggested that the pediveliger is well equipped both to select and to colonize a site on the bottom and to convert rapidly to the adult modes of feeding and locomotion; the bases for these adaptations are discussed.

I should like to record my thanks to Professor G. Thorson, in whose laboratory this work was carried out, and to Cand. mag. K. Ockelmann for many helpful discussions. I am grateful also to The University of Wales for financial support.

REFERENCES

ALLEN, J. (1961). The development of *Pandora inaequivalvis* (Linne). *J. Embryol. exp. Morph.* 9, 252–68.

ANSELL, A. D. (1962). The functional morphology of the larva, and the post-larval development of *Venus striatula* (da Costa). *J. mar. biol. Ass. U.K.* 42, 419–43.

BAYNE, B. L. (1964). The responses of the larvae of *Mytilus edulis* (L.) to light and to gravity. *Oikos* 15, 162–74.

BAYNE, B. L. (1965). Growth and the delay of metamorphosis of the larvae of *Mytilus edulis* (L.). *Ophelia* 2, 1–47.

BERRILL, N. J. (1947). Metamorphosis in ascidians. *J. Morph.* 81, 249–67.

BROWN, C. H. (1952). Some structural proteins of *Mytilus edulis*. *Q. Jl microsc. Sci.* 93, 487–502.

BRUCE, J. R., KNIGHT, M. & PARKE, M. W. (1940). The rearing of oyster larvae on an algal diet. *J. mar. biol. Ass. U.K.* **24**, 337–74.

CADDY, J. F. (1969). Development of mantle organs, feeding and locomotion in post-larval *Macoma balthica* (L.) (Lamellibranchiata). *Can. J. Zool.* **47**, 609–17.

CARRIKER, M. R. (1961). Interrelation of functional morphology, behaviour and autecology in early stages of the bivalve *Mercenaria mercenaria*. *J. Elisha Mitchell scient. Soc.* **77**, 168–242.

COLE, H. A. (1937). Metamorphosis of the larva of *Ostrea edulis*. *Nature, Lond.* **139**, 413.

COLE, H. A. (1938). The fate of the larval organs in the metamorphosis of *Ostrea edulis*. *J. mar. biol. Ass. U.K.* **22**, 469–84.

COLE, H. A. & KNIGHT-JONES, E. W. (1949). The setting behaviour of larvae of the European flat oyster, *Ostrea edulis* L., and its influence on methods of cultivation and spat collection. *Fishery Invest. Lond.*, Ser. II **17** (3).

CREEK (GRANT), G. A. (1960). The development of *Cardium edule* L. *Proc. zool. Soc. Lond.* **135**, 243–60.

CRISP, D. J. (1955). The behaviour of barnacle cyprids in relation to water movement over a surface. *J. exp. Biol.* **32**, 569–90.

DREW, G. A. (1899). Some observations on the habits, anatomy and embryology of members of the Portobranchia. *Anat. Anz.* **15**, 493–519.

DREW, G. A. (1901). The life history of *Nucula delphinodonta*. *Q. Jl microsc. Sci.* **44**, 349–52.

ERDMANN, W. (1935). Über die Entwicklung und die Anatomie der 'ansatzreifen' Larve von *Ostrea edulis*, mit Bemerkungen über die Lebensgeschichte der Auster. *Wiss Meeresunters., Abt. Helgoland*, N.F. **19**, 1–25.

FIELD, I. A. (1922). Biology and economic value of the sea mussel, *Mytilus edulis*. *Bull. Bur. Fish., Wash.* **38**, 125–259.

JØRGENSEN, C. B. (1946). Lamellibranchia. In *Reproduction and larval development of Danish marine bottom invertebrates*. (G. Thorson.) *Meddr. Kommn Danm. Fisk.-og. Havunders.*, Kbh., Ser. Plankton **4**, 277–311.

MILLAR, R. H. (1955). Notes on the mechanism of food movement in the gut of the larval oyster, *Ostrea edulis*. *Q. Jl. microsc. Sci.* **96**, 539–44.

MILLAR, R. H. & SCOTT, J. M. (1967). The larva of the oyster *Ostrea edulis* during starvation. *J. mar. biol. Ass. U.K.* **47**, 475–84.

OCKELMANN, K. (1965). Developmental types in marine bivalves and their distribution along the Atlantic coast of Europe. *Proc. 1st Eur. Malac. Congr.* (1962), 25–35.

PRYTHERCH, H. F. (1934). The role of copper in the setting, metamorphosis and distribution of the American oyster, *Ostrea virginica*. *Ecol. Monogr.* **4**, 49–107.

PUJOL, J. P. (1967). Formation of the byssus in the common mussel (*Mytilus edulis* L.). *Nature, Lond.* **214**, 204–5.

PUJOL, J. P., ROLLAND, M., LASRY, S. & VINET, S. (1970). Comparative study of the amino acid composition of the byssus in some common bivalve molluscs. *Comp. Biochem. Physiol.* **34**, 193–201.

QUAYLE, D. B. (1952). Structure and biology of the larva and spat of *Venerupis pullastra* (Montagu). *Trans. R. Soc. Edinb.* **62**, 255–97.

RAVEN, C. P. (1958). *Morphogenesis: the Analysis of Molluscan Development*. London: Pergamon Press.

REES, C. B. (1950). The interpretation and classification of lamellibranch larvae. *Hull Bull. mar. Ecol.* **3**, 73–104.

RICE, E. L. (1908). Gill development in *Mytilus. Biol. Bull. mar. biol. Lab. Woods Hole* **14**, 61–77.

RICHARDS, O. W. & KAVANAGH, A. J. (1945). The analysis of growing form. In *Essays on growth and form*, pp. 180–230 (ed. W. E. Le Gros Clark and P. B. Medawar). Oxford: Clarendon Press.

RYDER, J. A. (1884). The metamorphosis and post-larval stages of development of the oyster. *Rep. U.S. Commnr. Fish.* **10**, 779–91.

SIGERFOOS, C. P. (1907). Natural history, organisation and late development of the Teredinidae, or shipworms. *Bull. Bur. Fish., Wash.* **27**, 191–231.

SMYTH, J. D. (1954). A technique for the histochemical demonstration of polyphenol oxidase and its application to egg-shell formation of helminths and byssus formation in *Mytilus. Q. Jl microsc. Sci.* **95**, 139–52.

STAFFORD, J. (1913). *The Canadian Oyster: its Development, Environment and Culture*. Ottawa.

SULLIVAN, C. M. (1948). Bivalve larvae of Malpeque Bay, P.E.I. *Bull. Fish. Res. Bd Can.* No. 77.

THORSON, G. (1950). Reproductive and larval ecology of marine bottom invertebrates. *Biol. Rev.* **25**, 1–45.

THORSON, G. (1966). Some factors influencing the recruitment and establishment of marine benthic communities. *Neth. J. Sea Res.* **3**, 267–93.

TRUEMAN, E. R. (1950). Observations on the ligament of *Mytilus edulis. Q. Jl microsc. Sci.* **91**, 225–36.

TRUEMAN, E. R. (1951). The structure, development and operation of the hinge ligament of *Ostrea edulis. Q. Jl microsc. Sci.* **92**, 129–40.

WERNER, B. (1939). Über die Entwicklung und Artunterscheidung von Muchellarven des Nordseeplanktons, unter Gesonderer Beruchsichtigung der Scholenentwicklung. *Zool. Jb. Abt. Anat.* **66**, 237–70.

WILLIAMSON, H. L. (1906). The spawning, growth and movement of the mussel (*Mytilus edulis* L.), horse mussel (*Modiolus modiolus* L.) and the spoutfish (*Solen siligna* L.). Rep. Fishery Bd Scotl. **25**, Part III.

YONGE, C. M. (1962). On the primitive significance of the byssus in the Bivalvia and its effects in evolution. *J. mar. biol. Ass. U.K.* **42**, 113–25.

ZEUTHEN, E. (1949). Body size and metabolic rate in the animal kingdom. *C. r. Trav. Lab. Carlsberg Ser. chim.* **26**, 17–161.

THE HISTOLOGY OF THE LARVA OF
OSTREA EDULIS DURING METAMORPHOSIS

R. W. HICKMAN* AND LL.D.GRUFFYDD

*N.E.R.C. Unit of Marine Invertebrate Biology, Marine Science Laboratories,
Menai Bridge*

INTRODUCTION

A large volume of literature has been written on the biology of the oyster, largely because of its economic importance. Most of this work has been reviewed by Korringa (1952) and Galtsoff (1964). Much less information is available concerning the metamorphosis of the oyster larva and the greater part of our knowledge relates to the American Atlantic oyster, *Crassostrea virginica*, Gmelin, and is due to the work of Ryder (1884), Jackson (1888) and Stafford (1913). These authors were perhaps influenced to a great extent by earlier descriptions by Davaine (1853) and Huxley (1883). The only detailed work on metamorphosis in *Ostrea edulis* is that of Cole (1938), who reviewed the previous work in this field and drew attention to the considerable gap in the detailed information available on stages between the fully developed 'ansatzreifen' larva (Erdmann, 1934) and the settled spat figured by Yonge (1926). Cole dealt, in the main, with the fate of the characteristic organs of the larva, namely the velum, foot, adductor muscles and eyespots. The present investigation extends and supplements Cole's descriptions by investigating at various stages in metamorphosis changes in organ systems that he had omitted. Cole did not present complete descriptions of the gross morphology at any particular stage as has been done here.

MATERIALS AND METHODS

The oyster larvae used were obtained from the Ministry of Agriculture, Fisheries and Food Laboratory at Conway and reared to metamorphosis at constant temperature (see Gruffydd & Baker, 1969) according to the method developed by Walne (1956, 1966). Surfaces for settlement consisting of No. 50 Whatman filter papers which had previously been coated with oyster extract were inserted into the culture vessels when the proportion of eyed to non-eyed larvae reached 3:1. It was thought that by using such a substratum, the spat could be induced to settle readily (Crisp, 1967; Bayne, 1969) and could be sectioned *in situ*. However, it was found that

* Present address: Fisheries Research Division, Marine Department, P.O. Box 19062, Wellington, NZ.

they could be brushed or scraped off the filter paper without damage. Six hours after the settlement surface was added, the larvae still swimming were removed from the culture vessel so that only attached and crawling larvae remained. The larvae which had settled on the filter paper were removed from the culture vessels at known intervals and transferred to a solution of 0.1 % nembutal in sea water which acted as a narcotic by preventing the larvae from closing the valves tightly on subsequent immersion in fixative.

Such small relaxed larvae could be conveniently handled by pipetting them into 7 mm glass tubing closed at one end with nylon mesh. These proved to be very suitable vehicles for transferring the larvae from one reagent to the next without loss and with the minimum of damage.

Although conventional embedding in wax blocks was tried, the method eventually used was one adapted from standard electron microscope preparative techniques. The material was fixed in 5 % unbuffered glutaraldehyde solution (Sabatini, Bensch & Barnett, 1963) for 24 h at room temperature. It was dehydrated in alcohols and then placed in a 50/50 Araldite/absolute alcohol mixture for 2 h followed by 100 % Araldite for a further 2 h. The larvae were finally embedded in fresh Araldite contained in gelatin capsules, which were then placed in an oven at 48 °C for at least 48 h to harden. The larvae could not be orientated in the gelatin capsules, hence large numbers had to be embedded in order that sufficient would be found in the appropriate plane for sectioning. The embedded specimen was clearly visible within the Araldite block. Sections were cut at thicknesses varying from 0.5 to 5 μm on an ultramicrotome, the sections being floated on to droplets of water on glass slides which were then dried on a hotplate. The drying caused the Araldite to flatten out and adhere to the slide.

The sections were stained according to the method of Richardson, Jarrett & Finke (1960) using Mallory's azur II methylene blue stain, dried in air by reheating after the excess stain had been washed off with distilled water, and finally mounted in D.P.X.

One major advantage of the Araldite embedding method is that the larva is retained in a non-distorted condition. Moreover, decalcification of the shell is unnecessary because in hardened Araldite the shell presents no difficulties during the section cutting stage. Fixation in Bouin's solution and embedding in wax did produce considerable distortion of the larvae, presumably because the fixative, being acidic, had a considerable decalcifying action on the larval shell. The glutaraldehyde solution used in this study as a fixative was not buffered as is normal practice in electron microscopy and it may therefore have had some slight decalcifying action on the shell. However, the shell valves retained their normal configuration during sectioning and mounting. Glutaraldehyde has a distinct advantage over Bouin's fixative in that it preserves much more of the interstitial and connective tissue in its natural position.

Using serial sections of embedded larvae and whole mounts, pictures of the whole larvae at three selected points during metamorphosis were constructed. Twelve intermediate stages between those selected for detailed study were also examined in order that changes could be explained more fully. All the intermediate stages were found to show a degree of change or development consistent with the rate of development suggested by these three stages that were examined in detail. The three stages selected were the pediveliger, the 2-day-old spat and the 6-day-old spat.

The larvae and early spat used by Cole were obtained from the M.A.F.F. outdoor oyster breeding tanks at Conway (Cole, 1936). These were fixed in Bouin's solution and embedded in paraffin wax. Spat were collected on oyster shells and he found that they could be removed easily from the shell after the surface layers of the shell had been decalcified by the fixative.

DESCRIPTIONS

The terminology referring to the stages of bivalve development is confused by the fact that European and North American authors have tended to use different names to describe the same stage. Despite obvious contradictions, the adjectives 'adult' and 'mature' are frequently used for the advanced stages of larval development just prior to settlement. The term 'pediveliger' proposed by Carriker (1961) to describe the swimming–creeping stage of the clam *Mercenaria mercenaria* L. is considered preferable to either of the terms 'adult' or 'mature' and is applicable to oyster larvae in which the foot and the velum are functional at the same time. It will therefore be used in this report when referring to the larval stage just before settlement. The term 'spat' will be used to refer to the larva after it has attached to the substratum. Larval life is usually considered to cease at attachment, settlement, setting or spatfall; these terms have been used synonymously in the literature.

Thus, the pediveliger constitutes the last truly larval stage; it is fully developed, ready to settle and attach, and equipped to select a suitable substratum. Its behaviour changes from planktonic swimming, characteristic of the veliger, to alternating periods of swimming just above, and crawling upon, the substratum. The organs characteristic of the swimming larva are still present as well as those specifically developed for the transitory crawling phase which precedes attachment.

The size of the larva at this or any other time cannot be used as a criterion for determining the stage of development reached, since there can be great variation at a given stage in development (see Loosanoff, Davis & Chanley, 1966). A morphological examination is necessary to establish the stage of development.

General descriptions of the pediveliger and subsequent stages will refer to the diagrammatic reconstructions of the whole larvae (Figs. 1–3). Where the descriptions confirm or contradict the work of previous authors, reference is made to these authors. Points of detail will be illustrated by plates.

The pediveliger

In general, the pediveliger studied agreed closely with Erdmann's (1934) description of the 'ansatzreifen' or fully developed larva.

When the pediveliger stage is reached (Fig. 1), the shell valves are distinctly unequal, the left valve being the more convex. Pronounced umbones are present to allow for the attachment of the larval muscles and the hinge line, at this stage, is relatively long.

The velum which is the most characteristic organ of the swimming larva is still large and functional in the pediveliger. The crown of the velum is fringed with extremely long cilia (Plate 1 a) but the central area, the apical plate, is covered by relatively short cilia. In the very centre is a small region having a different cellular arrangement from the rest of the velar tissue, and although it was difficult to interpret the structure of this apical organ in detail, it corresponds in position to the structures figured by Erdmann (1934) as the 'scheitel organ'. Cole (1938) does not make a distinction between the apical plate, which is the area of the velum covered by short cilia, and the apical organ itself. Galtsoff (1964) refers to this organ as the 'apical sense organ' although he admits that its function is unknown. It seems likely, however, that it has a sensory function since it is positioned directly over the cerebral ganglion. The precursor of the upper labial palps can be seen as a small, darkly stained area of the apical plate adjoining the oral side of the apical organ. In the fully retracted velum shown in Plate 1 a, the apical plate, part of the edge of the velum and the cerebral ganglion all lie close to the oesophagus. The large velar retractor muscles are attached chiefly to the right valve and serve to withdraw the velum into a position between this valve and the digestive diverticulum, the retractors passing between the lobes of the diverticulum. The two adductor muscles (Fig. 1) are roughly equal in size, the posterior muscle occupying a rather more dorsal position relative to the anterior one.

Characteristically, in the pediveliger the foot (Fig. 1) is a large tongue-

shaped organ which, when relaxed, projects out through the posterior edge
of the valves. The foot is covered by a single layer of ciliated cells enclosing
a matrix of various cell types interspersed with a few muscle fibres. Much
of the foot and its base is made up of the byssus gland which can be
differentiated into various cell types with different staining properties.
A large foot retractor muscle (Fig. 1, Plate 1 *b*) runs diagonally from the
base of the foot to its point of attachment in the dorsal region of the left
valve (Plate 1 *b*).

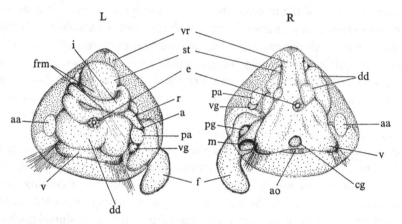

Fig. 1. A semidiagrammatic representation of the pediveliger, drawn from serial sections,
seen from the left side (L) and right side (R).

Key to abbreviations to Figs. 1–3. a, Anus; aa, anterior adductor muscle; ao, apical organ;
bg, byssus complex; cg. cerebral ganglion; dd, digestive diverticulum; e, eye; if, foot;
fr, remnants of foot; frm, foot retractor muscle; i, intestine; lp, labial palp; lg, left gill;
m, mouth; oe, oesophagus; pa, posterior adductor muscle; pg, pedal ganglion; r, rectum;
rg, right gill; st, stomach; v, velum; vr, velar retractor muscles; vg, visceral ganglion.

The actual process of attachment of the larva of *Ostrea edulis* has been
described in detail by Cole & Knight-Jones (1939) and there is no doubt
that the byssus gland provides the cement by means of which the larva
fixes the left valve to the substratum. The structure of the byssus gland
itself has not been described in detail by any author, but Erdmann (1934)
and Galtsoff (1964) show the gland as a simple, roughly spherical structure
occupying a large proportion of the base of the foot and opening by a duct
on to the 'sole' of the foot. The present study showed that it has certain
similarities to the byssus gland of *Mytilus edulis* as described by Brown
(1952), Pujol (1967) and Smyth (1954), but it is composed of a greater
variety of cell types. It is suggested that the byssus gland is likely to be

a complex of glands and storage areas, and a detailed study of this organ is in progress.

The mouth (Fig. 1) which lies just anterior to the base of the foot, between the foot and the velum, is a wide funnel-shaped structure leading into the thick walled, heavy ciliated, oesophagus (Plate 1 a). The oesophagus runs antero-dorsally to the stomach (Fig. 1) which lies on the left side of the larva and contains the gastric shield made up of fused cilia. The lumen of the stomach appears to be continuous with that of the digestive diverticulum, the latter being a large, many-lobed structure occupying much of the internal shell cavity and lying between and around the other viscera. It consists of a mass of large, highly vacuolate cells with numerous intercellular spaces and at this stage bears no resemblance to the tubule structure described by Yonge (1926), which he says is common to larva and adult. Both Yonge and Erdmann describe it as a two-lobed structure, each lobe being connected by a duct to the stomach. The stomach itself has a hemispherical shape, the flattened inner wall being formed by the diverticulum; the complex stomach structure described by Erdmann was not revealed by this study. No obvious caecum or style sac were seen.

The midgut or intestine (Fig. 1) leaves the stomach posteriorly and runs in a loop anteriorly around the left side of the stomach, becoming thin-walled and dilated before bending ventrally and passing backwards as the rectum to terminate at the anus. The latter lies in a postero-dorsal position adjacent to the posterior adductor muscle.

Of the three stages to be described here, it is the pediveliger that reveals the structure of the nervous system most clearly. In the pediveliger, it consists of a cerebral ganglion (Fig. 1), a pair of pedal ganglia and a visceral ganglion. The cerebral and visceral ganglia are not obviously paired, though Erdmann describes all three as paired structures. The large cerebral ganglion lies just below the apical sense organ (Plate 1 a) and, in common with the other ganglia, the cell bodies with the prominent nuclei surround an inner core of nerve fibres. The cerebral ganglion is connected by thickened nerve trunks to each eye (Fig. 1). These nerve trunks are the equivalent of Erdmann's (1934) pleural ganglia but here they did not appear as a typically ganglionic structure.

The eyes (Plate 1 b) lie roughly at the centre of each valve just beneath the shell. Each consists of a pigment cup surrounding a central amorphous mass termed the lens. The cells of the pigment cup have granular cytoplasm and darkly staining nuclei towards their base. The lens appears to be supported by thin strands of cytoplasm extending from the inner ends of the pigment cells. The section shown in Plate 1 b is slightly to one side of the central plane of the eye. In this central plane, the pigment cup is U-shaped, the open end being towards the exterior of the larva.

The paired pedal ganglia lie in the base of the foot (Fig. 1) and nerves from each run into the distal part of the foot. The visceral ganglion is small in the pediveliger but it lies in its definitive position just ventral to the posterior adductor muscle.

No gill rudiments, reported by Erdmann (1934) to be present in the pediveliger, were seen at this stage, and the heart and kidney rudiments, also seen by Erdmann, could not be positively identified, possibly due to the partially contracted state of the larvae.

The 48 h spat

The oyster larva settles by cementing the more convex left valve on to the substratum. At this time the larva displays a slight increase in the ratio of shell width to shell length in anticipation of the adult form.

The highly developed locomotory organs necessary for the pediveliger's active mode of life are much less prominent (Fig. 2). The velum is greatly reduced and is undergoing disintegration. The labial palp precursor lies anterio-dorsal to the mouth and adjacent to the oesophagus as in the pediveliger, and at this stage shows no obvious signs of developing into labial palps. That the apical area of the velum is the precursor of the labial palps was suggested originally by Meisenheimer in 1901 (in *Dreissensia polymorpha*) and again in *Ostrea edulis* by Cole (1938).

Cole maintained that immediately after attachment the apical area of the velum sank through the other tissues to reach the position occupied by the labial palps, but it is clear from Plate 1 *a* that the apical area of the velum, or apical plate, normally lies in this position when retracted. It seems likely therefore that the labial palp precursor is brought into this position as a result of the shrinkage of the velum and its retractor muscles. The latter, though still recognizable, are partly resorbed. It is certain that the labial palp precursor does not sink into the interior of the velum as stated by Cole.

The other highly characteristic organ of the pediveliger, the foot, is also much reduced (Fig. 2), the distal portion having been largely resorbed, and the retractor muscle lost. The two large pedal ganglia are still evident. Prominent connectives run forward from these to the centrally positioned cerebral ganglion which lies just anterior to the oesophagus. The visceral ganglion, lying on the ventral edge of the posterior adductor muscle, is large and elongate, running almost the entire length of the adductor. Here we have the beginning of a trend which results in the enlargement of the visceral ganglion relative to the cerebral and pedal ganglia.

The oesophagus (Fig. 2) running between the cerebro-pedal connectives, is very long and narrow and the mouth has now moved to a more anterior position and opens antero-ventrally. Cole (1938) considers that this change in position is due to the mouth's being closely bound up with the velar remnants as they sink more deeply into the body and he states that for a short while the opening faces inwards instead of outwards and is blocked by the velar remnants which may thus be ingested. The present work revealed no evidence that the remnants were in fact ingested. The shift of the position of the mouth also causes a sharp bend in the oesophagus so that the first part runs dorsally before turning forwards into the stomach. The

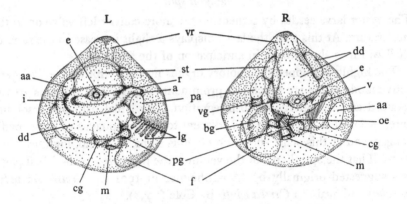

Fig. 2. A semidiagrammatic representation of the 48-h-old spat, drawn from serial sections, seen from the left side (L) and right side (R). For key see Fig. 1.

rest of the gut is very similar to that of the pediveliger. The digestive gland (or diverticulum), as a result of the distintegration of the velum and foot, occupies an even greater proportion of the body than it did in the pediveliger, and shows signs of definite division into small lobules separated from each other by a few strands of connective tissue. It is thus beginning to resemble the tubular arrangement described by Yonge (1926). At this stage the posterior adductor muscle is showing a very slight increase in size relative to the anterior one. The eyes are still prominent and they lie very close to the inner surface of the shell.

A newly developed structure in the 48 h spat is the gill. This is a paired structure in the adult, but, as has been noted by Prytherch (1934), the gills in the early spat are better developed on the left side than on the right. Not surprisingly, therefore, only one gill was seen at this stage, lying on the left side of the body (Plate 1 c) and formed from an outgrowth of the mantle wall. It is developed in a midlateral position and in the 48 h spat is composed of six processes. Erdmann figures a similar row of six extremely knob-like processes on the left side of the earlier pediveliger larva.

The 6-day-old spat

At this stage, the spat has almost lost the characteristic larval organs of locomotion (Fig. 3). The ciliated crown of the velum is highly degenerate, the remnants (Plates 1*d*, 2*a*, *b*) lying dorsally on the right side of the body. The cells of this residual tissue are very large and ciliated and the nuclei are not as distinct as in the functional velum. The complex system of basal bodies and rootlets (Galtsoff, 1964) which provide a system for ciliary co-ordination in the functional velum can be clearly seen in these large cells. Phagocytes can be seen in the vicinity of the degenerating velum; phagocytosis may therefore play a part in its removal.

The apical plate (Plate 2*a*) of the degenerate velum (Plates 2*a*, *b*) lies directly above the mouth and by this stage shows a definite development into a two-lobed structure constituting the developing upper labial palps. Few, if any, of the cells of the apical organ itself (Plate 2*a*) are incorporated in the developing labial palps; these latter are derived in the main from that region of the apical plate that lies between the apical organ and the mouth. A sagittal section through the developing labial palp (Plate 2*b*) shows this as a thickened fold of densely staining tissue. Sections made in planes lateral to the midline demonstrate that this tissue progressively invaginates and separates to form tentacular structures (Plate 2*c*) which Cole states will give rise to the outer palps of the adult oyster. Cole, however, found this development to take place in younger spat only 2–3 days after attachment.

Lying adjacent to the labial palps is the cerebral ganglion (Plate 2*b*) which is much less prominent than in previous stages (cf. Plate 1*a*). The small pedal ganglia (Plate 2*a*) which, according to Cole, disappear after about 24 h of attached life, are still present here in the base of the foot, and the visceral ganglion (Fig. 3), which is now the largest, lies below the posterior adductor muscle. The foot has very largely disappeared (Plate 2*a*), the remnants forming a mass in the midventral part of the body which contains numerous phagocytic cells, several intercellular spaces and the remains of the byssus complex.

The posterior adductor muscle has become slightly larger than the anterior and now occupies a more median position, heralding the monomyarian condition of the adult (see Fig. 3).

The length of the oesophagus, which is now short and straight, has decreased in relation to the rest of the gut, but the relative length of the narrow intestine has increased. The gills have enlarged, the left one (Plate 2a) now occupying a large proportion of the mantle cavity. The developing right gill is much smaller and is composed of only three lobes at this stage (Plate 2d).

The eyes are still present in the 6-day-old spat (Fig. 3) but are less prominent than in the younger spat.

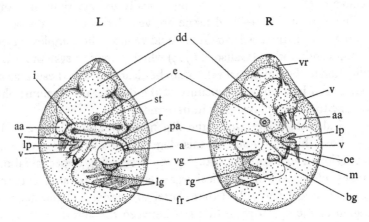

Fig. 3. A semidiagrammatic representation of the 6-day-old spat, drawn from serial sections, seen from the left side (L) and right side (R). For key see Fig. 1.

DISCUSSION

Cole (1938) chose to describe the fate of selected organs during metamorphosis. By means of a detailed analysis of his descriptions, it is possible to correlate the state of development of the various organs at specific instances in time so that his data can be related to this work.

Fig. 4 shows that where these studies overlap; the same general changes occurred at settlement and subsequently. There are differences, however, and the most marked perhaps is in the rate of metamorphosis. Metamorphosis generally took place more slowly in the larvae investigated in this study, kept under artificial laboratory conditions. For example, Cole found that the velum had disappeared completely at 48 h after settlement and that the foot was completely resorbed before the spat was 72 h old. In the present study, remnants of both foot and velum were still present 6 days after settlement and metamorphosis was therefore not completed even after 6 days.

These discrepancies are probably due to some extent to the reduced growth rate that occurs in spat settled and grown on filter paper. The reduction in growth rate and rate of metamorphosis that occurs in spat on filter paper has also been noted recently in rearing studies at the Fisheries Experimental Station, Conway (Helm, personal communication). Moreover, the dimensions of the larvae used by Cole and obtained from semi-natural habitats were certainly greater age for age than those of the larvae used here. Another factor that could have contributed to these differences is the temperature of development since the controlled temperature of 20 °C used in this work would have differed considerably from the presumably fluctuating and probably lower temperature of the external tanks

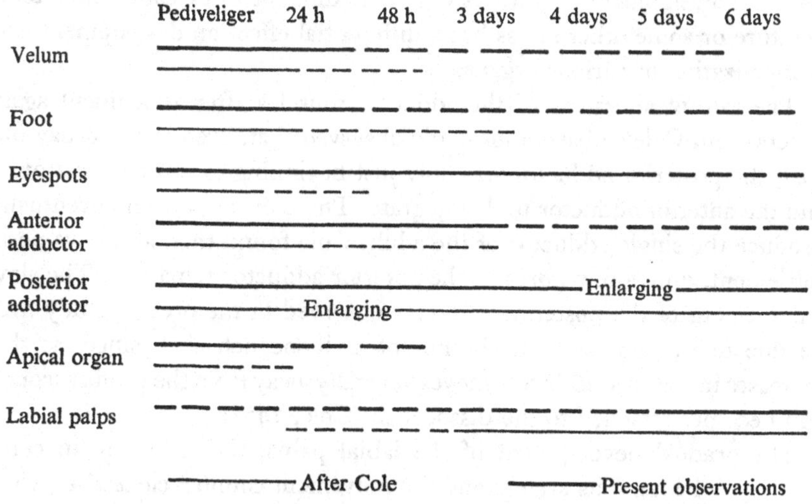

Fig. 4. A comparison of changes at metamorphosis between the present study and that of Cole (1938).

from which Cole obtained his larvae. However, it is difficult to explain the discrepancy on the basis of temperature alone, since the value of 20 °C is one that is considered to be roughly optimal for growth under laboratory conditions (Walne, 1956, 1966). It is possible of course that different processes have different optimal temperature requirements, and a fluctuating temperature might possibly promote better growth.

A further indication of the slow growth rate on filter paper comes from the poor development of the dissoconch shell in the spat. Cole observed the beginnings of dissoconch formation 48 h after settlement. A similar time for the initial development was observed in this work, but in contrast with Cole's observations, very little increase in the size of the dissoconch occurred subsequently.

The structure of the eyespots agrees very closely with the descriptions given by Erdmann (1934) and confirmed by Cole (1938). There seems little doubt that Cole was correct in rejecting Prytherch's (1934) view that these structures were not eyes, since their structure, position and innervation suggest that they are. However, both Prytherch and Cole found that the eyes were lost within a day or so of setting, whereas we found them still present 6 days after setting when both the foot and velum had disappeared. Furthermore, Cole's descriptions show that they were lost before the velum and foot disappeared and before any evidence of labial palp development was seen; we found exactly the reverse. Such a discrepancy in the relative rates of development of various organs is difficult to explain unless temperature or some other stress has a differential effect on development and reorganization of various organs.

The rate of alteration of the adductor muscles after attachment again differs from Cole's observations. We observed that, even in the 6-day-old spat, the posterior adductor was only just beginning to show hypertrophy and the anterior adductor to disintegrate. These changes would eventually produce the single adductor of the adult. Cole found that, at 3 days post-settlement, only a few fibres of the anterior adductor remained. The slow enlargement of the posterior adductor observed in the present study may be due to the almost total absence of a dissoconch shell since, as this increases in size, the adductor moves ventrally away from the prodissoconch and becomes attached to the dissoconch (Cole, 1938).

The gradual development of the labial palps, the reduction in complexity of the nervous system and the movement counter-clockwise (when viewed through the right valve) of the mouth, all occur in roughly the same time sequence as that described by Cole, but at a later age, due to the generally reduced rate of development. Although labial palp development was generally slower than that observed by Cole, the initial stages of this process were observed at the earlier, pediveliger, stage. The slow rate at which metamorphosis occurred as compared to that recorded by previous authors may have been due to non-optimal conditions but may also be an indication of natural variation in rates of development of different broods of larvae (Loosanoff & Davis, 1963; Loosanoff et al. 1966).

Because of the differences mentioned earlier in the timing of changes in some organs relative to others, the question arises of the suitability of laboratory-reared larvae for this type of study. The advantages of using such material are that the problem of identifying the larvae in a plankton sample is removed and a far more precise estimation of the age of the spat used in the investigation is possible. On the other hand, the normal process of development may be affected in some way by the artificial environment in which they are reared, and by the artificial substrata on which they settled in the laboratory. It is possible, of course, that the small, but significant, differences that exist between the present and previous work fall within the limits of natural variation. If this is so, the degree of variation would seem to be rather too large to enable one to do what this work was intended to do, that is, to describe a particular stage in detail and to regard this description as being typical of all larvae reaching that stage. It would seem, therefore, that it is essential to extend this study to produce a comparison between wild larvae and ones reared and induced to settle artificially under various laboratory conditions, in order to establish how far such differences are due to natural variation on the one hand or to environmental differences on the other.

We are grateful to Mr B. S. Spencer and Mr M. Helm of M.A.F.F. Laboratories, Conway, for supplying larvae for this study, to Dr J. A. Nott for advice on preparative techniques and to Mr D. Williams for assistance with the photography. The work was carried out during the tenure of an N.E.R.C. Advanced Course Studentship held by R. W. H.

REFERENCES

BAYNE, B. L. (1969). The gregarious behaviour of the larvae of *Ostrea edulis* L. at settlement. *J. mar. biol. Ass. U.K.* **49** (2), 327–56.

BROWN, C. H. (1952). Some structural proteins of *Mytilus edulis*. *Q. Jl microsc. Sci.* **93**, 487–502.

CARRIKER, M. R. (1961). Interrelation of functional morphology, behaviour and autecology in early stages of the bivalve, *Mercenaria mercenaria*. *J. Elisha Mitchell scient. Soc.* **77**, 168–241.

COLE, H. A. (1936). Experiments in the breeding of oysters (*Ostrea edulis*) in tanks, with special reference to the food of the larva and spat. *Fishery. Invest., Lond. Ser. II.* **15** (4), 1–25.

COLE, H. A. (1938). The fate of larval organs in the metamorphosis of *Ostrea edulis*. *J. mar. biol. Ass. U.K.* **22**, 469–84.

COLE, H. A. & KNIGHT-JONES, E. W. (1939). Some observations and experiments on the setting behaviour of larvae of *Ostrea edulis*. *J. Cons. perm. int. Explor. Mer.* **14**, 86–105.

CRISP, D. J. (1967). Chemical factors inducing settlement in *Crassostrea virginica* (Gmelin). *J. Anim. Ecol.* **36**, 329–36.

DAVAINE, C. (1853). Recherches sur la génération des huîtres. *C. r. Séanc. Soc. Biol.* **4** (1) 1852, 297–339.

ERDMANN, W. (1934). Uber die Entwicklung und die Anatomie der 'Ansatzreifen' Larve von *Ostrea edulis*. *Wiss. Meeresunters, Abt. Helgoland*, N.F. **19**, 1–25.

294   R. W. HICKMAN AND LL.D. GRUFFYDD

GALTSOFF, P. S. (1964). The American oyster *Crassostrea virginica*, Gmelin. *Fishery Bull. Fish Wildl. Serv. U.S.* **64**, 1–480.

GRUFFYDD, LL.D. & BAKER, W. F. (1969). An integrated multiple unit controlled temperature system for sea water aquaria. *Lab. Prac.* **18**, 300–4.

HUXLEY, T. H. (1883). Oyster and the oyster question. *Engl. illust. Mag.* 47–55, 112–21.

JACKSON, R. T. (1888). Development of the oyster. *Proc. Boston. Soc. nat. Hist.* **23**, 531–56.

KORRINGA, P. (1952). Recent advances in oyster biology. *Q. Rev. Biol.* **27**, 266–308, 339–65.

LOOSANOFF, V. L., DAVIS, H. C. & CHANLEY, P. E. (1966). Dimensions and shapes of larvae of some marine bivalve molluscs. *Malacologia* **4**, 351–435.

LOOSANOFF, V. L. & DAVIS, H. C. (1963). Rearing of bivalve molluscs. *Adv. mar. Biol.* **1**, 1–136.

MEISENHEIMER, J. (1901). Entwicklungsgechichte von *Dreissensia polymorpha* Pall. *Z. wiss. Zool.* **69**, 1–137.

PRYTHERCH, H. F. (1934). The role of copper in the setting, metamorphosis and distribution of the American oyster, *Ostrea virginica*. *Ecol. Monogr.* **4**, 47–107.

PUJOL, J. P. (1967). Le complexe byssogene des Mollusques Bivalves. Histochimie comparée des secretions chez *Mytilus edulis* L. et *Pinna nobilis* L. *Bull. Soc. linn. Normandie* **8**, 308–32.

RICHARDSON, K. C., JARRETT, L. & FINKE, E. H. (1960). Embedding in epoxy resins for ultra-thin sectioning in electron microscopy. *Stain Technol.* **35**, 313–30.

RYDER, J. A. (1884). The metamorphosis and post-larval stage of development of the oyster. *Rep. U.S. Commnr Fish.* **10**, 779–91.

SABATINI, D. D., BENSCH, K. & BARNETT, R. J. (1963). Cytochemistry and electron microscopy. The preservation of cellular ultrastructure and enzymatic activity by aldehyde fixation. *J. Cell Biol.* **17**, 19.

SMYTH, J. D. (1954). A technique for the histochemical demonstration of polyphenol oxidase and its application to egg shell formation in helminths and byssus formation in *Mytilus*. *Q. Jl microsc. Sci.* **95**, 139–52.

STAFFORD, J. (1913). *The Canadian oyster*. Commission of Conservation, Ottawa.

WALNE, P. R. (1956). Experimental rearing of larvae of *Ostrea edulis* L. in the laboratory. *Fishery. Invest., Lond.*, Ser II, **20**, 1–23.

WALNE, P. R. (1966). Experiments in the large-scale culture of the larvae of *Ostrea edulis* L. *Fishery. Invest., Lond.*, Ser II, **25** (4), 1–53.

YONGE, C. M. (1926). Structure and physiology of the organs of feeding and digestion in *Ostrea edulis*. *J. mar. biol. Ass. U.K.* **14**, 295–386.

TRANSIENT LARVAL GLANDS
IN *PALAEMONETES*

J. H. HUBSCHMAN
Wright State University, Dayton, Ohio, USA

INTRODUCTION

In spite of the impetus provided by research on juvenile hormonal control in insects, there is no detailed information on the endocrine control of metamorphosis in crustaceans. This is, however, a potentially fruitful area for the study of fundamental processes of growth and development.

Passano (1960) reviewed the available evidence and finds it not unreasonable to assume, by analogy with known control mechanisms of insect metamorphosis, the existence of a substance similar to insect juvenile hormone.

Insect metamorphosis has been attributed to the interaction of secretions of the moulting gland (Prothoracic gland) and a decrease in the titre of a juvenile hormone secreted by the *corpora allata*. Stimulation of the prothoracic gland (by secretory activity of the central nervous system) in the presence of a sufficiently high concentration of juvenile hormone results in pupation. Activity in the absence of juvenile hormone results in metamorphosis to the adult form.

As yet there is no positive evidence of a juvenile hormone in crustaceans. Schneiderman & Gilbert (1958) surveyed a number of organisms in an attempt to detect possible sources of juvenile hormone. Their assay revealed no juvenile hormone activity in extracts of *Palaemonetes vulgaris*. If it is the absence of juvenile hormone that allows expression of the adult form, one would expect to find little or none in adult shrimp. Passano (1961) discussing the problem of crustacean metamorphosis suggested that a careful histological study of crustacean larvae might disclose glandular activity related to ecdyseal stages. Up to that time, all knowledge of crustacean developmental endocrinology was based upon the work of Pyle (1943) on *Pinnotheres* and *Homarus*, and the embryological study of *Crangon* by Dahl (1957). Since then the development and function of neurosecretory sites in larval *Palaemonetes* has been studied in detail (Hubschman, 1963).

The apparent lag between experimental work on insect and crustacean metamorphosis is due to several factors which relate to their size and availability. Decapod crustaceans exhibiting larval development are usually very small at the time of metamorphosis. Most post-larvae are but a few millimetres in total length.

Experimentation on larval development in shrimp was quite limited until post-larvae were successfully reared on a routine basis in the laboratory (Broad, 1957). Since then, a great many descriptive and experimental studies have been conducted.

The latter have been confined essentially to two approaches. One has been concerned with the influence of environmental (ecological) factors such as temperature, light and salinity. The other has been the experimental extirpation of larval organs. Usually this has involved the eyestalks, which are known to contain neurosecretory sites having relationship to several physiological functions in adult organisms. My earlier experiments have suggested that metamorphosis is not under eyestalk control in larval *Palaemonetes*, and that we must look elsewhere for potential sources of regulatory factors (Hubschman, 1963).

In adult crustaceans, several sources of moult-inhibiting hormone (x organs) and a release site (sinus gland) are found in the eyestalks. The absence of ganglionic x organs in the eyestalks of larval *Palaemonetes* and the failure of the eyestalk removal to interfere with either moulting or metamorphosis suggests that the control mechanisms are different in larvae and adults. The moulting process (limited only by the maximum rate of the combined physiological processes involved) apparently proceeds without inhibition in larval shrimp.

The process of metamorphosis may, on the other hand, be under the control of some substance produced within the body proper. Progressive development and metamorphosis may take place as inhibition is removed due to the decreasing activity of some gland or tissue during larval life. If the existence of a crustacean juvenile hormone may be hypothesized, then it may follow that *Palaemonetes* and most other decapods have a larval phase because of the presence of the hormone at hatching. In addition, prothoracic-like organs, the Y-organs, are known in many Malacostraca (Gabe, 1953). These are non-nervous secretory organs which are located in the lateral thoracic region and function in the control of moulting. No Y-organ is known in larval *Palaemonetes*. Among crustacea in which the Y-organ is present, it may occupy either of two locations. The Y-organ occupies the maxillary segment in crustaceans whose adult excretory structures are in the antennal segments (antennal glands). Among crustaceans having the adult excretory organs in the maxillary segment (maxillary glands) the Y-organ is found in the antennal segment.

The objective of the work described here was to determine if structures analogous to the insect prothoracic glands are present in larvae of the shrimp *Palaemonetes*. Five species of this genus inhabit eastern North America. These are *P. vulgaris* (Say), *P. pugio* Holthuis, *P. intermedius* Holthuis, *P. paludosus* (Gibbes), and *P. kadiakensis* Rathbun. The first three species are marine and the latter two inhabit fresh water. Larval development has been described for all five species (Broad, 1957; Broad & Hubschman, 1962, 1963; Faxon, 1879; Dobkin, 1963).

DESCRIPTION OF GLANDS

A pair of glands has been discovered in the midthoracic region of *Palaemonetes* larvae that exhibit rather striking appearance with various histological techniques. A most interesting feature of these glands is their transient nature. They develop early in larval life and appear active in all larval forms up to and including the early form VI larva (the last stage). During the last phase of the terminal larval form (VI), the glands begin to deteriorate. At metamorphosis (post-larva), the glands are absent, leaving an empty space where the glands once were. These glands are compact oval structures located in lateral ventral thorax near the cuticle at the base of the first maxilla. They are not to be confused with the maxillary (excretory) glands in the base of the second maxilla, which are also present in the larvae but are very different in appearance. The glands of the first maxilla (hereafter larval glands) lie at the point where the coxopodite promotor muscle crosses the coxopodite adductor. Each gland is vertically elongate and applied to the posterior aspect of the coxopodite adductor and coxopodite abductor muscles just above the point where they are crossed by the coxopodite promotor when viewed in frontal section. A small duct may be traced from the centre of the gland to the tip of the maxilla. The glands are pictured in Plate 1.

Allen (1893) briefly mentioned these glands in his description of nephridia of the larva of *P. varians*. His impression at that time was that these were simply another pair of 'spherical' glands abundant in the gill axis of adult shrimp.

These glands are present in all five North American species. The glands are well developed at the time of hatching and the history of progressive development with subsequent degeneration is essentially the same throughout the group. Naturally, in *P. paludosus*, which exhibits abbreviated larval development (Dobkin, 1963), the process is correspondingly shortened.

Since crustaceans are notorious for the occurrence of the so-called tegumental glands, I made detailed comparisons of the larval and tegumental glands. The latter are generally irregular in form in *Palaemonetes* while the larval glands have smooth oval outlines. Both are composed of short pyramidal cells. With various techniques (Hubschman, 1963) they may be compared as follows:

Tegumental glands

Azan:- cytoplasm, yellowish; nuclei, red; droplets, blue. PAF (Paraldehyde-Fuchsin):- cytoplasm, green; nuclei, orange; fine droplets, deep purple; large droplets, clear. CHP (Chrome-Haematoxylin-Phloxine):- cytoplasm, gray; nuclei, red; droplets, red. PAS (Periodic acid-Schiff):- cytoplasm, pink, red or purple.

Larval glands

Azan:- cytoplasm, pale blue; nuclei, red. PAF:- cytoplasm, orange; nuclei, deep orange; evenly distributed deep purple granules. CHP:- cell evenly stained deep purple throughout. PAS:- cell evenly stained deep purple throughout.

DISCUSSION

The role that these glands play in larval life is still unknown. The detailed work of Stevenson (1964, 1967) on isopod crustaceans suggests that certain glands function in cuticular tanning and others may assist the feeding process. In this case, the occurrence of a single pair of larval glands in *Palaemonetes* does not offer much support for the notion that they function in cuticular tanning. On the other hand they may well play an important role in the feeding process. With the various histological techniques used, however, I could not demonstrate activity related to this function. Likewise, I could not establish a correlation between glandular activity and the moulting cycle. The larval glands may well be related to the process of metamorphosis. The fact that they appear active in the first larval phase (Form I) and disappear at metamorphosis, strongly suggests a functional relationship to larval development or at least some process essential to larval life.

SUMMARY

Structures analogous to the prothoracic glands of insects responsible for secreting a juvenizing hormone were sought in the larvae of *Palaemonetes*.

A pair of compact oval glands, which develop early in larval life but disappear after stage VI, were found in the midthoracic region. These transient larval glands differed histologically and by their staining reactions from tegumental glands, and showed no cycle of activity correlated with the moult. It is suggested that they have some function essential to the larval life.

This work was supported in part by National Science Foundation Grants no. 4965 and no. 8314. This support is gratefully acknowledged. The marine shrimp studied in this work were reared at the Duke Marine Laboratory, Beaufort, North Carolina. I thank Dr C. G. Bookhout and Dr J. D. Costlow, Jr., and their staff for the assistance offered during my visits to Beaufort. I am indebted to Dr L. S. Putnam, Director of the F. T. Stone Laboratory, Put In Bay, Ohio for the use of his facilities in the preparation of this manuscript.

REFERENCES

ALLEN, E. J. (1893). Nephridia and body-cavity of some decapod crustacea. *Q. Jl microsc. Sci.* (NS) **34**, 403–26.

BROAD, A. C. (1957). Larval development of *Palaemonetes pugio* Holthius. *Biol. Bull. mar. biol. Lab., Woods Hole* **112**, 144–61.

BROAD, A. C. & HUBSCHMAN, J. H. (1962). A comparison of larvae and larval development of species of Eastern US Palaemonetes with special reference to the development of *Palaemonetes intermedius* Holthuis. *Am. Zool.* **2**, 394–5.

BROAD, A. C. & HUBSCHMAN, J. H. (1963). The larval development of *Palaemonetes kadiakensis* M. J. Rathbun in the laboratory. *Trans. Am. microsc. Soc.* **82**, 185–97.

DAHL, E. (1957). Embryology of the *x*-organs in *Crangon allmanni*. *Nature, Lond.* **179**, 482.

DOBKIN, S. (1963). The larval development of *Palaemonetes paludosus* (Gibbes, 1850) (Decapoda, Palaemonidae), reared in the laboratory. *Crustaceana* **6**, 41–61.

FAXON, W. A. (1879). On the development of *Palaemonetes vulgaris*. *Bull. Mus. comp. Zool. Harv.* **5**, 303–30.

GABE, M. (1953). Sur l'existence, chez quelques Crustacés Malacostraces, d'un organe comparable a la gland de la mue des Insectes. *C. r. hebd. Séanc. Acad. Sci., Paris* **237**, 1111–13.

HUBSCHMAN, J. H. (1963). Development and function of neurosecretory sites in the eyestalks of larval *Palaemonetes* (Decapoda:Natantia). *Biol. Bull. mar. biol. Lab., Woods Hole* **125**, 96–113.

PASSANO, L. M. (1960). Molting and its control. In *Physiology of Crustacea*, Vol. I (ed. T. H. Waterman). New York: Academic Press.

PASSANO, L. M. (1961). The regulation of crustacean metamorphosis. *Am. Zool.* **1**, 89–95.

PYLE, R. W. (1943). The histogenesis and cyclic phenomena of the sinus gland and *x*-organ in crustacea. *Biol. Bull. mar. biol. Lab., Woods Hole* **85**, 87–102.

SCHNEIDERMAN, H. A. & GILBERT, L. I. (1958). Substances with juvenile hormone activity in crustacea and other invertebrates. *Biol. Bull. mar. biol. Lab., Woods Hole* **115**, 530–5.

STEVENSON, J. R. (1964). Development of the tegumental glands in the pillbug, *Armadillidium vulgare* in relation to the molting cycle. *Trans. Am. microsc. Soc.* **83**, 252–60.

STEVENSON, J. R. (1967). Mucopolysaccharide glands in the isopod crustacean *Armadillidium vulgare*. *Trans. Am. microsc. Soc.* **86**, 50–7.

SPICULE FORMATION AND CORROSION IN RECENTLY METAMORPHOSED *SYCON CILIATUM* (O. Fabricius)

W. C. JONES

Department of Zoology, University College of North Wales, Bangor, Wales, UK

INTRODUCTION

At the start of the present century Maas (1900, 1904, 1906, 1907, 1910) published the results of some pioneering experiments involving the larvae (and adults) of the calcareous sponges *Sycon* (= *Sycandra*) *setosa* and *S. raphanus*. At the appropriate season the larvae emerge in great numbers from the osculum of the parent sponge and Maas pipetted sea water containing the larvae into watchglasses in sufficient amount to enable the sea water to come into contact with the undersurface of a coverslip resting on its four corners in each watchglass. He then floated the watchglass on sea water contained in a light- and air-tight aquarium tank (forming a primitive, but effective, constant-environment cabinet) and left them for a day or so, after which many larvae had settled on the coverslips and could be examined with the highest powers of the optical microscope merely by transferring the watchglass to the microscope stage. By means of this simple technique Maas contributed much to our knowledge of the settlement, metamorphosis and development of calcareous sponges. Maas's most significant observations, however, concern his experiments with carbonate-free and lime-free sea waters. Initially he prepared the carbonate-free sea water by evaporating normal sea water to dryness and adding distilled water to the residue (Maas, 1904). In the process calcium carbonate was precipitated and did not redissolve, but calcium sulphate remained in the resulting solution. Later Maas (1906) made up artificial sea waters lacking either carbonate or both calcium and carbonate by mixing salts (NaCl, KCl, $MgCl_2$, $MgSO_4$, $CaSO_4$) in the required proportions. Using the method outlined above he found that in these artificial media the metamorphosing larvae and juvenile spongelets produced no spicules composed of calcite. Instead, organic 'spiculoids' were formed by the sclerocytes, and because some of these exhibited the normal regular spicule form, Maas (1904) concluded that calcite played no part in the determination of spicule form. The form was the result of an organic process, independent of the nature of the inorganic material of which the spicules are largely composed.

[301]

When the spongelets were transferred from normal sea water to the artificial media lacking carbonate or lime, Maas observed that the existing spicules became corroded, and that those in contact with living tissue tended to dissolve at a faster rate than spicules that had fallen out, or were otherwise isolated from the spongelets. The spongelets generally underwent involution (retrogression) such that their tissues shrank inwards leaving an exposed framework of interlocking spicules. Such isolated spicules, and spicules which had fallen out, maintained their uncorroded appearance much longer than the spicules within the shrinking tissues or in more normal juveniles alongside. Maas hence introduced the concept of 'spiculoclastic activity'. The process of spicule corrosion was a physiological one involving cellular activity, and not a direct chemical action of the ambient medium on the spicule calcite. This conclusion was supported by the additional observation that the cells tended to acquire granules stainable with aniline blue as the spicules corroded away (Maas, 1906); the granules first appeared in the pinacocytes and later in the choanocytes.

It was clear that in carbonate-free sea water the spongelets were unable to furnish sufficient carbonate for spicule production from the carbon dioxide produced by their own respiration, but Maas also concluded that only traces of calcium and carbonate were required in the water surrounding the spongelets.

Last year (June, 1968) I decided for two main reasons to reinvestigate the effect of rearing spongelets in artificial media of various compositions. First, the nature and source of the organic matter forming Maas's spiculoids was problematical because I had not been able to confirm the existence of organic matter within the spicule calcite (Jones, 1967; Jones & James, 1969). Should the spiculoid material be considered as the equivalent of the organic spicule sheath, perhaps overabundantly secreted by the sclerocytes in the carbonate- and lime-free media? Maas himself had observed in these media a thickening and hardening of the mesohylial matrix, the material between the pinacoderm and choanoderm which embeds the three-rayed spicules (triacts). Secondly, Maas was unable to state precisely just how much calcium and carbonate were required for normal spicule production. The volumes and concentrations used were much too small for quantitative analysis at that time. But what are the threshold levels of calcium and carbonate needed? Is there a calcium-pumping mechanism as powerful as the sarcoplasmic reticulum of a vertebrate striated muscle fibre? This can concentrate calcium from solutions as dilute as 0.02 micro-molar and less (Weber, Herz & Reiss, 1963).

MATERIALS AND METHODS

Large and medium-sized specimens of *S. ciliatum* (O. Fabricius) which were collected during the period from 26 June to 15 July in 1968 and 1969 were observed to liberate larvae when brought back to the laboratory. They were brushed well to remove adhering detritus, transferred to clean sea water and cut transversely into 1–2 mm thick slices. These were distributed at the bottom of small bowls containing 150 ml sea water, whereupon the larvae were rapidly released in large numbers. They swam upwards to the surface at first and could be dispersed in fresh sea water by decanting the surface layers in which most of the larvae swam. The fresh sea water containing larvae was then stirred and poured into a series of plastic Petri dishes (diameter 9 cm) on the bottoms of which were laid coverslips (22 × 22 mm) covering as much of the area as possible without overlap. The lids of the dishes were then replaced and the dishes stored in complete darkness in a constant temperature cabinet at 14 °C. Complete darkness ensured that the larvae maintained a random distribution in each dish. After 48 h some larvae had settled on the coverslips, which could be carefully removed if required. Larvae might remain swimming for several more days, however, before settling. The attachment of the already settled larvae progressively became stronger and this lessened the risk of dislodgment as the coverslips were removed. Sometimes coverslips were floated on the surface, because metamorphosis of larvae settled in the surface film frequently occurred. The coverslips were mostly soaked in sea water for several weeks beforehand, but it is doubtful if this made them more suitable for settlement than the clean coverslips also tried. Doubtless there is an optimal concentration of larvae, too many quickly fouling the water and thereby inhibiting settlement, and too few resulting in insufficient numbers per coverslip for the purposes of the experiment. However, little difficulty was experienced in obtaining many coverslips with numbers of attached spongelets ranging from some six to 50 or more per coverslip. The spongelets could be kept for many weeks simply by decanting the water and replacing it with fresh periodically. Juveniles a few weeks old survived a period of 3 weeks at 16 °C without change of water.

Three types of experimental procedure were adopted. For the first, pairs of coverslips with attached, metamorphosing larvae lacking spicules were employed. Each pair was drained against blotting paper and floated, spongelets downwards, on the surface of 20 ml of an experimental solution contained in a small (6 cm in diameter) pyrex glass Petri dish. The dishes were covered and maintained in a constant temperature cabinet at 14 °C for

48 h. One coverslip from each pair was then immersed in absolute alcohol to fix the spongelets, while the other was rinsed in sea water and then transferred to fresh sea water for a further 48 h before fixing. The two coverslips were washed in distilled water (buffered to pH 7.4), immersed in ammoniacal picrocarmine for 30 min, dehydrated, cleared and mounted on a slide in DPX. Later the spongelets were examined under the microscope and compared with controls fixed at the start of the experiment. The immersion of one coverslip in sea water after the experimental solution enabled one to test whether the secretory mechanism had been permanently harmed.

The second type of experiment was similar, except that older juveniles which already possessed spicules were employed. Pairs of coverslips bearing these were first immersed in a bicarbonate-deficient medium (pH about 7) for 24 h in order to corrode strongly the existing spicules. The pairs were then rinsed in sea water and floated on the experimental solutions as before. Again one coverslip of a pair was given a subsequent immersion in sea water after 48 h in the experimental solution. As will be seen below the spicules forming during this experiment were readily distinguishable from the remnants of pre-existing spicules.

The third type of experiment was designed to test whether spicule corrosion did occur faster in spicules when in contact with living tissue. Pairs of coverslips bearing juveniles with spicules were selected and one member of each pair was immersed in absolute alcohol for a few minutes, then rinsed in sea water and floated alongside its partner on the experimental solution at 14 °C. After a suitable period of time (12 h to 4 days depending on the solution) the two coverslips were immersed in absolute alcohol, and stained and mounted as before. In this way the rate of spicule corrosion in fixed juveniles could be compared with that in living juveniles of roughly similar stage of development.

The experimental media were prepared by mixing in the requisite proportions solutions of sodium chloride, potassium chloride, calcium chloride, magnesium chloride, sodium sulphate and sodium bicarbonate, all isotonic with sea water. For artificial sea water the proportions used were the same as those given by Pantin (1946, p. 64), except that an equal amount of sodium chloride was added in place of the sodium bromide. For respectively calcium-deficient and bicarbonate-deficient solutions the proportions used were as in Table 1. Certain volumes of respectively calcium chloride and sodium bicarbonate were omitted and replaced by the same volumes of sodium chloride.

Two separate sets of stock mixtures were made up, one in 1968 and the

Table 1. Composition of the experimental solutions

Description	Solution no.	Volumes (ml) of isotonic solutions/1000 ml						Molarity (approximate)		pH			
		NaCl	KCl	$CaCl_2$	$MgCl_2$	Na_2SO_4	$NaHCO_3$	Ca^{2+}	HCO_3^-	1968 (20.5 °C)	16 July 1969 (23 °C)	29 July 1969 (19 °C)	After use 1969 (17.5 °C)
Normal sea water								10^{-2}	25×10^{-4}	7.81	7.70	7.75	7.80
Artificial sea water	1	740.65	18.05	28.0	145.7	63.0	4.6	10^{-2}	25×10^{-4}	7.70	7.77	7.77	7.82
$CaCl_2$ replaced by NaCl	2	740.65 + 28.0	18.05	0	145.7	63.0	4.6	0	25×10^{-4}	7.80	7.60	7.64	7.92
	3	740.65 + 27.72	18.05	0.28	145.7	63.0	4.6	10^{-4}	25×10^{-4}	7.80	7.77	7.79	7.95
	4	740.65 + 26.6	18.05	1.4	145.7	63.0	4.6	5×10^{-4}	25×10^{-4}	7.80	7.86	7.84	7.92
	5	740.65 + 25.2	18.05	2.8	145.7	63.0	4.6	10^{-3}	25×10^{-4}	7.75	7.68	7.69	7.92
	6	740.65 + 14.0	18.05	14.0	145.7	63.0	4.6	5×10^{-3}	25×10^{-4}	7.70	7.65	7.69	7.85
$NaHCO_3$ replaced by NaCl	7	740.65 + 4.6	18.05	28.0	145.7	63.0	0	10^{-2}	0	5.65	5.59	5.21	5.50
	8	740.65 + 4.4	18.05	28.0	145.7	63.0	0.2	10^{-2}	10^{-4}	6.65	6.50	6.46	6.49
	9	740.65 + 3.68	18.05	28.0	145.7	63.0	0.92	10^{-2}	5×10^{-4}	7.25	7.08	7.04	6.90
	10	740.65 + 2.76	18.05	28.0	145.7	63.0	1.84	10^{-2}	10^{-3}	7.55	7.31	7.32	7.42

other in 1969. The mixtures were stored in polypropylene bottles in a refrigerator when not in use. The pH of the 1969 stock solutions was determined on two occasions. The pH of the solutions immediately after use in one experiment was also determined (see Table 1). There was relatively little variation.

In 1968 a mixture of equal parts of 2 % osmium tetroxide and sea water was employed as a fixative. This gives better results, but is much less convenient to use.

RESULTS

Normal metamorphosis and development

Before describing the results of the experiments a brief account of the metamorphosis and development of *S. ciliatum* will be given. The larva is an amphiblastula with an anterior flagellated half and a posterior half of granular, non-flagellated cells (Fig. 1a). The inner ends of the flagellated cells contain a dark brown pigment such that there appears to be a pigmented, hollow dome at the centre of the larva. The sides seem somewhat more pigmented than the ceiling of the dome. When the larva is about to settle it comes to rest with usually its flagellated pole in contact with the substratum. The granular cells then move round the flagellated cells and spread between the latter and the substratum, and outwards also along the attachment surface. According to Schulze (1878) and Hammer (1909) the larvae undergo invagination whereby the layer of flagellated cells moves inwards to line the cavity enclosed by the dome-shaped granular cell layer, and it is the ring of granular cells around the opening of the 'gastrula' that makes contact with the substratum. However, their observations were made on larvae occurring in the surface film of hanging drops of sea water. There seems little doubt that when a hard substratum is concerned the larva adheres first by either its flagellated or its granular cells before the invagination process takes place. In the process the pigmented dome widens and the brown pigment becomes progressively less discernible as the flagellated cells round off. The attached larva can spread out to cover a considerable area (Fig. 1c), forming a thin plaque (as Minchin, 1900a, depicted for *Clathrina blanca*), but it can also remain more hump-shaped, like that of *Leucosolenia variabilis* (Minchin, 1896). Sometimes a distinct, funnel-shaped depression is visible at the apex of the projecting surface.

The granular cells spreading out at the borders of the plaque form a double layer of flat, extended pinacocytes. An abundant secretion of mesohylial matrix then occurs between the surface pinacoderm and the previously flagellated cells now aggregating closely to form a rounded central

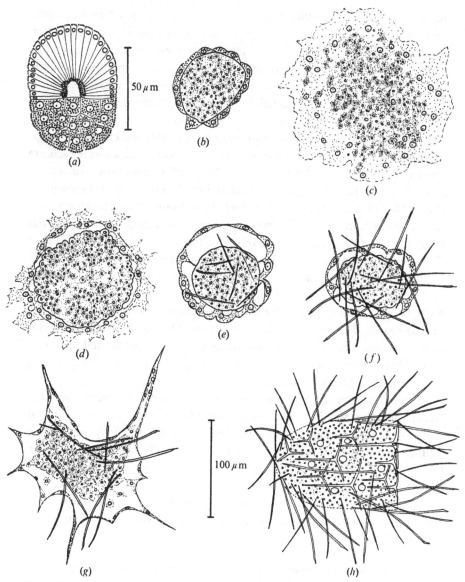

Fig. 1. Stages in the normal development of *Sycon ciliatum*. (a) Optical section of an amphiblastula attached laterally to a coverslip (somewhat diagrammatic). The flagella have vanished. Inclusions in the anterior cells are not shown. (b) Optical section of a later stage. Posterior granular cells now envelope the anterior cell mass. (c) A thin, greatly extended pupa seen in surface view. (d) Pupa withdrawing transversely and projecting more longitudinally. The anterior cells are aggregating to form a ball and a clear mesohyl has become evident in places between this and the outer pinacoderm. (e) Later stage seen in optical section. Distinct mesohyl and internal monacts are present. (f) Pre-olynthus. Mesohyl narrower, monacts project, cavity (not shown) present within ball of anterior cells. (g) Stellate pupa. (h) Young olynthus attached laterally to coverslip. Osculum (not visible), choanocytes and pores are present. Triacts and a distal tetract have appeared, but there are no long slender monacts around the osculum.

All drawings depict picrocarmine-stained, osmic acid fixed stages, excepting (g) and (h) for which absolute alcohol was used as fixative. All from (b) to (h) inclusive were drawn at approximately the same magnification.

mass (Fig. 1 *d*, *e*). A cavity next appears within this mass and the cells arrange themselves into a single layer of choanocytes lining this cavity. The mesohyl is crossed in places by inwardly projecting pinacocytes and by processes linking the two cellular layers together. It also contains sclerocytes derived from immigrated pinacocytes. At first only single-rayed monacts are formed, and these are internal, but later they pierce the pinacoderm and project far from the surface (Fig. 1 *e*, *f*). Meanwhile the young spongelet has changed shape, contracting somewhat transversely and extending longitudinally. The mesohyl becomes less obvious, pores and a distal osculum open and the three-rayed triacts, which have beforehand begun to appear, now tend to acquire an orientation related to the direction of the osculum (Fig. 1 *h*). Tetracts form at first around the osculum, their fourth ray projecting inwards in the membranous diaphragm enclosing the osculum, but later they are to be found lower down in the wall of the young olynthus (Fig. 2 *a*). The monacts formed at first are slender, and straight or gently curved. Some bear a tiny 'lance-head' distally. Later appear numerous very long, slender monacts to form an oscular crown. The later development, which Maas (1900) observed, of exceptionally stout monacts in the oscular crown, did not occur with *S. ciliatum*. The usually wide mesohyl is another difference, which doubtless Maas would have attributed to starvation. This, however, could not have applied in the case of very young spongelets living in fresh sea water.

Thus one can recognize several distinct stages of development. First, the larva; then the attached, metamorphosing larva; then the pre-olynthus (with pinacoderm, mesohyl, choanoderm and central cavity, but lacking a functional water conducting system (Duboscq & Tuzet, 1937); and then the olynthus (equipped with osculum and pores). Later, diverticula would appear and the asconoid olynthus would assume the form of the syconoid adult. Minchin (1900 *b*) used the term 'pupa' to denote the stages from settlement up to the olynthus. For Maas (1904) the 'pupa' would equal the 'pre-olynthus'.

Abnormal development in certain of the experimental solutions

When the pupae or olynthi were left in some of the experimental solutions they underwent retrogressive changes to an extent depending upon the degree of deficiency of calcium or bicarbonate. The process has been termed involution, and two types may be distinguished. The first is shown by pupae which have few spicules, and by olynthi in which the spicules are rapidly corroded by the experimental solution (Fig. 2 *b*). The spongelet contracts and rounds off, its cavity becomes slit-like and difficult to detect

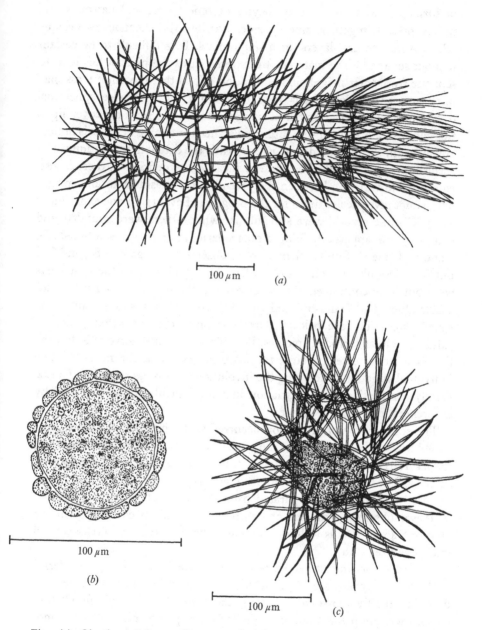

Fig. 2(a). Olynthus of *Sycon ciliatum* attached laterally to a coverslip. Note the long slender monacts around the osculum and the few tetracts amongst the triacts distributed throughout the wall. (b) Involuted spongelet produced by 46.5 h immersion in solution 9 at 12 °C (Osmic acid – picrocarmine). (c) Involuted olynthus. The living tissues have withdrawn inwards leaving a framework of isolated spicules.

in surface view, while its choanocytes become longer and narrower and tightly pressed together, side by side. Possibly the choanocytes become piled up into several layers, or a single mass. It is difficult to be certain without sectioning the material. In extreme cases the pinacocytes tend to separate from their neighbours and round off, so that the inner mass may burst out and isolated cells may lie around the otherwise spheroidal organism. Doubtless a number of spongelets fall away from the coverslip in this condition. A plate of cementing substance containing detritus is noticeable beneath the involuted spongelets. While the extent of involution varies according to the extent of calcium or bicarbonate deficiency, there is also considerable variation between the juveniles in the same solution, so that one cannot give precise times for the progress of events. In one experiment at 14 °C some olynthi were still tubular while others were rounded and compact, with apparently injured pinacoderm, after 12 h in solutions, (S), 2 and 7 (Table 1). Similar changes were observed after 20 h in S3 and S8, but were less obvious after 27 h in S9. After 70 h in S10 the spongelets were but little contracted. Those in S6 and S1 still had open pores and oscula after 96 h. Hand-in-hand with this process of involution proceeds spicule corrosion (see below). The least involuted spongelets generally exhibit the least corroded spicules. Possibly the spicules serve to buffer the tissues to some extent against the ionic deficiency of the medium. The individual variation concerning the involutionary process may therefore be merely a reflection of the variation in initial spicule content, caused by differences in age, for example.

The second type of involution occurred with olynthi that already possessed triacts in quantity. Here the pores and osculum closed and the living tissues retreated inwards leaving a framework of exposed, interlocking spicules. The tissues rounded off, producing a gemmule-like spherical body enclosed by, but mostly separated from the spicule framework (Fig. 2c). This was observed with one or two of the olynthi on coverslips left for 96 h in S4 and S6, as well as with some olynthi left for several weeks in unchanged sea water.

If one excludes the ultimate pathological changes (disrupted pinacoderm) of the former type it seems probable that the two involutionary processes described briefly above are essentially the same, differing only in whether or not there are present uncorroding and overlapping spicules that cannot be carried inwards by the shrinking organism.

A brief reference to Maas's studies would not be out of place here. Maas observed several types of degenerative change. First, when *S. setosa* larvae were allowed to metamorphose in carbonate-free sea water, the pupae lacked a skeleton, were much flatter in shape, and tended to round off and lose their cavity. The pinacocytes spread out on the coverslip and the choanocytes disaggregated and likewise spread, so that a single sheet of cells was obtained. Further spreading resulted in a loose network of cells which finally broke up into separate cells. It was in the early stages of this process that Maas observed the spiculoids. They consisted of large irregular triacts and projecting, flexible, shadowy-granular monacts. He also noted that the hyaline mesohyl at times appeared more abundant and contained membranous formations similar to cuticular thickenings between the pina-coderm and choanoderm (Maas, 1904). I have observed that the mesohyl appears to be strongly refringent in some spongelets, particularly after immersion in S7 and S8.

Secondly, Maas (1907, 1910) described a hunger response. When adult calcareous sponges were kept in filtered sea water the osculum closed and certain parts of the sponge died while the remainder underwent reorganiza-tion to form a smaller organism equipped with a functional water con-ducting system. Hunger thus resulted in a step-wise diminution in size. Sometimes 'blisters' appeared on the tube wall of *Leucosolenia lieberkühnii* and there was a considerable increase in the mesohyl. No spicule corrosion occurred. I have also observed 'blisters' on a spongelet after 27 h in S9. Earlier (Jones, 1956) they appeared on the inner surface of tubes of *L. botryoides* and as 'dermal blow-outs' (a term used by Huxley, 1921) on healing membranes developed across the cut ends of the tubes, when these were left in slightly acidified sea water.

The third type of degenerative change noted by Maas (1906, 1907, 1910) occurred when juvenile and adult sponges were transferred gradually into calcium- and carbonate-free sea water. The living tissue withdrew inwards leaving the spicules exposed, but united by strands of mesohylial matrix. The living substance formed at first narrow branching tubes, but even-tually these broke up into ovoid or round gemmule-like bodies. The process involved the movement of the porocytes into the cavity of the flagellated chamber, the piling up of choanocytes into several layers, and the engulf-ment of choanocytes by the porocytes. The resulting gemmule consisted of a pinacoderm enveloping a mass of archaeocytes derived from the original porocytes. Spicule corrosion occurred, but only of the small spicules carried inwards by the withdrawing tissues. The mesohyl was much reduced. If the spongelets were immersed directly in the calcium- and carbonate-deficient sea water (i.e. an immediate, rather than a gradual, change), the spicules underwent rapid corrosion or fell out, contraction occurred, the cavity became obliterated and the pinacoderm suffered damage in that pinacocytes separated and the inner mass tended to burst out in places. The mesohyl was feebly developed. Thus again it would appear that the speed of spicule corrosion is the main factor distinguishing the two types of involution, if one neglects the ultimate disintegration of the tissues following prolonged immersion in an unfavourable medium. The corrodability of the spicules is important, for I have shown (Jones, 1956) that direct transfer of tubes of *Leucosolenia* to isotonic sodium chloride solution results in involution to form gemmule-like bodies (possibly lacking pinacoderm) within the slowly corroding spicule framework, in contrast to Maas's observations on the direct transfer of juvenile *Sycon* to lime-free sea water. Spicule corrosion is presumably slower with the adults, either because there are more spicules initially, or because the spicules are better protected by their organic sheath and by the mesohylial matrix.

Spicule corrosion

As would be expected the spicules already present in pupae and olynthi tend to corrode in the experimental solutions containing less calcium or less bicarbonate than sea water, and the corrosion is faster the lower the concentration of either of these ions. For the solutions 7–10 the acidity (Table 1) is an additional factor favouring spicule dissolution. The following data give an indication of the speed of corrosion in the different solutions at 14 °C. Only markedly corroded spicules remained after 14 h in S2 and S7, after 20 h in S3 and S8, after 27 h in S9, and after 70·5 h in S10. There was very little corrosion, if any, in S4, S5, S6 and S1 after 4 days.

A difference was noticed between the spicules corroded in calcium-deficient media and those corroded in bicarbonate-deficient media. In both the corrosion proceeded with the formation of pits on those surfaces that were more or less parallel with the optic axis of the spicule calcite. Also the axial parts of the rays tended to dissolve from the tips inwards. Thus the monacts and the rays of tri- and tetracts tended to dissolve distally into paired 'splints'. However, in the calcium-deficient medium these were smooth throughout their length (Plate 1a), whereas in the bicarbonate-deficient medium the corrosion seemed to be coarser and more localized, so that the corroded rays appeared like ladders or girders with cross-bars joining the two splints together (Plate 1b). With further corrosion these cross-bars severed, leaving projections on the inwardly facing surfaces of the two splints, and eventually the remnants appeared to break up each into a row of granules, before finally disappearing completely. Precisely the same differences in corrosion by the two types of media can be seen on living as on fixed spongelets, which suggests strongly that the corrosion is not a physiological process, but caused by the direct action of the medium on the spicule calcite.

In both types of deficient media one observes considerable differences in the rate of corrosion between different spicules in one and the same spongelet, as well as in different spongelets on the same coverslip. Quite possibly this is to be correlated with the age of the spicules, for it is known that the organic sheath enveloping the spicules tends to thicken as the spicules of *Clathrina coriacea* grow (Minchin, 1898), and it seems likely that the sheath would protect the spicule calcite to some extent from the corrosive action of the medium. Moreover the sclerocytes completely envelope the smaller, growing spicules and might be expected to hinder corrosion in consequence. Thus the oldest and the youngest spicules would be expected to corrode more slowly than those recently formed. Older spongelets on the same coverslip would in general possess a greater proportion of older spicules than the younger. The position of the spicules in the spongelets would also be a factor. One commonly finds in olynthi that corrosion of triacts is first noticeable at the oscular edge, a site where the mesohylial matrix is more liable to swell in potassium nitrate solution than elsewhere in the oscular tube of *Leucosolenia* (Jones, 1956). The variation in corrosion rate between spicules and individuals was also noted by Maas (1906).

When the corrosion rate in living spongelets was compared with that in alcohol fixed spongelets it became obvious that, if anything, the corrosion proceeded faster in the latter. In several of the experiments, involving either calcium- or carbonate-deficiency, all traces of calcite had vanished from the fixed material after a time when there were still traces of spicules in the living spongelets. For example, in an experiment with pupae there were 24 living and six fixed juveniles placed in S2. After 14 h the fixed juveniles exhibited no spicule remnants, whereas all the living had traces of spicules (short internal paired splints and nodules of calcite). In S7 there were no traces in the seven fixed pupae whereas most of the 28 living pupae exhibited spicule remnants after the same time (14 h). In a later experiment in which olynthi were used the seven fixed olynthi exhibited only a few slivers of calcite after 20 h in S3, whereas the five living had less corroded remnants. After 20 h in S8 there were obvious remnants in the four living, but of the six fixed, two had no trace of calcite, one had only traces and the remaining three exhibited remnants. In a similar experiment conducted in 1968 at 12 °C, after 24 h in S2 all seven living spongelets had spicule remnants, whereas five of the 15 fixed spongelets had only the merest traces left and the remaining ten had none whatsoever. After 10.5 h in S7 all 13 living juveniles had remnants and these were abundant in seven, whereas 29 of the 32 fixed spongelets had no trace of spicules at all.

In only two cases, involving S9 and S10 in a single experiment, was the corrosion apparently more rapid in the living than the fixed spongelets, but this result is capable of other interpretations and was not repeated. The protruding monacts of living pupae tend to fall out of the contracting masses (Maas, 1904), whereas they remain *in situ* in the fixed material. The smaller amount of spicular remnants in the former may thus not indicate a faster corrosion. Moreover, this experiment used pupae, and it was difficult to ensure that their ages were the same on the two coverslips. If, as suggested above, the age is a factor in spicule corrosion, it is obvious that careful controls are needed for reliable results. In the experiment referred to above all 19 living spongelets exhibited only a few remnants after 27 h in S9, whereas of the 13 fixed juveniles, some had no remnants, others only slivers, and others again had abundant remains. After 27 h in S10 there was noticeable corrosion in the 13 living, but in general only slight corrosion in the 14 fixed juveniles, but once again variation between the latter was observed.

It may be concluded that there is certainly no good evidence for faster corrosion in living than in fixed juveniles. On the contrary, in many cases the reverse appeared to be true. The results do not support Maas's contention that spicules in contact with living tissues corrode more rapidly because of spiculoclastic activity. Nor was there any noticeable appearance of granules in the cells, either pinacocytes or choanocytes, as corrosion proceeded. Granules were just as likely to be present in the control juveniles reared in sea water as in any of the other experimental solutions. The appearance of stainable granules could merely reflect the occurrence of nutritive processes, for example, uptake of food particles. Controlled experiments were not apparently carried out by Maas.

Spicule formation

The results of many experiments carried out during July–August of 1968 and 1969 indicate quite unequivocally that spicule formation can occur in solution 6, but not in 5. Thus when bicarbonate is at the concentration found in sea water the amount of calcium needed for spicule production is between 0.001 and 0.005 M. Normal sea water contains 0.01 M Ca^{2+}.

The determination of the threshold concentration of bicarbonate required when calcium is present at the concentration found in sea water is more difficult. Firstly, the bicarbonate ions dissociate to form carbonate and hydrogen ions. Secondly, carbon dioxide dissolves from the air and furnishes additional bicarbonate and hydrogen ions. The effect is less the more sodium bicarbonate is added to the mixture. Finally, the amounts of sodium bicarbonate added in solutions 7–10 were too small to enable spicule production to be detected with certainty. There was certainly no spicule production in solutions 7–9, but in 10, which only slowly corrodes the spicules, it is possible that thin-rayed monacts and triacts were formed in some spongelets. One cannot be certain, because the thin-rayed structures seen could well have been the corroded remains of spicules existing at the start of the experiment. However, for normal, thicker-rayed spicule production one must certainly add sodium bicarbonate to an extent which makes the molarity of bicarbonate plus carbonate greater than 0.001 M. The combined molarity of these ions in normal sea water is 0.0025 M.

It can be concluded that the concentrations of calcium and bicarbonate-carbonate needed for spicule production are greater than would be expected from Maas's assertion that only traces were required. Almost half the normal concentrations in sea water seem in fact to be needed. Further experiments are planned to enable the threshold levels to be determined more precisely. For the present it should be noted that spicule production occurs only in media which do not appreciably corrode the existing spicules within 4 days. This is helpful in that, were corrosion and production to occur in the same medium, one would not be able to determine accurately the threshold concentrations required for the latter, because the former would tend to increase locally the concentrations of spicule-forming ions.

Recovery of spicule production in normal sea water

When juveniles were returned to sea water after 48 h immersion in the experimental solutions, normal spicule production was resumed in all cases excepting those which had been in solution 7. Thus, even though spicule production was inhibited in S2–5 and S8–10 inclusive, the spicule-secreting mechanism had not been permanently harmed by the 48 h immersion in these media. Presumably the insufficient supply of essential ions, or the acidity in some of the solutions, prevented spicule formation. Even the juveniles from S7 showed some signs of recovery (for example, regeneration of pinacoderm), so that spicule production might have been observed had they been left in the sea water for a longer period (> 48 h). However, Maas (1907) reported that after 3 days in a mixture lacking both calcium and carbonate no spicule production occurred on restoration of the spongelets to normal sea water.

Restoration of spicule production does not have to await the reconstruction of a functional water-conducting system in the juveniles. As described above, the spongelets underwent contraction and retrogression in the solutions tending to corrode the spicules, and they had not recovered their normal asconoid form after 48 h in normal sea water. Only those juveniles transferred from solutions 6 and 1 showed open pores and oscula after the 24 h in S8, followed by the 48 h in S5 or S6 and the 48 h in sea water. The others showed varying degrees of recovery, from an inflated closed bag down to a compacted, rounded mass enveloped by a regenerated pinacoderm. Spicule formation had none the less occurred in all except the latter extreme, which had suffered retrogression in S7. Thus, just as the pupa can produce spicules (monacts mainly) before the advent of a functional water-conducting system, so the involuted juveniles can do likewise (Plate 1c, d).

The existing spicule remnants are apparently of no use in the formation of new spicules. There was no evidence of any repair of previously corroded spicules, and the remnants persisted apparently unchanged when the juveniles were replaced in sea water. In time, no doubt, they would be shed from the surface of the sponge (Jones, 1964), or phagocytosed by porocytes (Minchin, 1898; Prenant, 1925).

DISCUSSION

The results published above contradict the three assertions of Maas concerning spiculoclastic activity, the formation of organic spiculoids and the need for only traces of calcium and carbonate in the medium for spicule production. The absence of spiculoclastic activity is shown by: (1) the often faster corrosion of spicules embedded in fixed rather than living spongelets, (2) the same characteristic type of corrosion in both living and fixed material in respectively calcium-deficient and bicarbonate-deficient media, and (3) the lack of correlated change (granule formation) in the cells as corrosion proceeds. How then can Maas's observation that corrosion occurs faster in living material be explained? The answer would appear to be that he was comparing spicules in living tissue with spicules that had fallen out, or were united to form isolated frameworks, whereas in the experiments described above, the comparison was made between spicules of similar ages *in situ* in living and fixed material. It is known that the spicule sheath becomes progressively thicker as the spicule grows (in *Clathrina* at least; Minchin, 1898), and that the outermost parts of the mesohyl of *Leucosolenia* are less easily softened by colloid-dispersing salt solutions (Jones, 1956). Also, the growing spicules of *Clathrina, Leucosolenia* and *Sycon* lie in close contact with the innermost cell layer (Woodland, 1905; Minchin,

1908). One would hence expect the outermost triacts to corrode least rapidly, for these would be the oldest and the most protected by their sheath and by the mesohylial matrix. Moreover, the innermost, growing spicules would be the ones that would be carried inwards by the involuting tissues when these were withdrawing from the spicule framework. Thus Maas's observations are explicable in terms of the variable protective effect of the sheath, mesohyl and formative cells.

Photographic records of growing oscular tubes of *Leucosolenia variabilis* have indicated that even the fragments of broken spicules remain uncorroded, and that spicules are removed by protrusion through the pinacoderm rather than by spiculoclastic activity (Jones, 1964). The observations of Minchin (1898), Prenant (1925, fig. 15) and Maas (1910) on the occurrence of spicule fragments within porocytes need not indicate spiculoclastic activity, but merely phagocytic uptake by scavenging cells of already corroded remnants. The concept of spiculoclastic activity by cells of calcareous sponges must thus be abandoned, together with Maas's conclusion that the unruly cells of *Sycon* spongelets do not subserve the needs of the whole organism, because they dissolve away the spicules when they are still required.

With regard to Maas's spiculoids one can only suppose that the triacts (which are mentioned only in the 1904 paper) were merely the cavities left by spicule corrosion *after* fixation, for this is what the depicted example appears to be. Care must be taken in fixation, staining and dehydration to ensure that no corrosion occurs. The flexible monacts (mentioned in 1904, 1906), on the other hand, could well have been the projections of stellate spongelets. Some juveniles have a markedly stellate appearance (Fig. 1*g*) and when undergoing involution the projections tend to become detached distally and withdraw only slowly as the organism rounds off. Such projecting, unattached extensions of the organism, consisting of a thin strand of mesohyl enveloped by pinacoderm, have been seen alongside others that were still attached to the coverslip on some of my preparations.

Finally, as stated earlier, it is clear that more than traces of calcium and bicarbonate must be supplied in the medium. Between 0.001 and 0.005 M Ca^{2+} and more than 0.001 M HCO_3^- and CO_3^{2-} must be present. Either the carbon dioxide liberated by the respiration of the spongelets is insufficient, or the associated acidity precludes its use for spicule production. It would be interesting to test the effect of media lacking bicarbonate but buffered to a higher pH than those actually used.

Glover & Sippel (1967) have suggested that organisms have merely to bring sea water and bicarbonate ions into contact in order to produce skeletons of magnesian calcite. They obtained a calcite containing 14.3 mole % Mg^{2+} by adding sodium bicarbonate to sea water at about 0 °C. Kinsman & Holland (1969), however, using conditions giving lower degrees of supersaturation and slower rates of precipitation, obtained calcite containing 8–10 mole % $MgCO_3$ at 16 °C, and less than 1 mole % at 3 °C. When Jones & Jenkins (1970) analysed the spicules of adult *S. ciliatum* they found them to contain 5.2–5.4 mole % $MgCO_3$ (sea water temperature 7–16 °C). The Mg^{2+} content of spicules of various species varied, however, even when the sponge specimens were collected from the same pool at the same time of year, so that the mechanism controlling composition is ion-selective. Glover & Sippel (1967) have demonstrated that both the temperature and the concentration of sodium chloride influence the Mg^{2+} content of the calcite precipitated, in addition to the MMg^{2+}/MCa^{2+} ratio.

SUMMARY

1. Recently metamorphosed spongelets of *Sycon ciliatum* were immersed in artificial 'sea waters' in which either the $CaCl_2$ or the $NaHCO_3$ had been replaced by NaCl. Spicule formation occurred only when the 'sea water' contained between 0.001 and 0.005 M Ca^{2+} and more than 0.001 M HCO_3^-.

2. Spicule corrosion occurred within 4 days in 'sea waters' containing (a) 0.001 M, or less, HCO_3^-, or (b) less than 0.0005 M Ca^{2+}, namely, in solutions in which spicule formation could not take place.

3. The appearance of the corroding spicules in Ca^{2+} deficient media differed from that in HCO_3^- deficient media, for both living and alcohol-fixed spongelets, suggesting that the corrosion was caused by direct chemical action of the medium, rather than by physiological, spiculoclastic activity of cells.

4. Spicule corrosion did not in general occur at a faster rate in living than in alcohol-fixed spongelets of comparable age, again indicating the absence of spiculoclastic activity. Maas's observation that isolated spicules corrode more slowly than spicules in contact with living tissue is explicable in terms of differences in the protective effect of the spicule sheath and enveloping matrix.

5. Spicule production was rapidly restored by return of the spongelets to sea water, after 48 h in solutions lacking sufficient Ca^{2+} or HCO_3^-, but containing at least 10^{-4} M HCO_3^-. A functional water-conducting system was not essential for spicule formation to recommence.

6. Organic spiculoids were not formed in place of spicules in any of the media employed.

7 The spongelets underwent retrogression (involution) in unfavourable media.

8. Aspects of normal development are briefly described.

REFERENCES

DUBOSCQ, O. & TUZET, O. (1937). L'ovogénèse, la fécondation et les premiers stades du développement des Éponges calcaires. *Archs Zool. exp. gén.* **79**, 157–316.

GLOVER, E. D. & SIPPEL, R. F. (1967). Synthesis of magnesium calcites. *Geochim. cosmochim. Acta* **31**, 603–13.

HAMMER, E. (1909). Neue Beiträge zur Kenntnis der Histologie und Entwicklung von *Sycon raphanus*. *Arch. Biontol.* **2**, 289–334.

HUXLEY, J. S. (1921). Further studies on restitution-bodies and free tissue-culture in *Sycon*. *Q. Jl microsc. Sci.* **65**, 293–322.

JONES, W. C. (1956). Colloidal properties of the mesogloea in species of *Leucosolenia*. *Q. Jl microsc. Sci.* **97**, 269–85.

JONES, W. C. (1964). Photographic records of living oscular tubes of *Leucosolenia variabilis*. II. Spicule growth, form and displacement. *J. mar. biol. Ass. U.K.* **44**, 311–31.

JONES, W. C. (1967). Sheath and axial filament of calcareous sponge spicules. *Nature, Lond.* **214**, 365–8.

JONES, W. C. & JAMES, D. W. F. (1969). An investigation of some calcareous sponge spicules by means of electron probe micro-analysis. *Micron* **1**, 34–9.

JONES, W. C. & JENKINS, D. A. (1970). Calcareous sponge spicules: a study of magnesian calcites. *Calcif. Tiss. Res.* **4**, 314–29.

KINSMAN, D. J. J. & HOLLAND, H. D. (1969). The co-precipitation of cations with $CaCO_3$. IV. The co-precipitation of Sr^{2+} with aragonite between 16 °C and 96 °C. *Geochim. cosmochim. Acta* **33**, 1–17.

MAAS, O. (1900). Die Weiterentwicklung der Syconen nach der Metamorphose. *Z. wiss. Zool.* **67**, 215–40.

MAAS, O. (1904). Ueber die Wirkung der Kalkentziehung auf die Entwicklung der Kalkschwämme. *Sber. Ges. Morph. Physiol. Münch.* **50**, 4–21.

MAAS, O. (1906). Uber die Einwirkung Karbonatfreier und Kalkfreier Salzlösungen auf erwachsene Kalkschwämme und auf Entwicklungsstadien derselben. *Arch. EntwMech. Org.* **22**, 581–602.

MAAS, O. (1907). Ueber die Wirkung des Hungers und der Kalkentziehung bei Kalkschwämme und anderen kalkausscheidenden Organismen. *Sber. Ges. Morph. Physiol. Münch.* **23**, 82–9.

MAAS, O. (1910). Uber Involutionserscheinungen bei Schwämmen und ihre Bedeutung für die Auffassung des Spongienkörpers. In *Festschrift zum 60 Geburtstage Richard Hertwigs* pp. 93–130. Jena: G. Fischer.

MINCHIN, E. A. (1896). Note on the larva and post larval development of *Leucosolenia variabilis* with remarks on the development of other Asconidae. *Proc. Roy. Soc. Lond.* B **60**, 42–52.

MINCHIN, E. A. (1898). Materials for a monograph of the ascons. I. On the origin and growth of the triradiate and quadriradiate spicules in the family Clathrinidae. *Q. Jl microsc. Sci.* **40**, 469–588.

MINCHIN, E. A. (1900a). Éponges calcaires. In *Zoologie Descriptive des Invertébrés* Tome 1, Chap. 5, pp. 107–47. Louis Boutan. Paris: Octave Doin.

MINCHIN, E. A. (1900b). Sponges – Phylum Porifera. In *A Treatise on Zoology*, Pt. II, Ch. III, pp. 1–178 (ed. E. Ray Lankester). London: Adam and Charles Black.

MINCHIN, E. A. (1908). Materials for a monograph of the ascons. II. The formation of spicules in the genus *Leucosolenia*, with some notes on the histology of the sponges. *Q. Jl microsc. Sci.* **52**, 301–35.

PANTIN, C. F. A. (1946). *Notes on Microscopical Technique for Zoologists*. Cambridge: University Press.

PRENANT, M. (1925). Observations sur les porocytes de *Clathrina coriacea* Mont. *Trav. Stn. zool. Wimereux* **9**, 198–204.

SCHULZE, F. E. (1878). Untersuchungen über den Bau und die Entwicklung der Spongien. V. Die Metamorphose von *Sycandra raphanus*. *Z. wiss. Zool.* **31**, 262–95.

WEBER, A., HERZ, R. & REISS, I. (1963). On the mechanism of the relaxing effect of fragmented sarcoplasmic reticulum. *J. gen. Physiol.* **46**, 679–702.

WOODLAND, W. (1905). Studies in spicule formation. I. The development and structure of the spicules in Sycons: with remarks on the conformation, modes of disposition and evolution of the spicules in calcareous sponges generally. *Q. Jl microsc. Sci.* **49**, 231–82.

THE MEASUREMENT OF INSHORE SUBMARINE IRRADIATION

R. M. SMITH AND W. E. JONES

Marine Science Laboratories, Menai Bridge, Anglesey

INTRODUCTION

In earlier work (Smith, 1967) on sublittoral algae at sites on the western and eastern coasts of Anglesey, considerable differences were noted in the composition of the floras of the two coasts and also in the distribution of the plants in relation to depth. A number of species found on the west coast are absent or scarce on the east and plants of the same species usually occur to greater depths on the west coast.

One of the reasons for these differences may be the greater turbidity of water on the eastern side of the island which would result in a lower average level of submarine illumination and a more rapid attenuation with increasing depth as compared with the west coast water. In this paper an attempt to assess these differences quantitatively is described.

As part of studies on the life histories of some sublittoral Rhodophyceae, spores and mature plants have been cultured for extended periods in the laboratory (Jones & Smith, 1970). It was desired to maintain illumination levels and day length at values similar to those in the field at the same season; this provided a second reason for making observations of submarine illumination.

Jerlov (1953) described the use of a barrier-layer photocell for measuring the attenuation of radiation in oceanic waters. Levring (1959) applied these methods to determine the qualitative changes in illumination which occur with increasing depth. He related these to the photosynthetic efficiency of plants in light of different wavelengths and to the distribution of algae in the sublittoral.

MATERIALS AND METHODS

The instrument used was a 'Mini-photometer' manufactured by Evans Electroselenium Ltd. This was housed in a waterproof Perspex case to permit it to be used underwater. The spectral response of the instrument is shown in Fig. 1. In order to extend the upper range of the instrument beyond its calibration limit of 2000 lx a neutral density filter was used which transmitted 14% of the incident light. Other filters were used to

11   [321]   CFE

check the differences in the attenuation of various parts of the visible spectrum. These were Ilford spectrum filters nos. 601, 602, 604, 605 and 607; the spectral regions in which these filters transmit light are indicated in Fig. 1.

When in use, the filters were held close against the instrument case by means of a rubber sleeve; above the filter was placed a flat translucent cosine collector of the type described by Smith (1969). The filters could be readily interchanged by the diver whilst underwater.

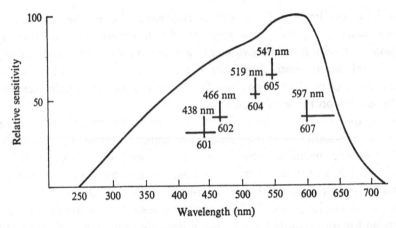

Fig. 1. Spectral response curve of the barrier layer photo cell. The peak transmission and half band width of each of the filters used in the work are indicated.

Attenuation through a vertical water column was measured along a plumb line supported by a small buoy at the surface and anchored to a heavy concrete block, which was laid at the station over which the measurements were made. Readings were taken above the surface and by a diver at 0.1 m below the surface and at 1 m intervals from the surface to the seabed. The readings were repeated on the ascent and the surface and 0.1 m subsurface values were compared with the initial readings to record any changes which might have occurred during the ascent and descent. It was usually obvious when changes of incident illumination, due to clouds, occurred during the dive. The series of readings was repeated if changes did occur. Under constant surface conditions two descents and ascents were made, hence four sets of figures were available to reduce error. The four readings for each depth were averaged to give mean values which were used to construct the attenuation curve. A note was made of the time, cloud cover, water surface conditions and the velocity of the wind. These observations were repeated as far as possible at weekly intervals throughout the year; in addition measurements were made of the attenuation at different times of the day, at dawn and dusk, and under various sea conditions from calm to storm.

Water samples were collected simultaneously with each set of light readings. These were used for investigation of the suspended matter. Four 250 ml samples were taken at the surface, at a depth of 4 m and at 9 m; these were mixed to provide a homogenous sample which was analysed for sediment content. A coulter counter was used to count particles in the size range 1–50 μm. Larger particles could not be estimated by this method; it was found best when large particles were present to evaporate the sample and weigh the washed residue.

During laboratory work involving the culture of benthic marine algae, irradiation levels and photoperiods were adjusted to approximate to illumination in the field. All readings, both in the field and in the laboratory, were taken with the photometer submerged below the water surface to standardize conditions as far as possible. The field values obtained using the spectrum filters were used directly to set up irradiation levels in the laboratory using Atlas 'Daylight' fluorescent tubes. For this the wavelengths selected were those transmitted by the blue filter 602, with a half band width of 23 nm and a peak transmission at 466 nm, and the 604 green filter with a half-band width of 12 nm and a peak transmission at 519 nm. This was the spectral region of least attenuation as measured in these inshore waters. In the laboratory, all the algae were irradiated with light of the same spectral composition, a recognized but unavoidable error, in an attempt to reproduce quantitatively the submarine irradiation recorded at different depths.

The culture of algae in the laboratory under irradiation regimes designed to recreate field conditions must take into account the change in day length during the period of the experiment; normally this would be satisfied by quite rough approximations. During our work, however, some difficulties have been experienced in culturing some summer annual species of Rhodophyceae beyond a phase of creeping filaments. Since the irradiation requirements of these species might be critical in initiating the formation of reproductive structures, a more refined method was used. The photoperiods and irradiation levels used in the laboratory during the culture experiments were calculated from the field data as follows. The mean day length for a month was taken as occurring about the 15th day and for this the average daily irradiation curve for that month was assumed to apply. This curve, derived from measurements at the surface, could be adjusted for depth by applying the appropriate attenuation as measured in the corresponding weekly dive. The equivalent laboratory irradiation level was calculated in two ways, the simplest giving a single level of irradiation throughout the day. In this case if

$$\int_{t_0}^{t} I_0 dt,$$

is the sum of irradiation at 0.1 m between dawn (t_0) and dusk (t), derived from the curve of diurnal irradiation (e.g. Fig. 2), then for a laboratory day length θ and a laboratory irradiation level I_1:

$$\theta I_1 = \int_{t_0}^{t} I_0 dt.$$

If E is the attenuation per metre, then, at a depth of h metres, the intensity will be a fraction E^h of the surface intensity I_0. The equivalent laboratory intensity should therefore be $E^h I_1$.

A second method giving a better approximation to field conditions used two levels of illumination, a second lamp being switched on for part of the day to increase the irradiance. In this case if the higher level of irradiation, I_2 lasts for a fraction $1/n$ of the laboratory daylength, so that the total will be

$$\theta I_1 + \frac{\theta}{n}(I_2 - I_1).$$

If the maximum value of I_0 recorded is I_{max}, and the laboratory value for I_2 is taken to be some fraction K of this, then $I_2 = K I_{max}$ and,

$$A = \theta I_1 + \frac{\theta}{n}(KI_{max} - I_1),$$

where

$$A = \int_{t_0}^{t} I_0 dt.$$

Therefore,

$$\frac{A}{\theta} = I_1\left(1 - \frac{1}{n}\right) + \frac{K}{n} I_{max},$$

and,

$$I_1 = \left[\frac{A}{\theta} - \frac{K}{n}.I_{max}\right] / \left(1 - \frac{1}{n}\right).$$

Thus for example, taking the data in Fig. 2, where $A = 85\,600$ lx h, $\theta = 11$ h $I_{max} = 14\,500$ lx, values of $K = 0.7$ and $n = 4$ (i.e. irradiating the plants $\frac{1}{4}$ of the daylength at 70% of the maximum field irradiation), this gives values of $I_1 = 6980$ lx and $I_2 = 10\,180$ lx.

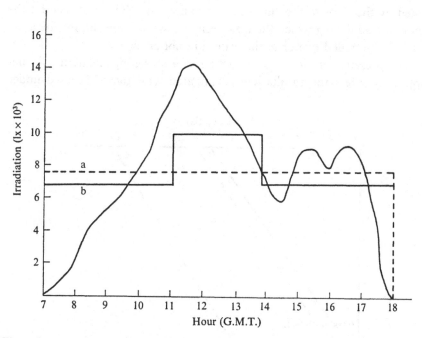

Fig. 2. Diurnal irradiation during February 1968 at 0.1 m depth. Readings were taken at ½ h intervals. The corresponding laboratory illumination regimes are shown. One step (*a*) and two step (*b*) cases for laboratory irradiation are illustrated. The method of calculating the laboratory regimes is explained in the text.

RESULTS AND DISCUSSION

Short term effects

In considering the results and their relevance to the floristic data, short-term changes in the attenuation pattern require attention as well as the seasonal aspects. Two factors were considered: (1) the effect of the sun's altitude and (2) the effect of water movement on the sediment load.

(1) Effect of the sun's altitude

The assumption seemed justified that, while the value of irradiation at the surface would vary with the sun's altitude, attenuation would not. This was tested by measuring the attenuation through the water column at dawn, ½ h after sunrise, at noon, ½ h before sunset and at sunset on days in October, February, May and July, noting the state of the water surface and cloud cover at the time. Fig. 3 shows the results for 8 July, 1968 which were typical. The surface value (the intercept on the irradiation axis) changes with the diurnal variation in the sun's altitude but the attenuation, indicated by the slope of the curves, remains the same. This was true for the short period during which the measurements were made and in which the load of suspended matter in the water did not change.

The presence of cloud cover, acting as a scattering medium, did not appear to affect the results when compared with those obtained under

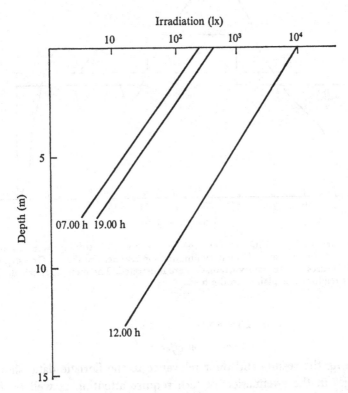

Fig. 3. The variation in attenuation at different times of day – 07.00, 12.00 and 19.00 h G.M.T. The small change in the slope of the curve indicates that the solar altitude has little effect upon attenuation, but only on the level of radiation.

cloudless conditions. This is probably because scattering at the water surface is more effective, except on the rare occasions when the surface is without ripples.

The pattern of attenuation with increasing depth shown in Fig. 3 agrees broadly with the results of Jerlov (1953), working on coastal waters. As will be seen, however, our results in close inshore waters are greatly affected by rapid changes in the sediment load.

(2) *Sediment load and the attenuation of light*

The effect of the changing sediment load is illustrated by an account of observations made over a period of 13 days in February 1968. The variation in attenuation through the water column is shown in Fig. 4. 10 February was the first day of a storm; an onshore wind increased rapidly in strength. The resulting increase in water movement raised more material into suspension and evidently caused a rapid decrease in transmittance, measured at a depth of 9 m, from 5% of the surface value (10 February) to 0.5%

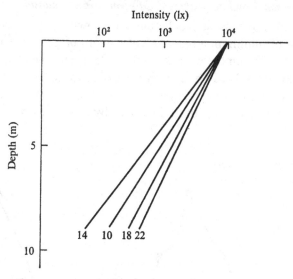

Fig. 4. Attenuation curves measured at the same station over a period of 13 days. Between 10 and 14 February 1968, onshore winds and heavy swells resulted in increased attenuation. After this period the seas subsided, sediment settled and the attenuation decreased. Numbers against the curves indicate dates.

(14 February). During the following 10 days the wind velocity, though variable, did not exceed 10 knots. Measurements were made of attenuation through the water column on alternate days up to an including 22 February; the curves obtained on 10, 14, 18 and 22 February are shown in Fig. 4. The intercept on the intensity axis is the value of the surface irradiation calculated from the reading taken at 0.1 m below the surface.

Similar storms have been recorded at all months of the year at this station and the results of Fig. 4 are typical of conditions when a storm with heavy swells was followed by a calm period during which the swells decreased in magnitude. The gradual decrease in attenuation as the sea moderated was due to the reduction in sediment load as water movement decreased.

This particular series of readings illustrated the difficulty of correlating attenuation in the inshore waters with the local weather conditions. The local wind strength and direction was very variable and was not responsible for the development of the relevant swells. The latter probably originated from an active depression approaching the Bay of Biscay during 9 and 10 February (*Weather*, 1968). Winds associated with this centre would have been sufficient to cause the arrival of heavy swells between 10 and 16 February. By 16 March the depression had filled and the swells were decreasing.

Seasonal and geographical differences in attenuation

In Fig. 5 monthly means of irradiation are shown as measured in the spectral region of minimum attenuation at a depth of 4 m at two stations, one at Trearddur Bay on the west coast of Anglesey and the other at Fedw Fawr on the east coast. It will be seen that there are considerable differences and that, in general, the attenuation is greater on the eastern coast. Floristic differences at these sites have been described by Smith (1967).

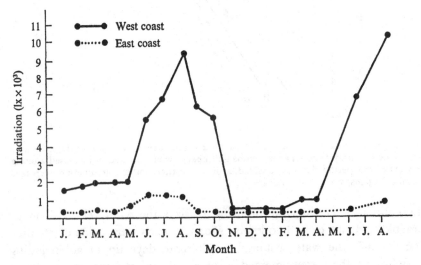

Fig. 5. Monthly means of irradiation at a depth of 4 m at stations on the west and east coasts of Anglesey.

Subjective estimates had suggested that the east coast water was more turbid. The relation between suspended load and attenuation is borne out in the results presented in Fig. 6, in which the monthly means of irradiation at a depth of 4 m at the western (Fig. 6a) and eastern (Fig. 6b) stations are plotted, together with the monthly means of sediment load measured at the

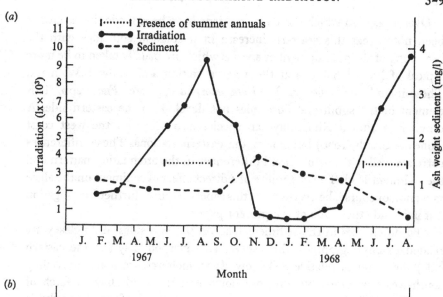

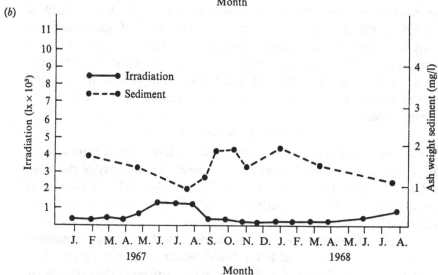

Fig. 6(a). Monthly means of irradiation received at 4 m on the west coast of Anglesey and the mean monthly dry weight of suspended sediment per litre of sea water sampled at the same depth. (b) Monthly means of irradiation received at 4 m on the east coast of Anglesey and the mean monthly dry weight of suspended matter per litre of sea water sampled at the same depth.

same place and depth. It will be seen that, on the west coast, there is a rapid increase in irradiation in May–June. This is apparently due to (1) the seasonal increase in incident radiation, coinciding with some reduction in cloud cover and (2) the onset of calmer weather with reduced water movement and hence smaller sediment load.

Over 3 years in which these observations have been made, it has always been noted that this seasonal increase in irradiation coincides with the beginning of the development of species which are usually taken to be more typical of littoral habitats at the western station 4 m below L.W. mark. These include *Ulva lactuca* L., *Enteromorpha* spp. and *Fucus* spp. This element of the sublittoral flora does not develop on the eastern side. A number of annual Rhodophycean species also occur on the west coast (Jones & Smith, 1970) but not at the eastern stations. These differences correlate well with the observed differences in the attenuation pattern and the sediment load. The possibility of a direct effect of sediment on the algae must not, of course, be overlooked; this aspect requires further investigation but is outside the scope of the present paper.

The differences in the sediment load on the two sides of Anglesey are related to physiographical differences. In Liverpool Bay (on the eastern side) the water is relatively shallow. It is sheltered from the prevailing south-westerly winds but onshore north-easterly winds have a fetch of 170 km (70 miles). The seabed close to the shore is of soft mud and this is easily disturbed by wave action. Strong tidal currents also carry sediment into the area. On the western coast the exposure to wave action is much greater and the accumulation of fine sediment is prevented; the local load of sediment was observed to be less effective in the reduction of transmittance.

The nature of the sediment load

The attenuation of radiation through a water column can be expressed as: attenuance = absorbance + scatter. Absorbance is a property of the water and is dependent upon the wavelength of the incident light. In the inshore waters measured by us absorbence resulted in the loss of the longer wavelengths of the visible light (Fig. 7). Scatter due to water molecules and very small particles is proportional to the density of the scattering medium and is inversely proportional to the fourth power of the wavelength of the incident radiation. Scatter and absorbance by particles of the same size or greater than the wavelength of the radiation is much less dependent upon wavelength. The scattering coefficient, S_d, of these particles has been expressed by Ångström (1929) as $S_d = \beta r^{-\gamma}$ where γ is proportional to the particle density and decreases with particle size. With particles much greater in size than 1 μm, $-\gamma$ tends to zero and the scattered radiation has the same spectral distribution as the incident radiation. It is these larger particles that are most effective in the attenuation of light radiation in inshore waters.

Particles between 1 and 50 μm were found to be evenly distributed

through the water column and showed consistent variations directly proportional to the attenuation in the water column. The particles above this
size were the most difficult to analyse. The variations in the dry weight of
sediment in a given volume of water sampled at the station showed some
correlation with the seasonal changes in attenuation (Figs. 6, 7). However,
the measurement of sediment by weight does not record particle shape or
numbers of particles. Thus the inorganic fraction will tend to have a high
density with a small cross-sectional area whilst the organic fraction will be

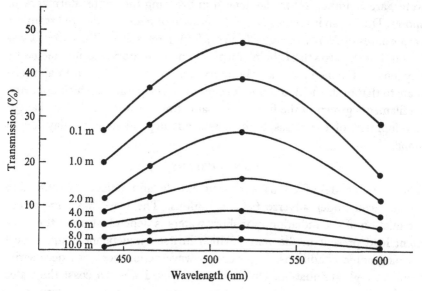

Fig. 7. Transmission of different wavelengths measured with the aid of four filters, characterized in Fig. 1. Readings were taken at depths of 0.1, 1, 2, 4, 6, 8 and 10 m, and are
expressed as percentage of the values recorded just above the surface.

of low density and the particles will tend to have a larger cross-sectional
area. Sometimes the suspended particles were large enough to be seen with
the naked eye. The turbulence at the seabed created by water movement
during a storm causes debris, particularly fragments of algae and low
density particles of shell gravel, to be raised into suspension. Whilst algal
fragments are held in suspension for long periods the shell gravel is in
constant oscillation between the seabed and suspension as the local eddies
vary in velocity. In summer, during calm weather, strands of mucilage
5–8 cm long have been observed in some quantity from the seabed to the
water surface. The strands were very tenuous and disintegrated rapidly in
contact with a solid object; attempts to collect samples of them have not

been successful. They occurred in particularly dense quantities above the forests of *Laminaria hyperborea* (Gunn.) Fosl. and it is believed that they originate from these plants as an extracellular exudate. Both *L. hyperborea* and the mucilage strands were absent at the east coast stations.

Use of laboratory illumination adjusted to match field measurements

Some results of experiments using the method of adjusting laboratory irradiation have appeared elsewhere (Smith & Jones, 1970) and suggest that other environmental conditions, such as the temperature of the sea water, are as important as day length in breaking the winter dormancy of spores. During an investigation of the growth of filaments derived from the carpospores of *Naccaria wiggii* (Turn.) Endl (Jones & Smith, 1970), it was found that the growth of the filaments in the laboratory was unaffected by day length. Filaments under laboratory illumination, adjusted to approximate to that in the field, did not develop the tetrasporangia which appeared on filaments grown in the field at the same time. It was concluded that the development of the tetrasporangia was not dependent upon day length alone.

CONCLUSIONS

The methods described have proved simple and reliable in obtaining results even under adverse field conditions. The work has shown that attenuation of the radiation in inshore waters is dependent upon the sediment load present in the water, which in itself is dependent upon local oceanographic conditions. Exposure to wave action does not necessarily result in high attenuation. On the more exposed western coast the water was found to have a lower attenuation, as sediment did not accumulate at these stations. In more sheltered regions, the accumulation of sediment in relatively shallow water resulted in a high attenuation of radiation through the water. It is concluded that the differences in the algal flora observed on the eastern and western coasts of Anglesey are due in some part to these contrasting conditions.

SUMMARY

The purpose of the work was to correlate the observed differences in the algal vegetation at two stations on the eastern and western coasts of Anglesey with the differences in the illumination regimes.

Irradiation was measured by means of a barrier layer photo cell in a Perspex waterproof case. Neutral filters were used to extend the range of the instrument. Spectrum filters of known transmission were used to determine the extinction curve of known wavelengths.

Values were measured *in situ* at metre intervals along a vertical plumb line. A measure of light intensity at the sea surface was taken before and after the irradiation readings. Corrections for filter density and instrument sensitivity were applied to the field readings before constructing the extinction curve.

The results confirmed that the regimes of water transparency on the west and east coasts of Anglesey were different. The differences were found to be due to the size and density of suspended particles in the water. Consistent quantitative values for the irradiation of the seabed under a wide range of inshore conditions were obtained throughout the year and could be correlated with differences in the populations of the benthic marine algae.

REFERENCES

ÅNGSTRÖM, A. (1929). On the atmospheric transmission of sun radiation and on the dust in the air. *Geogr. Annlr.* **11**, 156–66.

JERLOV, N. G. (1953). Optical studies of ocean water. *Rep. Swed. deep Sea Exped.* **3**, Fasc. II, 1–61.

JONES, W. E. & SMITH, R. M. (1970). The occurrence of tetraspores in the life history of *Naccaria wiggii* (Turn.) Endl. *Brit. phycol. J.* **5**, 91–95.

LEVRING, T. (1959). Submarines licht und die Algen Vegetation. *Botanica mar.* **1**, 67–75.

SMITH, R. C. (1969). An underwater spectral irradiance collector. *J. mar. Res.* **27**, 341–51.

SMITH, R. M. (1967). Sublittoral ecology of marine algae on the North Wales coast. *Helgoländer wiss. Meeresunters.* **15**, 467–79.

SMITH, R. M. & JONES, W. E. (1970). The culture of benthic marine algae in the laboratory and in the field. *Helgoländer wiss. Meeresunters.* **20**, 62–69.

WEATHER (1968). Weatherlog for February 1968, supplement. *Weather, Lond.* **23**, No. 4.

CONTINUOUS RECORDING OF UNDERWATER LIGHT IN RELATION TO *LAMINARIA* DISTRIBUTION

JOANNA M. KAIN (Mrs Jones)

Marine Biological Station, Port Erin, Isle of Man

INTRODUCTION

In western Europe where conditions are suitable, *Laminaria hyperborea* (Gunn.) Fosl. is the dominant plant below extreme low water spring tides (ELWS) on rocks and stable boulders. In some places the lower limit is at 30 m or more (Jorde, 1966; Ernst, 1966) but in others it is at considerably shallower depths (Kain, 1962). In the Isle of Man most of the subtidal rock gives way to sand or gravel at a depth of 5–15 m but there are a few sites with deeper rock; here the lower limit of *L. hyperborea* is inconstant (Kain, 1962, 1966). Where the rock ends at about 10 m below ELWS the lower limit of *L. hyperborea* is at about 6 m; where the rock ends at 15 m *L. hyperborea* extends to about 10 m and where the rock reaches 20 m or more the lowest plants of this species are found between 15 and 20 m. Thus there is always a zone of rock next to the sand or gravel which is bare of *Laminaria*. This zone has been shown in at least one case to be due to a higher density of *Echinus esculentus* L. which seems to congregate next to the sand and which destroys young *Laminaria* plants by grazing (Jones & Kain, 1967). There must be a depth, however, where *L. hyperborea* colonization is limited by light. The question is whether the deepest plants found off the Isle of Man are at the light limit or whether some other factor prevents deeper colonization there too.

Young stages of *L. hyperborea* and its main competitors in Britain have been grown in the laboratory and their light requirements determined (Kain, 1969). In order to relate this information to conditions in the sea, light measurements had to be made in the field locally. For some years this was done by suspending a photocell equipped with Schott BG 12, VG 9 and RG 1 filters at various depths from a boat (Kain, 1966). The two main disadvantages of this system are the relative sparsity of measurements and their prevention by rough seas. Accordingly, a continuously recording system was set up.

CONSTRUCTION AND CALIBRATION OF RECORDING SYSTEM

Two selenium barrier-layer cells are each housed in a watertight Perspex case. The upper part of the case is covered with black Perspex with a 5 cm window over the photocell. The window is encircled by a filter-holder containing a green Schott VG9 filter, a neutral filter and an opal glass, the last being flush with the top of the holder. A sensitivity curve for a VG9 filter and photocell is given by Sauberer (1962, Abb. 5). Because of trouble with neutral glass filters, a 'pair' of which did not remain identical underwater, one or two layers of black nylon chiffon glued to a rigid plastic ring are used. The cases are bolted to two shelves on a galvanized Dexion (angle-iron) and concrete construction (Fig. 1). The upper shelf is at 3.1 m below ELWS and the lower one exactly 2 m below that. Each case is provided with enough flexible wire for it to be brought to the surface without disconnecting. The other end of the flex is joined to a heavy cable laid in the angle of further Dexion and led across gravel and boulders to the lower end of Port Erin lifeboat slip. The cable is fixed to the side of the lifeboat slip through the littoral zone which is often subject to heavy wave action. The availability of the slip was the reason for siting the apparatus inside and not outside Port Erin Bay. The cable is then led to the Biological Station to a 2-channel recorder with three alternative scales of different sensitivities. A Kipp and Zonen solarimeter is installed on the cliff behind the Station and the cable from this led to a second recorder. The thermopile and light cells are about 230 m apart.

About twice a year the cell cases are brought to the surface for calibration. For this they are attached for at least a day to a floating frame in such a way that the opal filters are about 5 cm below the surface. Westlake (1965) has pointed out that a depth of water smaller than half the diameter of the opal may allow back reflection from the underside of the water surface. Calibrations can be carried out only when weather conditions are stable, so that the sea surface is not rough and irradiance records smooth and regular. From a comparison of the solar irradiance records (in the visible band) and the photocell records (green) a calibration factor can be calculated for each photocell. This gives, from any irradiance reading, the output that each cell would give if it were at the surface. This calibration factor is designed for estimates of energy as a percentage of that at the surface. The response of the cells decreased, amounting to 25% in 1 year. The calibration factor is appropriately increased from the date of one calibration to that of the next.

A common difficulty with permanently installed photocells is unwanted materials such as silt and diatoms ('dirt') accumulating on the opal glasses. Whenever feasible (in practice every 1–3 weeks) divers have cleaned the opals. Cleaning between the filters is less simple and done infrequently.

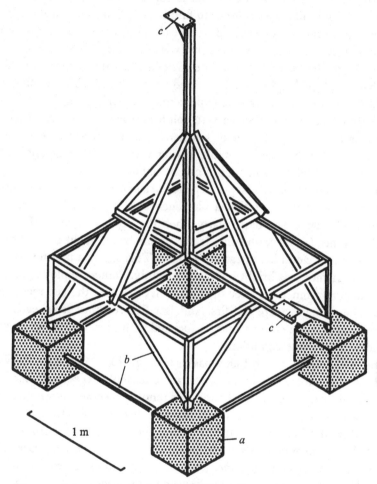

Fig. 1. Diagram of the construction bearing the photocell cases. (*a*) Concrete block weighing 124 kg. (*b*) Dexion galvanized angle iron 60 × 40 mm. (*c*) Shelf on which is bolted a Perspex case containing a photocell.

CALCULATION OF RESULTS

The two separate record charts are superimposed one upon the other and aligned exactly for each day. Four or five times when conditions were stable are selected and the simultaneous values read off each chart, one total visible energy value for the surface and two values from underwater.

The energy recorded from the thermopile is corrected by a factor of 0.46 to obtain a value for energy between 400 and 700 nm. Appropriate calibration and dirt correction factors are applied to the figures from the underwater cells so that the result from each cell is expressed as a percentage of the surface visible irradiance. In order to find the depth of water at the time of each reading a set of predicted tidal curves has been prepared, showing the height of water above the upper cell and with the time scale identical to that on the record charts. The appropriate one of these is then superimposed on the charts and the heights at the selected times read off. After adding 2 m for the lower cell the values for the percentage of surface irradiance can be plotted against depth. This gives 8–10 points for each day. When conditions are stable all day the number of selected times for measurement can be increased, and if the tide is suitable, giving a wide range of height during daylight, a good plot against depth can be obtained.

As light attenuation with depth depends to some extent on the altitude of the sun one might expect that a plot such as this would be affected by variations during 1 day, particularly when the sky is cloudless. However, Smith & Jones (1971) found this factor unimportant. That it can be disregarded under cloud is indicated by Fig. 2. This shows a plot for 23 May 1968, a day which was bright but mainly cloudy. Points are plotted at half-hour intervals from 1.5 h after the first to 1.5 h before the last recordable light. The triangles were the first and last five readings from each cell. There is no divergence apparently associated with time, and therefore with altitude, under these conditions.

The extrapolated line in Fig. 2 passes through 100%. Because the VG9 filter used transmits a fairly wide waveband one would expect a slight curve, particularly near the surface, though this tendency would be minimized if this waveband included the wavelengths of greatest penetration in this water. More detailed measurements with a photocell operated from a boat have failed to demonstrate a curve in these waters so it seems reasonably safe to extrapolate a straight line. This can be useful. The problem of dirt on and between the filters has already been mentioned. If dirt is present the line has the same slope but extrapolates to less than 100%. An example is shown in Fig. 3 where the two cells had different amounts of dirt on them. This can be used to determine a correction factor for dirt at any one time. However, it is not often possible to obtain as good a plot as that shown in Fig. 3 and extrapolation may be inaccurate. A better way of making this correction is to take readings before and after cleaning on the same day. The points after cleaning can be extrapolated to 100% and the divergence from this line of the readings before cleaning used to produce a correction factor. It has to be assumed that dirt accumulates quite regularly on and between the filters and that the correction factor to be applied changes linearly with time between one cleaning and the next. The factors for

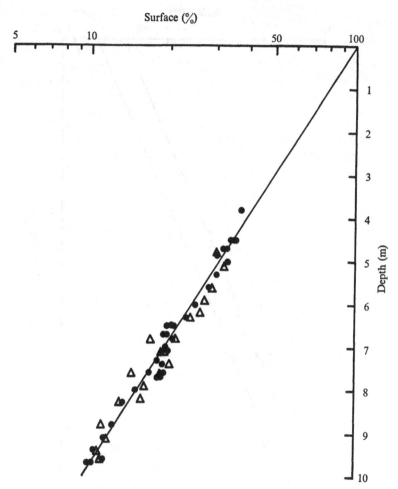

Fig. 2. Plot of green irradiance as a percentage of surface recorded by the two underwater photocells. △ 1.5–4 h from first or last recordable light. ● More than 4 h from first or last recordable light.

between-filter dirt, cleaned off less regularly, and dirt on the top filter, cleaned more often, are treated separately throughout the year but combined with the calibration factor for each day.

The results are a measure of only the green part of the spectrum, and need to be converted to total visible energy (effective in photosynthesis).

Some information can be obtained from the photometer lowered from the boat and equipped with blue, green and red filters, though a set of narrower band filters would have been better. The total visible irradiance in this case

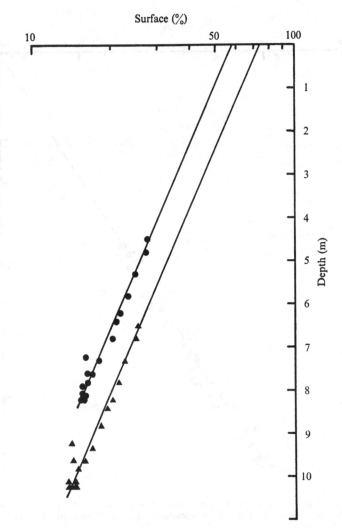

Fig. 3. Plot of green irradiance as a percentage of surface recorded by underwater cells partly covered by dirt. ● Upper cell. ▲ Lower cell.

has been calculated using the weighting factors 0.3, 0.35 and 0.35 respectively for the readings from the filters BG 12, VG 9 and RG 1 (Talling, 1957). In Fig. 4 is plotted the green value as percentage surface against the calculated total visible as percentage surface for depths of 1–20 m on 13 different occasions including a wide range of conditions. The curve produced has been used to convert green percentages to total visible percentages.

Four depths have been chosen for the final calculations of irradiance: 0, 5, 10 and 15 m below ELWS. The mean tide height at Port Erin is about 3 m, but the addition of 3 m to the above depths would not give a mean water depth for light penetration because of the logarithmic change of

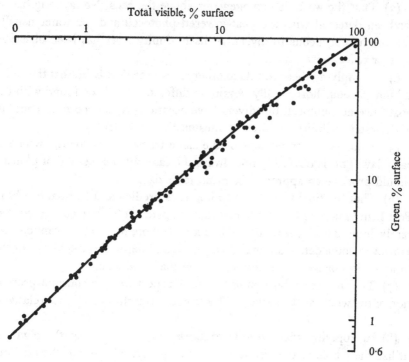

Fig. 4. Percentage of surface irradiance recorded through a Schott VG9 filter plotted against the percentage of surface irradiance between 400 and 700 nm.

light with depth. The height to be added to allow for the tide should vary with the variation in transparency but this is impracticable. A good compromise for these waters is 2.5 m.

For each day, therefore, available values for the green as a percentage of surface are plotted against depth. A line is fitted, passing through 100% at the surface. Values are read off at 2.5, 7.5, 12.5 and 17.5 m and converted to total visible percentages using the curve in Fig. 4. Each percentage value is then multiplied by the integrated mean surface visible solar irradiance for that day obtained from the record from the solarimeter.

This method involves the following assumptions.

(1) That cell deterioration and dirt accumulation are linear with time. They are more likely to be irregular.

(2) That the irradiance at the thermopile is the same as at the sea surface over the cells. Some anomalous values must result from cloud edges but should not cause systematic errors. There may be some shading of the underwater cell site by the land in winter.

(3) That the water is homogeneous above the cells. No layering has so far been detected with the boat-lowered photocell and the water usually appears homogeneous to divers. A discontinuity layer has been observed once or twice, however.

(4) In applying these results to other areas near Port Erin, that the water is homogeneous horizontally. Again no difference has been found with the boat-lowered photocell but divers have occasionally noticed differences in underwater visibility in different sites on the same day.

(5) That the water remains of the same transparency for the whole of each day. This is obviously not always the case but the method of plotting should produce an approximate mean for 1 day.

(6) That the tidal height and timing is as predicted. The tidal height in Port Erin in fairly good weather conditions (when it has been observed) has rarely differed from the predicted by more than 0.5 m, usually by much less. However, some detailed plots of irradiance and depth show evidence of the time of high or low water differing from that predicted.

(7) That the green portion of the visible spectrum is a constant proportion of the whole at the surface. This is not so but the deviation is relatively small.

(8) In applying these results to *Laminaria* growth, that the alga can utilize the whole of the visible spectrum equally. This is not the case but in the absence of accurate knowledge of the action spectrum of *L. hyperborea* this approximation is unavoidable.

In spite of these assumptions the method has the obvious advantage that extrapolation is possible, to determine deep water values. For this sort of study great precision is not necessary.

RESULTS AND DISCUSSION

This is a preliminary report and the results available are restricted to two contrasting months, June and December 1968. Fig. 5 shows the calculated daily percentages of the surface irradiance for the 2 months at 0, 5, 10 and 15 m below ELWS. Much of June was remarkably constant. This is often the case in the summer when there are periods of calm. The first part of December was also fairly constant and similar to June. In this case it was because the wind was easterly, from which direction Port Erin Bay is sheltered. In the second half of the month, however, the wind was mainly west or northwest and the waves stirred up bottom sediments and reduced light penetration. This is probably the more usual winter situation. It clearly makes a considerable difference to the irradiance in deeper water.

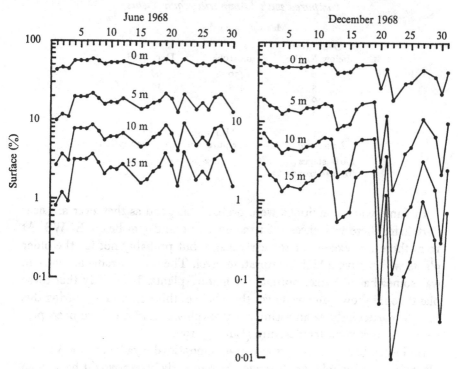

Fig. 5. Mean daily percentage of surface visible irradiance at 4 depths calculated from observations during June and December 1968.

Mean daily irradiance figures for each depth have been combined to give a mean figure for the whole of each month and are shown in Table 1. The requirements of the early stages of *L. hyperborea*, *L. digitata* (Huds.) Lamour., *L. saccharina* (L.) Lamour. and *Saccorhiza polyschides* (Lightf.) Batt. are 1–2 μgcal/cm^2/s as a minimum below which growth cannot take place and 50–60 μgcal/cm^2/s as light saturation for growth (Kain, 1969). These figures are for continuous irradiance from fluorescent ('Northlight') tubes at 10 °C. The difference in spectrum from the light in the sea will cause some inaccuracy in comparison. The mean figures for whole months are equivalent to continuous irradiance only if all the irradiance can be utilized, that is, if none was above saturation for photosynthesis. In addition, these figures apply to open water at each depth. The energy received at the rock surface may be reduced by shading by cliffs, other rocks, the *Laminaria* canopy, other algal forms and possible silt deposition.

Table 1. *Calculated visible irradiance for 2 months
compared with Laminaria requirements*

Mean μgcal/cm^2/s

Depth (m) below ELWS	June 1968	Dec. 1968
0	1450	68
5	455	17
10	165	5.3
15	64	1.9

Laminaria	Minimum requirement	Saturation
Early stages	1–2	50–60
Mature fronds		400?

In June, when conditions were probably as good as they ever are near Port Erin, there was excess of irradiance at 0 and 5 m below ELWS. At 10 m there was excess for the early stages but probably not for the older plants which have a higher saturation level. The mean irradiance at 15 m was somewhat less than optimal for mature plants. It is likely that these plants could grow quite well over the whole of this depth range during this month, particularly as an assimilatory surplus should not have been prevented by increased temperature (Lüning, 1971).

In December, however, there was suboptimal irradiance for mature plants over the whole depth range. At 0 m early stages would have been saturated during the (short) day in open water but under the canopy of the *Laminaria* forest they too would probably have been limited. As this was not a bad December for winds the winter is clearly a rigorous time. At 15 m the early stages of *Laminaria* would have been unable to grow. The maximum reproductive period of *L. hyperborea* is in December and January and although the spores can survive for long periods in the dark they could be subjected to *Echinus* grazing before an irradiance level sufficient for growth occurred. The establishment of this species at this depth seems likely to be chancy. In a good winter it might be possible and in a bad one perhaps not. This probably explains the predominance of particular age groups at this depth (Kain, 1963) and the fact that few plants occur below it. The relatively high summer figure for 15 m probably explains why plants once established can grow reasonably well and are not small and stunted.

More results are necessary to confirm these conclusions.

SUMMARY

Around the south end of the Isle of Man the vertical distribution of *Laminaria hyperborea* and its main competitors is known, as are the light requirements of their early stages. It seems likely that light only sometimes limits their depth penetration. In order to measure the light environment at various depths and seasons in the area some method of continuous recording is desirable.

Two selenium barrier layer photocells have been maintained in Port Erin Bay at 3 and 5 m below ELWS for 2 years. They are contained in Perspex waterproof boxes and provided with green, neutral and opal filters. A cable runs to a continuous two-channel recorder. Another recorder is linked to a solarimeter. The photocells are brought to the surface for periodic calibration against a thermopile. More detailed single observations of light penetration have been made with a photocell with blue, green and red filters lowered to various depths from a boat.

From the continuous underwater records, the light reaching various depths can be calculated approximately by extrapolation, using the following factors. (1) The height of the water over each photocell, varying with the tide. (2) The relationship between the green part of the spectrum, as a percentage of the surface, and the total visible energy, as a percentage of the surface. (3) Calibration factors for the photocells. (4) Neutral filter factors. (5) Corrections for dirt on or between the filters (in spite of cleaning).

Mean irradiance values for June and December 1968 have been calculated for the depths 0, 5, 10 and 15 m below ELWS. In June there was excess light at 0 and 5 m and enough for early stages but not mature plants at 15 m. In December all depths received suboptimal light, with growth impossible at 15 m, which is near the lower limit of *L. hyperborea* in the Isle of Man.

I am very grateful for assistance underwater, in setting up the apparatus and in cleaning the cell covers, from my husband Dr N. S. Jones, Dr R. G. Hartnoll, Mr T. P. Bell and Mr M. J. Bates.

REFERENCES

ERNST, J. (1966). Données quantitatives au sujet de la répartition verticale des Laminaires sur les côtes nord de la Bretagne. *C. r. hebd. Séanc. Acad. Sci., Paris* **262**, 2715–17.

JONES, N. S. & KAIN, J. M. (1967). Subtidal algal colonization following the removal of *Echinus*. *Helgoländer wiss. Meeresunters.* **15**, 460–6.

JORDE, I. (1966). Algal associations of a coastal area south of Bergen, Norway. *Sarsia* **23**, 52 pp.

KAIN, J. M. (1962). Aspects of the biology of *Laminaria hyperborea*. I. Vertical distribution. *J. mar. biol. Ass. U.K.* **42**, 377–85.

KAIN, J. M. (1963). Aspects of the biology of *Laminaria hyperborea*. II. Age, weight and length. *J. mar. biol. Ass. U.K.* **43**, 129–51.

KAIN, J. M. (1966). The role of light in the ecology of *Laminaria hyperborea*. In *Light as an Ecological Factor*, pp. 319–34 (ed. R. Bainbridge, G. C. Evans and O. Rackham). Oxford: Blackwell.

KAIN, J. M. (1969). The biology of *Laminaria hyperborea*. V. Comparison with early stages of competitors. *J. mar. biol. Ass. U.K.* **49**, 455–73.

LÜNING, K. (1971). Seasonal growth of *Laminaria hyperborea* under recorded underwater light conditions near Helgoland. This volume, 347–61.

SAUBERER, F. (1962). Empfehlungen für die Durchführung von Strahlungs-messungen an und in Gewässern. *Mitt. int. Verein. theor. angew. Limnol.* **11**, 77 pp.

SMITH, R. & JONES, W. E. (1971). Measurement of submarine illumination. This volume, 321–33.

TALLING, J. F. (1957). Photosynthetic characteristics of some freshwater plankton diatoms in relation to underwater radiation. *New Phytol.* **56**, 29–50.

WESTLAKE, D. F. (1965). Some problems in the measurement of radiation under-water: a review. *Photochem. Photobiol.* **4**, 849–68.

SEASONAL GROWTH OF
LAMINARIA HYPERBOREA UNDER
RECORDED UNDERWATER LIGHT
CONDITIONS NEAR HELGOLAND

K. LÜNING

Biologische Anstalt Helgoland (Meeresstation),
Helgoland, Germany (FRG)

INTRODUCTION

Laminaria species, like many other marine perennial algae in Western Europe, have their period of maximum growth during the cold season. This fact has been interpreted in the past by the classical theory of Kniep (1914) and Harder (1915). These authors reported that, with increased temperature, respiration of marine algae (e.g. *Laminaria saccharina* (L.) Lamour., *Fucus serratus* L.) increases much more than photosynthesis and concluded that there is no assimilatory surplus in these algae in summer because of their high respiration at this time, and that the highest net gain of photosynthesis occurs in winter and early spring (see discussion in Gessner, 1955). These suggestions were confirmed by an early paper of Ehrke (1931) working with *F. serratus, Plocamium* sp., and *Delesseria* sp. Lampe (1935) and Montfort (1935), on the other hand, found that seasonal adaptation of respiration and potential photosynthesis occurs in a variety of perennial species (e.g. *L. saccharina, L. hyperborea* (Gunn.) Fosl., *F. serratus*). This possibility has been admitted by Ehrke (1934) and Levring (1947). Kanwisher (1966) reported that partial seasonal adaptation occurs in *Chondrus* and *Ascophyllum*. For instance, *Chondrus crispus* Stackh. respires in winter at the same rate as at a 10 °C higher temperature in the summer. According to Lampe (1935), Montfort (1935) and Kanwisher (1966) species of *Fucus* and *Laminaria* have their highest net gain of photosynthesis during the warm season, since underwater irradiance is strongest in summer and respiration does not consume all the assimilatory surplus which occurs at that time, as respiration rates in summer are only slightly above those measured in winter. Kniep (1914), Harder (1915) and Ehrke (1931) had conducted only short-termed measurements of respiration and photosynthesis, hence they had overlooked the effect of seasonal adaptation.

In this paper attempts are made to interpret the rhythm of seasonal growth of the brown alga *L. hyperborea* on the basis of data on continuously measured underwater irradiance and rates of respiration and photosynthesis measured at different seasons.

MATERIAL AND METHODS

The rhythm of seasonal growth of *L. hyperborea* was followed by transplanting specimens, grown under natural conditions at a depth of 2.5 m below mean low water of spring tides, on to PVC plates fixed to subtidal growth stations. The SCUBA diving technique was employed, as was described in an earlier paper (Lüning, 1969a). The algae were raised to the surface and photographed on board a boat at intervals of 2–4 weeks. Measurements of underwater irradiance at 2.5 m water depth were performed according to the method of Johnson & Kullenberg (1946), which has also been used by Levring (1947). This method involves multiplication of the photocurrent of one individual photoelement, used without colour filters at a certain water depth, with a factor k. This factor is a function of water depth and of the ratio between the photocurrents at the given water depth and above the water surface. In the present study one photoelement (Type S 50; manufacturer: Dr B. Lange, Berlin, West Germany) was mounted in a watertight housing, made of plexiglas, and fastened to a subtidal station at 2.5 m water depth, near the stations to which *L. hyperborea* specimens had been fixed. The signals of the photoelement were conducted to a recorder in the laboratory by underwater cable. Another photoelement of the same type was mounted on the roof of the laboratory, some 300 m distant from the underwater station. Opal glass filters (1:100 in summer, 1:10 in winter) were used, as well as a protective plexiglas cap on the underwater light sensor, preventing by its smooth surface the settling of benthic diatoms. The constancy of the light-exposed photoelements was checked at intervals of some months by comparison with the response of a photoelement kept in the dark, using a standardized light source. Deviations of sensitivity of the light-exposed photoelements were below 5% after 1 year exposure. The photoelements were calibrated in the range between 380 to 720 nm against a Kipp and Zonen thermopile (Delft, Holland) by means of solar radiation, according to the procedure recommended by Johnson & Kullenberg (1946).

A total of 58 measurements of respiration rates and curves of photosynthesis *v.* irradiance was performed in the laboratory. Attempts were made to simulate natural conditions of underwater light in the laboratory. A tungsten filament lamp (Osram 12 V, 150 W; with ellipsoidal reflector bulb) was supplied with Schott filters KG3 (4 mm), BG22 (3 mm) and BG18 (1 mm). Optical arrangement employed was similar to that described by Ziegler, Ziegler & Schmidt-Clausen (1965). Values of relative spectral emission (Table 1) of this combination (measured with a Kipp and Zonen thermopile supplied with a 'Schott-Interferenzverlauffilter') are comparable with the transmission curve of type 6 of coastal water, as characterized by Jerlov (1954), except for a higher relative spectral emission at 500–600 nm, occurring in the combination of lamp and filters employed.

Table 1. *Values of relative spectral emission*

Of a tungsten filament lamp (Osram 12 V, 150 W; with ellipsoidal reflector bulb), supplied with Schott filters KG 3 (4 mm), BG 22 (3 mm), BG 18 (1 mm).)

nm	400	450	500	550	600	650	700
%	10	32	71	100	74	29	3

Experiments on photosynthesis and respiration were carried out in March (4 °C), May (8 °C), August (16 °C), and November/December (10 °C). Fronds of *L. hyperborea* were taken from the sea immediately before the start of the experiment. A circular disc of a diameter of 8 cm was punched out of a frond. The tissue was transferred into a plexiglas chamber filled with filtered sea water collected from the sea immediately before use. The chamber was closed with a plexiglas cover sealed with an 'O' ring. Water was pumped from the chamber, at a rate of 3.8 l/min, passing a polarographic oxygen electrode (manufacturer: Hydrobios, Kiel, West Germany) and recirculated into the chamber. The total volume of water in the closed system was 335 ml (volume of algal tissue about 5 ml). The whole assembly was immersed in a water bath kept at constant temperature ($\pm$0.1 °C) by a thermostat. Water temperature was within $\pm$1 °C of the temperature of the sea at the time of the experiment. After equilibration between water temperatures inside and outside the chamber, rates of oxygen consumption in the dark were read for 30 min, at intervals of 5 min, on a Microva AL4 galvanometer (manufacturer: Kipp and Zonen, Delft/Holland), until the oxygen content of the water had decreased to about 80%. saturation. The tissue was then irradiated from above. The highest irradiance from the lamp, supplied with filters, as described above, was about 700 μgcal/cm²/s on a circular area of 15 cm diameter. Irradiance decreased with age of the lamp and was measured after each experiment with a Kipp and Zonen thermopile. Rates of gross photosynthesis were measured at seven different irradiances. Neutral glass filters (Schott NG11) with decreasing thickness were employed. Five minutes duration at each light intensity was sufficient to get a significant increase of oxygen content of the water. After the light was turned off, rates of dark respiration were measured again for 30 min. The experiment was stopped after about 2 h. Bacterial activity in the water could be neglected for a 2 h duration of one experiment, as was proved by test runs without algal tissue in the chamber. Dry weight of algal tissue was determined after drying at 103 °C.

A method similar to that employed by Parker (1963, 1965) was used for labelling the old frond of *L. hyperborea* with C^{14}. A plexiglas reaction chamber containing 400 ml sea water, to which was added a few ml of a $NaHC^{14}O_3$ solution (800 μCi), was immersed in running sea water (temperature 10 °C) to within 1.6 cm of the open top. The old frond of a small specimen of *L. hyperborea* was placed into the reaction chamber, the young frond and stipe into the running sea water, permitting the exposure to air of about 3 cm of the tissue connecting the old to the new frond. The old frond was illuminated from above for 14 h. Discs of frond 22 mm in diameter and sections of stipe about 1 mm thick were cut, treated with 0.1 N-HCl, dried and counted for radioactivity, using a Tracerlab FDA-50 GS/SC-511 AL gas flow instrument. Counting was considered to be at infinite thickness. Artificial translocation can be excluded, as was proved by a control specimen placed outside the reaction chamber in running sea water, this showing no signs of radioactivity at the end of the experiment.

RESULTS AND DISCUSSIONS

The growth records of a specimen of *L. hyperborea* are illustrated in Fig. 1. This specimen was collected from a community with a low population density at a depth of 2.5 m and transplanted to a subtidal growth station, situated at the same water depth, in October 1966. From comparison of the frond area and stipe length of this plant at the time of transplantation with the size of sporophytes of *L. hyperborea* whose growth under the same natural conditions had been followed since they first appeared, it is suggested that the transplanted alga arose in the early months of 1965, and Fig. 3 demonstrates growth in the third, fourth and fifth years. The new frond is produced during the first half-year, maximum increase of frond area occurring from late April until the end of May in all years. The area attained by the new frond at the end of the period of rapid growth (June) increases from year to year (by a factor of 1.6–1.7 in this example). During the second half-year frond tissue is lost more or less continuously at the distal end. In December/January, when growth of the new frond commences, the old frond area has been reduced by about half. It seems most likely that the increase in frond area from year to year, which can be valid, of course, only for the early years of *L. hyperborea*, is partly due to reserve materials assimilated by the old frond during the preceding period of slow growth, supporting the growth of the new frond (Lüning, 1969*a*). Growth rate of the new frond decreases if the old frond is amputated at the beginning of the year. Complete specimens produce a small new frond during the first half-year even in complete darkness (Lüning, 1969*a*). It is sug-

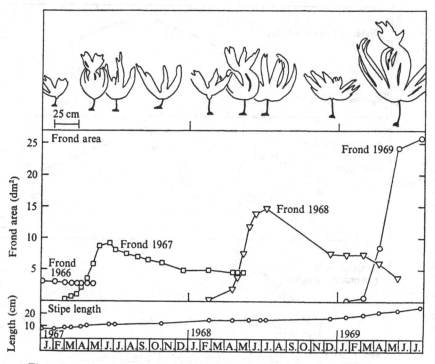

Fig. 1. Records of the growth of *L. hyperborea* growing on a subtidal growth station at a depth of 2.5 m.

gested that during the early years of *L. hyperborea* about 30% of the new frond area is due to reserve materials, stored mainly in the old frond (Lüning, 1970).

The seasonal variation in respiration rates of *L. hyperborea* fronds (Fig. 2) shows some remarkable features. Respiration rates per frond area (Fig. 2*b*) show great variation, but generally follow the shift of water temperature *in situ* (Fig. 2*a*) throughout the year. Dry weight per unit frond area increases from March to August by about three times. If respiration rates are calculated on dry weight basis (Fig. 2*c*) it appears that the respiration rate of the new frond is highest in May and decreases significantly from May to August, even though water temperature *in situ* increases from 8 to 16 °C during the same period. The seasonal shift of respiratory activity may partly be due to a large increase in metabolically inactive components occurring in *L. hyperborea* fronds during summer. According to Black (1948, 1950) the laminarin content in the fronds increases from May to November from 0 to about 30% on dry weight basis. A similar

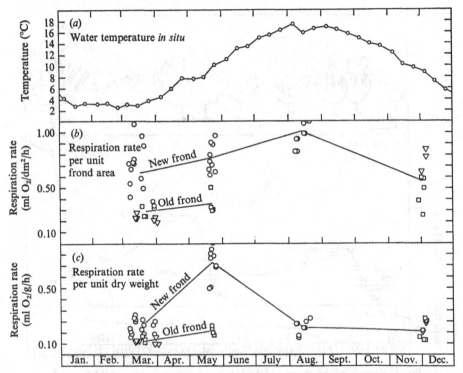

Fig. 2. (*a*) Water temperature *in situ*; (*b*) and (*c*) respiration rates of *L. hyperborea* fronds, measured in the laboratory at different seasons at temperatures corresponding to water temperature *in situ*. ○, Young frond; ▽, immature old frond; □, fertile old frond.

increase is reported to occur in the mannitol content of the fronds. The decrease in respiratory activity is, however, also noticed, if respiration rates are referred to the crude protein content, as determined by Black (1948, 1950) at different seasons (Table 2). It is concluded, hence, that an adaptation to higher temperatures in summer takes place in *L. hyperborea* fronds, apart from the influence of seasonal variation in the content of metabolically inactive components. The influence of growth activity on respiration rate is inferred from the following findings. Maximum growth activity, occurring in May, is paralleled by maximum respiration rate per year. Respiration of the old frond (Fig. 2b, c) is only about 50% of respiration of the new frond in March and 27% in May.

Curves of gross photosynthesis v. irradiance (P v. I curves), measured in *Laminaria hyperborea* fronds at different seasons, are presented in Fig. 3. In spite of the scattering of values, referring to several algal samples, familiar also to other investigators measuring photosynthesis (Kanwisher, 1966), the well-known temperature dependence of light-saturated rates is clearly demonstrated (Lampe, 1935; Montfort, 1935; Talling, 1957, 1961; Jørgensen & Steeman-Nielsen, 1966; Kanwisher, 1966). Since high water temperature and high underwater irradiance are positively correlated, the possibility cannot be excluded that adaptation to irradiance also causes the shift of the P v. I curves (Talling, 1957).

Table 2. *Respiration rate of Laminaria hyperborea fronds*

(Rate during the first life year on crude protein basis (crude protein content of fronds according to Black, 1950).)

	March (4 °C)	May (8 °C)	August (16 °C)	Nov./Dec. (10 °C)
Respiration rate (ml O$_2$/g crude protein/h)	1.7	6.9	2.8	1.9

The net gain of photosynthesis per unit frond area, calculated at two irradiances (100 and 400 μgcal/cm²/s) and at different seasons, is illustrated in Fig. 4. Net photosynthesis at 400 μgcal/cm²/s, which is about saturation irradiance, increases from March to August during the frond's first year of life. This increase is caused by the seasonal shift of the P v. I curve (Fig. 3). The small increase of respiration rate per unit frond area (Fig. 2b), occurring in summer, is rather unimportant in regard to the net gain of photosynthesis in summer. The decrease of light-saturated rates in autumn and winter (Fig. 3) causes at the same time the decrease of net photosynthesis at 400 μgcal/cm²/s. Net photosynthesis of the old frond in May (second year of life) is smaller than at the same month, during the first year of life, at 400 μgcal/cm²/s. An ontogenetic change in potential photosynthesis is reported also to occur in Macrocystis (Sargent & Lantrip, 1952; Clendenning, 1961). Quite another picture is to be seen, if the net gain of photosynthesis per unit frond area is calculated at a lower irradiance (100 μgcal/cm²/s, Fig. 4). In this case the maximum net photosynthesis occurs during the cold season, since respiration consumes the greatest part of gross photosynthesis in summer. Kniep (1914), Harder (1915) and Ehrke (1931) employed only low irradiances, when measuring photosynthesis, so they concluded that the highest assimilatory surplus occurs in winter and early

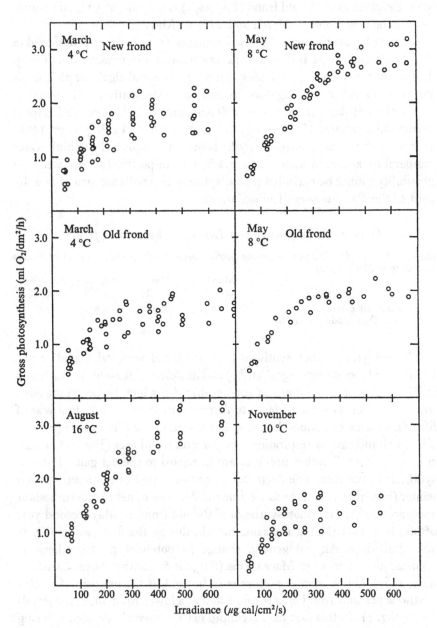

Fig. 3. Relation between gross photosynthesis and irradiance in *L. hyperborea* fronds, measured at different seasons at temperatures corresponding to water temperature *in situ* (see Fig. 2). Results of 3–7 experiments, all performed under similar conditions are represented.

spring. Continuous measurements of underwater irradiance at a depth of
2.5 m (Fig. 5) revealed, however, that irradiance was below compensation
irradiance for most of the time from October until March. On the other
hand, irradiance at the same water depth was near or above saturation
irradiance for most of the time from June to August. Hence, it is concluded
that the highest assimilatory surplus in *L. hyperborea* fronds occurs in
summer, part of the stored material being consumed in respiration and the
formation of the new frond in winter and early spring, when waters become
turbid and underwater irradiance is reduced to levels below compensation

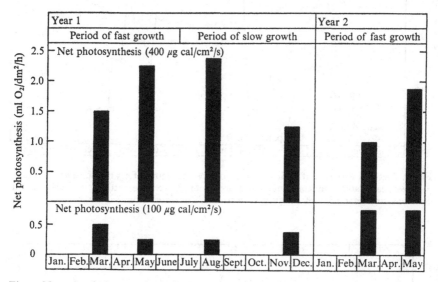

Fig. 4. Net gain of photosynthesis during the first and second year of life of *L. hyperborea*
fronds, calculated (from Figs. 2, 3) at two irradiances.

irradiance for long periods. Similar conclusions were drawn by Kain (1966)
in regard to the growth conditions for mature *L. hyperborea* fronds off the
Isle of Man. It should be mentioned that compensation irradiance is much
lower in early sporophytes (Kain, 1965) than in mature fronds.

Fig. 6 (solid curve) shows underwater irradiance at 2.5 m water depth as
percentage portion of irradiance above water surface. High percentage
values (15–27%) prevail in June/July, very low percentage values (below
1%) in March, October and December. Among the numerous factors
causing the seasonal variation in light penetration (Clarke, 1938; Strickland,
1958; Sauberer, 1962; Kain, 1966) the effects of wave action stirring up
sediment are very important in regard to underwater light conditions near
Helgoland. The broken curve in Fig. 6 shows seasonal variation in seston

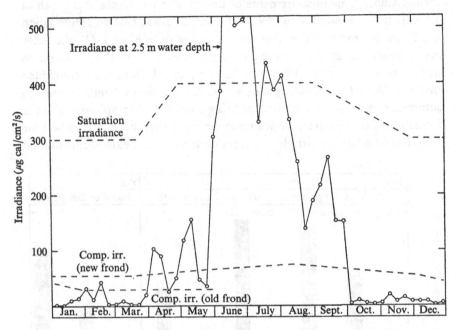

Fig. 5. Seasonal variation of underwater irradiance at a depth of 2.5 m. Solid line, average per week, calculated from continuous measurements. Broken line, saturation irradiance and compensation irradiance of *L. hyperborea* fronds, calculated from Figs. 2, 3.

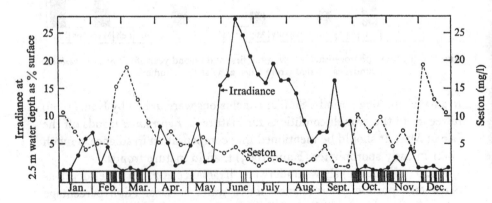

Fig. 6. Solid line, underwater irradiance at 2.5 m water depth as percentage of irradiance above water surface (average per week, calculated from continuous measurements). Broken line, Seston; dry particulate matter, > 1 μm; mean values for 10 days obtained from three samples per week taken from surface waters at the *Kabeltonne* station, about 800 m distant from the underwater light measuring station (Hagmeier, unpublished). Vertical lines in lower part of diagram indicate days with winds of Beaufort force 6 or more.

content (organic and inorganic particulate matter) of surface waters near Helgoland (Hagmeier, unpublished). The vertical lines in the lower part of the diagram indicate days on which wind velocity reached Beaufort numbers of six and more. The shift to low values of transparency, occuring in March and October to December, is accompanied by periods of rough weather and strong wave action resulting in a high seston content of the sea

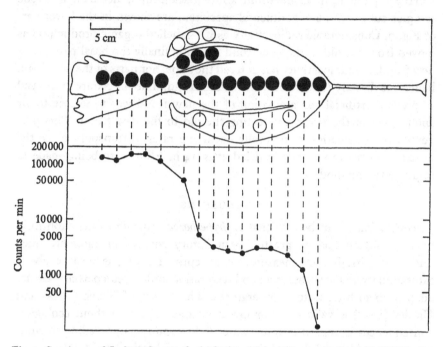

Fig. 7. Specimen of *L. hyperborea*, of which the old frond has assimilated NaHC^{14}O$_3$ for 14 h (water temperature 10 °C). Circles in upper part of diagram indicate samples taken for counting of radioactivity (●, radioactive samples; ○, inactive samples). Curve in lower part of diagram indicates counts/min above normal background.

water. Plankton blooms occurring in May and July/August (Hagmeier, unpublished) seem to have less importance in reducing transparency of the sea water near Helgoland than the sediment stirred up by wave action.

Since the old frond of *L. hyperborea*, if compared to the new frond, respires at a reduced rate (Fig. 2), it has also a lower compensation irradiance (Fig. 5). In consequence, net photosynthesis is higher in the old than in the new frond from January until May (see Fig. 4, low irradiance level). It seems likely that the growth of the new frond is supported not only by reserve materials stored in the old frond and assimilated during the preceding season of slow growth, but also by organic material assimilated by the old frond during the period of forming the new frond. Parker (1963, 1965, 1966) demonstrated that newly synthesized organic products of photosynthesis (mainly D-mannitol) are translocated through the stipe tissue of *Macrocystis* from mature to young blades at the actively growing apical region. Fig. 7 shows the results of an experiment with a small specimen of *L. hyperborea* of which the old frond was allowed to assimilate $NaHC^{14}O_3$ for 14 h. Counts/min above background, detected in tissue samples cut from the experimental specimen, are shown in the lower part of Fig. 7. Considerable radioactivity as C^{14}-labelled organic product(s) has moved from the old to the new frond reaching finally the basal part of the new frond. No radioactivity has entered the stipe, nor parts of the new frond being not directly in contact with the old frond. The procedure employed to prevent artificial contamination of the new frond was to expose to air about 3 cm of the tissue connecting the old to the new frond. This prevented the calculation of naturally occurring rates of transport, since the air-exposed tissue was nearly dried after 14 h. Efforts are being made to improve the method.

CONCLUSIONS

In conclusion it can be said that *L. hyperborea* growing in the sublittoral near Helgoland has its greatest assimilatory surplus in summer, since underwater irradiation, measured at a depth of 2.5 m, is near or above saturation irradiance in summer and respiration undergoes a partial seasonal adaptation to high water temperatures. The theory of Kniep (1914) and Harder (1915) is valid only for low irradiances and is without ecological consequences in regard to underwater light conditions prevailing in summer in the sublittoral. Part of the organic material assimilated by the fronds in summer is stored, mainly in the frond itself, and used up both in respiration and in supporting the new frond's growth in winter and early spring, when underwater irradiance is below the compensation point for long periods, even at moderate depths. The growth of the young frond seems to be supported not only by reserve materials assimilated by the old frond during the preceding season of slow growth, but also by organic products assimilated by the old frond during the formation of the new frond.

SUMMARY

(1) Respiration rate of *Laminaria hyperborea* fronds, measured in the laboratory by means of an oxygen electrode and calculated on a dry weight basis, attains a maximum in May, when growth of the frond is most active.

Respiration decreases towards summer, even though temperature *in situ* increases. Respiration rate of the young frond is two to three times faster than that of the old frond.

(2) Light saturated rates of gross photosynthesis of *L. hyperborea* fronds are higher in summer than during the cold season.

(3) The net gain of photosynthesis, calculated at 400 μgcal/cm^2/s, which is about saturation irradiance, is higher in summer than in winter. The reverse is true, if net photosynthesis is calculated at a low irradiance (100 μg/cal/cm^2/s).

(4) Continuous measurements of underwater irradiance at 2.5 m water depth revealed that saturation irradiance of *L. hyperborea* fronds was reached for most of the day from June until August. Irradiance at the same water depth was below compensation irradiance for most of the period from October until March. It is concluded that *L. hyperborea* has its greatest assimilatory surplus in summer.

(5) The old frond (connected to the young frond until May) has a higher net gain of photosynthesis than the young frond. Labelling the old frond with NaHC^{14}O$_3$ demonstrated the translocation of organic material from the old to the young frond.

I sincerely thank Professor O. Kinne, Hamburg, for help and encouragement during my studies and for valuable criticism. Advice by Professor F. Gessner, Kiel, is gratefully acknowledged. I would like to thank Dr J. M. Jones, Port Erin, for helpful criticism of the manuscript. This work has been supported by grants Ki 41/10 and Ki 42/21 of the Deutsche Forschungsgemeinschaft.

REFERENCES

BLACK, W. A. P. (1948). Seasonal variation in chemical constitution of some of the sublittoral seaweeds common to Scotland. Part I. *Laminaria cloustoni. J. Soc. chem. Ind., Lond.* **67**, 165–8.

BLACK, W. A. P. (1950). The seasonal variation in weight and chemical composition of the common British Laminariaceae. *J. mar. biol. Ass. U.K.* **29**, 45–72.

CLARKE, G. L. (1938). Seasonal changes in the intensity of submarine illumination off Woods Hole. *Ecology* **19**, 89–106.

CLENDENNING, K. A. (1961). Photosynthesis and growth in *Macrocystis pyrifera. Int. Seaweed Symp.* (Biarritz), **4**, 55–65.

EHRKE, G. (1931). Über die Wirkung der Temperatur und des Lichtes auf die Atmung und Assimilation einiger Meeres und Süßwasseralgen. *Planta* **13**, 221–310.

EHRKE, G. (1934). Die Assimilation in ihrer Abhängigkeit vom Licht, von der Temperatur, der Kohlensäuremenge und der Zeit. *Int. Revue ges. Hydrobiol. Hydrogr.* **31**, 373–420.

GESSNER, F. (1955). *Hydrobotanik. I. Energiehaushalt*, 517 pp. Berlin: VEB Deutscher Verlag der Wissenschaften.

HARDER, R. (1915). Beiträge zur Kenntnis des Gaswechsels der Meeresalgen. *Jb. wiss. Bot.* **56**, 254–98.

JERLOV, N. G. (1954). Colour filters to simulate the extinction of daylight in the sea. *J. Cons. perm. int. Explor. Mer.* **20**, 156–9.

JOHNSON, N. G. & KULLENBERG, B. (1946). On radiant energy measurements in the sea. *Svenska hydrog.-biol. Kommn., Skr., Ser.* 3 **1**, 1–27.

JØRGENSEN, E. G. & STEEMANN-NIELSEN, E. (1966). Adaptation in plancton algae. *Memorie. Ist. ital. Idrobiol.* **18** (Suppl.), 39–46.

KAIN, J. M. (1965). Aspects of the biology of *Laminaria hyperborea*. 4. Growth of early sporophytes. *J. mar. biol. Ass. U.K.* **45**, 129–43.

KAIN, J. M. (1966). The role of light in the ecology of *Laminaria hyperborea*. In *Light as an Ecological Factor*, pp. 319–34 (ed. R. Bainbridge, G. C. Evans and O. Rackham). Oxford.

KANWISHER, J. W. (1966). Photosynthesis and respiration in some seaweeds. In *Some Contemporary Studies in Marine Science*, pp. 407–20 (ed. H. Barnes). London: George Allen and Unwin Ltd.

KNIEP, H. (1914). Über die Assimilation und Atmung der Meeresalgen. *Int. Revue ges. Hydrobiol. Hydrogr.* **7**, 1–18.

LAMPE, H. (1935). Die Temperatureinstellung des Stoffgewinns bei Meeresalgen als plasmatische Anpassung. *Protoplasma* **23**, 534–78.

LEVRING, T. (1947). Submarine daylight and the photosynthesis of marine algae. *Göteborgs K. Vetenst.-o Vitterh-samh. Handl., Ser. B* **5**, 1–89.

LÜNING, K. (1969 a). Growth of amputated and dark-exposed individuals of the brown alga *Laminaria hyperborea*. *Mar. Biol.* **2**, 218–23.

LÜNING, K. (1970). Cultivation of *Laminaria hyperborea in situ* and in continuous darkness under laboratory conditions. *Helgoländer wiss Meeresunters.* **20**, 79–88.

MONTFORT, C. (1935). Zeitphasen der Temperatur-Einstellung und jahreszeitliche Umstellungen bei Meeresalgen. *Ber. dt. bot. Ges.* **53**, 651–74.

PARKER, B. C. (1963). Translocation in the giant kelp *Macrocystis*. *Science, N.Y.* **140** (3569), 891–2.

PARKER, B. C. (1965). Translocation in the giant kelp *Macrocystis*. 1. Rates, direction, quantity of C^{14}-labelled products and fluorescein. *J. Phycol.* **1**, 41–6.

PARKER, B. C. (1966). Translocation in the giant kelp *Macrocystis*. 3. Composition of sieve exudate and identification of the major C^{14}-labelled products. *J. Phycol.* **2**, 38–41.

SARGENT, M. C. & LANTRIP, L. W. (1952). Photosynthesis, growth and translocation in a giant kelp. *Am. J. Bot.* **39**, 99–107.

SAUBERER, F. (1962). Empfehlungen für die Durchführung von Strahlungsmessungen an und in Gewässern. *Verh. Int. Verein. theor. angew. Limnol.* **11**, 1–77.

STRICKLAND, J. D. H. (1958). Solar radiation penetrating the ocean. A review of requirements, data and methods of measurement, with particular reference to photosynthetic productivity. *J. Fish. Res. Bd Can.* **15**, 453–93.

TALLING, J. F. (1957). Photosynthetic characteristics of some freshwater plankton diatoms in relation to underwater radiation. *New Phytol.* **56**, 29–50.

TALLING, J. F. (1961). Photosynthesis under natural conditions. *Rev. Pl. Physiol.* **12**, 133–54.

ZIEGLER, H., ZIEGLER, I. & SCHMIDT-CLAUSEN, H. J. (1965). Der Einfluß der Intensität und Qualität des Lichtes auf die Aktivitätssteigerung der NADP+-abhängigen Glycerinaldehyd-3-phosphat-Dehydrogenase. *Planta* **67**, 344–56.

THE EFFECT OF LIGHT ON THE
GROWTH OF ALGAL SPORES

W. E. JONES AND E. SUZANNE DENT
University College of North Wales, Marine Science Laboratories,
Menai Bridge, Anglesey

INTRODUCTION

It is known that certain species of red algae are more tolerant of high light intensity than others. There have been several reports on the effect of light on red algae both from the point of view of the intensity of the light they receive, and its wavelength. Plants of intertidal habitats are more light-tolerant than sublittoral species and within the intertidal region itself there are different degrees of tolerance; some plants are confined to shaded habitats, while others are able to tolerate high intensities.

Biebl (1952) found that certain sublittoral species of algae, and intertidal species occupying shaded habitats were killed by exposure to direct sunlight for 2 h. Algae from the intertidal zone and tide pools showed little or no injury after exposure for the same period of time. Similar results were obtained using continuous artificial illumination. Boney (1959) compared the tolerance to light of the sporelings of sun and shade plants and found that the tolerance of bright illumination varies according to the habitat of the test species. Sporelings of *Polysiphonia lanosa* (L.) Tandy., a plant of sunny habitats, showed 85 % survival when exposed to high light intensities for 9 days, whereas sporelings of the mainly sublittoral species *Antithamnion plumula* (Ellis) Thur in Le Jol. were killed. Arasaki (1953) showed that the growth of red algal sporelings was better at low intensities, and Jones (1959) has shown that sporelings of *Gracilaria verrucosa* (Huds.) Papenf. were killed by very bright illumination. Boney & Corner (1962) measured the growth rate of *Plumaria elegans* (Bonnem.) Schm., and found that sporelings grew best at low intensities. The same authors (1963) found that sporelings of *A. plumula* also showed better growth at low intensities, and that *Brongniartella byssoides* (Good et Woodw.) Schm., an exclusively sublittoral species, required even less light for growth.

The reason for the present investigation was to extend the knowledge of the way in which light intensity affects red algae, and to relate the findings to the distribution of plants on the shore. Observations have also been made on the way in which light affects the pigmentation of red algae. Possible explanations of the observed differences between species are given.

Because of the difficulty of working with adult plants it was decided to carry out experiments on algal sporelings. In addition to the practical advantages sporelings represent a critical stage in the life history of the plant.

METHODS

Methods for obtaining experimental material

Species used for experiments on sporeling growth included the following: *Rhodymenia palmata* (L.) Grev., *Dilsea carnosa* (Schmidel.) Kuntze., *Dumontia incrassata* (O. F. Müll.) Lamour., *Naccaria wiggii* (Turn.) Endl., *Bonnemaisonia asparagoides* (Woodw.) C. Ag., and *Halarachnion ligulatum* (Woodw.) Kutz.

Fertile plants were brought in from the field and laid on glass slides. Mature spores were shed on to the slides. As soon as settlement and attachment of the spores had taken place they were used in growth experiments. Usually, plants had to be left for no more than 24 h before sufficient spores had been shed for experimental purposes. While still at the one-cell stage the spores were transferred to different light regimes.

Culture conditions

Slides bearing sporeling cultures were placed in plastic dishes 5 cm in depth. Into these was passed filtered, running sea water, supplied by a recirculating sea water system. Details of this system and methods of filtration are given elsewhere (Jones & Dent, 1970).

Plants were illuminated by Atlas 'Daylight' fluorescent tubes for a daylength of 16 h; Fig. 1 shows the spectral distribution of light emitted by these tubes. The dishes were placed in a series of separate chambers. Variations in light intensity were produced by varying the number of fluorescent tubes in each chamber, and by adjusting the distance of the tubes from the culture dishes. Illumination was measured using an E.E.L. selenium barrier layer photocell scaled in lumens/ft². It was not necessary to make corrections to the measurement on the basis of spectral response, as the sensitivity of this type of photocell corresponds closely to the spectral composition of the light used in the experiments. Using the data in Fig. 1 and the results of photometric measurements, the radiant energy equivalent to a reading of 1 lx on the photometer was calculated to be 0.01 erg/s/mm².

Measurements of illumination are expressed as lx rather than in units of light energy, since actual measurements were made in terms of light intensity.

It was not possible to keep the temperature of the laboratory sea water constant throughout the year (Jones & Dent, 1970), but seasonal changes were gradual and during each experimental period the variation in water temperature was never more than 5 deg.C. A constant flow of sea water was maintained to avoid excessive temperature rise in the culture dishes, particularly those receiving high illumination. With maximum flow, water temperature was never more than 0.5–1.0 deg.C higher in the strongly illuminated dishes than in those kept at low intensities.

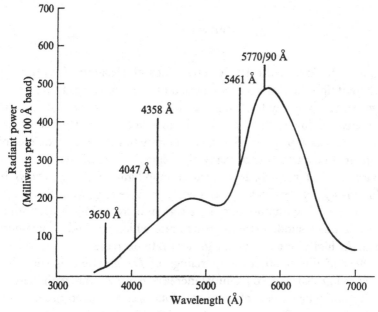

Fig. 1. Spectrum of the radiation emitted by an Atlas 2 ft 40 W 'Daylight' tube (data supplied by manufacturers). The 'line spectra' of emission over five restricted spectral ranges are shown by co-ordinates on the same scale as the curve.

Growth measurement

Any approach to the measurement of algal sporelings can be criticized; their small size combined with a multicellular structure make it impossible to use the more conventional methods suitable for larger plants. In this case it was decided to measure the linear dimensions of sporelings under the microscope. Where the initial spore development produced disc-shaped plates of cells the diameter of the discs was measured. Where filaments were produced, filament length was measured.

Measurements were carried out using a microscope fitted with an eye-piece graticule engraved with a linear scale. A magnification of ×400 was used.

Weekly measurements of sporeling size at each light intensity were made over a period of 3–4 weeks. The number of sporelings measured was between 10 and 25 per sample. The same sporeling population was used throughout each experiment, so that it was necessary to avoid prolonged exposure to light from the microscope lamp. Results are expressed as the mean increase in size (either disc area or filament length) per sporeling during the whole of the experimental period.

GROWTH

Results

Fig. 2 shows the increase in size of sporelings of *Dumontia incrassata* at three different light intensities over a period of about 6 weeks. Except at an intensity of 6620 lx, growth during the first week was slow. The rate of growth increased during the second and third week of development, and after this there was a progressive decrease in growth rate. This sequence of development is characteristic of many algal sporelings studied.

The variation in total growth with variation in light intensity is shown in Fig. 3 for *Rhodymenia palmata*. Sporelings were able to grow at low intensities; with increasing illumination there was an increase in growth until at 5650 lx there was a sudden rise in growth rate which reached a maximum at 6620 lx. At higher intensities the growth rate decreased.

The effect of illumination on sporelings of *D. incrassata* is shown in Fig. 4. *Dumontia* also showed an initial increase in growth rate with increase in intensity in the lower part of the range, followed by a steep rise over the intermediate range. The amount of growth taking place continues to increase up to 10760 lx. The variation between values in the range 5650 to 10760 lx may be due to experimental error. A Student's t test has shown that there is no significant difference between values for 5650, 6620, 9150 and 10760 lx ($P > 0.05$). The variation in the gowth in area of sporelings is due to variation within samples at these intensities.

The growth of sporelings of *Dilsea carnosa* is shown in Fig. 5. Growth occurred at very low intensities, and the same rate of growth was maintained between 100 and 2690 lx beyond which the growth taking place during the experimental period decreased. At 9150 and 10760 lx few sporelings were able to survive.

Fig. 6 summarizes the results obtained with an exclusively sublittoral species *Naccaria wiggii*. Here there is a steady rise in rate of growth with increase in light intensity until a maximum is reached at around 1000 lx; after this growth rate decreases rapidly.

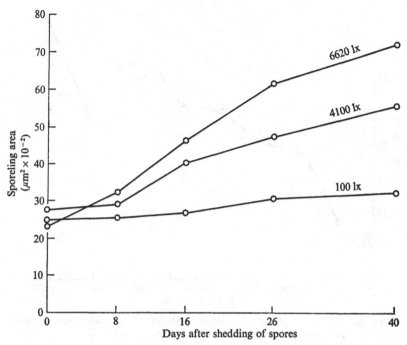

Fig. 2. Increase in size of *Dumontia* sporelings grown at different light intensities. June–August 1968. Temperature 15.5–19.6 °C.

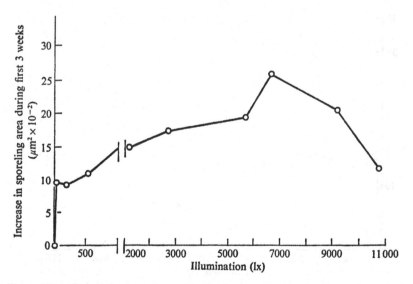

Fig. 3. The effect of the intensity of illumination on the growth of sporelings of *Rhodymenia palmata*. March 1968. Temperature 8.0–11.0 °C. In this and subsequent figures the lower part of the illumination range (0–1000 lx) is plotted on a scale twice as great as that used for the remainder in order to improve legibility.

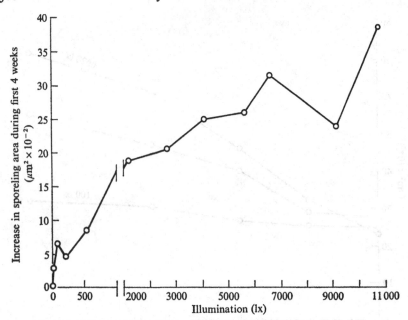

Fig. 4. The effect of the intensity of illumination on the growth of sporelings of *Dumontia incrassata*. June–July 1968. Temperature 15.5–19.5 °C.

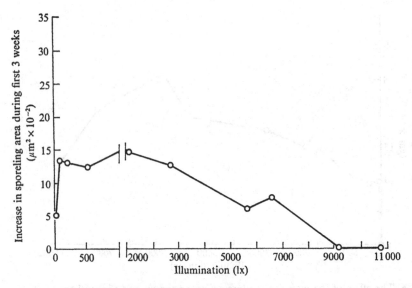

Fig. 5. The effect of the intensity of illumination on the growth of sporelings of *Dilsea carnosa*. March 1968. Temperature 8.0–9.25 °C.

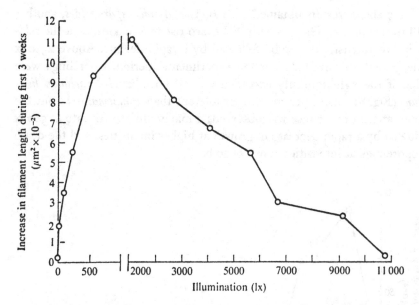

Fig. 6. The effect of the intensity of illumination on the growth of sporelings of *Naccaria wiggii*. September–October 1968. Temperature 15.1–19.25 °C.

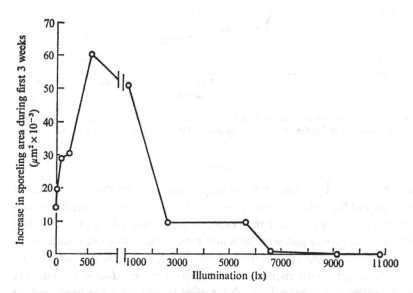

Fig. 7. The effect of the intensity of illumination on the growth of sporelings of *Bonnemaisonia asparagoides*. July–August 1968. Temperature 17.5–19.6 °C.

Fig. 7 shows results obtained with *Bonnemaisonia asparagoides*, another sublittoral species. There is an initial sharp rise to a maximum at the relatively low intensity of 540 lx, followed by a rapid decline. Above 2500 lx little growth occurred during the experimental period. Sporelings were killed if the light intensity exceeded around 6620 lx. *Halarachnion ligulatum* (Fig. 8) was rather more light tolerant than *Bonnemaisonia* but the same pattern of response was observed; a maximum growth rate at 540 lx followed by a rapid decrease in growth at higher intensities, and the death of sporelings at intensities above 6620 lx.

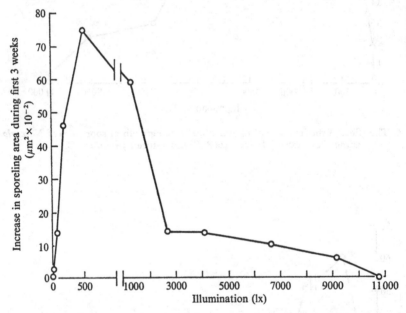

Fig. 8. The effect of the intensity of illumination on the growth of sporelings of *Halarachnion ligulatum* September 1968. Temperature 15.25–19.25 °C.

Conclusions

From Fig. 2 it is clear that a relationship exists between the growth rate of sporelings and the light intensity at which they are grown. From the subsequent graphs it appears that the response varies according to the species involved. *Rhodymenia palmata* has a low growth rate at low intensities with a maximum, under the type of illumination used, at an intensity of 6620 lx, and a falling rate with increase above this intensity. *Dilsea carnosa* has a lower optimum for growth, at about 2690 lx and does not grow well at higher intensities. Both *Rhodymenia* and *Dilsea* are plants of the immediate sublittoral, and sublittoral fringes, with a limited range into the littoral region. *Dilsea* however, when found in the littoral, grows in shaded regions; *Rhodymenia* appears to be more light tolerant.

Dumontia incrassata is a littoral summer annual, occupying open habitats, and therefore likely to receive more light in the field than the other species discussed; this fits in well with its maintaining a high growth rate at high intensities. The low growth rate at low intensities corresponds to the limited range of *Rhodymenia* and *Dilsea* downward into the less well-illuminated depths, and the limitation of *Dumontia* to the littoral region.

In contrast, the purely sublittoral species, *Bonnemaisonia asparagoides*, *Halarachnion ligulatum* and *Naccaria wiggii*, show their maximum growth at much lower intensities; at 540 lx in the first two species and at 1000 lx in *Naccaria*. Since these are species which grow at some depth, and therefore in poorly illuminated habitats, these results are not surprising. However, they are damaged and eventually killed at intensities around 6620 lx, which produce good growth in species occupying habitats receiving better illumination; this observation requires further exploration and will be considered in the next section.

PIGMENTATION

Results

In addition to causing variation in growth rate, the other most noticeable effect of light is to cause bleaching of red algal sporelings (Jones, 1959). In the summer bleaching occurs in mature plants in nature due to the loss of phycoerythrin at high light intensities. Bleaching was found to occur both in the intertidal and the sublittoral species studied, but it followed a different course in each case.

In intertidal plants loss of phycoerythrin did not appear to affect growth until all the phycoerythrin had been lost and the plants appeared pale yellow in colour. Even then, in the case of *Rhodymenia* and *Dumontia*, it was possible for some increase in size to take place. Sporelings of intertidal species were well pigmented at intensities up to 1075 lx. After this, phycoerythrin was lost with increase in light intensity and at 9150 lx and above they had entirely lost their red colouration. The morphology of the plants was only slightly affected by variation in intensity; higher light intensities tended to stimulate the production of hairs.

Sporelings of the sublittoral species *Naccaria wiggii*, *Halarachnion ligulatum*, and *Bonnemaisonia asparagoides* showed a different response. They also lost their phycoerythrin, but once the pigment had been lost, the plants became colourless. There was no stage at which they had the yellow appearance found in sporelings of intertidal species. Moreover, once phyco-erythrin has been lost the plants die or show abnormal development. Bleaching of *Bonnemaisonia* and *Halarachnion* sporelings began at 540 lx, as compared to 1075 lx in intertidal species, and at 4100 lx most of the phycoerythrin was lost and the plants were very pale in colour. *Naccaria* was more tolerant, and although sporelings began to bleach at 540 lx phycoerythrin was still present at 6620 lx although the amount of growth taking place was small. In both the intertidal and the sublittoral species studied, bleaching of sporelings was fully established after 1 week under experimental conditions. As already stated the amount of bleaching which occurred was dependent on the intensity of illumination received by the sporelings.

Conclusions

This marked difference in the pattern of bleaching, and in growth response, suggests that different species of red algae may have different systems for the utilization of light. The intertidal species investigated, even when most of the phycoerythrin was lost, were able to grow and presumably to photo-synthesize normally.

It appears then that these algae are able to photosynthesize without utilizing phycoerythrin. Intertidal Rhodophyceae have a darker colour than the clear, bright red of the sublittoral species. In addition to the phyco-erythrin–chlorophyll system for the utilization of light they may have a second, possibly carotenoid–chlorophyll system which functions most efficiently at higher intensities of illumination. These intertidal species therefore have a capacity to occupy more strongly illuminated habitats. The system is better developed in purely littoral species like *Dumontia*, but is also seen in species such as *Gracilaria verrucosa* (Jones, 1959) in which growth occurs under conditions of high illumination when plants are bleached. Such species, bleached to a straw yellow colour, recover their red pigmentation when the light intensity is reduced.

In the case of sublittoral algae, however, there is no evidence of the presence of a second system – once the phycoerythrin had been destroyed, the growth rate of the sporelings decreased rapidly and at the highest intensities no growth occurred. The indication is that these sublittoral species have a phycoerythrin–chlorophyll system only. This functions well at low intensities, but under stronger illumination the system is damaged, and the plants become colourless and die.

The observations of Boney & Corner (1962) on *Plumaria*, a shade-loving species, seem to support this hypothesis. They suggest that phycoerythrin has a different function in submerged and intertidal plants. Submerged plants use phycoerythrin as an accessory pigment in photosynthesis, whereas in intertidal plants it is utilized as a means of protection against green light.

Strain (1958) has investigated the components of pigment systems of a large number of red algae, and has found considerable variation between species as to the pigments present.

An alternative possibility is that the intertidal species are able to adapt to conditions of high light intensity whereas sublittoral species are not. Red algae contain relatively large concentrations of chlorophyll and if it is assumed that intertidal plants can utilize this chlorophyll when in a bleached condition, the ability to grow without the presence of phycoerythrin would be explained. Yocum (1951) has shown that certain algae can become adapted to utilizing certain wavelengths of light, and Haxo & Blinks (1950) produced evidence that the chlorophyll of red algae may be activated in bleached plants.

As a preliminary test of the differences in the pigment systems of the species investigated, the thallus absorption spectra were investigated. It is hoped to continue these investigations but until this has been done, only the tentative conclusions previously described can be offered as to the nature of the differences observed between species.

SUMMARY

The growth of sporelings of the intertidal algae *Rhodymenia palmata*, *Dilsea carnosa* and *Dumontia incrassata*, and the sublittoral algae *Naccaria wiggii*, *Halarachnion ligulatum* and *Bonnemaisonia asparagoides* was measured at light intensities ranging from 0 to 10750 lx. Measurements were made over the first 3–4 weeks of sporeling development, as the growth rate of most species studied is maximal during this period under laboratory conditions.

The amount of growth which occurred was dependent on the light intensity at which the sporelings were grown.

The intertidal species studied were able to grow at intensities below 100 lx. Growth increased with increase in light intensity until a maximum was reached, the optimum intensity for growth being dependent on the species concerned. Sporelings of the sublittoral species were also able to grow at low intensities (below 100 lx). *Bonnemaisonia asparagoides* and *Halarachnion ligulatum* showed maximum growth at 540 lx and *Naccaria wiggii* at 1075 lx. At intensities above the maximum, growth decreased rapidly in all sublittoral species studied, and at the higher intensities used death of the sporelings occurred.

Increase in light intensity caused loss of phycoerythrin and bleaching in all species studied. Sporelings of intertidal plants bleached to a pale yellow colour, and were still able to grow when all the phycoerythrin had been lost. Sporelings of sublittoral species became colourless and died when the light intensity was high enough to destroy their phycoerythrin content.

The results indicate that the sublittoral species depend on a phyco-erythrin–chlorophyll system for the utilization of light, and can grow only at intensities at which phycoerythrin is not destroyed. Intertidal species appear to have a second system which can function without phycoerythrin, and at high intensities of illumination. The distribution of the species studied is discussed in the light of these results.

REFERENCES

ARASAKI, S. (1953). An experimental note on the influence of light on the development of spores of algae. *Bull. Jap. Soc. scient. Fish.* **19**, 466–70.

BIEBL, R. (1952). Ecological and non-environmental constitutional resistance of the protoplasm of marine Algae. *J. mar. biol. Ass. U.K.* **31**, 307-15.

BONEY, A. D. (1959). *The Ecology and Biology of certain Intertidal red Algae.* Ph.D. Thesis. University of London.

BONEY, A. D. & CORNER, E. D. S. (1962). The effect of light on the growth of sporelings of the intertidal red alga *Plumaria elegans* (Bonnem.) Schm. *J. mar. biol. Ass. U.K.* **42**, 65–92.

BONEY, A. D. & CORNER, E. D. S. (1963). The effect of light on the growth of sporelings of the red algae *Antithamnion plumula* and *Brongniartella byssoides*. *J. mar. biol. Ass. U.K.* **43**, 319–25.

HAXO, F. T. & BLINKS, L. R. (1950). Photosynthetic action spectra of marine algae. *J. gen. Physiol.* **33**, 389–422.

JONES, W. E. (1959). Experiments on some effects of certain environmental factors on *Gracilaria verrucosa* (Huds.) Papenf. *J. mar. biol. Ass. U.K.* **38**, 153–67.

JONES, W. E. & DENT, E. S. (1970). Culture of marine algae using a re-circulating sea water system. *Helgoländer wiss. Meeresunters.* **20**, 70–8.

STRAIN, H. H. (1958). *Chloroplast pigments and Chromatographic analysis.* 32nd Annual Priestley Lectures. Penn. State Univ., Univ. Park, Penn.

YOCUM, C. S. (1951). Some experiments on photosynthesis in marine algae. *Abstr. Diss. Stanford Univ.* 166–68.

LIGHT QUALITY AND THE PHOTOMORPHOGENESIS OF ALGAE IN MARINE ENVIRONMENTS

M. J. DRING

Department of Botany, The Queen's University, Belfast BT7 1NN, Northern Ireland

INTRODUCTION

The changes in spectral composition which occur as solar radiation penetrates sea water have been fully analysed and discussed by investigators of photosynthesis in marine algae, with particular reference to the hypothesis of 'complementary chromatic adaptation' (e.g. Levring, 1966). Now that this hypothesis has been shown to be substantially correct, and submarine light quality can be accepted as an important factor in the control of algal zonation in littoral regions, attention can perhaps be shifted to another aspect of the biology of marine algae which may be substantially affected by the quality of the underwater light. Physiological work with algae in recent years has revealed a considerable number and variety of responses to light which appear to be mediated by pigments other than those involved in photosynthesis. These non-photosynthetic pigment systems have been shown to differ from the photosynthetic pigments in their absorption spectra and energy requirements, and must, therefore, be affected differently by changes in the quality of the incident light. It is possible, therefore, that light quality could influence the distribution of certain algal species through its effects on non-photosynthetic responses to light, and that such responses may be of considerable ecological significance. On the other hand, all of the physiological work in which these responses have been demonstrated has been carried out in highly artificial culture conditions, in which the spectral composition of the incident light has certainly been very different from that which the plant would experience in nature, and it has been suggested that some of these responses or pigment systems can have little or no ecological significance under natural conditions (e.g. West, 1968). They may, indeed, merely be artifacts of the conditions which the physiologists have been using.

In this paper, an attempt is made to resolve some aspects of this problem by examining the available evidence both on these non-photosynthetic responses of algae to light, and especially on the spectral and energy requirements of the pigment systems which are thought to be responsible, and on the spectral composition of underwater light in various marine environments. It is hoped that this approach will direct the attention of physiologists to those responses which appear to have greater ecological significance, and of optical oceanographers to those aspects of the sub-marine light environment which physiologists – or, at least, algal morpho-geneticists – would find most interesting.

NON-PHOTOSYNTHETIC LIGHT RESPONSES AND PIGMENT SYSTEMS IN ALGAE

Phytochrome-controlled responses

The importance of phytochrome as a photomorphogenic pigment in higher plants is now well established. Its activity has been demonstrated in numerous species of flowering plants, as well as in pteridophytes and bryo-phytes, and it appears to control a wide variety of photomorphogenic and photoperiodic responses (e.g. Salisbury, 1963, table 7–1). The presence of phytochrome has been inferred, as a result of physiological work, in four species of algae (Table 1), and the pigment has been extracted from one of these species (*Mesotaenium*; Taylor & Bonner, 1967). The algal species in which phytochrome activity has been demonstrated are representatives of widely separated taxonomic groups, and have little in common, either morphologically or ecologically. In addition, the responses which appear to be controlled by phytochrome cover a wide range – behaviour, orientation and photoperiodism. The distribution and range of these responses suggest that phytochrome may be widespread among algal species, and that it may be as important as a photomorphogenic pigment in algae, as it is in higher plants. Photoperiodic responses have invariably been shown to be associated with the activity of phytochrome, and the evidence for the existence of photoperiodic responses in several algal species, in addition to *Porphyra tenera* (Dring, 1970), indicates that further reports of phytochrome activity in algae may soon be forthcoming.

Table 1. *Phytochrome responses in algae*

Alga	Response	Reference
Mougeotia sp.	Positive light-induced chloroplast movement	Haupt, 1959
Mesotaenium sp.	As *Mougeotia*	Haupt & Thiele, 1961
Gyrodinium dorsum	Reactivation of 'stop response' to brief blue irradiation	Forward & Davenport, 1968
Porphyra tenera	Inhibition of conchosporangium formation in short days	Dring, 1967; Rentschler, 1967

In higher plants, phytochrome responds to low energies of red (660 nm) radiation, and the energy requirements which have been determined for three of the algal responses (Table 2) indicate that algal phytochrome is also sensitive to low energies at similar wavelengths. Since phytochrome exists in two interconvertible forms, P_R and P_{FR}, which absorb maximally at 660 and 730 nm respectively, the response of a plant to red light can be reversed by exposure to far-red (Hendricks & Borthwick, 1965). In radiation of mixed spectral composition, the ratio of red to far-red light will determine the photostationary state of the pigment, and this, in turn, will determine the physiological response of the plant (Cumming, Hendricks & Borthwick, 1965). The ecological significance of this ratio will be discussed below. Both forms of phytochrome also have minor absorption peaks in the blue region of the visible spectrum (Butler, Hendricks & Siegelman, 1965), and it is these peaks which are now thought to be responsible for the effects of prolonged or high-intensity irradiation with blue light, known as the high energy reaction (HER; Hartmann, 1966; Hillman, 1967). The action spectrum results obtained by Rentschler (1967) and by Dring (1967) for the inhibitory effects of light breaks on sporangium production in *P. tenera* indicate that the phytochrome system in this alga is sensitive to high energies of blue (447 nm) radiation, and this suggests that HER-type responses may also occur in algae (see discussion in Dring, 1970). This aspect of phytochrome activity, therefore, should also be considered in relation to the submarine light environment.

Red–green reversible responses of Nostoc

When the blue–green alga, *Nostoc muscorum*, is grown in darkness, it forms amorphous colonies of cells, but exposure to low energies of light results in the development of the filamentous colonies which are generally regarded as more typical of the genus (Lazaroff, 1966). The action spectrum for photoinduction of development has a sharp peak at 650 nm. The energy requirement at this wavelength is low (Table 2), and both the spectral and energy requirements of the response are similar to those of phytochrome-controlled responses. The effect of red light on this response can also be reversed by subsequent exposure to another wavelength, but here the similarity to phytochrome ends since the action spectrum for reversal shows a broad peak at 500–600 nm. The photoreceptor pigment for red-light induction has been shown to be allophycocyanin, and phycoerythrins are thought to be responsible for the antagonistic effect of green light. The effects of mixtures of red and green light on the response have not been investigated, but the red/green ratio could possibly have the same signifi-

Table 2. *Non-photosynthetic pigment systems in algae, responding to red light*

(Energy (erg/cm^2) required to produce a 50% response.)

Alga	Type of response	Photoreceptor	Wavelength	Energy	Reference
Mougeotia	Chloroplast orientation	Phytochrome	679 nm	$3-7 \times 10^3$	Haupt, 1959
Mesotaenium	Chloroplast orientation	Phytochrome	665 nm	$6-14 \times 10^4$	Haupt & Thiele, 1961
Porphyra	Photoperiodic	Phytochrome	662 nm	3.24×10^5	Dring, 1967
Nostoc	Photomorphogenic	Allophycocyanin	650 nm	1.6×10^5	Lazaroff & Schiff, 1962

cance for this response as the red/far-red ratio has for phytochrome-controlled responses.

Although *N. muscorum* is not a marine alga, the genus *Nostoc* does include marine species, and this response is of great interest as the only well-documented example of a photomorphogenic response in a blue-green alga. Observations of morphological responses to changes in light intensity by *Calothrix scopulorum* (W. D. P. Stewart, unpublished), and the plasticity of the morphology of blue–green algae in general, indicate that photomorphogenic responses may be important in the Cyanophyta.

Photomorphogenic response of Acetabularia

Plants of *A. crenulata* and *A. mediterranea* form caps when grown in blue (400–550 nm) or white light, but not when grown in red light (600–750 nm). Red light may (Richter & Kirschstein, 1966) or may not (Terborgh, 1965) support the growth and elongation of the stalks. This appears to be a complex response which has yet to be fully investigated, and the specific effects of red and blue light on the morphogenesis of *Acetabularia* are far from clear. According to Terborgh (1965), stalk elongation and cap formation occur in both white and blue light, but in blue light cap formation is delayed, so that the caps are formed on longer stalks. In red light alone, there is little growth and no cap formation, but in mixtures of red and blue light, cap formation does occur, the stalk lengths recorded being shorter with higher ratios of red to blue light. Thus, red light appears to stimulate cap formation in the presence of blue light, but inhibits cap formation in the absence of blue. The morphology of the caps formed in blue light also differed from those formed in white light. The latter were irregular with free rays, whereas the former were more regular and resembled the caps formed on plants growing in natural conditions.

Although no action spectrum has been determined for this response, and the energy requirements of the photoreceptor have yet to be investigated, it is clear that changes in the spectral composition of the incident light have important effects on the morphology of *Acetabularia*.

Photomorphogenic responses to blue light

The responses to blue light, which have been reported for various algal species, include a number of phototopotactic and chloroplast orientation responses, which are reviewed by Haupt (1965). Responses to blue light which do not involve changes in the orientation of organisms or organelles are listed in Table 3. They may not all appear to deserve the adjective 'photomorphogenic', but most of the authors have postulated that blue

Table 3. *Photomorphogenic responses of algae to blue light*

Alga	Response to blue	Active wavelength (nm)	Photoreceptor pigment proposed	Reference
Chlorophyta				
Chlorella spp.	Induction of cell division	485	Carotenoid(s)	Senger & Schoser, 1966
	Protein synthesis from exogenous glucose	450–490	Carotenoid	Kowallik, 1966
	Inhibition of respiratory decline in dark	550	?	Kowallik & Gaffron, 1966
	Increased photosynthetic rate and changes in distribution of carbon fixed	480 ± 150	?	Hauschild *et al.* 1962–5
Prototheca zopfii	Inhibition of cell division	Near u.v., 420	Cytochrome	Epel & Kraus, 1966
Scenedesmus acuminatus	Changes in distribution of photosynthetically fixed carbon	480 ± 150	?	Hauschild *et al.* 1962b
Chrysophyta				
Coccolithus huxleyi	Coccolith formation	440	?	Paasche, 1966
Phaeophyta				
Fucus spp.	Induction of polarity in germinating zygote	250, 455, 254, 370, 439	Carotenoid Riboflavin	Bentrup, 1963
Cyanophyta				
Microcystis aeruginosa	As *Scenedesmus*	480 ± 150	?	Hauschild *et al.* 1962b
Rhodophyta				
Nitophyllum punctatum	Maintenance of spore discharge rhythm	Blue	?	Sagromsky, 1961

light induces the synthesis of a specific enzyme which results in the responses observed, whether morphogenetic (e.g. Senger & Schoser, 1966; Paasche, 1966) or physiological (Hauschild *et al.* 1962–5). The similarity between the various hypotheses about different responses suggests that the primary effect of blue light is to stimulate a biochemical response, which may underlie a variety of physiological responses. In this way, a single pigment system sensitive to blue light could be responsible for several types of response, resembling the phytochrome system in this respect, and more responses of algae to blue light may remain yet to be discovered. The taxonomic, morphologic and ecologic diversity of the algae in which such responses have already been observed, supports the idea that the effects of blue light on algae may cover a wide range.

Detailed action spectra are available for some of these responses (e.g. *Fucus, Prototheca, Coccolithus*), but the minimum specific energy requirements are difficult to determine, because long irradiation periods are often required, and the energy used in the blue-light response cannot always be distinguished from that used in photosynthesis. Little is known, therefore, of the energy requirements for any of these responses, except that in *Fucus* zygotes. The primary photoresponse in the induction of polarity in these cells involves a carotenoid pigment and requires 4.3×10^4 erg/cm^2 (i.e. 10^{16} quanta/cm^2 at 455 nm) in order to produce a 50% response, whereas the second, high-intensity, response involving riboflavin requires 10^6 erg/cm^2 (Bentrup, 1963).

Blue light with wavelengths around 450 nm clearly has important non-photosynthetic effects on a variety of algae, and the ecological significance of these effects must be considered in discussions of the submarine light environment.

UNDERWATER LIGHT SPECTRA IN RELATION TO NON-PHOTOSYNTHETIC ALGAL PIGMENT SYSTEMS

The review of the non-photosynthetic responses of algae to light which has been presented above has shown that the wavelengths of light which are of greatest importance in relation to the photoreceptor pigments are red (660 nm) and far-red (730 nm) for the phytochrome-controlled responses, and blue (450 nm) for most of the other systems. A quantitative analysis of the penetration of these wavelengths has been carried out in order to determine the irradiant energy at each wavelength at various depths in different water types. Since the minimum energy required to activate the photoreceptor pigment has been determined for many of the responses, it should be possible to calculate the maximum depth at which such systems can operate.

Calculations have been based on the values for irradiance transmittance at the different wavelengths in different water types, which are given by Jerlov (1968). These are listed in Table 4. No direct measurements appear to have been made of the penetration of wavelengths greater than 700 nm, and transmittance values for 730 nm have been determined by extrapolating curves of percentage irradiance transmittance per metre against wavelength from 700 nm to a value of about 10% at 750 nm. The latter figure is derived from the laboratory studies of pure water by James & Birge (1938) and by Clarke & James (1939). The values obtained for transmittance at 730 nm (Table 4) show little variation with water type, and any in-accuracies in their estimation should therefore be slight. Nevertheless, direct readings of irradiance transmittance in this region of the spectrum would be valuable.

The minimum energy available to a plant at any one wavelength is the irradiation at the maximum depth which will support plant growth in each water type (i.e. the irradiant energy at the lower limit of the photic zone). This limit is marked by the depth at which the surface irradiance (350–700 nm) is reduced to 1% (Jerlov, 1968). In oceanic waters, the photic zone extends down to 30–100 m, but in coastal waters the lower limit occurs at as little as 6 m (Table 4; Jerlov, 1968). Calculation of irradiation at the limit of the photic zones (Table 4) is based directly on the transmittance data and assumes that the transmittance of each wavelength does not vary with depth. Although this assumption will result in a slight underestimation of some of the irradiance values, particularly those for 450 nm radiation in the clearer waters, the errors involved are unlikely to be significant.

The irradiant energy at each of the principal wavelengths at the lower limit of the photic zone has been calculated as a percentage of the energy at that wavelength at the surface (Table 4). The absolute energy values will vary according to the lighting conditions at the surface, and the width of the waveband absorbed. Therefore, in order to calculate absolute energy values for submarine habitats, it is necessary to hypothesize specific surface conditions, so that the irradiation at the surface can be evaluated, and to assume a value for the width of the absorption band of the photoreceptor pigments. In these calculations, an attempt has been made to approximate to the optimum daily lighting conditions during the algal growing season in various regions of the world by calculating the irradiation at a sea surface which is exposed to zenith sun for 10 h under cloudless conditions. The

Table 4. Penetration of blue, red and far-red radiation through oceanic and coastal waters

Water type	Irradiance transmittance (%/m) Wavelength (nm)			Depth of lower limit of photic zone (m)	Irradiance at lower limit of photic zone as % of surface irradiance. Wavelength (nm)		
	450	660	730		450	660	730
I	98.1	68.4	24.0	105	13.34	$< 10^{-8}$	$< 10^{-8}$
II	94.0	65.9	24.0	55	3.35	$< 10^{-8}$	$< 10^{-8}$
III	88.5	63.4	24.0	33	1.78	2.95×10^{-5}	$< 10^{-8}$
1	84.0	63.4	24.0	28	0.75	2.88×10^{-4}	$< 10^{-8}$
3	75.0	60.0	24.0	18.4	0.52	8.20×10^{-3}	$< 10^{-8}$
5	60.0	55.7	23.0	11.5	0.28	1.18×10^{-1}	4.58×10^{-6}
7	42.0	50.0	23.0	8.1	8.70×10^{-2}	3.67×10^{-1}	6.76×10^{-4}
9	21.0	44.5	17.5	6.0	8.58×10^{-3}	7.76×10^{-1}	2.87×10^{-3}

width of the absorption band has been taken as 50 nm, centred on each of
the principal wavelengths. Since it is the orders of magnitude of the energy
values which are of importance, substantial corrections could be made to
these assumptions without affecting the final figures significantly. The calculations of the irradiant energy (in erg/cm²) at the lower limit of the photic
zone (Table 5) are based on values for solar radiation at zero air mass given
by Johnson (1954), which are reduced by 25 % for attenuation by passage
through the atmosphere (Gates, 1962) and integrated over 10 h.

Table 5. *Irradiation in three wavebands at the lower limit of the photic
zone in oceanic and coastal waters*

(Surface conditions assumed: zenith sun for 10 h.)

Irradiation at lower limit of photic zone (erg/cm²).
Waveband (nm)

Water type	425–475	635–685	705–755
I	3.94×10^8	< 1	< 1
II	9.88×10^7	< 1	< 1
III	5.25×10^7	6.34×10^2	< 1
1	2.21×10^7	6.19×10^3	< 1
3	1.53×10^7	1.76×10^5	< 1
5	8.32×10^6	2.54×10^6	8.24×10^1
7	2.57×10^6	7.89×10^6	1.22×10^4
9	2.53×10^5	1.67×10^7	5.17×10^4

Calculations based on figures for solar radiation at zero air mass: 450 nm 0.220 w/cm²/μm;
660 nm 0.159 w/cm²/μm; 730 nm 0.134 w/cm²/μm (Johnson, 1954).

The review of non-photosynthetic responses above (p. 376) also demonstrated that certain wavelength ratios are important in determining the
physiological response or the morphology of certain algal species. Thus, the
red/far-red ratio influences the photostationary state of phytochrome, and
the ratio of blue to red light has been shown to affect the morphology and
rate of cap formation in *Acetabularia*. Because of the differential absorption
of these three wavelengths by sea water, the ratios between them will show
considerable variation with depth. The red/far-red ratio has been analysed
by calculating the irradiant energy at both 660 and 730 nm at various
depths as a percentage of the irradiant energy at 730 nm at the surface. The
ratio between the two values at each depth gives the red/far-red ratio in
terms of incident energy. Since pigments absorb light in discrete quanta,
which have different energies at different wavelengths, the ratio between
any two wavelenths has been calculated in terms of the numbers of incident
quanta. The energy ratio has thus been converted to a quanta ratio by
means of a factor which expresses the ratio of the energy content of individual quanta at each wavelength, and this ratio has then been converted

to a percentage of quanta at 660 nm relative to the total number of quanta at 660 and 730 nm. These values are plotted against depth in Fig. 1. The percentage of red quanta rises rapidly with increasing depth in all water types, although the rise is most rapid in type I, oceanic water and least rapid in type 7, coastal water. The ratio always exceeds 99.9% below 9 m. The blue/red ratio has been analysed in the same way (Fig. 2). The pattern

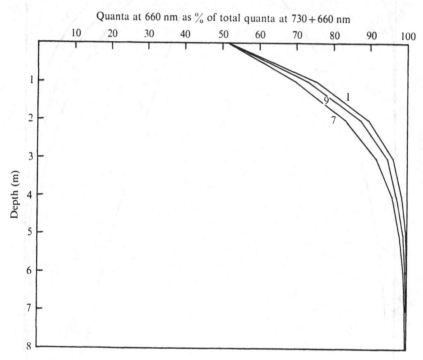

Fig. 1. The quantum ratio of red (660 nm) to far-red (730 nm) irradiation as a function of depth in oceanic water (type I) and in two types of coastal water (types 7, 9).

of variation of this ratio with depth changes markedly in the different water types. The percentage of blue quanta rises rapidly with increasing depth in all oceanic waters and in the clearer coastal waters, but in type 5, coastal water there is relatively little change with depth, and in types 7 and 9 the percentage of blue quanta falls rapidly.

This analysis shows that considerable variation in both the red/far-red and the blue/red ratios may result from changes in depth of only a few metres (Figs. 1, 2). Changes in depth of this order occur regularly in the lower littoral and upper sublittoral zones of coastal regions as a result of normal tidal activity, and, therefore, variation in these ratios will not only

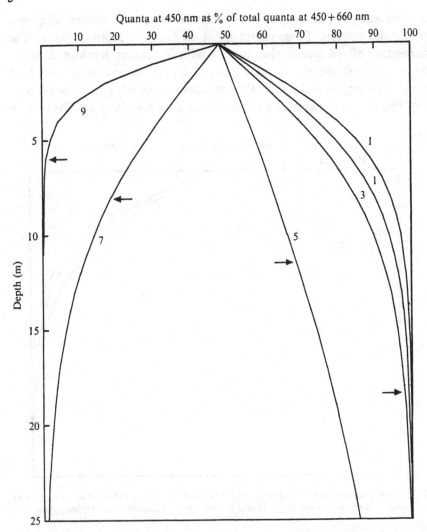

Fig. 2. The quantum ratio of blue (450 nm) to red (660 nm) irradiation as a function of depth in oceanic water (type I) and in five types of coastal water (types 1, 3, 5, 7, 9). The arrows indicate the lower limit of the photic zone for waters in which this limit occurs at depths of less than 25 m.

occur from one place to another, with changes in depth, but will also occur in the same place from one time to another, with changes in the level of the tide. Variation in these two ratios will always be accompanied by changes in the spectral composition of the incident light in general. The spectral composition of the incident radiation at any point in littoral and upper sublittoral zones must, therefore, be dependent, not on its vertical depth from some hypothetical mean water level, but on the depth of water over-lying that point at any particular time, and this depth is entirely dependent on the state of the tide. The change in spectral composition of the radiation incident upon individual plants, which must occur in response to the con-tinual change in the state of the tide, is an aspect of the littoral environment which has so far received little attention. The possible significance of such changes for the biology of littoral algae will be discussed below.

DISCUSSION: THE ECOLOGICAL SIGNIFICANCE OF PHOTO-MORPHOGENIC PIGMENT SYSTEMS IN MARINE ALGAE

Phytochrome

West (1968) argues that 'a red/far-red reversible photoperiodic process could not reasonably be expected in the sublittoral algae because neither type of light is present at depths where they grow'. This argument appears to assume that light is suddenly cut off at certain depths, instead of being gradually reduced on a logarithmic scale, and also takes no account of the extreme sensitivity of the phytochrome system to low energies of red radiation. A comparison of the energy requirements of the phytochrome-controlled responses in algae (Table 2) with the irradiant energy within the 635–685 nm waveband at the lower limit of the photic zone in various water types (Table 5) shows that there is sufficient red light to satisfy the require-ments of phytochrome throughout the photic zone in all coastal waters, except the clearest (type 1). This implies that the majority of benthic algae, and certainly all littoral and sublittoral forms, will receive sufficient energies of red radiation to enable a phytochrome system to operate by a mechanism similar to that in flowering plants. In littoral and upper sublittoral regions, the situation may be complicated by tidal fluctuations in the red/far-red ratio (Fig. 1). However, the only phytochrome-controlled response which has so far been reported for a benthic alga from these zones is the photoperiodic response of *Porphyra tenera*, and photoperiodic responses are unlikely to be influenced by tidal variations in the red/far-red ratio since the ratio can never decrease below its value in full sunlight (Fig. 1). The fluctuations will always serve, therefore, to reinforce the light–dark rhythm of the terrestrial environment. As and when other types of phytochrome-controlled responses are reported for littoral algae, it will be necessary to investigate them under a range of red/far-red ratios before the responses can be accepted as ecologically significant.

Benthic or planktonic algae which are growing in relatively deep, clear waters (e.g. near the lower limit of the photic zone in type 1, coastal water, or in any oceanic waters) will receive much greater energies of irradiation in the blue than in the red (Tables 4, 5). In these environments, it is possible that the minor absorption bands of phytochrome around 450 nm may assume greater significance than in terrestrial or shallow-water environments, and HER-type responses could occur. Theoretically, such responses would enable the phytochrome system to act as a photomorphogenic pigment in algae throughout the full range of marine environments, but the responses of the known algal phytochrome systems to various energies of red and blue radiation, and to various ratios of red to far-red and blue to red radiation, need to be investigated before this theoretical possibility can be confirmed.

Pigment systems in blue–green algae

The majority of marine blue–green algae are found in the upper littoral zone or in the 'littoral fringe' (Lewis, 1964), and they will, therefore, rarely be covered by sufficient water to affect the spectral composition of the incident light significantly. Thus, any pigment which responds to wavelengths which are abundant in solar radiation will be able to function adequately in these algae under natural conditions. However, the energy requirements for red light of the response of *Nostoc muscorum* (Table 2) are similar to those for phytochrome-controlled responses, and the conclusions reached in the previous section concerning the depths at which the red-response of phytochrome could operate, can also be applied to pigment systems of the type found in *Nostoc*.

Pigment system in Acetabularia

It has been suggested (Terborgh, 1965) that the similarities in cap morphology and stalk length between plants grown in blue light and those grown under natural conditions, which contrast with these characters in plants grown in white or red light, is due to the fact that 'the portion of the spectrum which penetrates the sea most readily lies between 450 and 550 nm, the energy reaching a depth of a few metres at 650 nm being only 11 % of that at 500 nm'. According to Taylor (1960), however, *Acetabularia* is usually found in shallow waters in which there is little tidal variation and little wave action, and so the conditions 'at a depth of a few metres' are not relevant to this alga. At 3 m in type 1, coastal water, the 450/660 nm ratio, on a quantum basis, will be 2.20, and in type 3, it will be 1.84. These ratios are close to the experimental ratios in which significant differences from blue-grown cells were observed, and blue light as 'pure' as that in which cells resembling natural plants were obtained, will only be found at depths of 9–10 m or more. The morphology of *Acetabularia* in nature cannot, therefore, be attributed entirely to the spectral composition of the incident light, and the response of this alga to changes in light quality requires more detailed investigation before its ecological significance can be established. If the effects of changes in blue/red ratio on stalk length and cap morphology are confirmed, however, it must be concluded that plants growing at different depths will show considerable morphological divergence, and the taxonomic criteria in this genus may require drastic revision.

Pigment systems responding to blue light

The wide range of algae in which these systems appear to operate, and the variety of the responses controlled by them (Table 3) make it difficult to draw any general conclusion about their significance under natural conditions. Quantitative estimates of the energy required are lacking for most of the responses, but these details are available for the response of *Fucus*. The energy of blue radiation required to induce polarity in *Fucus* zygotes (see *Photomorphogenic responses to blue light*) is less than that available at the limits of the photic zone in all water types, and the requirements can certainly be satisfied at the relatively shallow depths at which *Fucus* species are found. Since polarity can also be induced by a variety of other environmental factors, such as gravity, temperature and chemical gradients (Jaffe, 1968), it is by no means certain that blue light does serve this function in nature. Nevertheless, the calculations presented here clearly indicate that it could do so.

The reports (Table 3) that blue light may induce cell division in *Chlorella*, but inhibits this process in *Prototheca*, appear to be contradictory and require confirmation. An effect of light quality on cell division may have some ecological significance, however, since it could lead to the synchronization of the cell division cycle, which is commonly observed in natural populations of phytoplankton. This hypothesis is discussed by Epel & Kraus (1966) in relation to the *Prototheca* response. Their conclusion that 'in nature this phenomenon is probably of widespread occurrence, since natural sunlight and skylight are very rich in those wavelengths (320, 420 nm) most active in promoting this phenomenon', can be accepted for planktonic organisms in the upper layers of the sea, where the spectral composition of the light will be little different from that of solar radiation, but it will require modification for algae in deeper coastal waters, in which blue light is subject to rapid attenuation. This conclusion can be extended to cover all responses to blue light in planktonic organisms (e.g. *Chlorella*, *Scenedesmus*, *Coccolithus*, *Microcystis;* Table 3).

The effects of changes in light quality on the products of photosynthesis, which have been reported for various planktonic algae (Table 3, Hauschild *et al.* 1962–5) could assume ecological significance if they also occurred in benthic algae which are subject to large tidal variations in the blue/red ratio. In turbid coastal waters (e.g. type 9), the percentage of blue light falls below 4% (the proportion used in the experiments of Hauschild *et al.* 1962) at depths greater than 4.5 m (Fig. 2), and the incorporation of photosynthetically fixed carbon into certain amino acids (e.g. aspartic and glutamic acids) may be inhibited in these conditions. This reduction in the rate of synthesis of amino acids could influence protein synthesis and ultimately produce a morphological or physiological response. A mechanism of this type could account for the effects of blue light on the spore discharge rhythm of *Nitophyllum punctatum* (Sagromsky, 1961), and might also be involved in the maintenance of other tidal rhythms. Such photosynthetic responses to light quality deserve further detailed investigation, and the effects of different blue/red ratios should also be examined in order to assess their possible importance in natural environments.

CONCLUSIONS

The spectral composition of the incident light in marine environments is subject to much greater variation than in terrestrial environments, and any conclusions about the physiological activity of any pigment systems in marine algae must be based on exact quantitative data about the spectral composition of the light to which the plant is exposed, about the response of the pigment system to specific lighting conditions and about the ecology of the alga concerned. Complete data of this type are rarely available and most of the conclusions about each pigment system which have been derived from the results of the present analysis are subject to reservations because of lack of data. Nevertheless, it has been possible to show, on the basis of the data that are available, that most of the non-photosynthetic responses of algae to light can occur in their natural habitats, and it is worth considering how to extend our knowledge of these effects of light on algal morphology. Further direct measurements of underwater spectra, which should include wavelengths up to at least 750 nm, are required for shallow coastal waters, in order to test the hypothesis that the spectral composition of the incident light will vary with the state of the tide. More detailed action spectra are required for most of the responses, together with exact determinations of the minimum energy requirements, and this approach could be supplemented by field experiments or by laboratory experiments in which critical features of the natural light environment are accurately reproduced.

I wish to thank Professor E. W. Simon for his encouragement, and Mr R. E. Parker for valuable discussion and criticism of this manuscript.

REFERENCES

BENTRUP, F. W. (1963). Vergleichende Untersuchungen zur Polaritätsinduktion durch das Licht an der *Equisetum*-spore und der *Fucus*-zygote. *Planta* **59**, 472–91.

BUTLER, W. L., HENDRICKS, S. B. & SIEGELMAN, H. W. (1965). Purification and properties of phytochrome. In *Chemistry and Biochemistry of Plant Pigments*, pp. 197–210 (ed. T. W. Goodwin). London: Academic Press.

CLARKE, G. L. & JAMES, H. R. (1939). Laboratory analysis of the selective absorption of light by sea water. *J. opt. Soc. Am.* **29**, 43–55.

CUMMING, B. G., HENDRICKS, S. B. & BORTHWICK, H. A. (1965). Rhythmic flowering responses and phytochrome changes in a selection of *Chenopodium rubrum*. *Can. J. Bot.* **43**, 825–53.

DRING, M. J. (1967). Phytochrome in red alga, *Porphyra tenera*. *Nature, Lond.* **215**, 1411–12.

DRING, M. J. (1970). Photoperiodic effects in microorganisms. In *Photobiology of Microorganisms*, pp. 345–68 (ed. P. Halldal). London: Wiley.

EPEL, B. & KRAUS, R. W. (1966). The inhibitory effect of light on growth of *Prototheca zopfii* Kruger. *Biochim. biophys. Acta* **120**, 73–83.

FORWARD, R. & DAVENPORT, D. (1968). Red and far-red light effects on a short-term behavioural response of a dinoflagellate. *Science, N.Y.* **161**, 1028–9.

GATES, D. M. (1962). *Energy Exchange in the Biosphere*. New York: Harper.

HARTMANN, K. M. (1966). A general hypothesis to interpret 'high energy phenomena' of photomorphogenesis on the basis of phytochrome. *Photochem. Photobiol.* **5**, 349–66.

HAUPT, W. (1959). Die Chloroplastendrehung bei *Mougeotia*. I. Uber den quantitativen und qualitativen Lichtbedarf der Schwachlichtbewegung. *Planta* **53**, 484–501.

HAUPT, W. (1965). Perception of environmental stimuli orienting growth and movement in lower plants. *A. Rev. Pl. Physiol.* **16**, 267–90.

HAUPT, W. & THIELE, R. (1961). Chloroplastenbewegung bei *Mesotaenium*. *Planta* **56**, 388–401.

HAUSCHILD, A. H. W., NELSON, C. D. & KROTKOV, G. (1962a). The effect of light quality on the products of photosynthesis in *Chlorella vulgaris*. *Can. J. Bot.* **40**, 179–89.

HAUSCHILD, A. H. W., NELSON, C. D. & KROTKOV, G. (1962b). The effect of light quality on the products of photosynthesis in green and blue-green algae, and in photosynthetic bacteria. *Can. J. Bot.* **40**, 1619–30.

HAUSCHILD, A. H. W., NELSON, C. D. & KROTKOV, G. (1964). Concurrent changes in the products and rate of photosynthesis in *Chlorella vulgaris* in the presence of blue light. *Naturwissenschaften* **51**, 274.

HAUSCHILD, A. H. W., NELSON, C. D. & KROTKOV, G. (1965). On the mode of action of blue light on the products of photosynthesis in *Chlorella vulgaris*. *Naturwissenschaften* **52**, 435.

HENDRICKS, S. B. & BORTHWICK, H. A. (1965). The physiological functions of phytochrome. In *Chemistry and Biochemistry of Plant Pigments*, pp. 405–436 (ed. T. W. Goodwin). London: Academic Press.

HILLMAN, W. S. (1967). The physiology of phytochrome. *A. Rev. Pl. Physiol.* **18**, 301–24.

JAFFE, L. F. (1968). Localisation in the developing *Fucus* egg and the general role of localising currents. *Adv. Morphogenesis* **7**, 295–328.

JAMES, H. R. & BIRGE, E. A. (1938). A laboratory study of the absorption of light by lake waters. *Trans. Wis. Acad. Sci. Arts Lett.* **31**, 1–154.

JERLOV, N. G. (1968). *Optical Oceanography*, 194 pp. Amsterdam: Elsevier.

JOHNSON, F. S. (1954). The solar constant. *J. Met.* **11**, 431–9.

KOWALLIK, W. (1966). Einfluss verscheidener Lichtwellängen auf die Zusammensetzung von *Chlorella* in Glucosekultur bei ehemmter Photosynthese. *Planta* **69**, 292–5.

KOWALLIK, W. & GAFFRON, H. (1966). Respiration induced by blue light. *Planta* **69**, 92–5.

LAZAROFF, N. (1966). Photoinduction and photoreversal of the Nostocacean developmental cycle. *J. Phycol.* **2**, 7–17.

LAZAROFF, N. & SCHIFF, J. (1962). Action spectrum for developmental photoinduction of the blue-green alga, *Nostoc muscorum. Science, N.Y.* **137**, 603–4.

LEVRING, T. (1966). Submarine light and algal shore zonation. In *Light as an Environmental Factor*, pp. 305–318 (ed. R. Bainbridge, G. C. Evans and O. Packham). Oxford: Blackwell.

LEWIS, J. R. (1964). *The Ecology of Rocky Shores.* London: English Universities Press.

PAASCHE, E. (1966). Action spectrum of coccolith formation. *Physiologia Pl.* **19**, 770–9.

RENTSCHLER, H. G. (1967). Photoperiodische Induktion der Monosporenbildung bei *Porphyra tenera* Kjellm. (Rhodophyta–Bangiophyceae). *Planta* **76**, 65–74.

RICHTER, G. & KIRSCHSTEIN, M. J. (1966). Regeneration und Photosynthese-Leistung kernhaltinger Zell-Teilstücke von *Acetabularia* in blauer und roter Strahlung. *Z. Pfl. Physiol.* **54**, 106–17.

SAGROMSKY, H. (1961). Durch Licht-Dunkel-Wechsel induzierter Rhythmus der Entleerung der Tetrasporangien von *Nitophyllum punctatum. Pubbl. Staz. zool. Napoli* **32**, 29–40.

SALISBURY, F. B. (1963). *The Flowering Process.* Oxford: Pergamon.

SENGER, H. & SCHOSER, G. (1966). Die spektralabhängige Teilungsinduktion in mixotrophen Synchronkulturen von *Chlorella. Z. Pfl. Physiol.* **54**, 308–20.

TAYLOR, A. O. & BONNER, B. A. (1967). Isolation of phytochrome from the alga, *Mesotaenium*, and liverwort, *Sphaerocarpus. Pl. Physiol.*, Lancaster **42**, 762–6.

TAYLOR, W. R. (1960). *Marine Algae of the Eastern Tropical and Subtropical Coasts of the Americas.* Ann Arbor: University of Michigan Press.

TERBORGH, J. (1965). Effects of red and blue light on the growth and morphogenesis of *Acetabularia crenulata. Nature, Lond.* **207**, 1360–3.

WEST, J. A. (1968). Morphology and reproduction of the red alga *Acrochaetium pectinatum* in culture. *J. Phycol.* **4**, 89–99.

STUDIES ON THE RESPONSES OF MARINE PHYTOPLANKTON TO LIGHT FIELDS OF VARYING INTENSITY

F. O. QURAISHI AND C. P. SPENCER

University College of North Wales, Marine Science Laboratories,
Menai Bridge, Anglesey

INTRODUCTION

The elegance and convenience of the C^{14} uptake method has made available a considerable amount of data concerning the effects of light intensity on the photosynthesis of marine unicellular algae under natural conditions and in the laboratory. Since, however, over short periods of time protoplasm formation and cell division may proceed more or less independently, it is difficult to infer growth from carbon dioxide uptake data. The term 'growth' as applied to cultures of unicellular algae in this paper implies increase in cell number by cell division.

The effect of the lighting conditions on the growth of these organisms has been relatively little studied. Previous laboratory experiments include work reported by Barker (1935a, b), Curl & McLeod (1961), Kain & Fogg (1958a, b, 1960), Maddux & Jones (1964), Jitts, McAllister, Stephens & Strickland (1964), and Thomas (1966). These studies have not, in general, included detailed measurements of the growth kinetics obtained under a range of light intensities.

Tamiya et al. (1953) and Gastenholtz (1964) have described the effect of variations in the day length on the growth in the laboratory of some planktonic algae. Cell division, or the release of daughter cells, has been reported to occur in some species during dark periods (Tamiya et al. 1953; Leedale, 1959; Ducoff, Butler & Geffon, 1965; Edmunds & Funch, 1969) but in others during the light periods (Spencer, 1954; Gastenholtz, 1964; Hastings & Sweeney, 1964; Paasche, 1966; Eppley, Holmes & Paasche, 1967). In most cases these studies involved the use of light of constant intensity during the period of illumination, conditions which are very different from the diurnal variation which occurs in nature. Little attention has been given to the effect of day length *per se* on the growth of phytoplankton.

The work described in this paper includes an investigation of the detailed growth kinetics of a number of marine unicellular algae when grown under various illumination regimes, including diurnal cycles which are similar in pattern to those of the natural light field.

[393]

MATERIALS AND METHODS

Five species of marine unicellular algae have been used in these studies; *Phaeodactylum tricornutum* Bohlin, *Chlorella ovalis* Butcher, *Brachiomonas submarina* Bohlin, *Dunaliella primolecta* Butcher and *Monochrysis lutherii* Droop. Axenic cultures of these organisms were established by antibiotic treatment of unialgal cultures (Quraishi, 1968) except in the case of *M. lutherii* which was obtained in an axenic state from Dr M. R. Droop of the Millport Laboratory of the Scottish Marine Biological Association.

The experimental cultures were grown at 15 °C in the apparatus and using the media described elsewhere (Quraishi & Spencer, 1971). The culture apparatus was set up in a room from which natural light was excluded. A fluorescent source of light (80 W, Osram 'daylight' tubes) was used throughout the work. The maximum light intensities incident on the culture vessels was controlled by varying both the number and the distance of the fluorescent tubes from the culture vessels. The continuously varying intensities used in some experiments were obtained using the rotating shutter assembly shown in Fig. 1*a*. The shutters were rotated mechanically at any desired constant angular velocity to produce a light field which varied from zero intensity (shutters closed) to the maximum output of the lamp array used (shutters fully open). The fluorescent tubes were switched off automatically as soon as the shutters had moved into the fully closed position and were switched on again immediately before the shutters started to open at the commencement of a light period. The variation with time of the light intensity incident on each side of the culture flasks obtained with this apparatus is shown in Fig. 1*b*. Light intensities were measured using an Hilger-Schwartz thermopile fitted with a 2 mm diameter silica window. The radiant energy reaching the thermopile was filtered by passage through 30 cm of water in a tank with optically plane glass sides and also through a filter array consisting of two Chance glass filters ('2y' and 'Solvicalor') mounted in front of the window of the thermopile. For comparative purposes, illumination measurements were made using a light meter (Evans Electroselenium Ltd) incorporating a selenium barrier layer cell protected by the same filter array. This set of filters acted as a band pass filter with a flat transmission between 400 and 700 nm and a transmittance of < 1 % at wavelengths below about 480 nm and above about 980 nm.

All the experimental work using constant illumination described here was performed at two light intensities. When the light periods included variations in the intensity of the light field, these same two intensities were used as the maximum intensities. The high intensity of 6000 lx (equivalent to 300 μgcal/cm^2/s or 1.26 × 10^{-3} watts cm^{-2} in the spectral band selected) was chosen because it is well above the intensity at which the growth rate of all the organisms used becomes independent of light intensity. The low light intensity of 1000 lx (equivalent to 50 μgcal/cm^2/s or 0.21 × 10^{-3} watts cm^{-2} is

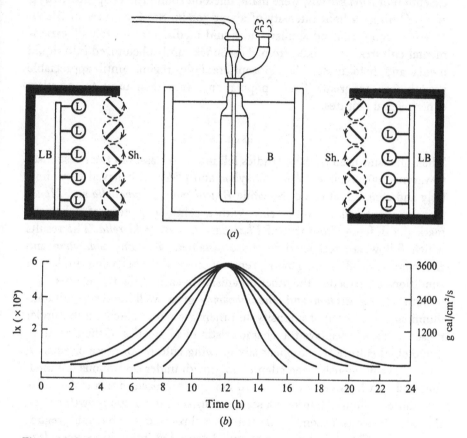

Fig. 1. (*a*) Culture apparatus used showing the arrangements of shutters to simulate the natural diurnal variation of light intensity. B, Constant temperature bath. LB, Light tight box. L, Fluorescent light sources. Sh., Shutters. Broken circles show direction of rotation of shutters. (*b*) Variation with time of the light intensity and incident radiant energy for several day lengths obtained with the apparatus shown in Fig. 1*a*.

in the range where the growth rate of the organisms is proportional to the light intensity (Quraishi & Spencer, 1971).

Cell concentrations were measured using a Coulter counter. On those occasions when counts could not be performed immediately after sampling, the cell suspensions were preserved by storage in the dark in the presence of 50 mg/l of actidione. This treatment stopped all cell division and prevented changes in cell volume for periods of up to 1 week. None of the experimental samples were, however, preserved for longer than 24 h before counting.

Stock cultures of the organisms used in the experimental work, with the exception of *Monochrysis*, were maintained on solid media (Spencer, 1954) at 15 °C under a light intensity of about 500 lx. Stock cultures of *Monochrysis* were maintained similarly in liquid media. Inocula for all experimental cultures were taken from these stocks and subcultured into liquid media and held under the appropriate light regime until appreciable growth was apparent. These populations were then used to inoculate experimental cultures.

RESULTS

The organisms used in these studies fell into two classes in their responses to various light regimes: *Phaeodactylum* and *Chlorella* behaved very similarly and constituted one class, while *Brachiomonas*, *Dunaliella* and *Monochrysis* constituted the other class with a common pattern of behaviour markedly different from that of *Phaeodactylum* and *Chlorella*. The results which follow are restricted to those obtained with *Phaeodactylum* and *Brachiomonas* and are, in general, representative of the behaviour under the conditions described of the other organisms in each of the two classes.

Both *Phaeodactylum* and *Brachiomonas* grow well under continuous illumination of constant intensity and their growth in batch culture under these conditions displays the characteristic ideal kinetics of the growth of a microbial culture. If exponentially growing cultures of these organisms, which have previously been adapted to growth under continuous illumination and in which cell divisions are not synchronized, are transferred to intermittent illumination of constant intensity, continuous growth ceases, the cell divisions becoming partially synchronized to the light regime (Fig. 2a, b). The growth curves obtained consist of intermittent periods of growth during either the light periods (*Phaeodactylum*, Fig. 2a) or the dark periods (*Brachiomonas*, Fig. 2b). Very similar patterns of growth occur whether light of constant intensity is used during the light periods or whether the cells are exposed to light of continuously changing intensity similar in pattern to the natural diurnal cycle. However, under the latter conditions, the maximum rate of cell division of *Phaeodactylum* is not reached until a short time after illumination has commenced. The growth of *Brachiomonas* is, in contrast, little different under the two types of intermittent illumination. The differences in the response of *Phaeodactylum* and *Brachiomonas* in this respect presumably reflect the much closer

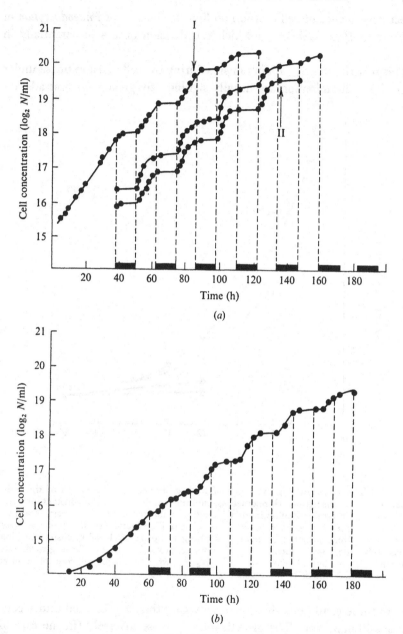

Fig. 2. (a) The progress of the induction of rhythmic growth by *Phaeodactylum* when transferred from continuous illumination to a light regime of 12 h dark and 12 h light of intensity
6000 lx. The bars enclosed by broken lines indicate the dark periods. The arrows I and II
indicate the time of subculturing. (b) The progress of the induction of rhythmic growth by
Brachiomonas when transferred from continuous light to a light regime of 12 h dark and
12 h light of intensity 6000 lx. The bars enclosed by broken lines indicate dark periods.

direct dependence of cell division on light in the case of *Phaeodactylum* in contrast to *Brachiomonas*, in which cell division occurs preferentially in the dark.

The growth rates of both organisms during the cell division phase under intermittent illumination are usually significantly greater on first adapta-

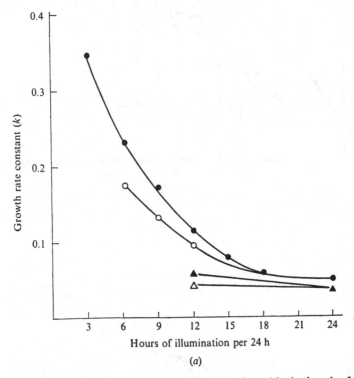

(*a*)

Fig. 3. (*a*) The variation of growth rate of *Phaeodactylum* with the length of the light period under various light–dark regimes. ●, Intermittent illumination of constant intensity (6000 lx). ○, Intermittent illumination including a diurnal cycle of intensity reaching a maximum intensity of 6000 lx. ▲, Intermittent illumination of constant intensity (1000 lx). △, Intermittent illumination including a diurnal cycle of intensity reaching a maximum intensity of 1000 lx. Growth rates expressed in terms of the growth rate constant k defined by the relationship $N = N_0 . e^{k \cdot t}$ where N_0 is the cell concentration at time $t = 0$ and N the cell concentration at time t, in hours.

tion to these conditions than the steady growth rates obtained under continuous illumination. The growth rates increase inversely (in the case of *Phaeodactylum*, Fig. 3*a*) and directly (in the case of *Brachiomonas*, Fig. 3*b*) with the length of the light period. However, these increased growth rates are limited in their duration, there being in general an inverse relationship between the magnitude of the increase in the growth rate and its duration.

It is also noticeable that during the cell division phase the growth rate of *Brachiomonas* is less affected by continuously changing the light intensity than is that of *Phaeodactylum*. In the latter case the growth rates obtained are, nevertheless, still markedly greater than those obtained under continuous illumination of constant intensity. When a light intensity of 1000 lx, which is suboptimal for growth under continuous illumination, is used in

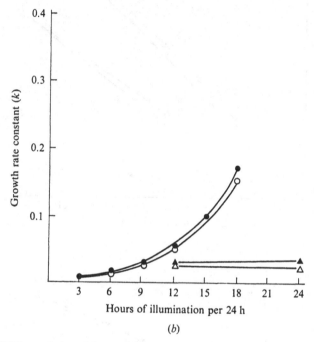

Fig. 3. (b) The variation of the growth rate of *Brachiomonas* with the length of the light period under various light–dark regimes Conditions, symbols and units as in Fig. 3 a.

intermittent light regimes the growth rates exhibited during the division phases in the light or dark periods are only slightly, if at all, greater than those observed under continuous illumination for 24 h/day (Figs. 3 a, b).

When a high constant light intensity is used in a light/dark cycle the yield of cells in each complete period of 24 h increases as the length of the light period increases (Figs. 4 a, b) but proportionately smaller increases are obtained for the longer light periods. The individual organisms differ somewhat in this respect. Thus, whereas the daily yield of cells of *Phaeodactylum* displays only a slight deviation from strict proportionality to the length of the light period, the daily yield of *Brachiomonas* approaches a maximum when the duration of the light period reaches 24 h. The species

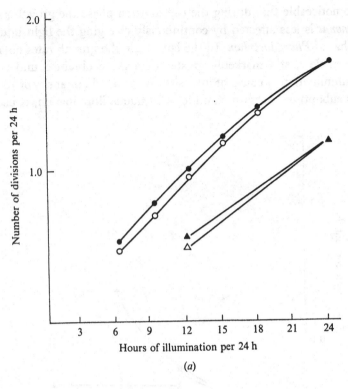

(a)

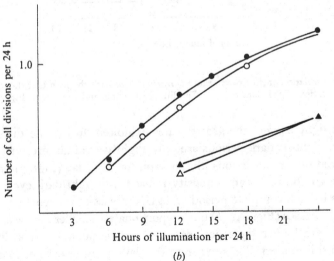

(b)

Fig. 4. (a) The yield of cells of *Phaeodactylum* obtained under various light–dark regimes. Conditions and symbols as in Fig. 3 a. (b) The yield of cells of *Brachiomonas* obtained under various light–dark regimes. Conditions and symbols as in Fig. 3 a.

of *Chlorella* used in these studies displayed no further increases in daily yield when the light period exceeded about 12 h. The use of continuously changing light intensity during the light periods of an intermittent light regime has little further effect on the total yield of cells obtained per 24 h with any of the organisms studied. It is possible that the interspecific variations observed may imply that the day length is an ecologically important factor for phytoplankton and plays some part in the control of the succession of species in nature.

The organisms used in these studies adapt readily to a change from continuous to intermittent illumination (Figs. 2a, b). If actively growing cells adapted to a particular regime of intermittent illumination are transferred to continuous light, the repeating pattern of intermittent growth is quickly lost by *Phaeodactylum*, cell divisions becoming asynchronous (Fig. 5a). In contrast, *Brachiomonas* maintains its intermittent growth pattern under continuous light (Fig. 5b), at the same time showing a greater yield of cells per 24 h then the immediate parent culture, as can be seen from the longer time during which the cell division phase continues. If, however, the culture is allowed to enter the senescent phase, a subculture quickly loses the rhythmic pattern of growth.

A particular pattern of intermittent growth once induced in *Brachiomonas* is, however, very labile to changes in the phasing of the particular light–dark regime. If an additional light period is introduced into such a regime so that the timing of light and dark periods are interchanged, the growth pattern quickly readjusts to synchronize with the new phasing of the light and dark periods (Fig. 5c).

DISCUSSION

All the organisms studied responded to intermittent illumination by confining the cell division stages of their growth to a limited period. The bursts of growth occurred at 24 h intervals and in the case of *Brachiomonas* the rhythmic pattern of growth persisted under suitable conditions after the stimulus which produced the behaviour was removed (Fig. 5b). Thus there is a superficial similarity in the response of these organisms to the well recognized rhythmic behaviour patterns of many higher organisms which are often interpreted in terms of inherent biological clocks which can be synchronized by exposure to an oscillating environmental factor.

The cells in a microbial culture which is growing regularly under constant conditions have an inherent rhythmic pattern of development and cell division. The period of this rhythm is the generation time of the cells.

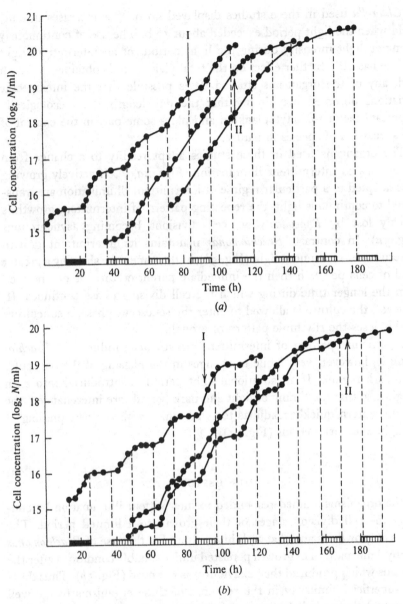

Fig. 5. (a) The progress of the loss of rhythmic growth in three sequential cultures of *Phaeodactylum* when transferred from intermittent illumination to continuous illumination after 24 h. Arrows I and II indicate the times of subculturing. The time scale is arbitrary for the two subcultures but the hatched bars indicate phasing of dark periods under the initial light regime. Illumination during initial light periods at a constant intensity of 6000 lx. Intensity during continuous illumination also 6000 lx. (b) The progress of the loss of rhythmic growth in three sequential cultures of *Brachiomonas* when transferred from intermittent illumination to continuous illumination. Conditions and symbols as in Fig. 5 a.

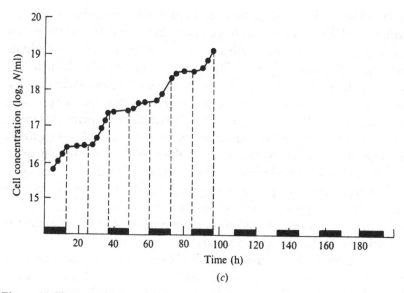

Fig. 5. (c) The progress of adaptation of *Brachiomonas* to rephasing of the light–dark periods. The bars enclosed by broken lines indicate the timing of the dark periods. Illumination during light periods was at constant intensity of 6000 lx.

In the case of exponentially growing cultures the population presumably consists of cells with a range of generation times and the growth rate of such cultures reflects the mean generation time of the population.

Under constant illumination at 15 °C the mean generation times of *Phaeodactylum* and *Brachiomonas* were 14.1 and 19.7 h respectively. These generation times will vary inversely with temperature and within restricted ranges of temperature will be expected to show a Q_{10} of about 2. The dependence on temperature and the marked difference in the period from that of the natural diurnal cycle distinguishes the inherent rhythm of cell development and division in these organisms from the circadian rhythms of activity in higher organisms.

The rhythmic growth patterns produced by intermittent illumination of cultures of unicellular algae is a result of two effects. First, the period of the cycle of cell development and division is lengthened to 24 h. Secondly, the cell division stages of all the population are confined to a particular restricted interval in each 24 h period.

If an obligate phototrophic organism is to grow under intermittent illumination it must be able to synchronize those metabolic activities which are dependent on light energy with the times of illumination. In any case, a period of darkness would be expected to extend the generation time of the cells. It is inevitable therefore that such a population will be forced into a rhythm which has a period which is equal to an integral ratio of the period of the incident light cycle. Furthermore, a repeating cycle of rhythmic bursts of growth will only persist if the generation times of individual cells are equal to the period of the cycle of illumination. It seems likely that this is the explanation of the diurnal rhythms observed rather than the presence of an inherent circadian rhythm. Results of preliminary experiments using intermittent illumination with different periods confirm this suggestion since under these conditions the period of the rhythm of cell development and division shows marked variation if the period of the illumination cycle is not an integral ratio of 24 h. In contrast, light cycles with a period of 12 h produced rhythmic growth patterns with a period of 24 h.

If the rhythmic growth behaviour of these organisms under intermittent illumination is dependent upon their ability to synchronize certain of their metabolic activities with the period of oscillations of the external stimulus, then only those cells in the population which are able to modify the phasing of their inherent rhythmic cycle of development will be able to proliferate. It is generally considered that newly formed daughter cells are the most active photosynthetically so that, for example, a *Phaeodactylum* cell which divides late in a light period will produce daughter cells which will be denied light at critical stages in their development. If, as seems probable, the generation time of such cells is increased, it is likely that they will not be able to complete their development and division before the end of the subsequent light period. The first dark period will thus tend to discriminate against such cells and their progeny as are unable to effect an appropriate shift of phase in their cycle of development. In the same way, such a mechanism will tend to establish and maintain a population of cells with a more limited range of individual generation times than is presumed to be general in exponentially growing populations under continuous illumination. Essentially the same mechanism can account for the development of rhythmic growth in *Brachiomonas*, but in this organism it seems that it is the light period following the first dark period which discriminates in favour of those cells in the population which happen to be at the optimum stage of their life cycle to exploit the light energy during the limited period when it is available.

The behaviour of *Brachiomonas* when returned from intermittent to continuous illumination suggests that the generation time of individual cells of this organism in the culture under intermittent illumination shows little variation and hence there is little tendency to revert to a heterogeneous population when continuous illumination is established. In this organism it is only after cultures have been allowed to enter the stationary phase that the population growing from a subsequent subculture is found to have lost its rhythmic growth pattern. During the senescent phase of a culture it is to be expected that a population will develop a wide spectrum of developmental stages and on subculture any synchronization of the growth will quickly disappear. The loss of rhythmic growth in this case therefore seems more likely to result from a randomization of the distribution of developmental stages in the population.

In contrast, the synchrony in the division stages of *Phaeodactylum* cultures is quickly lost when these are returned to continuous illumination (Fig. 5*a*). It appears therefore that cultures of this alga rapidly develop a population in which the individual cells have a wide range of generation times. Similar tendencies are apparent in the time scale of the development of rhythmic growth patterns in *Phaeodactylum* and *Brachiomonas*. The slower acquisition of rhythmic growth by the former organism again suggests a more heterogeneous population under continuous illumination.

It is also possible that the greater the difference between the mean generation time of a culture under continuous illumination and the period of the intermittent illumination cycle forced upon it, the more difficult it is for the organism to achieve synchrony with the external stimulus. This could clearly be a factor in the speed of adaptation to the conditions imposed and on the stability of the rhythmic growth patterns after return to continuous illumination.

The growth kinetics of cultures under intermittent illumination require several parameters to be specified if they are to be characterized completely. If the duration of the time interval during which cell division occurs is T and the period of the rhythm is P, then $1-(T/P)$ is an index of the synchrony in the timing of cell division. The experimental utility of synchronized microbial cultures is the possibility they provide for studying variations in the chemical composition of cells at different stages in their development. Such cultures are also attractive material for cytological investigations. The interpretation of results and the practical convenience of the material depends very much upon the proportion of the cells in the population which are in an identical condition at any one time. The percentage of cells in the population which divide in unit time, $100(N-N_0)/N_0 t$ where N_0 is the cell concentration at time $t = 0$ and N is the cell concentration at time t, is a practically useful index of this characteristic of the behaviour of the population.

Table 1 shows the value for these two parameters for the five species included in these studies and covers a range of illumination conditions.

Table 1. *Degree of synchrony obtained with four species of*
algae under various light regimes

(Degree of synchrony as the % cells dividing per hour. Illumination of constant intensity
of 6000 lx during the light periods. Figures in parenthesis are the indices of synchrony in
the timing of the cell division phases of the cultures.)

	Species			
Light regime	*Phaeodactylum*	*Chlorella*	*Brachiomonas*	*Monochrysis*
24 h light 0 h dark	4.9 (0)	3.1 (0)	3.5 (0)	3.6 (0)
18 h light 6 h dark	5.7 (0.50)	6.3 (0.58)	17.3 (0.90)	15.4 (0.90)
15 h light 9 h dark	7.7 (0.73)	7.7 (0.71)	9.9 (0.83)	8.9 (0.81)
12 h light 12 h dark	11.5 (0.86)	9.9 (0.84)	5.8 (0.75)	6.0 (0.75)
9 h light 15 h dark	17.2 (0.94)	13.9 (0.92)	3.1 (0.65)	3.2 (0.62)
6 h light 18 h dark	23.1 (0.97)	19.8 (0.96)	1.0 (0.54)	2.0 (0.54)

The degree of synchrony in the timing of cell division increases with the length of the time in any complete period of the cyclical light regime in which cell division is repressed. Even when the highest degree of synchrony in the timing of cell division is achieved, the proportion of cells in the population dividing in unit time does not show a very great increase compared with the situation which exists under continuous illumination. Changes in the biochemical composition of cells from such cultures cannot therefore be considered as representative of the composition of individual cells at unique stages in their life cycle although the trends should be similar. From this point of view it therefore seems better to refer to such cultures as showing rhythmic growth since the pronounced synchrony is only in regard to the timing of the division stages of those cells that complete their division in a particular period. Under such conditions this is not a high proportion of the cell population.

It is not easy to suggest a mechanism which accounts satisfactorily for the difference in growth behaviour of the organisms typified by *Phaeodactylum* or *Brachiomonas*. It seems that organisms such as *Phaeodactylum* have an obligate connection between photosynthesis (and/or some other photochemical reactions) and cell division whereas organisms such as *Brachiomonas* are able to separate some of the processes of cell development from those of cell division. The intimate biochemical connection between cell division and photochemical reactions in those organisms which are obligate 'light dividers' and the nature of the mechanism which allows a separation of these processes in the faculative 'light dividers' is uncertain. It is clear, however, that light does not exert a completely inhibitory effect on some or all the developmental processes of those organisms which in intermittent lighting conditions confine their cell division stages to the dark periods. The possibility that light exerts a retarding effect on one or more of these processes in these organisms cannot, however, be excluded at the present time.

REFERENCES

BARKER, H. A. (1935a). Photosynthesis in diatoms. *Arch. Mikrobiol.* 6, 141–56.
BARKER, H. A. (1935b). The culture and physiology of marine dinoflagellates. *Arch. Mikrobiol.* 6, 177–81.
CURL, H. & MCLEOD, G. C. (1961). The physiological ecology of a marine diatom *Skeletonema costatus* Cleve. *J. mar. Res.* 19, 70–88.
DUCOFF, H. S., BUTLER, B. D. & GEFFON, E. J. (1965). The effects of radiation on replication in *Brachiomonas submarina* Bohlin. *Radiat. Res.* 24, 563–71.
EDMUNDS, L. N. & FUNCH, R. (1969). Effects of skeleton photoperiods and high frequency light-dark cycles on the rhythm of cell division in synchronized cultures of *Euglena*. *Planta* 87, 134–63.
EPPLEY, R. W., HOLMES, R. W. & PAASCHE, E. (1967). Periodicity in cell division and physiological behaviour of *Ditylum brightwellii*, a marine planktonic diatom, during growth in dark and light cycles. *Arch. Mikrobiol.* 56, 305–23.
GASTENHOLTZ, R. W. (1964). The effect of day length and light intensity on the growth of littoral marine diatoms in culture. *Physiologia Pl.* 17, 951–63.
HASTINGS, J. W. & SWEENEY, B. M. (1964). Phased cell division in marine dinoflagellates. In *Synchrony in Cell Division and Growth* (ed. E. Zeuthen). New York: Interscience Publishers.
JITTS, H. R., MCALLISTER, C. D., STEPHENS, K. & STRICKLAND, J. D. H. (1964). The cell division rate of some marine phytoplankters as a function of light intensity and temperature. *J. Fish. Res. Bd Canada* 21, 139–57.
KAIN, J. M. & FOGG, G. E. (1958a). Studies on the growth of marine phytoplankton. I. *Asterionella japonica* Gran. *J. mar. biol. Ass. U.K.* 37, 397–413.
KAIN, J. M. & FOGG, G. E. (1958b). Studies on the growth of marine phytoplankton. II. *Isochrysis galbana* Parke. *J. mar. biol. Ass. U.K.* 37, 781–8.
KAIN, J. M. & FOGG, G. E. (1960). Studies on the growth of marine phytoplankton. III. *Prorocentrum micans* Erenberg. *J. mar. biol. Ass. U.K.* 39, 33–50.
LEEDALE, G. F. (1959). Periodicity of mitosis and cell division in Euglenineae. *Biol. Bull. mar. biol. Lab., Woods Hole* 116, 162–74.
MADDUX, S. W. & JONES, R. F. (1964). Some interaction of temperature, light intensity and nutrient concentrations during continuous culture of *Nitzschia closterium* and *Tetraselmis* sp. *Limnol. Oceanogr.* 9, 79–86.
PAASCHE, E. (1966). Phytoplankton growth in light dark cycles. In *Research on Marine Food Chains*, pp. 10–12. University of California Institute of Marine Research.

QURAISHI, F. O. (1968). *The reactions of phytoplankton to different illumination regimes*. Ph.D. Thesis. University of Wales.

QURAISHI, F. O. & SPENCER, C. P. (1970). Studies on the growth of some marine unicellular algae under different artificial light sources. *Marine Biology* **8**, 60–5.

SPENCER, C. P. (1954). Studies on the culture of a marine diatom. *J. mar. biol. Ass. U.K.* **33**, 265–90.

TAMIYA, H., SHABITA, K., SASA, T., IWAMURA, T. & MORIMURA, Y. (1953). Effects of diurnally intermittent illumination on the growth and some cellular characteristics of *Chlorella*. In *Algal Cultures. From Laboratory to Pilot Plant*, pp. 76–84 (ed. J. S. Burley). Washington Carnegie Institute. Publ. No. 600.

THOMAS, W. H. (1966). Effects of temperature and illumination on cell division rates of three species of tropical phytoplankton. *J. Phycol.* **2**, 17–22.

LIGHT AND THE DISTRIBUTION OF ORGANISMS IN A SEA CAVE*

T. A. NORTON, F. J. EBLING AND J. A. KITCHING

Department of Botany, University of Glasgow,
Department of Zoology, University of Sheffield,
and School of Biological Sciences, University of East Anglia

INTRODUCTION

In studies of the vertical zonation of sublittoral marine organisms it is extremely difficult to separate the effects of light intensity from those related to changes in the quality of light. The penetration of light into marine caves and the zonation of organisms in caves is therefore of interest since caves present natural situations in which there is a gradient of light intensity without the concomitant changes in wavelength which occur with increasing depth.

During the summers of 1967 and 1968 a detailed study was made of a sea cave on Bullock Island near to Lough Ine, Co. Cork, Eire. A short account of the algal vegetation of this cave has already been published by Grubb & Martin (1937). The mouth of the cave faces almost due west. Although close to the open sea the cave is sheltered from the full force of the Atlantic swell by the small inlet in which it occurs and because it faces at right angles to the swell. Even so, some tidal surge funnels into the cave on all but the calmest days.

The mouth of the Bullock Island Cave is 5–6 m in width and about 9 m from roof to sea floor, with the bottom submerged by about 3 m at low water of spring tides. The cave runs into the side of Bullock Island as a straight tube or tunnel 97 m long, without any other outlet. The width diminishes gradually, and the sandy floor also gradually rises but is permanently submerged for all except the last 10 m. The nearly vertical walls of the cave are composed of slate dipping nearly perpendicularly.

A longitudinal profile and cross-sections of the cave are shown in Fig. 1. For survey purposes horizontal distances (H) are given with respect to a position immediately below the end of the roof of the cave, over the cave mouth – positive for outside and negative for inside the cave. Vertical distances (V) are referred to the outer (west) end of the top of a central rock,

* Paper 17 in the series *The ecology of Lough Ine.*

[409]

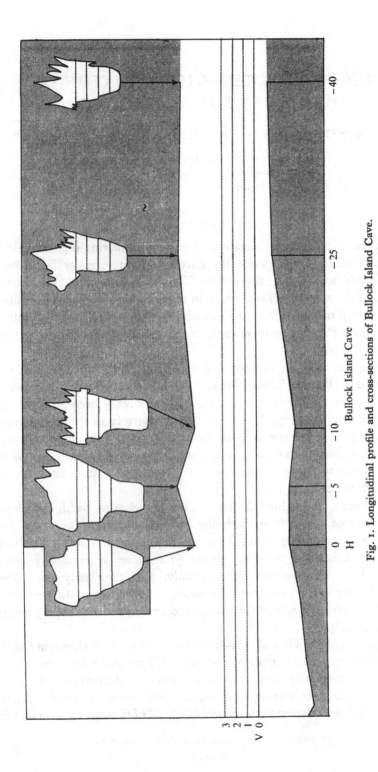

Fig. 1. Longitudinal profile and cross-sections of Bullock Island Cave.

at -14 m (H), submerged except at low water of ordinary spring tides, which this level approximately represents. Positive readings are upwards from this level.

Our readings of light intensity and our detailed records of flora and fauna, described below, have all been made on the north side of the cave. The south side is smooth and almost uniformly steep, has few ledges upon which photocells can be placed, and appears somewhat poorer in flora and fauna, although the distribution of species on the two sides in general appears similar.

ILLUMINATION

In order to obtain information for comparison with normal unshaded situations in the photic zone of the open coast, it was necessary to measure illumination at several wavelengths. Moreover, isolated readings would have been of little value; information was needed over the whole of the day, and at different stages of the semilunar cycle. Accordingly we used a number of photocells of narrow spectral sensitivity fitted with long cables, so that the cells could be mounted at suitable positions in the cave and read at frequent intervals from an observation post on the hillside above.

Methods

'Clairex' photocells, numbers CL 7021, CL 705 HL, CL 707 M, were used, sensitive to 515, 550 and 615 nm respectively. The choice was determined in part by availability of cells of narrow selectivity, although it is also related to the spectral absorption of important algal pigments. Each photocell was mounted in a waterproof container covered with a removable cap carrying a flashed opal filter within which could be mounted neutral perspex filters of individually measured transmission. The cells sensitive to 515 nm had a dark green (number 24) 'cinemoid' filter mounted next to the photocell in order to sharpen the band of sensitivity.

All the photocells were calibrated out of doors in daylight at their appropriate wavelengths against a spectroradiometer. The sensitive area of the latter carried an opal glass. The photocell containers were submerged under a few cm of water beside the spectroradiometer. For 'field' observations the photocells were mounted on brass carriers of adjustable angle so that the sensitive surfaces could be held at any angle from vertical to horizontal. The brass carriers were attached to concrete blocks to hold them in position in the cave. The angle of the carrier was adjusted to expose the cell to maximum illumination; it is given (Table 1) as between the opal glass and the horizontal.

Table 1. Daily illumination in Bullock Cave at 550 nm in μW h/cm²/nm

Date (1967)	Period of observations (h)	Photocell in relation to water level	Hilltop control (o°)	A −4 m (H) +1.5 m (V) 15° close to inner limit of *Fucus* spp.	B −5 m (H) +0.93 m (V) 70° above inner limit of *Laminaria*	C −17 m (H) +0.65 m (V) 75° near inner limit of *Lomentaria**	D −22 m (H) +1.08 m (V) 80° inner limit of *Plumaria* and *Pseudendoclonium*	E −42 m (H) +1.97 m (V) 80° no algae
6 July	04.45–22.15	Uncovered	—	185	17.7	2.55	1.82	0.057
		Covered	—	164	9.5	0.47	0.10	0.000
		Total	629	349	27.2	3.0	1.92	0.057
8 July	08.00–21.00	Uncovered	—	—	23.8	3.56	2.31	0.086
		Covered	—	—	5.53	0.50	0.015	0.0011
		Total	1025	—	29.3	4.1	2.325	0.087
10 July	09.45–21.45	Uncovered	—	—	37.9	5.9	3.26	0.115
		Covered	—	—	4.3	0.61	0.95	0.002
		Total	803	—	42.2	6.5	4.21	0.117
12 July	09.00–21.45	Uncovered	—	—	36.6	6.1	3.44	0.088
		Covered	—	—	5.25	0.52	0.70	0.004
		Total	965	—	41.9	6.6	4.14	0.092
14 July	05.45–22.15	Uncovered	—	131	73	7.7	4.24	0.080
		Covered	—	31	8.2	0.19	0.19	0.000
		Total	744	162	81	7.9	4.43	0.080
Total illumination for the 6 days			4166	(1560)	222	28.1	17.0	0.43
Total as percentage of that in open			100	37	5	0.7	0.4	0.01

Photometer positions are specified from the cave entrance (−) and above Centre Rock (+) which is at low water of a good but not extreme Spring tide.

* *Lomentaria articulata*, *Membranoptera alata*, *Phyllophora crispa* and *Cladophora rupestris* end at −16 m (H).

Readings were normally taken every quarter of an hour throughout the day. Control readings were taken from photocells submerged in a basin near the hill top.

Results

Illumination at 550 nm at five positions in the cave

Single cells sensitive to 550 nm were stationed as in Table 1 which also summarizes the results for the 5 days on which readings were taken. The results for 6 July and 14 July are shown in Figs. 2 and 3.

Illumination at limit of Laminaria at 515, 550 and 615 nm

A trio of photocells – one for each of the three wavelengths – was mounted on a single-hinged base plate fitted to a heavy weight, and was suspended from steel pins driven into the north rock face of the cave so as to lie just below low water at the inner limit of *Laminaria*, but not shaded by these algae, 6 m within the cave. The base plate was clamped at about 10° from the horizontal so as to point the photocells through the cave mouth at the sky. The results for 16 and 23 August are shown in Figs. 4 and 5, and summarized in Tables 2 and 3.*

Interpretations and conclusions

Photocells stationed at positions far into the cave (stations C, D and E in the first series) were directed towards the cave mouth. As was expected the readings obtained were very low as compared with outside. Note that in Figs. 2 and 3 the scale for these stations is expanded. Moreover, the period of significant illumination coincided with that during which the photocells were uncovered by the tide. There are several reasons why immersion should drastically reduce the illumination at the photocells. (*a*) With the rise of the tide the cave mouth is effectively reduced; light entering the cave through the water will have a long way to travel under water. (*b*) The smaller the angle made by the incident light with the water surface the more light will be reflected. Thus while the tide is out the photocells will receive both direct and reflected light, but while immersed they will only receive light which escapes reflection at the surface. (*c*) Light passing through the water surface will be refracted very considerably, so that the angle of incidence on the nearly vertical face of the photocell will deviate considerably from normal. The percentage collected by the opal will thereby be reduced.

In spite of (*c*), we must conclude that illumination is negligible at rock surfaces far within the cave except while they are uncovered, and even then it is very weak.

---

* The letters A–C and D–F used in Figs. 4 and 5 and in Tables 2 and 3 to refer to two trios of photocell readings, one at the control station and the other at the inner limit of *Laminaria* bear no relation to the stations designated A–E in Table 1 and Figs. 2 and 3.

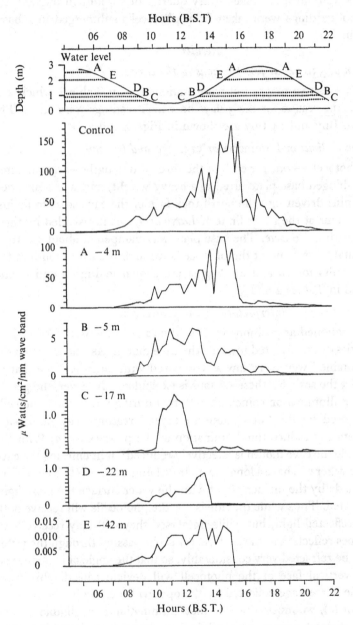

Fig. 2. Illumination at 550 nm at various positions in Bullock Island Cave on 6 July 1967. The levels of the photocells and the times of their submergence are shown in the top panel.

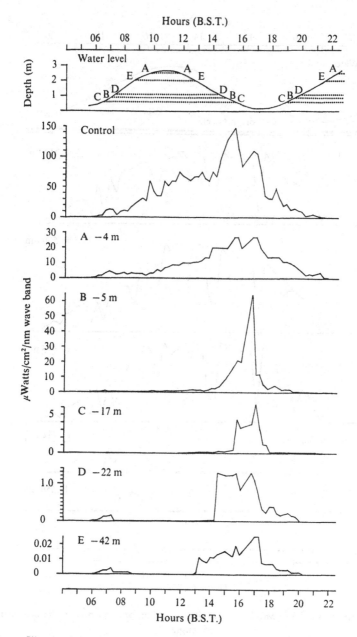

Fig. 3. Illumination at 550 nm at various positions in Bullock Island Cave on 14 July 1967. The levels of the photocells and the times of their submergence are shown in the top panel.

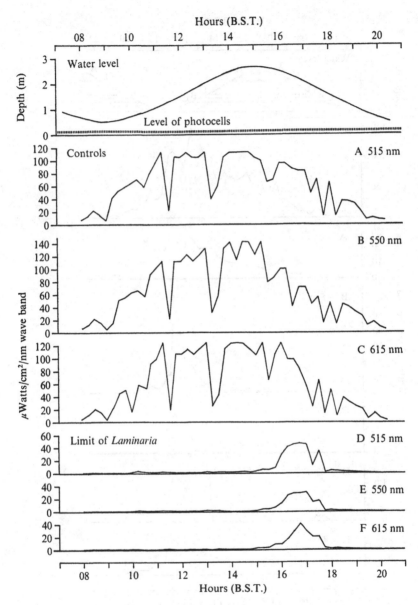

Fig. 4. Illumination at three wavelengths at the inner limit of *Laminaria* in Bullock Island Cave on 16 August 1967.

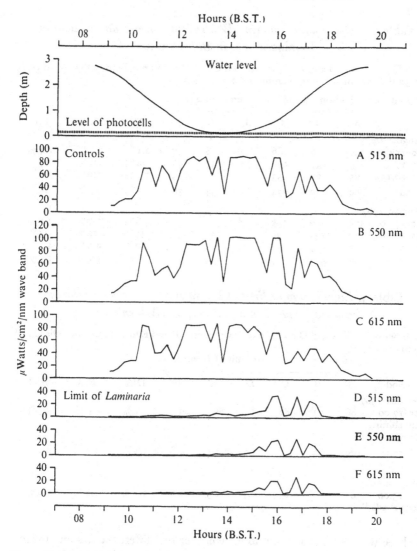

Fig. 5. Illumination at three wavelengths at the inner limit of *Laminaria* in Bullock Island Cave on 23 August 1967.

The second series of observations, with photocells sensitive to three very narrow wavebands, was designed to describe the illumination at the limit of penetration of laminarian algae in the cave. The percentage of illumination received at the inner limit of laminarian algae amounted to 6–12 % of that in an unshaded situation, with the values regularly higher for 515 nm than for 550 and 615 nm. Nevertheless the results are of the same order for

Table 2. *Total illumination (in $\mu W\,h/cm^2/nm$) over 1 day at limit of*
Laminaria *spp. in Bullock Cave*

(A and D at 515 nm, B and E at 550 nm, and C and F at 615 nm. During the period 15.15 to 17.30 the cave mouth was open to direct sunshine.)

Period	Hilltop control			At limit of *Laminaria*					
	A	B	C	D	E	F	D/A	E/B	F/C
16 August 1967									
08.00–15.00	530	576	486	11	8	7	2.1%	1.4%	1.4%
15.15–17.30	205	197	195	70	43	46	34%	22%	24%
17.45–20.15	65	66	53	5	3	3	7.7%	4.6%	5.7%
Total	800	839	734	86	54	56	11%	6.5%	7.6%
23 August 1967									
09.10–15.00	374	410	355	18	11	11	4.8%	2.7%	3.1%
15.15–17.30	140	153	130	46	36	31	33%	22%	24%
17.45–19.45	43	47	37	3	2	1.5	7.0%	4.3%	4.1%
Total	557	610	522	67	49	44	12%	8.0%	8.4%

Table 3. *Illumination at limit of* Laminaria *when sun was shining into cave and when it was not, in $\mu W\,h/cm^2/nm$*

(One h period; cells A and D respond to 515 nm, cells B and E to 550 nm, and cells C and E to 615 nm.)

Period	Hilltop control			At limit of *Laminaria*					
	A	B	C	D	E	F	D/A	E/B	F/C
16 August 1967									
16.15–17.00 (Sun shining into cave)	88	71	82	45	28	29	51%	39%	35%
12.15–13.00 (Cave shaded)	109	122	115	1.5	1.1	0.8	1.4%	0.9%	0.7%
23 August 1967									
12.15–13.00 (Cave shaded)	85	93	85	2.7	1.5	1.6	3.2%	1.6%	1.9%

the three wavelengths and we doubt whether the difference has any meaning. In sunny weather on 16 August 1967, 70% or more of the light energy reaching the inner limit of *Laminaria* reached this position during the few hours in the afternoon while the sun was shining into the mouth of the cave, even though the tide was high during this period. On the other hand, during short intervals in the afternoon, while the sun was covered with cloud (Fig. 5), or during periods when the sun was out but not shining into the cave (Table 3), the illumination at the limit of *Laminaria* was much less, both absolutely and in comparison with that on the hill top. Light reaches the limit of *Laminaria* from a severely limited area of sky, determined by the aperture of the cave; so that under cloudy conditions, or when the sun is in some other part of the sky, the light energy reaching the *Laminaria* amounts only to a small proportion of that reaching a fully open situation.

DISTRIBUTION OF ORGANISMS

Records

The presence or absence of as many species as possible was recorded – without regard for vertical distribution – at one metre intervals from a distance of $+10$ m (outside mouth of cave) to a distance of -20 m (inside). Further within the cave the species composition changed only gradually and records were made only at -26, -40 and -80 m. Upper and lower limits of the various species were determined at $+9$, $+1$ m (outside the mouth of the cave), and -10, -25 and -40 m (inside the cave). The information for some plants and animals common inside the cave is summarized in Figs. 6–10. More complete records are given in the Appendix.

Horizontal distribution

Outside the cave mouth the intertidal zone was dominated by maritime lichens, fucoids, and the small red algae *Laurencia pinnatifida* and *Gigartina stellata*, all at their appropriate levels. In the shallow sublittoral zone the laminarians predominated, with a rich epiflora and underflora of *Dictyota dichotoma* and a variety of red algae. All these continued for the first 6 m into the cave, although fucoids became so stunted that it was impossible to distinguish the species.

At -8 m the flora was quite different. Both fucoids and laminarians were absent. The intertidal zone was dominated by *Plumaria elegans*, *Ptilothamnion pluma* and *Lomentaria articulata* – which are normally found in crevices and as undergrowth on the open shore. In the shallow sublittoral the most abundant algae were *Phyllophora crispa* and *Cryptopleura ramosa*, the normal undergrowth of the *Laminaria* forest.

All these species became progressively sparser as they penetrated further into the cave, and at -20 m the only algae were *Plumaria elegans*, *Ptilothamnion pluma*, *Cladophora* sp., *Pseudendoclonium marinum*, *Hildenbrandia prototypus* and the calcareous lithothamnia. *Cladophora* sp., *Ptilothamnion*, *Hildenbrandia* and lithothamnia penetrated to -26 m but no algae were found at -40 m. Certain species normally associated with algae followed these in their distribution and limitation: *Elachista fucicola* on *Fucus* spp., *Polysiphonia lanosa* on *Ascophyllum*, and herbivorous gastropods such as *Littorina littoralis* which feeds on fucoids.

Fig. 6. Horizontal distribution of plants in Bullock Island Cave. Shaded bands represent actual records of presence in one or both summers of 1967 and 1968.

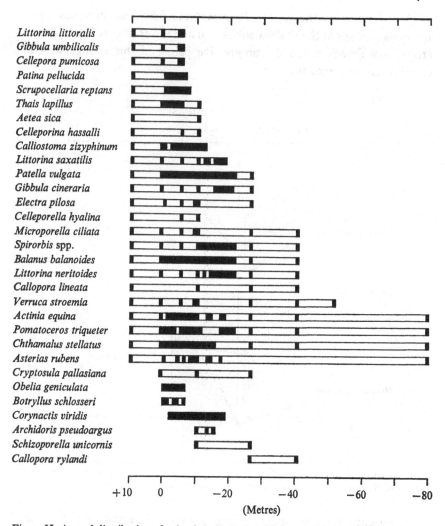

Fig. 7. Horizontal distribution of animals in Bullock Island Cave. Shaded bands represent actual records of presence in one or both summers of 1967 and 1968.

The greatest numbers of animal species were found between the cave mouth and − 10 m. We did not in fact especially investigate the rich fauna of the overhanging sublittoral cliff face in this region, which would doubt-less yield many more species than we have recorded. *Corynactis viridis* (Anthozoa) was found in large numbers on this sublittoral rock-face, and also on upward-facing surfaces in intertidal rockpools in the cave, where it was free from competition with algae as well as from the direct effects of bright illumination. We expect to report on this more fully elsewhere.

Slightly further in, the sponge *Pachymatisma johnstonia* becomes a conspicuous element of the shallow sublittoral and lower littoral fauna. Renouf (1931) has already noted the progressive paling of this sponge with increasing distance into the cave.

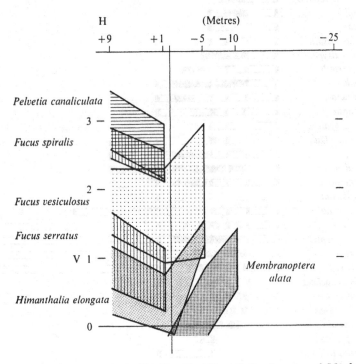

Fig. 8. Vertical distribution of various fucoids, *Himanthalia elongata* and *Membranoptera alata* based on observations made at +9, +1, −5, −10 and −25 m in relation to the mouth of the cave. For more exact horizontal distribution see Fig. 2.

The decline in the number of animal species with increasing distance into the cave was less steep than for algae. A number of filter feeders or carnivores were found to penetrate to −40 m, including sponges, the anemone *Actinia*, serpulid worms, and barnacles. *Actinia equina*, *Tealia felina*, *Pomatoceros triqueter*, *Chthamalus stellatus*, and certain sponges extended to −80 m.

Vertical distribution

The upper and lower limits of some plants and animals were higher outside the cave (+10 m) than in the cave mouth. Examples are shown in Figs. 8 and 9. It is possible that there is greater splash at +10 m in the entrance of the gully which leads into the cave than in the cave mouth.

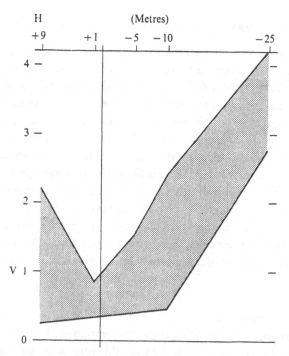

Fig. 9. Vertical distribution of *Hildenbrandia prototypus*, based on observations made at +9, +1, −5, −10, −23 m in relation to mouth of the cave. *Hildenbrandia* extends inwards to −27 m.

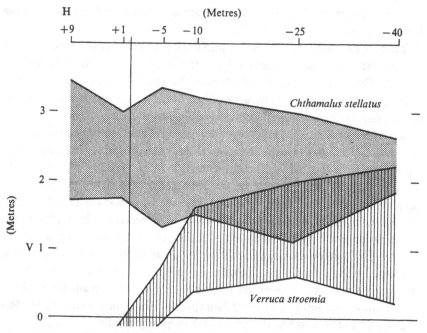

Fig. 10. Vertical distribution of *Chthamalus stellatus* and *Verruca stroemia*, based on observations made at +9, +1, −5, −10, −25 and −40 m in relation to the mouth of the cave. *Chthamalus* extends inwards to −80 m, and *Verruca* to −52 m.

For many algae and a few animals the upper or lower limits (or both) rose with increasing distance into the cave. (Occurrence in rock pools is excluded.) *Hildenbrandia prototypus* (Fig. 9), *Pseudendoclonium marinum*, and *Ulva lactuca* are notable examples. Several species normally confined to the sublittoral or lowest littoral levels, *Membranoptera alata*, *Phyllophora crispa*, the barnacle *Verruca stroemia* (Fig. 10), and the gastropod *Gibbula cineraria*, were found high in the intertidal region within the cave. At − 10 m several specimens of the nudibranch *Archidoris pseudoargus* were found in the upper littoral, ranging up to 3 m above low water of spring tides. The occurrence high between tidemarks of the Sponge *Pachymatisma johnstonia* and of a number of Polyzoa (*Microporella ciliata*, *Callopora lineata*, *Cryptosula pallasiana*, *Schizoporella unicornis*, *Cellepora pumicosa*, and others) is also noteworthy; particulars are given in the Appendix.

DISCUSSION

Although illumination is the most obvious factor to be considered in the ecology of caves, there are other important variables. Far into the cave the humidity of the air must remain in equilibrium with sea water, so that intertidal organisms will not be subject to desiccation, and they will also be safe from rises in temperature resulting from direct insolation on a calm day in the open. The surge of waves is also likely to build up as they drive up the cave, and there is likely to be more silt and debris in suspension at the innermost extremity of the cave.

There is no doubt that reduced illumination determines the inner limit of the various algae and lichens within the cave. The stunted growth of fucoids, *Gigartina stellata*, *Phyllophora crispa*, and *Plumaria elegans* towards their limits support this conclusion. The order of susceptibility is related to the normal habitat of these various algae on the open coast. Fucoids followed by laminarians dominate their habitats and receive all the insolation available and they are the first to be cut off within the cave. The numerous red algae which penetrate further into the cave are essentially members of the undergrowth of the open coast, and no doubt find comparable conditions of illumination in unshaded positions within the cave. *P. elegans* penetrates so far into the cave as to receive only 0.4 % of the illumination reaching an unshaded situation, although it is stunted beyond − 17 m (H), where the illumination amounts to 0.7 %. This alga has been reported by Boney & Corner (1962) as occurring at Church Reef, Wembury, in situations where it received only 0.8 − 8 % of the available daylight illumination. No doubt the *Cladophora* spp. and the rock-encrusting *Hildenbrandia* and lithothamnia also tolerate very shady conditions.

The lower limit of laminarian algae (*Laminaria hyperborea*) on the open coast off the nearby headland of Carrigathorna was found to be at 17 m below low water by Dr A Larkum and divers working with us in August 1967. Light transmission values in summer in Lough Ine and in the nearby open sea in calm weather range around 80–85 %/m at 550 nm. This would give an illumination at the lower limit of *Laminaria* of 2–6% of that at the surface (at 550 nm). The total daily illumination (at this wavelength) at the limit of *Laminaria* within the cave amounted to 5 and 8% of that in an unshaded position, and was of the same order at other wavelengths. However, the days on which readings were taken were sunnier than average days, and for periods when the cave mouth was shaded (Table 3) 1–4% is a better estimate of the proportion of light within the wavelengths measured which reaches the *Laminaria* at its limits within the cave. Thus the total illumination with green light is similar at the limits of *Laminaria* within the cave and in the open sea, although the intensity of red light will be much greater in the cave. According to Levring (1947) and Haxo & Blinks (1950) the greatest area of absorption of light by various dark-brown algae, including *Laminaria*, is in the blue and green, thanks to the presence of the pigment fucoxanthin, so that agreement of the results for green light is particularly satisfactory. We may note, however, that for short periods on sunny afternoons *Laminaria* in the cave may be illuminated at or above saturation level.

The vertical distribution of plants and animals at various distances within the cave probably depends on the interplay of a number of factors. High humidity and freedom from heating by the sun probably allows various organisms – both plant and animal – to extend up from the sublittoral to the littoral. The upper and lower limits of plants are likely to be elevated by the better illumination higher up in the littoral. The examples already given can be explained in these terms. We do not know to what extent the behaviour of settling larvae, known in many cases to be influenced by light (Thorson, 1964), is modified by the darkness of the cave.

SUMMARY

The total daily illumination for certain days during July and August has been estimated for certain specified wavelengths (but mainly at 550 nm) at several critical positions in the Bullock Island Cave. Far within the cave illumination is negligible except while the site in question is exposed to air, since light striking at a low angle is mainly reflected by the water surface.

The horizontal and vertical distribution of the commoner plants and animals within the cave has been determined. Fucoids penetrate a short distance within the cave, and laminarian algae only a little further. Shade-tolerant intertidal algae and algae normally forming an undergrowth to *Laminaria* forest penetrate further, and encrusting coralline algae penetrate the furthest. Herbivorous animals living on macrophytes in general die out with their algal food, but filter feeders and carnivores penetrate very far into the cave.

At the inner limit of *Laminaria* the total daily incidence of green light is of the same order as that at the lower limit of *Laminaria* in the open sea (1–8% of that received at an unshaded situation). For the inner limit of *Cladophora rupestris*, *Membranoptera alata*, *Phyllophora crispa* and *Lomentaria articulata* the value is 0.7%, and stunted *Plumaria elegans* persists to a position at which the total incidence amounts to 0.4%. No algae were found at 40 m within the cave (0.01%).

The upper and lower limits of some plants and animals are elevated within the cave. A continuously high humidity of the air probably assists by abolishing evaporation, while for some algae the greater illumination at higher intertidal levels may lead to elevation of the limits of vertical distribution.

REFERENCES

BONEY, A. D. & CORNER, E. D. S. (1962). The effect of light on the growth of sporelings of the intertidal red algae *Plumaria elegans* (Bonnem.) Schm. *J. mar. biol. Ass. U.K.* **42**, 65–92.

GRUBB, V. M. & MARTIN, M. T. (1937). The algal vegetation of a cave. *J. Bot., Lond.* **75**, 89–93.

HAXO, F. T. & BLINKS, L. R. (1950). Photosynthetic action spectra of marine algae. *J. gen. Physiol.* **33**, 389–422.

LEVRING, T. (1947). Submarine daylight and the photosynthesis of marine algae. *Göteborgs K. Vetensk.-o. VitterhSamh. Handl. Ser.* B **5**(b), 1–88.

MARINE BIOLOGICAL ASSOCIATION (1957). *Plymouth Marine Fauna*. Plymouth, 457 pp.

PARKE, M. & DIXON, P. S. (1968). Check-list of British marine Algae – second revision. *J. mar. biol. Ass. U.K.* **48**, 783–832.

RENOUF, L. (1931). Preliminary work of a new biological station (Lough Ine, Co. Cork, I.F.S.). *J. Ecol.* **19**, 410–38.

THORSON, G. (1964). Light as an ecological factor in the dispersal and settlement of larvae of marine bottom invertebrates. *Ophelia* **1**, 167–208.

APPENDIX

List of species in or immediately outside Bullock Cave with horizontal and vertical limits of distribution. All measurements are in metres; reference points are described on page 419. Horizontal limits (H) are given to the nearest metre, and vertical limits (V), all positive, to the nearest 0.1 m. Either the extreme vertical limits are given, or the upper limit (UL) and lower limit (LL) for horizontally specified positions. Names are as in Parke & Dixon (1968) or in the M.B.A. *Plymouth Marine Fauna* (1957), unless the authorship is given.

LICHENS

Caloplaca marina   +10 to −6H; 3.6 to over 5.0V

Lichina pygmaea   +1 to −2H; 2.1 to 2.7V

Verrucaria maura   +10 to −10H; UL 4.2 at +10, 3.9 at +1, 4.7 at −5; LL 0.8 at +10H, 1.6 at +1, 2.85 at −5.

CHLOROPHYTA

Chaetomorpha sp.   +10H

Cladophora rupestris   +10 to −16H; UL 2.1 at +10H, 1.7 at −10; LL 0.1 at +10H, 1.6 at −10

Cladophora sp.   −5 to −27H; < 0 to 2.6V

Enteromorpha sp.   +10 to −6H; 0.4 to 2.3V

Pseudendoclonium marinum   +10 to −22H; UL 2.9 at +10H, 2.6 at +1, 3.1 at −5, 4.0 at −10; LL 0 at +10H, 0.8 at +1, 1.0 at −5, 1.7 at −10

Ulva lactuca   +9 to −11H; < 0 to +2.6V, limits irregular

PHAEOPHYTA

Ascophyllum nodosum   +10 to −6H; 1.4 to 2.6V

Dictyota dichotoma   +10 to −11H; < 0V throughout

Ectocarpus fasciculatus   +10H

Elachista fucicola   +10 to 0H; 0.3 to 1.7V

Fucus serratus   +10 to −3H; 0.2 to 1.7V

F. spiralis   +10 to −6H; 2.1 to 2.9V for outside stations

PHAEOPHYTA (*cont.*)

F. vesiculosus	+10 to −6H; 0.9 to 2.3 V for outside stations
Giffordia hincksiae	+10H
Halopteris filicina	+1 to −11H; UL < 0 throughout
Himanthalia elongata	+10 to −6H; Fig. 8
Laminaria digitata	+10 to −7H; UL < 0 throughout
L. hyperborea	+10 to −7H; UL < 0 throughout
L. saccharina	+1H
Pelvetia canaliculata	+10 to −2H; 2.2 to 3.4 V for outside stations
Saccorhiza polyschides	+10 to −6H; UL < 0 throughout
Sphacelaria cirrosa	−5H
S. radicans	−5 to −11H; +2.0V
Spongonema tomentosum	+10H

RHODOPHYTA

Acrosorium reptans	+10H
A. uncinatum	+1H; UL < 0 throughout
Apoglossum ruscifolium	+1 to −11H; UL < 0 at +1H, 1.0 at −5H; LL < 0
Calliblepharis ciliata	+10 to −6H; < 0V
Callophyllis laciniata	+10 to −6H; UL < 0V at +10 and +1H, and +0.2 at −5H
Catanella repens	+1 to −11H; 1.4 to 2.9V
Ceramium flabelligerum	+10H
C. rubrum	+10 to 0H; UL < 0 at +10H, 1.5 at +1H
Chondrus crispus	+10 to −8H; < 0 to 0.6V
Corallina officinalis	+10 to −11H; UL < 0 at −10H, 0.8V at +1 and −5H; LL < 0 throughout
Cryptopleura ramosa	+10 to −11H; UL < 0 at +10, +1, −10, and +1.0 at −5H; LL < 0 throughout
Delesseria sanguinea	+10 to −6H; UL < 0 throughout
Dilsea carnosa	+1 to −11H; < 0 throughout
Epilithon membranaceum	+10 to −11H; < 0 throughout

RHODOPHYTA (*cont.*)

Gigartina stellata	+10 at −11H; UL 1.4 at +10H, 0.8 at +1H, 1.4 at −5, 1.6 at −10; LL<0 at +10 and +1H, 0.6 at −5, 1.0 at −10H
Gonimophyllum buffhami	+10H
Heterosiphonia plumosa	+1H
Hildenbrandia prototypus	+10 to −27H; Fig. 9
Hypoglossum woodwardii	+10 to −6H; UL < 0 throughout
Laurencia pinnatifida	+10 to −8H; UL 1.7 at +10H, 0.95 at +1, 1.9 at −5; LL < 0 throughout
Lithothamnia	+10 to −27H; < 0 to 3.1 V
Lomentaria articulata	+10 to −16H; UL 1.4 at −10H, 0.8 at +1, 1.4 at −5, 1.6 at −10; LL < 0 at +10 and +1, 0.3 at −5, 0.85 at −10H
Membranoptera alata	+10 to −16H; Fig. 8
Phycodrys rubens	+10 to −6H; UL < 0 throughout
Phyllophora crispa	+10 to −16H; UL < 0 at +10 and +1H, 0.8 at −5, 1.3 at −10H. LL < 0 throughout
P. membranifolia	+1 to −15H; UL < 0 throughout
Pleonosporium borreri	−10H
Plumaria elegans	+10 to −22H; UL 1.35 at +10H, 0.8 at 0H, 1.5 at −5, 1.9 at −10; LL < 0 throughout
Polyneura hilliae	+10 to −6H; UL < 0 throughout
Polysiphonia fibrata	+10H
P. lanosa	+10 to +1H; 0.8 to 2.4 V
Pterosiphonia complanata	+10H
P. parasitica	+10H
Ptilothamnion pluma	+1 to − 27H; 0.4 to 3.3 V, limits irregular
Rhodochorton purpureum	+1 to −11H; < 0 to +1.0 V
Rhodophyllis divaricata	+10 to −6H; UL < 0 throughout
Rhodymenia palmata	+10 to −6H; UL < 0 throughout
R. pseudopalmata	+1H
Schizymenia dubyi	−5H

PORIFERA

*Pachymatisma johnstonia	−8 to −41H; < 0 to 0.9V
*Clathrina coriacea (Montagu)	(Leucosolenia in Plymouth Marine Fauna) −5 to −50H; UL 0.3 at −10H and 1.2 at −40

HYDROZOA

Obelia geniculata	0 to −7H; on Laminaria

ANTHOZOA

Actinia equina	+10 to −80H; 0.4 to 2.9V, limits irregular
Caryophyllia smithi	−15H
Corynactis viridis	−2 to −19H; abundant within the cave on overhanging rock in sub-littoral and upward-facing in rock pools
Tealia felina	−80H

POLYCHAETA

Pomatoceros triqueter	+10 to −80H; < 0 to 1.5V
Spirorbis spp.	+10 to −80H; < 0 to 1.7 at +10H and UL rising to 2.8 at −26 and 2.7 at −40H

CRUSTACEA

Balanus balanoides	+10 to −41H; 0.2 to 2.7V, limits irregular
Chthamalus stellatus	+10 to −80H; Fig. 10
Verruca stroemia	+10 to −52H; Fig. 10
Cancer pagurus	+1H; < 0V

GASTROPODA

Patella vulgata	+10 to −27H; 0.6 to 3.0V, limits irregular
Patina pellucida	+10 to −7H; on Laminaria
Acmaea virginea	+10H; UL < 0
Calliostoma zizyphinum	+10 to −13H; < 0 to 1.0V

* Kindly determined by Miss S. M. Stone of the British Museum (Natural History).

GASTROPODA (*cont.*)

Gibbula cineraria
+10 to −27H; UL 0.9 at −10H, < 0 at +1, −5, and −10H, and 1.7 at −26H

G. umbilicalis
+10 to −6H; 0.6 to 2.9 V

Littorina littoralis
+10 to −6H; UL 1.8 at +10H, 2.3 at +1, 2.9 at −5; LL 1.0 at +10H, 0.4 at +1, 2.2 at −5

L. saxatilis
+10 to −19H; 1.7 to 3.1 V

L. neritoides
+10 to −41H; 2.3 to 5.3 V

Nucella (= Thais) lapillus
+10 to −11H; 0.8 to 2.0 V

Archidoris pseudoargus
−10 to −41H; 2.7 to 3.0 V, occasional specimens

LAMELLIBRANCHIA

*Musculus marmoratus
+10H, < 0 V

Mytilus edulis
−5H, UL < 0

Kellia suborbicularis
−10H, UL < 0

*Hiatella arctica
+10H, UL < 0

POLYZOA

Crisidia cornuta
−10H; 0.3 V

Crisia eburnea
+10H; UL < 0

Aetea sica
+10 to −11H; < 0 to 0.3 V

Electra pilosa
+10 to −40H; UL < 0 throughout except for 1.8 at −40

†Callopora lineata
+10 at −41H; UL < 0 at +10 and −5H, 0.3 at −10, 2.25 at −26, 4.5 at −40; LL < 0 at −10H, 0.4 at −26 and 1.0 at −40

†C. rylandi Bobin and Prenant
−26 to −41H; 0.4 to 1.6 V

Scrupocellaria scruposa
+10H, UL < 0

S. reptans
+10 to −8H; UL < 0

Hippothoa divaricata
−10H; UL < 0

Celleporella hyalina
+10 to −10H; < 0 to 0.3 V

†Escharella immersa
−26H; 0.8 to 0.9 V

†E. immersa var. labiosa

* Kindly determined by Dr N. Tebble of the British Museum (Natural History).
† A sample was kindly determined by Dr J. S. Ryland. Our previous records of *Membraniporella nitida* were probably *Callopora rylandi*.

POLYZOA (*cont.*)

Schizoporella unicornis	−10 to −27H; 1.1 to 1.5V
Schizomavella linearis	−10H; 0.3V
Escharina spinifera	−26H; 0.8 to 1.6V
Cryptosula pallasiana	+1 to −27H; UL 0.3 at +1H and 2.0 at −26H
Microporella ciliata	+10 to −41H; UL < 0 at +10 and −5H, 0.3 at −10, and 0.9 at −26; LL < 0 at −10H and 0.4 at −26
Escharoides coccineus	−5H, 0V
Cellepora pumicosa	+10 to −6H; UL < 0 throughout
Celleporina hassalli	+10 to −11H

ASTEROIDEA

Asterina gibbosa	−5 to −11H; UL < 0
Asterias rubens	+10 to −80H; UL < 0
Marthasterias glacialis	+10H; UL < 0

OPHIUROIDEA

Ophiothrix fragilis	−5 to −11H; UL < 0

ECHINOIDEA

Echinus esculentus	+10 to −1H; UL < 0

TUNICATA

Botryllus schlosseri	0 to −7H; UL < 0
Botrylloides leachi	+10H; UL < 0

LIGHT CONDITIONS AND
SHADE SEEKING POPULATIONS AMONG
ALGAL SETTLEMENTS

D. ZAVODNIK

Centre of Marine Research, Institute 'Rudjer Bošković', Rovinj, Yugoslavia

INTRODUCTION

The conditions of illumination affect the vertical distribution of many benthic organisms and of several communities, especially those of photophilous sea weeds and of sea grasses. The angle of inclination of the substratum and the cover of sessile plant and animal organisms is of utmost importance. Pérès & Picard (1964) and Riedl (1966) showed that isolated enclaves of communities distributed in the circalittoral zone can be found in the infralittoral but only in shade. Beneath large projecting rocks and in submarine caves, the illumination is considerably lessened, even at depths of only a few metres, and the light intensity can fall below 0.01 % of the incident light at the sea surface. The topography of submarine caves greatly influences the light penetration; at the same depths, straight and wide caves are always better illuminated than are narrow and curved ones.

Similarly, the illumination of the substratum is greatly dependent on the density of the sea weeds and sea grass populations. Beneath luxuriously branched algal thalli or dense sheaves of sea grass leaves, the light intensity can be reduced by more than 90 %, enabling many sciophilous organisms to colonize and inhabit the shaded substrata, even near the surface. However, in the upper part of the infralittoral zone, shade seeking organisms are not widely distributed because the favourable habitats provided by submarine caves and very dense plant cover are discontinuous. Consequently, true sciophilous populations extend into the better illuminated upper infralittoral zone mainly in the form of enclaves. Furthermore, only those sciophilous organisms that are not dependent on pressure or on deep types of sediment can extend to these upper levels.

METHODS AND RESULTS

Sea weed communities are luxuriously developed in temperate seas all over the world, predominantly on hard substrata at depths of 0–30 m. In European waters, the greatest mean of algal biomass at depths of 5–10 m

434 D. ZAVODNIK

was estimated as 10.5 kg (wet weight) per m² (Munda & Zavodnik, 1967). Algal settlements are usually polyspecific, with one or more species dominant, the so called 'Pilot species' (Bellan-Santini, 1967), on which the physiognomical aspect of the community is dependent. Pilot species are usually large species of macroalgae, which form an upper canopy or raised stratum (strate élevée). At the level of the basal parts of algal fronds, that is

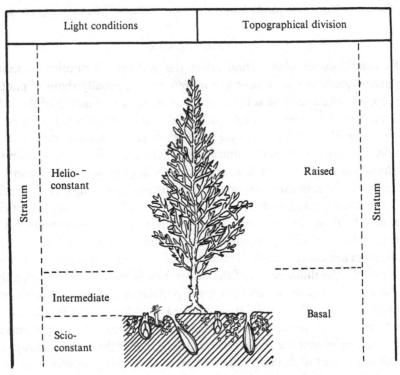

Fig. 1. Differentiation of the strata in algal settlement according to the conditions of illumination and the topography.

close to the substratum, a great variety of plant and animal species live, forming a shaded zone or lower stratum (sous strate). Some of these creatures live on basal parts of algal fronds as epiphytes, others crawl or swim among them, others bore into hard substratum (usually rock) or construct tubes. The illumination in these borings or tubes of petrobionts (endolithon) is insignificant even compared with that of the shade zone. Consequently, three main zones of illumination can be distinguished in a typical algal settlement (Fig. 1). (1) Helioconstant zone, comprising raised stratum described above, together with all its heliophilous organisms, plant and animal. (2) Intermediate zone, for which the epibionts of the lower stratum are characteristic. (3) Scioconstant zone, which comprises almost exclusively endobiont animals living in borings or crevices (*Clionidae*, *Lithophaga*, *Gastrochaena* and others).

Unfortunately, because of technical problems, exact data on light intensities in such complex algal settlements are poorly known. The standard photometers are usually too crude to enable precise registrations in any typical phytal settlement, as for example the settlements of *Padina*, *Fucus*, *Cystoseira* and others. However, in well developed and dense settlements of giant kelps and other big sea weeds, such as *Macrocystis* and *Sargassum*, classical photometers can be successfully used. A device suitable for light measurement in typical algal settlements was developed recently by Ott & Svoboda (1967). With small photosensitive cells, measurements can be made by this device in any desirable part of an algal frond, and automatic registrations enable observations to be made over a longer period. Lacking this equipment, we tried to measure the light intensity in the shaded zone with a standard GM Manufacturing Comp. Photometer combined with a frame of 400 cm² with nylon network of 2 cm mesh. Fronds of several pilot species were removed carefully from the bottom together with their basal discs and mounted on the network and placed by skin divers in the same position in which they were growing in the algal community (Fig. 2). Light intensities were recorded on the coast or from a boat; control measurements were made by removing the frame mounted with algae from the photosensitive unit.

The results obtained show that the branching of algal thalli and the density of the populations of pilot species sometimes reduces the illumination of the shaded zone by more than 95 % in relation to the illumination above the canopy (Table 1). When the shading effect of the rest of the vegetation and of luxuriously growing animal colonies in the shaded zone are also taken into consideration, the light intensities must presumably be still lower. For this reason sea-weed communities are believed to represent a transition between photophilous populations of higher horizons, and sciophilous populations of greater depths and of submarine caves.

A serious problem encountered during the measurement of light intensity in algal settlements at higher levels is the continual moving of fronds due to wave action. Also, on bright sunny days, variable refraction of the sun's rays can upset the measurements. But such rapid fluctuations in illumination are part of the environmental conditions of the habitat to which plants and animals are well adapted.

Detailed studies on phytal populations were initiated only 40 years ago,

436   D. ZAVODNIK

and afterwards they were continued in Europe only sporadically (Segerstråle, 1927, 1943; Colman, 1940; Dahl, 1948; Delamare-Deboutteville & Bougis, 1951; Wieser, 1959; Makkaveeva, 1959, 1960, 1964; Bellan-Santini, 1961, 1962, 1963, 1967; Zavodnik, 1962, 1967, 1969; Herberts, 1964; Hagerman, 1966; Jansson, 1966). The communities of the following pilot species were studied : *Acetabularia mediterranea, Ascophyllum nodosum, Cladophora glomerata, Corallina mediterranea, C. officinalis, Cystoseira abrotanifolia, C. barbata, C. crinita, C. granulata, C. spicata, C. stricta, Fucus serratus, F. spiralis, F. vesiculosus, F. virsoides, Gigartina stellata, Halopteris scoparia, Laminaria digitata, Laurencia pinnatifida,*

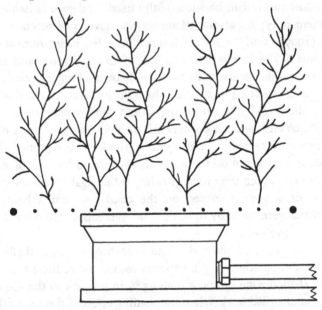

Fig. 2. Apparatus to measure reduction in transmittance due to algal canopy. The position of the photosensitive unit of the photometer and of the frame mounted with algae is shown.

Table 1. *Mean obscurity values for some littoral algae from north Adriatic*

(The obscurity value in column 4 is the percentage reduction in transmittance on passing through the canopy formed by the species given in Column 1.)

Species	Depth (m)	Number of observations	Mean obscurity (%)
Dictyota fasciola	0.4–1.0	12	97.27
Fucus virsoides	0.1–0.3	11	95.82
Cystoseira abrotanifolia	0.2–0.4	9	98.90
Cystoseira barbata	0.3–1.2	5	94.80
Cystoseira crinita	0.3	13	97.68
Cystoseira spicata	0.3–0.6	12	97.06
Laurencia obtusa	0.5–2.2	11	91.35

Lichina pygmaea, Lithophyllum tortuosum, Padina pavonia, Pelvetia canaliculata, Petroglossum nicaeense, Phyllophora nervosa, Plocamium coccineum, Polysiphonia lanosa, Ulva lactuca, and a special facies of *Mytilus galloprovincialis*. Nearly all algae cited are typical photophilous species of the midlittoral or upper infralittoral zones. Unfortunately, for various reasons, the flora and fauna have not been studied fully in any of the cited references. Usually the microflora and the microfauna have been neglected, in spite of their undoubted importance to the metabolism of the community. In the communities cited above, about 230 plant and 1100 animal species were identified in European waters. But this number is likely to be far below the true total; the presence of at least 2500–3000 animal species can be expected in European communities of photophilous sea weeds.

The ecological preferences of most species has not yet been established. But, thanks to the thorough observations of Dr Bellan-Santini, in the northwestern Mediterranean several ecological classes were established among phytal communities. The class of strictly photophilous organisms is represented by about 10% of the species present. In the sciophilous communities of *Petroglossum nicaeense* and *Plocamium coccineum* the total number of characteristic photophilous species is greatly reduced; in spite of this, the photophilous species participate in the populations with approximately 63 representatives, and a partial dominance value of 93%. Usually, in algal communities, the class of photophilous organisms is represented by more than 70%. On the other hand, Bellan-Santini (1967) reports only 16 typically cavernicolous species found in the algal communities in the region of Marseille, which is about 5% of the total number of species identified. The class of cavernicolous species is obviously poorly represented in photophilous algal settlements of the mid- and upper-infralittoral zones, and its partial dominance value is usually less than 1%. In the northern Adriatic near Rovinj, the number of cavernicolous animals in phytal communities was found to be only six, that is less than 2%.

Consequently, typically shade seeking species are not constant dwellers in communities of photophilous sea weeds. Generally they do not form dense populations and they can be regarded as guests in photophilous phytal communities. An exception to this rule is provided by the characteristic sciophilous endobionts living in the substratum (clionid sponges, boring bivalves) which are always to be found within the class of phytal communities, with which they are closely linked by their localization and by their trophic relations. Simultaneously, because of the topographical characteristics of their borings, these species have to be treated also as cave dwellers (Riedl, 1966).

DISCUSSION

It would be interesting to compare the light preferences of populations of photophilous algal communities, and of the sciophilous populations of submarine caves. But any such comparisons are very difficult to make at present for two reasons: first, the studies of both communities have not been carried out in the same geographical regions, and secondly the material has not been studied with equal care, especially in regard to the microflora and microfauna. Whereas detailed observations in phytal communities in Europe were carried out in the Baltic, in the Northern Sea, in the channel, in the north-west Mediterranean, in the Adriatic and in the Black Sea, exhaustive investigations of submarine caves have been undertaken almost exclusively along the European coasts of the Mediterranean Sea (Riedl, 1966; Bellan, 1968; Zibrowius, 1968). Hence, comparisons between the phytal and cave populations have to be restricted to the Mediterranean area. The lists of plant and animal species hitherto cited for phytal and cave populations respectively in this region (Bellan, 1968; Bellan-Santini, 1967; Makkaveeva, 1964; Riedl, 1966; Zavodnik, 1967, 1969; Zibrowius, 1968) indicate that 218 species are common to phytal and cave communities. This is 25.4% of total plant and animal species found in the biocoenose of photophilous algae, and 24.2% of the species which have as yet been reported from submarine caves in the Mediterranean. Whereas most typically cavernicolous species are absent from infralittoral algal communities, presumably because of excessive illumination and an inadequate substratum for such organisms as sponges and tunicates, there is a relative abundance of typically photophilous species (32), which were to be found in submarine caves, even in their darkest parts. The majority of these species are vagile animals, to which the darkness and semidarkness of the habitat resemble the environmental conditions in phytal settlements during the night. Their presence in the caves can therefore result from migration. Riedl (1966) has shown similarly that several sciophilous decapods and fishes leave the cavernicolous habitat during the night. The presence of photophilous sessile animals in the caves, such as hydroids, bivalves and the tube-worm *Spirorbis pagenstecheri*, indicates that the illumination is probably not the principal limiting factor for them.

When comparing the general structures of phytal and cave communities, their resemblances become evident (Fig. 3). In algal settlements, as well as in the caves, two main strata can be distinguished. The canopy or raised stratum of the submarine cave populations is formed by branched sponges, hydroids, corals and bryozoans, while the majority of other sessile organisms form a lower stratum, just as the raised stratum in phytal communities is composed almost exclusively of sea weeds, with the other sessile non-epiphytic organisms distributed in the lower stratum. Furthermore, in submarine caves endolithic sponges and bivalves are often contained within the substratum which, in phytal communities, was referred to as the scioconstant zone. Obviously, the different light conditions prevent the terminology used for the littoral phytal communities, and based on illumination levels, being used for the communities of submarine caves.

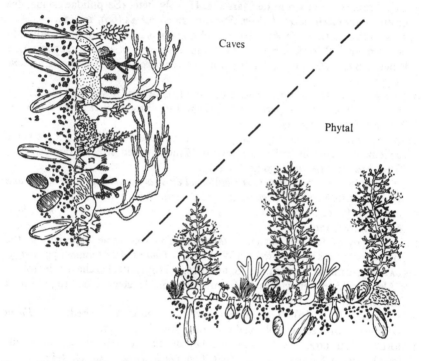

Fig. 3. Structural similarities of cave and phytal communities (partly after Riedl, 1966).

SUMMARY

Because of the density of the sea-weed settlements, and due to the ramification and the consistency of algal fronds, the quantity of incident light which reaches the bottom is reduced. According to illumination conditions within the settlements, three main strata are distinguished: the helioconstant stratum, the intermediate stratum, and the scioconstant stratum. The total reduction in transmittance is not the same for all algal species but usually the larger sea weeds reduce the light by more than 95%. For this reason sea-weed communities are believed to represent a transition between the

heliophilous populations of higher horizons, and the shade seeking populations of greater depths and of submarine caves. In fact, about 25 % of algal
and animal species is common to sea-weed communities and to the cave
populations. There are structural similarities between these two types of
communities.

REFERENCES

BELLAN, G. (1968). Contribution à l'étude des Polychètes des substrats solides
circalittoraux des environs de Marseille. II. Polychètes (Serpulidae exclus) des
grottes sous-marines. *Recl Trav. Stn mar. Endoume* **44** (60), 109–23.

BELLAN-SANTINI, D. (1961). Note préliminaire sur la faune et la flore du
peuplement à *Petroglossum nicaeense* (Duby) Schotter et sur ses rapports avec
le peuplement a *Cystoseira stricta* (Mont.) *Sauv. Recl Trav. Stn mar. Endoume*
23 (37), 19–30.

BELLAN-SANTINI, D. (1962). Étude du peuplement des 'dessous de blocks non
ensables' de la partie supérieure de l'Etage Infralittoral. *Recl Trav. Stn mar.
Endoume* **27** (42), 185–96.

BELLAN-SANTINI, D. (1963). Comparaison sommaire de quelques peuplements
rocheux de l'Infralittoral supérieur en Manche et en Méditerrannée. *Recl
Trav. Stn mar. Endoume* **30** (45), 43–75.

BELLAN-SANTINI, D. (1967). *Contribution à l'étude des peuplements Infralittoraux
sur substrat rocheux (Étude qualitative et quantitative)*. Thése, Marseille.

COLMAN, J. (1940). On the faunas inhabiting intertidal seaweeds. *J. mar. biol.
Ass. U.K.* **24**, 129–83.

DAHL, E. (1948). On the smaller *Arthropoda* of marine algae, especially in the
polyhaline waters off the Swedish West Coast. *Unders. över Öresund* **35**, 1–193.

DELAMARE-DEBOUTTEVILLE, C. & BOUGIS, P. (1951). Recherches sur le trottoir
d'Algues calcaires effectuées à Banyuls pendant le stage d'été 1950. *Vie et
Milieu*, **2** (2), 161–81.

HAGERMAN, L. (1966). The macro- and microfauna associated with *Fucus
serratus* L., with some ecological remarks. *Ophelia* **3**, 1–42.

HERBERTS, CH. (1964). Contribution à l'étude du peuplement rocheux sessile
dans la zone a *Fucus serratus* L. *Bull. Lab. marit. Dinard* **49–50**, 5–61.

JANSSON, A. M. (1966). Diatoms and Microfauna-producers and consumers in
the *Cladophora* belt. *Veröff. Inst. Meeresforsch. Bremerh.* **2**, 281–8.

MAKKAVEEVA, E. B. (1959). Biocenoz *Cystoseira barbata Ag.* (*Wor.*) pribrežnego
učastka Černego morja. *Trudy sevastopol. biol. Sta.* **12**, 168–91.

MAKKAVEEVA, E. B. (1960). K ekologii i sezonnym izmenenijam diatomovyh
obrastanij na cistozire. *Trudy sevastopol. biol. Sta.* **13**, 27–38.

MAKKAVEEVA, E. B. (1964). Zaroslevye biocenozy Adriatičeskogo morja. *Trudy.
sevastopol. biol. Sta.* **17**, 39–47.

MUNDA, I. & ZAVODNIK, D. (1967). Algenbestände als Konzentrationen organ-
ischen Materials im Meer. *Helgoländer wiss. Meeresunters.* **15** (1–4), 622–9.

OTT, J. & SVOBODA, A. (1970). Messungen der qualitativen Lichtverteilung in
Phytalbeständen. Int. Colloquium 'The actual problems of the phytal communities', Rovinj, 5–6 October 1967. *Thalassia jugosl.* **6**, 185–8.

PÉRÈS, J. M. & PICARD, J. (1964). Nouveau manuel de bionomie benthique de la Mer Méditerranée. *Recl. Trav. Stn. mar. Endoume*, **31** (47), 3–137.

RIEDL, R. (1966). *Biologie der Meereshöhlen*, Berlin u. Hamburg: 636 pp. Verlag Paul Parey.

SEGERSTRÅLE, S. G. (1927). Quantitative Studien über den Tierbestand der *Fucus*-Vegetation in den Schären von Pellinge (an der Südküste Finnlands). *Commentat. biol.* **3** (2), 1–14.

SEGERSTRÅLE, S. G. (1943). Weitere Studien über die Tierwelt der *Fucus*-Vegetation an der Südküste Finnlands. *Commentat. biol.* **9** (4), 1–28.

WIESER, W. (1959). Zur Ökologie der Fauna mariner Algen mit besonderer Berücksichtigung des Mittelmeeres. *Int. Revue ges. Hydrobiol. Hydrogr.* **44** (2), 137–80.

ZAVODNIK, D. (1962). Prelimary observations on the phytal populations of the rocky shore near Rovinj (Northern Adriatic). *Publ. staz. zool. Napoli* **31** (47), 3–137.

ZAVODNIK, D. (1967). Dinamika litoralnega fitala na zahodnoistrski obali (Dynamics of the littoral Phytal on the west coast of Istria). *Raspr. slov. Akad. Znan. Umet.* **10** (1), 5–67.

ZAVODNIK, D. (1969). La communauté de *Acetabularia mediterranea* Lamour. dans l'Adriatique du Nord. *Int. Revue ges. Hydrobiol. Hydrogr.* **54** (4), 543–51.

ZIBROWIUS, H. (1968). Étude morphologique, systématique et écologique des Serpulidae (Annelida Polychaeta) de la région de Marseille. *Recl Trav. Stn mar. Endoume* **43** (59), 81–252.

RESPONSES OF THE LARVAE OF *DIPLOSOMA* *LISTERIANUM* TO LIGHT AND GRAVITY

D. J. CRISP AND A. F. A. A. GHOBASHY

Marine Science Laboratories, Menai Bridge

INTRODUCTION

The majority of ascidians are sublittoral, being found in shaded situations such as the undersides of rocks and boulders. Early work on the light responses of ascidian tadpoles showed that those with well-developed eyes were photonegative at the time of settlement, though the great majority of such larvae, if not all, had an initial photopositive phase just after liberation from the adult or emergence from the egg. Examples are: *Amaroucium constellatum* (Verrill) and *A. pellucidum* (Leidy) (Grave, 1920; Mast, 1921); *Botryllus schlosseri* (Pallas) (Castle, 1896; Grave & Woodbridge, 1924; Grave, 1932); *Symplegma viride, Polyandrocarpa tincta* and *P. gravei* (Grave, 1935); *Ascidia nigra* (Grave, 1935; Goodbody, 1963), and *Ciona intestinalis* (Berrill, 1947; Millar, 1953; Dybern, 1962). One species only, *Perophora viridis* Verrill, is reported as remaining predominantly photopositive up to the time of settlement (Grave & McCosh, 1923), but even so, 40% of the larvae became photonegative. Moreover, this is a species occurring in well-illuminated as well as dark situations in the intertidal zone (McDougall, 1943). It thus confirms Thorson's (1964) generalization that intertidal species may remain photopositive up to the time of settlement.

Ascidians with well-developed light responses possess a single asymmetrically placed eye with pigment cup and lens. The angle of acceptance of light is strictly limited. Mast (1921) describes in detail how the clockwise rotation of the larva as it swims results in regular changes in the light received by the eye when the larva is differentially illuminated from one side. At each rotation the tail responds in the opposite sense to increasing or decreasing illumination received by the eye. This mechanism causes the larva to move towards the light in its initial state and away from it when it later becomes photonegative.

Some ascidian larvae are indifferent to light, but may nevertheless swim upwards in response to gravity. Examples are *Dendrodoa grossularia* (Berrill, 1950), *Distaplia* sp. (Berrill, 1948) and larvae of the Molgulidae which are eyeless (Grave, 1926, 1932). The eye of *Cynthia partita* is poorly developed and the directional responses of the larvae to light are weakly developed in

[443]

comparison with the geotactic response (Grave, 1941, 1944). The majority of ascidian larvae show a strong tendency to swim upwards, either spontaneously, or more commonly in response to shading (Grave, 1920; Mast, 1921; Woodbridge, 1924). Even larvae endowed with poorly developed eyes (*C. partita*) respond vigorously to shade (Grave, 1944), indicating that the eye is functional as a receptor as in other ascidian tadpoles (Mast, 1921). Larvae without eyes (*Molgula citrina*) also swim upwards on resuming activity after a period of rest (Grave, 1926).

Although the general behaviour of ascidian tadpoles is well established, there are considerable differences between species in regard to the duration of the free living larval stage and the relative length of the initial photopositive and the later photonegative phases, though interspecific differences are sometimes masked by individual variation. Early workers have paid little attention to the possible ecological significance of responses to light and gravity in guiding the organism to its destination or maintaining it at a preferred depth. Thus a modification of the photic and gravitational responses by environmental variables such as pressure, temperature, and the intensity of illumination could well determine the level and habitat where the larva eventually metamorphoses.

The didemnid, *Diplosoma listerianum* was a suitable species for such an investigation; it is very common in the lower part of the littoral zone and is an important fouling organism (Stubbings & Houghton, 1964) with a long summer breeding season (Orton, 1914; Millar, 1952).

PROVENANCE OF LARVAE

Colonies of *D. listerianum* are generally to be found in North Wales encrusting all kinds of surfaces, such as the fronds of algae, mussel shells, the underside of boulders, and pilings, wherever there is sufficient shelter to enable them to grow. The colonies usually contain eggs and larvae between June and late October. To obtain larvae, colonies were transferred to the laboratory and laid on the bottom of large dishes to which they reattached within a few hours. Phenol formaldehyde panels mounted in racks on a raft quickly become covered with this species and form a convenient source of mature colonies. In the laboratory the colonies were supplied with a continuous stream of aerated sea water at a temperature of 19–20 °C and were kept in the dark. When larvae were required light from a 200 W bulb was directed towards one end of the bowl and after a few minutes larvae appeared and swam towards the illuminated end. Larvae could be obtained most readily during the daytime with maximum liberation at approximately mid-day. Only newly liberated larvae were used in experiments in which time was a factor under investigation; the age of the larvae was measured from the time of liberation.

It was noticed that when larvae emerged from the parent colonies they swam upwards and towards a source of diffuse illumination; if a strong beam of light was shone across the tank, larvae were observed to move upwards until they came into the light beam when they moved horizontally towards the light source though still swimming upwards. If a light source was placed beneath or at the sides of the tank the larvae were always found to swim towards the surface, indicating that initially they are influenced mainly by a strong negative response to gravity. Larvae were rarely observed to swim downwards; they sank passively.

After the initial period of upward swimming the larvae tended to sink to the bottom, remaining motionless on the bottom for increasingly long periods interrupted from time to time by spells of active swimming. During this phase of behaviour the larvae were seen to be extremely sensitive to shadow. If the hand were passed over the tank or the intensity of illumination decreased, large numbers of larvae would react by swimming upwards from the bottom of the tank. Soon after reaching the top, or sometimes before reaching the top, they would sink again. Thus, if a larva which had been recently liberated was caught in a pipette while swimming towards the light and replaced in the darkest region of the container it would immediately resume its journey towards the light. If this procedure were repeated some 20 or 30 times the larva would eventually adopt a different pattern of behaviour – active swimming towards the light at the surface of the water alternating with periods of rest when the larva sank passively. Throughout the latter part of the larval life periods of rest progressively replaced those of activity, the larva sinking to the bottom of the tank.

In studying the responses of the larvae of *Diplosoma* to light and gravity care was taken to control both temperature and illumination. Very high levels of illumination, comparable with those experienced in sunlight, were obtained by the use of a 500 W focussing theatre lamp. The heat produced by the lamp was dissipated by using an apparatus similar to that employed by Ryland (1962) in which the experimental vessel was surrounded by a bath of continuously renewed water at constant temperature through which the rays of light passed. Measurements of light intensity were made by means of an E.E.L. photometer.

The majority of observations were made by noting at intervals of time the positions of swimming and settled larvae in the experimental apparatus. The percentage in each section was then calculated and represented by the width of a histogram. Each histogram represented the distribution of the larvae in relation to the applied stimulus at successive intervals of time, any change in distribution being evident from successive histograms. The experiments were generally left until all the larvae had metamorphosed and on no occasion did the larvae still swimming at the end of the experiment constitute more than 10 % of the total number used. At least 25 larvae were used in each experiment and all had settled within 12 h of being placed in the apparatus. When the number of metamorphosed larvae was less than three the result was ignored because of the large sampling error involved. The results of experiments carried out in the dark were obtained by inspecting the experiment under a deep red light at the lowest intensity that could be appreciated by the observer when he had become fully dark adapted. The red light had no obvious influence on the larvae though Mast (1921) found ascidian larvae sensitive to red light. For this reason critical experiments were replicated a number of times to enable examinations to be made once only after a period of complete darkness.

RESPONSES TO LIGHT

Larvae can respond to light either by directed movements towards or away from a light source, termed by Fraenkel & Gunn (1940) 'phototactic behaviour', or they may accumulate in regions of high or low light intensity as a result of changes in the pattern of activity related to the intensity of light but not to its direction; the latter form of behaviour is termed 'photokinesis'. Phototactic behaviour will be tested if the larvae are exposed to a parallel beam of light passing through a transparent medium, since the light flux will be constant without change of intensity. To demonstrate photokinetic behaviour the light intensity must be varied but the light flux must be minimized in the direction in which the intensity change occurs. The simplest method of presenting this situation is to expose the animals to a broad beam of parallel light with a graded neutral density filter interposed so that changes in intensity are produced at right angles to the direction of the light flux. By carrying out the experiments in a horizontal plane the effect of gravity is eliminated.

The apparatus used for both these purposes was similar to that described by Ryland (1960) in which a pair of U-shaped plastic troughs were illuminated along their lengths or at right angles to their lengths by means of a horizontal beam of light. The troughs were divided into six equal sections and placed on a black surface to show up the larvae in the light beam.

Response to a parallel beam of light

These experiments tested phototactic behaviour since the rays of light passed through the long axis of the trough with little attenuation. A wide range of light intensity was used in order to include levels which the larvae would experience in the field. The highest intensity (27000 lx) was obtained by direct illumination, the lower intensities by placing appropriate

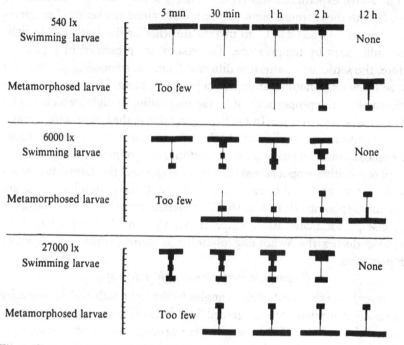

Fig. 1. Histograms to show the distribution of larvae at 20 °C in a horizontal beam of light of the intensity and at the time intervals stated. The light direction was from the top to the bottom of each histogram. Larvae which were free swimming or resting are distinguished from attached and metamorphosed larvae.

neutral density filters in the path of the light beam. The experiments were carried out at 20 °C and the results are shown in Fig. 1. The behaviour of the larvae before metamorphosis is shown in the upper of each pair of histograms, the distribution of larvae after metamorphosis is shown in the lower of the pair.

In the absence of any gravitational influence, the larvae before metamorphosis were positively phototactic at all three intensities tested, and were most strongly phototactic at 540 lx, the lowest intensity measured. At this intensity and temperature the majority of larvae settled towards the light, but at the higher intensities there was, just before metamorphosis, a clear reversal in response to light, with the result that the metamorphosed larvae settled predominantly away from the light source. A reversal of response to light at metamorphosis is characteristic of the majority of experiments carried out. It will be observed from the figure that there is no gradual change in the distribution of the free-swimming larvae. The reversal in phototaxis must occur suddenly, just before metamorphosis.

The above experiment was repeated using a constant light intensity of 6000 lx with the surrounding water bath maintained at a series of temperatures between 11 and 29 °C, to investigate whether the responses to light were influenced by temperature. The results are presented in Fig. 2. As before, the settlement pattern is different from the swimming pattern, the larvae becoming photonegative just before settlement. However, there are differences in the proportions of larvae responding to light, which can be related to the temperature. In the swimming phase, the larvae show a maximum positive reaction to light at 16 °C. As the temperature rises a higher proportion of swimming larvae become photonegative or neutral until at 29 °C the photopositive response has almost disappeared, the larvae becoming indifferent to light at this temperature. At 11 °C the swimming larvae are less strongly phototactic than at 16 °C. At metamorphosis, the photonegative response predominates at all temperatures, though, like the photopositive response during the swimming phase, it is more pronounced at lower temperatures.

Response to variation in light intensity

The troughs were placed at right angles to the light path and its intensity was varied by means of neutral graded light filters. The intensities measured in successive sections of the troughs were 14000, 10000, 3000, 2000, 1300 and 500 lx respectively and the water temperature again was kept at 20 °C. To observe the behaviour at intensities below 500 lx, a separate experiment was carried out in which one half of the trough was maintained at 300 lx and the other half was in darkness; it was not possible in this experiment to observe the swimming larvae but the pattern of settlement was recorded. The results of the two experiments are shown in Fig. 3 a, b. Fig. 3 a shows that the swimming larvae were more or less evenly distributed along the trough with no obvious preference for high or low light intensities. On metamorphosing, however, they aggregated in the least illuminated section at 500 lx. However, when offered a choice between 300 lx and darkness (Fig. 3 b) there was a slight tendency for them to settle in the dimly illuminated parts rather than in the dark. The larvae, at settlement, appear to have a preferendum of the order of 300 lx.

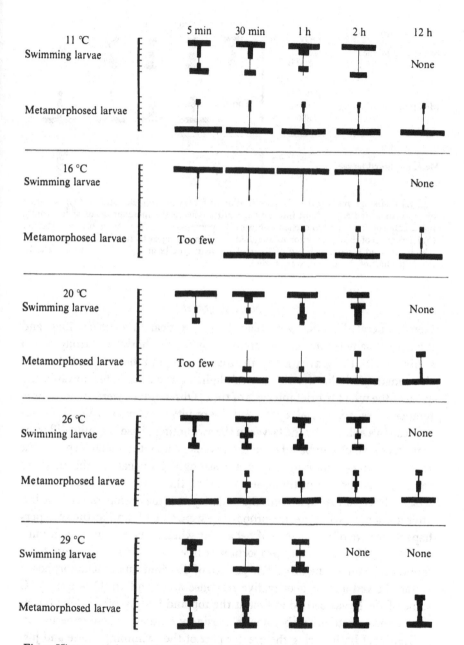

Fig. 2. Histograms showing the distribution of larvae in a horizontal light beam of 6000 lx at five different temperatures. The light direction is from the top to the bottom of each histogram.

450   D. J. CRISP AND A. F. A. A. GHOBASHY

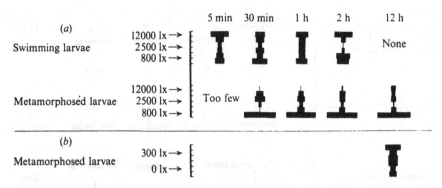

Fig. 3. (a) Histograms showing the distribution of free swimming and attached larvae when exposed to a gradient of light intensity at right angles to the incident beam of horizontal light. Intensities applied to certain points of the perspex trough are shown on the ordinate. (b) Histogram of the distribution of attached larvae in a perspex trough, half the long axis of which received a parallel beam of light of intensity 300 lx at right angles to its length; the other half of the trough was shaded.

Response to gravity

Newly liberated larvae were placed in a vertical tube 20 cm long and 2.5 cm in diameter marked in divisions as for the horizontal troughs. To study the effect of gravity alone, the tube was kept in the dark and observations made periodically with a weak light of 100 lx directed towards the sides of the tube for brief intervals of time. Observations were made at three temperatures, 12, 19 and 25 °C, and the results are shown in Fig. 4. Again the responses shown by the larvae in the swimming phase are quite different from those which can be inferred for larvae just about to metamorphose. At 12 °C the free-swimming larvae are strongly geonegative, the majority remaining at the top of the tube. At 19 °C, the larvae were at first somewhat indifferent to gravity with the majority near the bottom of the tube but after 2 h a somewhat greater proportion were to be found at the top, perhaps indicative of the change of behaviour when metamorphosis is imminent. At 25 °C, only a small proportion of free-swimming larvae were to be found at the top of the tube. The pattern of distribution of metamorphosed larvae showed a clear geonegative response at 12 and 19 °C but at 25 °C some of the larvae settled at first at the top and bottom of the tube but as time went on settlement became increasingly random. These experiments indicate that both during the greater part of the swimming phase and just prior to metamorphosis the upward swimming activity is greater at lower temperatures while the time spent in passive sinking increases with rise in temperature.

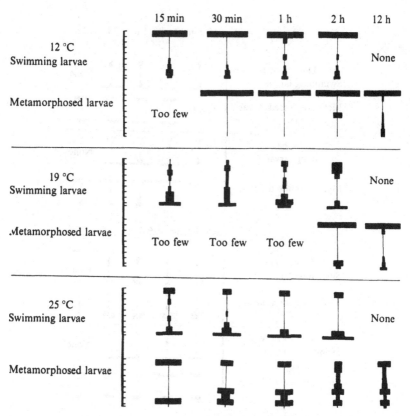

Fig. 4. Histograms showing the distribution of larvae at 12, 19 and 25 °C in a vertical tube in the dark at intervals of time stated. The top of the tube corresponds to the top of the histogram.

RESPONSES TO LIGHT AND GRAVITY COMBINED

In these experiments the larvae were again observed in a vertical tube, but with the effect of light superimposed on that of gravity. In one set of experiments a parallel light beam was shone from above so that the combination of the two stimuli resembled the natural situation. In the other series of experiments a parallel light beam was reflected on to the tube from below. The results of both types of experiment carried out at a series of different light intensities and at a temperature approximately constant between 19 and 20 °C are shown in Fig. 5. The behaviour of the larvae in the swimming phase was again clearly different from the distribution of metamorphosed larvae.

Whether the light was directed from above or below and whatever its intensity within the range employed, the free-swimming larvae collected at

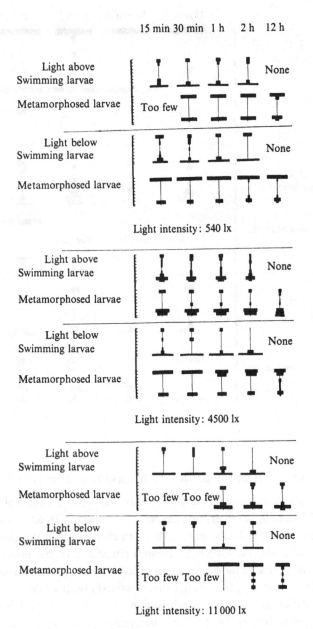

Fig. 5. Histograms to show the distribution of larvae when exposed at 19–20 °C in vertical tubes to a combined stimulus of gravity and a parallel light beam at three intensities of illumination. In the upper of each pair of histograms the light is from above, as in nature, in the lower of each pair the direction of light is from below.

RESPONSES OF *DIPLOSOMA* LARVAE
453

the bottom of the tube. Evidently at 19–20 °C the phototactic response shown in Fig. 1 is not capable of overcoming the tendency to sink to the bottom, nor, except perhaps in the experiment at 4500 lx, was there much evidence that reversing the direction of illumination had any influence on the pattern of distribution of the swimming larvae. Indeed, if one compares Fig. 5 with the results at 19 °C in the dark (Fig. 4), a higher proportion of larvae seem to remain at the bottom when the vertical tube is illuminated – whatever the direction of the light. This fact suggests that swimming activity is reduced by light independently of the direction of the light, a response which in Fraenkel & Gunn's (1940) terminology would be called 'low photokinesis'.

However, the settling larvae showed a clear response to the direction of light at all three intensities used, since when the light was directed from below a much higher proportion of larvae metamorphosed in the upper parts of the tube. At metamorphosis therefore the larvae become negatively phototactic as well as negatively geotactic.

A closer examination of the final settlement pattern, when the light was directed from below, reveals a result somewhat contrary to expectation. As the intensity of the light source increases the larval settlement in the upper part of the tube decreases. Table 1, which includes experiments carried out in the dark and at an additional light intensity of 22 500 lx, confirms that as the intensity of light increases the pattern of settlement is displaced lower down the tube. This effect can again be explained in terms of low photokinesis, the tendency to sink passively increasing at higher light intensities. Such an effect on behaviour is quite consistent with a negative phototaxis.

Figs. 6a, b give the results of a similar set of experiments with the light directed from above and from below at an intensity of 540 lx and at a series of different temperatures. Results shown for 19 °C are identical with those in Fig. 5 and are included here for comparison.

It will be seen that at all three temperatures reversal of the direction of light had very little effect on the larvae during the swimming phase.

Table 1. *Settlement of the larvae of* Diplosoma *in relation to the light intensity from below*

Light intensity (lx)	0	540	4500	11 000	22 500
Percentage in the upper two sections	76.5	68.5	59.0	54.5	48.0
Percentage in the middle two sections	0.5	0	14.5	13.5	8.5
Percentage in the lower two sections	23.0	31.0	25.5	31.5	44.5

However, a close examination of the histograms for 12 °C shows that a slightly higher proportion of larvae occur in the upper parts of the tube when light was directed from above than when light was directed from below, indicating a weak photopositive response. At this temperature there was also a slight change in the distribution of larvae with the passage of

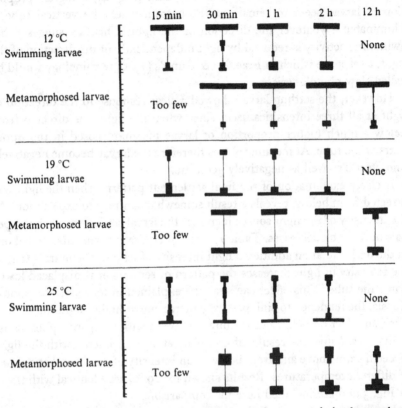

Fig. 6. Histograms showing the distribution of larvae when exposed simultaneously to gravity and to a parallel beam of light of intensity 540 lx in a vertical tube, at three different temperatures. (a) Light from above.

time, the proportion lying at or near the bottom of the tube gradually increasing. The influence of light on the distribution of metamorphosed larvae is much more pronounced, at all temperatures a higher proportion were found near the top of the tube when the light was directed from below. Evidently at this stage the larvae are negatively phototactic. However, even with the light directed from above, a considerable proportion of larvae must have swum upwards just before metamorphosis even though earlier in the free-swimming stage the majority were lying near the bottom. It follows that just before metamorphosis the larvae become strongly geonegative.

The influence of temperature on behaviour is very evident both in the free-swimming stage and just prior to metamorphosis. At low temperatures the larvae accumulate to a much greater degree near the top of the tube.

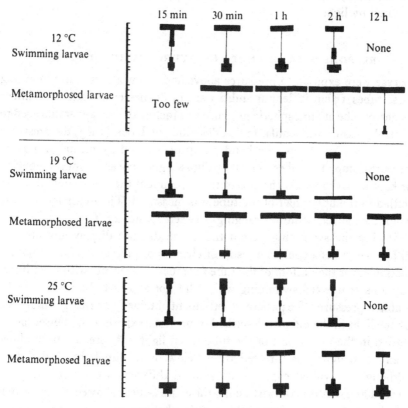

Fig. 6.(*b*) Light from below.

They are strongly geonegative even in the swimming phase at 12 °C, though not at 19 and 25 °C. Just before metamorphosis, they are strongly geonegative at 12 °C, somewhat geonegative at 19 °C, but at 25 °C they settle at the top of the tube only when the light is directed from below, that is when the geonegative response is reinforced by the photonegative response.

If we compare the settlement pattern after 12 h, with the light directed from below (Fig. 6*b*) with the settlement pattern in the dark (Fig. 4) it will be seen that despite the negatively phototactic behaviour which was evident when the light direction was reversed (see Fig. 6*a*) there was no decrease in the proportion of larvae settled near the bottom as a result of light being directed from below as compared with darkness. Indeed at 19 and 25 °C considerably more larvae settle near the bottom when lit from below than in the dark. This unexpected result can best be explained in terms of low photokinesis, upward swimming activity being partially inhibited or arrested by light.

RESPONSES TO CHANGES IN HYDROSTATIC PRESSURE

Larvae were exposed to pressures above and below atmospheric in strong glass tubes 15 cm in length and 2.5 cm in diameter; in each experiment a control tube at atmospheric pressure was included. The apparatus used to control pressure was similar to that described by Bayne (1963) the pressures being raised by means of a hand pump or lowered by means of a small vacuum pump. The tubes were lit by diffuse light from above, or from below, or kept in darkness. At the end of the experiment the proportion of larvae settled in the upper half of the tube was measured. The experiments were conducted at temperatures ranging between 18 and 20 °C.

During the swimming period the larvae did not display any obvious difference in behaviour as a result of changes of pressure. They remained motionless at the bottom of the tube for much of the time with brief interruptions of upward swimming which did not appear to be related to the ambient pressure. The pattern of settlement obtained was clearly influenced by the light in the manner described in previous experiments. Fewer larvae settled in the upper part of the tube when light was directed from above than when light was directed from below or in dim light or darkness. Although individual experiments showed differences between the pressurized and control compartments, these differences showed no consistent trend and it was therefore concluded that hydrostatic pressure within the range of 0.5–2.0 atmospheres produced no definite effect on the settlement behaviour of *Diplosoma*.

THE INFLUENCE OF LIGHT ON THE LENGTH OF THE FREE-SWIMMING PHASE

During the above experiments on the behaviour of the larvae of *Diplosoma*, a wide variation in the free-swimming periods of the tadpoles was evident. To investigate the possible influence of light on the length of the free larval stage, batches of recently liberated larvae of identical origin were kept in the dark, in the light and in intermittent illumination. The experiments were carried out at 18 °C (± 1 °C) with groups of 8–12 larvae in solid watch-glasses illuminated at 650 lx, or kept in complete darkness. The experiments in total darkness were arranged in such a way that the larvae were

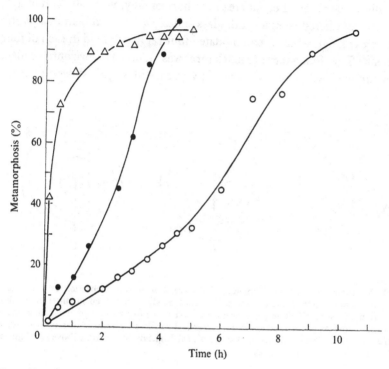

Fig. 7. Duration of the free-swimming period of *Diplosoma* larvae under different conditions of illumination. O, Continuous light of 650 lx; ●, darkness; △, intermittent light of 650 lx.

not exposed to light for examination at any time while the experiment was in progress. The larvae subjected to alternating light and dark periods were illuminated at an intensity of 650 lx every few minutes.

The result for experiments under each of the above conditions is illustrated in Fig. 7. Metamorphosis was clearly accelerated by rapid fluctuations of light and dark, and the length of the free larval life was much greater when illuminated than in total darkness.

THE SETTLEMENT OF DIPLOSOMA ON SURFACES AT
DIFFERENT ANGLES TO THE HORIZONTAL

In order to relate the behaviour of the larvae to the pattern of settlement on exposed surfaces, experiments were carried out in which *Diplosoma* larvae were offered surfaces of equal area held horizontally, vertically and at 45°. The surfaces offered were smooth glass slides 7.5 × 3.5 cm held in a plastic holder 13 cm high which accommodated four replicates held in each of four positions. The following nomenclature was adopted: horizontal slides: upper surface, 180°; lower surface, 0°; vertical slides: 90°; slides sloping at

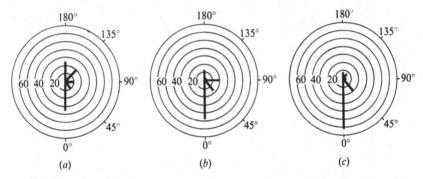

Fig. 8. Percentage of *Diplosoma* which settled in relation to the orientation of the surface when larvae were offered equal areas of transparent slides mounted at different angles. Temperature: 18–21 °C. The convention adopted is: upper surface of horizontal slides 180°; upper surface of inclined slides 135°; vertical surfaces 90°; lower surface of inclined slides 45°; lower surface of horizontal slides 0°. (*a*) Light of 650 lx from above; (*b*) light of 650 lx from below; (*c*) in darkness.

45° to the horizontal: upper surface, 135°; lower surface, 45°. The experiment was carried out in a dish containing colonies which were liberating larvae continuously, first in total darkness, then with a light of 650 lx above, and finally with a light of 650 lx directed up from below. The temperature was between 18 and 21 °C. The results are shown in the form of a polar diagram (Fig. 8) each polar co-ordinate representing an increment of 10% of the total settlement. It is clear that under all conditions the undersides of the slides (0°) are the most attractive surfaces for settlement but more especially so in total darkness. However, when the light was shining from above (Fig. 8*a*) a considerable proportion settled on the upper surfaces (180°) and on the upper side of the oblique slides (135°), surfaces which were avoided in total darkness (Fig. 8*c*) or when the light was shining from below (Fig. 8*b*).

These results are consistent with the observed behaviour of the larvae just before metamorphosis when they become strongly negatively geotactic and negatively phototactic. In the dark they will swim upwards coming into contact mainly with the underside of the horizontal and inclined surfaces. With light from below they will swim in a similar manner, but the element of low photokinesis may lead a small but significant proportion to drop on to the upper surface of the horizontal slides. When the light is shining from above, the photonegative response, possibly reinforced by photokinesis, will cause a fair proportion of the larvae to sink or swim downwards settling on the 180 and 135° surfaces. The generally low proportion of larvae settling on the vertical surfaces suggests that their main direction of movement under the conditions of this experiment are upwards and downwards.

SETTLEMENT ON HORIZONTAL SURFACES AT DIFFERENT DEPTHS

The glass surfaces used in the above experiments would not cause any significant shading. A further experiment was therefore carried out to compare the settlement on transparent and opaque slides. Four groups of square panels 2 × 2 cm were held horizontally from the four sides of a small rectangular column 20 cm high into which they were slotted. The panels were held one above another at distances of 8 cm apart and arranged alternately around the column at depths of 4, 8, 12 and 16 cm from the bottom. The column was placed vertically in a cylindrical perspex vessel 20 × 18 cm and the experiments were again carried out with light directed from above, from below and in darkness. In one set of experiments smooth square glass panels provided a transparent settling surface; in the other experiment square, slightly roughened, non-toxic PVC panels were used. The experiments were carried out at temperatures ranging from 16 to 20 °C. Fig. 9 shows the average percentage settlement of the larvae on the upper and lower surfaces of the slides at each of the four depths. The conditions are: (a) with light directed from above (b) with light directed from below and (c) in darkness. A stronger light source of 1500 lx was used in these experiments.

It can be seen that in general a greater proportion of larvae settled on the under surfaces. In all experiments together, the ratio of settlement on the lower sides to the upper was 2.8:1. However, the ratio increased to 9.6:1 on the uppermost group of slides and decreased to only 1.2:1 in the lowermost group. In fact, in the case of the transparent slides there was an apparent reversal of the behaviour of the larvae, the uppermost pair having a preponderance settled on the undersides while the lowermost pair had a preponderance settled on the upper sides.

There were also clear differences between the pattern of settlement on the transparent slides and that on the opaque slides. First, there was a much more pronounced difference between the settlement on the upper and on the under surfaces of the opaque slides. Secondly there was an increase in the density of settlement in going from the lowermost to the uppermost panels in the case of the opaque slides whereas the transparent slides showed more or less uniform settlement throughout the column.

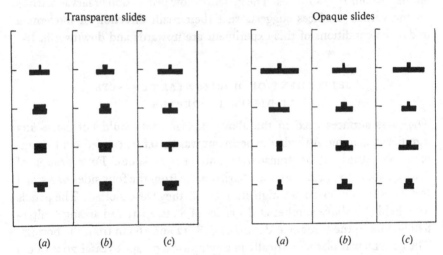

Fig. 9. Percentage settlement on upper and lower surfaces of a column of four slides 2 × 2 cm separated by a vertical distance of 4 cm from each other. The set on the left were transparent, those on the right dark and opaque. (a) Light of 1500 lx from above; (b) light of 1500 lx from below; (c) in darkness.

These results can readily be explained by visualizing the behaviour of the larvae as a result of their responses to light and gravity. At the time when metamorphosis is imminent larvae resting on the bottom become active and swim upwards towards the surface, sinking down after a short interval of time and perhaps repeating this behaviour several times. On touching the underside of the panel they may attach or they may continue to swim upwards and, presumably after spending a while near the surface, they will sink down again. Some of those which sink may alight on the upper sides of the panels and attach to them. The underside of the uppermost panels in the column can be reached by any larva which is swimming upwards in the main body of water beneath the panel whereas the upper side will be reached only by larvae sinking through the short head of water above it.

Hence the chances of the under surface receiving settlement are much greater. On the other hand, the upper surfaces of the lowermost series of panels can be reached by any larva that sinks through the whole length of the water column apart from the 4 cm of water below the panel, whereas its lower surface can be reached only by larvae swimming upwards through that 4 cm.

The differences in the pattern of settlement when opaque slides are used must result from the photic responses of the larvae. Surprisingly, it is when the light is shining from above that settlement is concentrated most on the upper panels and when light is shining from below the settlement is most evenly distributed. At first sight, this result is unexpected since the larvae are known to have become negatively phototactic at this time. The explanation must be sought in terms of the responses of the larvae to sudden shading; when they experience the shading due to an opaque panel they are stimulated to swim upwards. It is probable that this will take place whether they move into the shadow of a panel on their upward swimming excursion or when passively sinking. If the light is shining from above they will be constantly stimulated to swim upwards in the shade of the panel and attach to its under surface. In darkness they will swim upwards on account of their geonegative tendency. In the unnatural condition with the light shining from below, however, they will not be shaded to the same extent as they approach the under surface of a panel and will not therefore be drawn so strongly towards it. They will, instead, be shaded as they drop towards the upper surface of the panel where they may settle if the geonegative response does not come into play. Thus with light from above the majority settle on their way upwards but with light from below a higher proportion will probably settle while sinking down. Hence the greater numbers in the upper part of the column and on the lower side of the opaque panel when the light is directed from above.

SETTLEMENT ON LIGHT AND DARK SURFACES

In all experiments where a surface was shaded from the light – for example an opaque vertical panel placed in a light beam, or the inner surface of a cylinder with alternating opaque and transparent bands illuminated from above or from the side – the larvae settled on the darker surfaces, the ratios varying from 4:1 to 30:1.

To test the narrowest band of light and shade which a larva could detect, parallel black and white stripes of varying width were reproduced on matt photographic paper, the dark surface having about one-ninth the reflectance of the white surface. The results (Table 2) show that discrimination was impaired as the dark bands became narrower, but some differentiation was possible down to 1 mm band width. This is only three times the distance across the larval trunk which bears the eye. Evidently the larva must be capable of accurate orientation and navigation towards a dark stripe just before the moment of impact on the surface, or have some limited ability to migrate over the surface and locate a darker area after making contact.

Table 2. *Discrimination of narrow black and white bands by* Diplosoma *larvae*

Band width (cm)	Number settled on		Ratio Black/White
	Black surface	White surface	
2.5	67	9	7.45
1.0	52	12	4.33
0.5	22	7	3.14
0.2	22	9	2.44
0.1	95	45	2.11
0.05	34	31	1.10
0.025	11	13	0.85

In a further experiment, it was shown that differences in incident as well as in reflected light can influence settlement. When some 200 larvae were offered a choice between differentially illuminated light and dark bands 0.5 cm in width obtained by projecting an image on to a uniform white surface, five times as many larvae attached to the dark as to the light bands. The corresponding ratio shown in Table 2 for differentially reflecting surfaces of 0.5 cm band width was 3.14. This result confirms that larvae at settlement seek regions of low light intensity (see Fig. 3).

DISCUSSION AND CONCLUSIONS

In most respects the larvae of *Diplosoma* conform to the rather stereotyped behaviour pattern reported for other ascidians whose tadpoles possess an eye and have strongly directional vision: rhythmic liberation at dawn (Grave, 1920; Grave & Woodbridge, 1924); reversal of phototaxis; a strongly developed negative geotaxis; and an immediate swimming response to reduction in light intensity. The abbreviation of the free-swimming period by exposure to the activating effects of alternate periods of light and dark is probably general too, since it was observed by Grave using the larvae of *Symplegma viride* and *Polyandrocarpa tincta* (Grave, 1935). However, the lengthening of the free-swimming stage in continuous light as compared with darkness was not noted in Grave's experiments. If, as

RESPONSES OF *DIPLOSOMA* LARVAE 463

some of our experiments suggest, light is slightly inhibitory to swimming activity, this result may still be consistent with Grave's general conclusion that the more intense the activity after liberation, the earlier metamorphosis occurs.

In comparison with other ascidian tadpoles, those of *Diplosoma* are rather short lived, the majority settling within 6 h under the least favourable conditions of continuous light, and in less than an hour under favourable conditions of variable illumination. Indeed, the tadpoles may sometimes develop within the parent colony, omitting the free-swimming stage altogether (Millar, 1953).

Diplosoma larvae are strongly geonegative and weakly photopositive on first emergence, then display a phase of exploratory swimming behaviour with alternating periods of activity and of rest, but remaining photopositive and geonegative when swimming. It is only at the very end of the free-swimming stage that the sign of phototaxis is reversed and light strongly avoided. Even at this stage the tadpoles of *Diplosoma* remain geonegative, unlike the reported behaviour of *Amaroucium* (Grave, 1920); *Ciona* (Berrill, 1947); *Phallusia, Polyandrocarpa* and *Symplegma* (Grave, 1935). In retaining the tendency to keep near the surface of the sea for almost the whole duration of the swimming phase they resemble the larvae of another shallow water inter-tidal form *Perophora viridis*, but differ in avoiding illumination in excess of 300 lx.

Diplosoma larvae show greatly reduced upward swimming activity if the temperature rises above the optimum of 16–17 °C, and the tendency to sink also appears to be increased as the light intensity rises (low photokinesis). These modifications of the swimming behaviour apply both to the swimming–sinking phase, and to the immediate premetamorphosis excursions. They may well prevent the larvae from approaching and settling too close to a fully illuminated region near the water surface, while at the same time precluding too great a descent into deep cold water. Surprisingly, no comparable depth regulation mediated by pressure change could be detected.

SUMMARY

The larvae of *Diplosoma listerianum* swim actively on emergence and remain positively phototactic and strongly geonegative during the free-swimming stage. There follows an alternation of periods of swimming and sinking, until, just before settlement, they become strongly photonegative and seek dimly lit environments. Throughout the free-swimming stage they respond vigorously to shade by swimming upwards. The length of the free-swimming stage is greatest in light (*c.* 6 h) less in darkness (*c.* 2 h) and very short (*c.* 10–20 min) if the larvae are stimulated by fluctuating periods of light and dark.

As the temperature of the environment rises above the optimum of about 16 °C the upward swimming activity is reduced and the majority of larvae would settle at lower levels. There is also evidence that swimming activity is reduced in strong light. These modifications of the normal geonegative behaviour may serve as a means of regulating depth. Larvae settle preferentially on surfaces of low reflectance and in the shade. They congregate on the underside of objects, even in the dark. In the normal situation with light falling from above they settle most abundantly on the underside of opaque objects near to the surface of the water. This pattern of settlement agrees with observations in nature, and can be shown to result from the behaviour of the larvae as demonstrated in the laboratory.

REFERENCES

BAYNE, B. L. (1963). Responses of *Mytilus edulis* larvae to increases in hydrostatic pressure. *Nature, Lond.* **198**, 406–7.

BERRILL, N. J. (1947). The Development and Growth of *Ciona*. *J. mar. biol. Ass. U.K.* **26**, 616–25.

BERRILL, N. J. (1948). Budding and the reproductive cycle of *Distaplia*. *Q. Jl microsc. Soc.* **89**, 253–9.

BERRILL, N. J. (1950). *The Tunicata*. Ray Society 354 pp.

CASTLE, W. E. (1896). Early embryology of *Ciona intestinalis* Flem. *Bull. Mus. comp. Zool. Harv.* **27**, 203–80.

DYBERN, B. I. (1962). Biotope choice in *Ciona intestinalis* L. Influence of light. *Zool. Bidr. Upps.* **35**, 589–601.

FRAENKEL, G. & GUNN, D. L. (1940). *The Orientation of Animals. Monographs on Animal Behaviour*, 352 pp. Oxford.

GOODBODY, I. (1963). Biology of *Ascidia nigra*. *Biol. Bull. mar. biol. Lab.*, *Woods Hole* **124**, 31–44.

GRAVE, C. A. (1920). *Amaroucium pellucidum* (Leidy) form *constellatum* (Verrill). I. The activities and reactions of the tadpole larvae. *J. exp. Zool.* **30**, 239–59.

GRAVE, C. A. (1926). *Molgula citrina* (Alder and Hancock). Activities and structure of the free swimming larvae. *J. Morph.* **42**, 453–71.

GRAVE, C. A. (1932). The *Botryllus* type of ascidian larva. *Pap. Tortugas Lab.* **28**, 145–56.

GRAVE, C. A. (1935). Metamorphosis of ascidian larvae. *Pap. Tortugas Lab.* **29**, 211–91.

GRAVE, C. A. (1941). The eye spot and light responses of the larvae of *Cynthia partita*. *Biol. Bull. mar. biol. Lab.*, *Woods Hole* **81**, 287.

GRAVE, C. A. (1944). The larvae of *Styela* (*Cynthia*) *partita*. *J. Morph.* **75**, 173–91.

RESPONSES OF *DIPLOSOMA* LARVAE   465

GRAVE, C. A. & MCCOSH, G. (1923). *Perophora viridis* Verrill: The activities and structure of the free swimming larvae. *Wash. Univ. Stud. Scient. Ser.* 11, 89–116.

GRAVE, C. A. & WOODBRIDGE, H. (1924). *Botryllus schlosseri* (Pallas): The behaviour and morphology of the free-swimming larva. *J. Morph.* 39, 207–47.

MCDOUGALL, K. D. (1943). Sessile marine invertebrates at Beaufort, North Carolina. *Ecol. Monogr.* 13, 321–74.

MAST, S. O. (1921). Reactions to light in the larvae of the ascidians, *Amaroucium constellatum* and *Amaroucium pellucidum* with special reference to photic orientation. *J. exp. Zool.* 34, 149–87.

MILLAR, R. H. (1952). The annual growth and reproductive cycle in four ascidians. *J. mar. biol. Ass. U.K.* 31, 41–61.

MILLAR, R. H. (1953). *Ciona.* Liverpool Marine Biological Committee Memoirs on typical British marine plants and animals. No. 35. Liverpool University Press. 78 pp.

ORTON, J. H. (1914). Preliminary account of a contribution to an evaluation of the sea. *J. mar. biol. Ass. U.K.* 10, 312–26.

RYLAND, J. S. (1960). Experiments on the influence of light on the behaviour of polyzoan larvae. *J. exp. Biol.* 37, 783–800.

RYLAND, J. S. (1962). The effect of temperature on photic responses of polyzoan larvae. *Sarsia* 6, 41–8.

STUBBINGS, H. G. & HOUGHTON, D. R. (1964). Ecology of Chichester Harbour, S. England with special reference to fouling species. *Int. Revue ges. Hydrobiol. Hydrogr.* 49, 233–79.

THORSON, G. (1964). Light as an ecological factor in the dispersal and settlement of larvae of marine bottom invertebrates. *Ophelia* 1, 167–208.

WOODBRIDGE, H. (1924). *Botryllus schlosseri* (Pallas). The behaviour of the larvae with special reference to habitat. *Biol. Bull. mar. biol. Lab., Woods Hole* 47, 223–30.

FOOD SEARCHING POTENTIAL IN MARINE FISH LARVAE

J. H. S. BLAXTER* AND MARY E. STAINES

Natural History Department, University of Aberdeen

INTRODUCTION

Judged by the high fecundity of marine teleosts, there must be periods of serious mortality during the early stages of their life history. In the delicate planktonic larval phase both predation and starvation play a part. However, the rate of natural mortality is difficult to trace owing to the problems of sampling larval populations for any considerable period of time. Estimates of predation are usually anecdotal and only occasionally quantitative (e.g. Cushing, 1968; Fraser, 1969) and the feeding 'status' of marine teleost larvae, i.e. how near they are to starvation, has only been studied by a few workers. Shelbourne (1957) showed that plaice larvae were in poor condition, as judged by five arbitrary categories, in the early part of the year in the southern North Sea when plankton was scarce. Hempel & Blaxter (1963), Blaxter (1971) and Bainbridge & Forsyth (1971) measured the condition factor of herring larvae (weight/length^3) and related it to experimentally starved larvae in tanks and to the biomass of plankton of suitable size (the relevant biomass) available on the larval feeding grounds.

Another approach to the problem of viability is to measure the searching capacity for food under aquarium conditions and relate this to the relevant biomass in the natural environment. This demands estimates of: (a) distance swum in unit time; (b) longer term estimates of activity, including the possibility of a diurnal or light-dependent feeding rhythm; (c) perception distance for food.

These estimates may be used to calculate the volume searched in unit time.

Further information is then required: (d) success in taking food organisms; (e) the daily requirements per larva for normal growth; (f) the relevant biomass available.

Some initial attempts have been made to obtain these estimates, for example by Ivlev (1960) and Blaxter (1966) on young herring, by Rosenthal (1966) on sole larvae and by Rosenthal & Hempel (1969) on herring larvae.

* Present address: Dunstaffnage Marine Research Laboratory, P.O. Box 3, Oban, Argyll, Scotland and Biology Dept., Stirling University, Scotland.

[467]

The present paper describes a comparative study of volume searched per day and feeding success (items (a–d) above) in the larvae of four species, two clupeid and two flatfish, i.e. the herring *Clupea harengus* L., the pilchard *Sardina pilchardus* (Walbaum), the plaice *Pleuronectes platessa* L. and the sole *Solea solea* (L.). The larvae of these species were chosen as they were available from other rearing experiments and because of interesting differences in early development and behaviour. The herring hatches as a relatively large, active larvae with functional eyes; the pilchard is much smaller and the eyes develop some days after hatching. Both clupeids become more active as they grow. At hatching the two flatfish show differences parallel to those of the clupeids, the plaice being larger with functional eyes, the sole being smaller and eyeless. On the other hand they both become less active as they near metamorphosis and settle to the bottom. In particular the adult sole is well known as a night feeder in contrast to adults of the other three species.

METHODS

Herring gametes were obtained from ripe fish in the Firth of Clyde in February 1967 and 1968, and transported to Aberdeen for fertilization and rearing. Pilchard larvae were reared in Plymouth from naturally spawned eggs caught offshore there in July 1968. Plaice and sole eggs were taken from natural spawnings of tank stock at the New Hatchery, Port Erin in April 1967 and 1968 and transported to Aberdeen for hatching and rearing. The techniques used are described for flatfish by Shelbourne (1964), for herring by Blaxter (1968a) and for pilchard by Blaxter (1969a).

The temperature range used and details of the larvae are given in Table 1. The herring were held in rectangular black plastic tanks measuring 80 × 70 × 34 cm and holding about 200 l and the other species in cylindrical black tubs 42 cm in diameter and 25 cm deep holding about 25 l. Although it would have been desirable to hold all the larvae in the same size of tank for a comparative study of this sort, the smaller tanks were too small for herring once they had grown to a length of 12–15 mm. The larger tanks were, on the other hand, rather too large for concentrating food for the other species.

The usual food organisms used were *Artemia* nauplii though it was necessary to feed the pilchard larvae on small and fairly inactive natural plankton as they were too small to take *Artemia*. To this extent the pilchard results are not comparable with the other species.

Observations on the larvae were made under the same light conditions, a 60 W lamp with reflector being positioned above and to one side of the

Table 1. *Growth characteristics of larvae*

Species	Temp. range (°C)*	Hatching		Yolk resorption		Metamorphosis		Duration of observation (weeks)
		Days from fertilization	Body length (mm)	Days from hatching	Body length (mm)	Weeks from hatching	Body length (mm)	
Herring	8–12	15	6–8	9	8–11	16–24	30–35	10
Pilchard	15–18	3	3–4	4	4–5	16–24	30–35	3
Plaice	10–12	14	5–6	7	6–7	6–8	9–10	9
Sole	10–12	8	3–4	7	4–5	6–8	9–10	9

* Gradually rising temperature during period of experiment.

tank with the observer behind it. There were small rises of surface temperature of one degree or so as a result of this but they were the same for all experiments. The light intensity at the surface was of the order of 500 lx.

In some early experiments cine-photography was used but was abandoned because of the technical difficulties of following small and transparent larvae for any length of time in a tank, especially if they left the surface. A routine series of observation by eye was developed so that each species was investigated at least once per week, and more often twice, during the experimental period. In each series the following was done.

(a) Distance covered in unit time was measured by following actively searching individuals for periods of 1 min and estimating the distance swum by comparing their movements against a square frame graduated in centimetres floating at the surface. Initially a floating lattice of 1 cm squares was used but was found to obscure the tank surface too much. The results recorded in cm/min were only approximate and could not express the finer movements of the larvae. Estimates on 20 larvae were made before and after food was put in the water.

(b) The swimming activity of 20 individual larvae, selected randomly, was aggregated over a 5 min period by means of a stopwatch which 'accumulated' time. It was not considered feasible to observe activity of individuals for longer owing to the extreme concentration required. The number of seconds of swimming activity was then expressed as a % of 300 s. No attempt was made to distinguish between different types or speed of swimming, except that activity was measured only in larvae that were apparently searching for food. Estimates were also made before and after food was put in the water.

(c) Perception distances could be estimated only by watching the larvae in relation to their food and using such criteria as changes in direction of movement, changes in speed and apparent 'sizing-up' of food as a means of determining when it was first perceived. There seemed to be no alternative to this highly subjective technique and it must be considered the least precise of the estimates. It was not possible to obtain values for sole larvae which usually fed near the bottom of the tanks.

(d) Feeding success. The percentage of feeding movements which were successful, unfinished or failed was fairly easily measured over a 30 min period of active feeding in all species except sole.

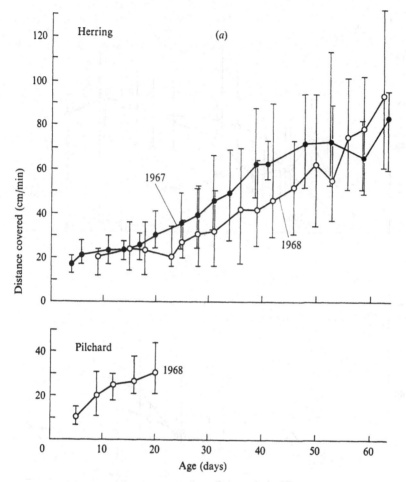

Fig. 1. (a) Distance covered in cm/min by herring and pilchard larvae, of different ages (from hatching). The year of each series of experiments is given. The vertical lines show the range of observations in each series of experiments.

RESULTS

(a) *Distance covered*

In Fig. 1 the distance covered in cm/min is shown for all four species with food present. In herring there was an increase from about 20 cm/min at the end of the yolk-sac stage to 80 cm/min 8 weeks later; in pilchard the increase was from 10 to 30 cm/min in 3 weeks, in plaice 10 to 60 cm/min over 7 weeks and in sole 5 to 40 cm/min over 7 weeks. The 90% reduction in distance covered at metamorphosis is clearly shown in the flatfish.

The influence of food is shown in Fig. 2. There seemed to be no change

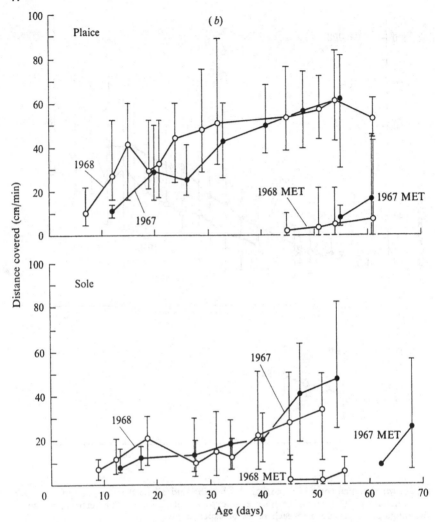

Fig. 1. (b) Distance covered in cm/min by plaice and sole larvae of different ages
(from hatching). As (a). MET means metamorphosing.

in distance covered by herring with or without food; in plaice the average
distance covered decreased when food was added, presumably due to the
time spent in actually feeding.

Rosenthal (1966) reported that sole larvae at first feeding covered about
28–32 cm/min and Rosenthal & Hempel (1969) that herring larvae 8 days
after hatching swam 160–300 cm/min, increasing to 600–1200 cm/min
about 6 weeks later. The results for herring are far in excess of the values
reported in this paper, a fact which is partly due to these authors including

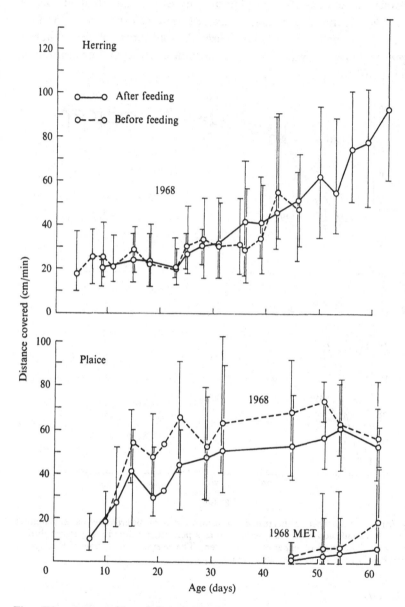

Fig. 2. Distance covered in cm/min by herring and plaice before and after feeding at different ages after hatching; 1968 only. The vertical lines show range of observations in each series of experiments. MET means metamorphosing.

undulatory movements of the head in their measurements. In other species Ivlev (1960) found that young bleak *Alburnus alburnus* (mean weight 250 mg) swam about 180 cm/min. Three-day-old larvae of the whitefish *Coregonus wartmanni* swam about 170 cm/min at 16 °C and 95 cm/min at 4 °C (Braum, 1964).

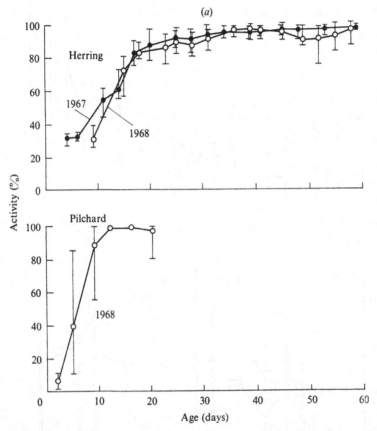

Fig. 3. (*a*) Per cent activity of herring and pilchard larvae of different ages after hatching (number of seconds active in 300 s, expressed as a percentage). Food was present during observations. The year of experiment is given. The vertical lines show the range of observations in each series of experiments.

(*b*) *Activity*

The results are shown in Fig. 3. In herring per cent activity (over 5 min periods) increased from 30 to 100% in 4 weeks and from 5 to 100% in pilchard in 2 weeks. In plaice activity remained at 80–100% until metamorphosis when it dropped to under 20%. In sole there was an increase from 25 to 80% over about 5 weeks with a similar sharp fall at metamorphosis.

In herring larvae observed by Rosenthal & Hempel (1969) per cent activity increased from 40% after hatching to 70% some 17 days later; activity also dropped when food was absent. Sole larvae (Rosenthal, 1966) were 70% active at yolk resorption.

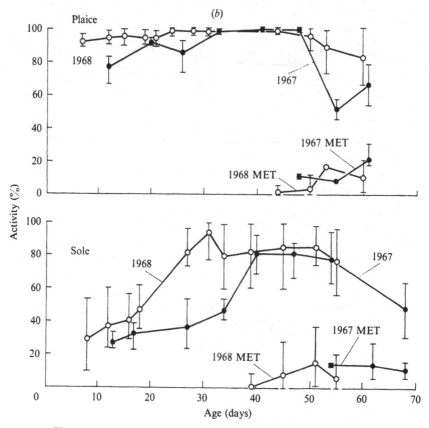

Fig. 3. (b) Per cent activity of plaice and sole larvae of different ages after hatching. As (a). MET means metamorphosing.

(c) *Perception distances*

These are shown in Fig. 4 for all species except sole. Over the rearing period the increase in perception distance was slight, from 3.5 to 5 mm in herring, 1 to 2.5 mm in pilchard and 3.5 to 5.5 mm in plaice.

According to Rosenthal (1966) early sole larvae have a perception distance of 3–4 mm, while herring larvae (Rosenthal & Hempel, 1969) in the yolk-sac stage have perception distances of 2–8 mm, increasing to as much as 30 mm a few weeks later. These values are also much in excess of the

476   J. H. S. BLAXTER AND M. E. STAINES

present findings. The latter authors also found greater perception distance during what they classed as 'search-swimming'. Braum (1964) estimated that the perception distances of early larvae of whitefish *Coregonus* and pike *Esox lucius* were about 10 mm.

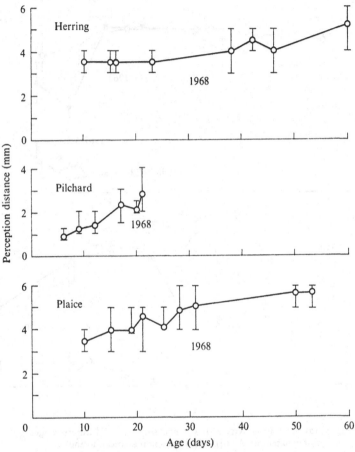

Fig. 4. Perception distances of herring, pilchard and plaice at different ages after hatching; 1968 only. The vertical lines show the range of observations in each series of experiments.

(d) *Feeding efficiency*

The proportions of successful, unfinished and failed feeding movements are shown for herring in some detail in Fig. 5. The improvement in efficiency is marked over 5 or 6 weeks, due possibly to a number of factors such as learning, increase in size of the mouth and death of the less viable larvae. The per cent success is shown for all species except sole in Fig. 6. The clupeids are inefficient in the early stages, under 5 % of feeding movements being successful at the end of the yolk-sac stage. In plaice, feeding is much more successful from the earliest stages. This may be due to the different type of locomotion, plaice being much more manoeuvrable in the water and able to 'back-up'. Clupeids have a more continuous type of serpentine swimming with less chance to correct an initial misjudgement.

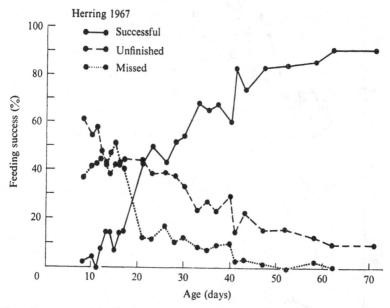

Fig. 5. Percentage of feeding movements of herring larvae of different age from hatching, classed as 'successful', 'unfinished' and 'missed'; 1967 only.

Rosenthal & Hempel (1969) found only 3–10% successful feeding in early herring larvae, increasing to 90% later. Ivlev (1960) calculated the feeding efficiency of young bleak was only 8% and Braum (1964) found an increase from 3–5 to 18–22% in coregonid larvae and from 30 to 80% in pike larvae between first feeding and about 2 weeks later.

DISCUSSION

The foregoing data may be used to calculate the volume searched per hour. First the shape of the tube searched visually by the larval eyes must be considered. Strictly it is elliptical in cross-section if the eyes can be used individually for food perception. There is at present no evidence of binocular vision in the larval stage of the species used, though Braum (1964) reported binocular fields of 45 and 80° in the early larvae of *Coregonus* and *Esox*

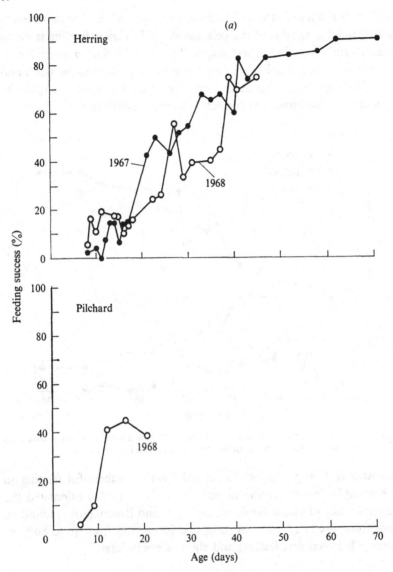

Fig. 6.(a) Percentage success in feeding of herring and pilchard of different ages after hatching. The year of experiment is given.

respectively. The following assumptions, based on observation, were therefore made. The eyes can be used individually, the maximum angle subtended by the visual field during feeding being 90° in the vertical plane and about 45° in the horizontal plane (see Fig. 7). Food is apparently not perceived much below the body axis so that the tube searched is elliptical in cross-section and offset above the axis of the body or swimming path of the larvae. Based on average distance between the eyes of 1 mm the cross-sectional area searched is approximately $\frac{2}{3}\pi x^2$ where x is the maximum perception distance.

The volume searched per hour is obtained from $\frac{2}{3}\pi x^2 \times$ distance swum/ min (Fig. 1) × min activity/h (Fig. 3). The reason for not extrapolating distance swum per minute to distance swum per hour was that the measurements shown in Fig. 1 were made on actively searching larvae, whereas

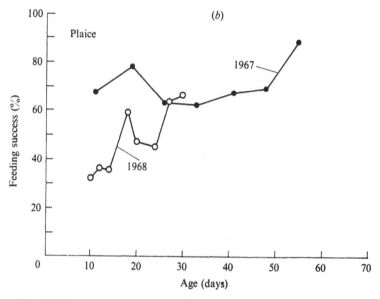

Fig. 6. (b) Percentage success in feeding of plaice larvae of different ages after hatching. As (a).

larvae selected at random in the activity observations showed that there were periods of inactivity which would reduce the path length per hour, especially in the younger larvae.

The results of calculating volume searched are plotted in Fig. 8 and show marked increases with age, from 0.1 to 2.4 l/h in herring over 8 weeks, 0.01 to 0.2 l/h in pilchard over 3 weeks and 0.1 to 2.3 l/h in plaice over 7 weeks. The lack of data on perception distances made it impossible to calculate similar volumes for sole larvae.

The values may be compared with those of previous authors. Rosenthal & Hempel (1969) found the volume searched per hour increased from about 1.5 l in early larval herring to about 10 l at an age of 4 weeks. The substan-

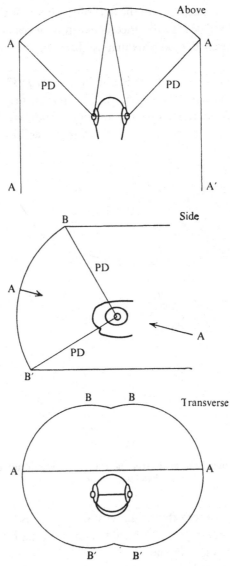

Fig. 7. Probable visual field of a fish larva viewed from above, from the side and from the front (transversely) PD means perception distance; other letters show equivalent axes in the three parts of the figure.

tial disagreement here lies in the greater perception distances and swimming performance per unit time found by these authors compared with the present results. Young whitefish *Coregonus wartmanni* larvae were found to search 14.6 l/h (based on a fairly high perception distance of 10 mm).

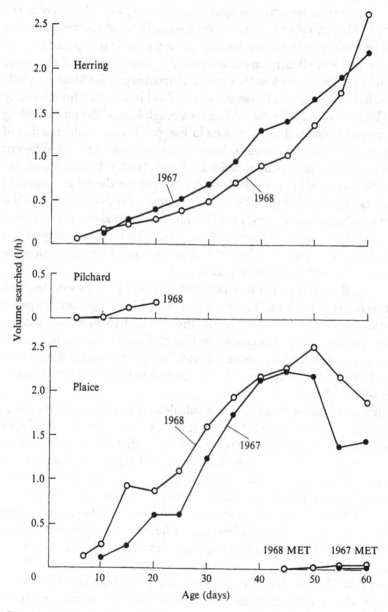

Fig. 8. Calculated volume searched in l/h by herring, pilchard and sole of different age from hatching. The year on which observations were based is given. MET means metamorphosing.

In making a further calculation for volume searched per *day*, the dependence of the visual response on light needs investigation. A review of the literature (Blaxter, 1969 b) shows that the feeding threshold for visual feeders (based on about a 10 % feeding level) is about 0.1 lx. Of the species investigated, herring and pilchard larvae were entirely visual feeders, plaice larvae could take food in the dark as they reached metamorphosis (Blaxter, 1968 b) but not before, while sole larvae could take food in the dark from an early stage (Blaxter, 1969 c). For visual feeders a rough idea of the hours feeding per day can be obtained by using data in Fig. 9. This shows the number of hours during which the surface illumination exceeds 0.1 lx at different seasons and latitudes based on tables by Brown (1952). By reading off the appropriate value of hours available for feeding per day the volume searched per day can be obtained from the volume searched per hour. This does not allow for reduced feeding levels at light intensities just above the threshold nor for smaller numbers of hours of suprathreshold light per day at any depth below the surface. Thus the values obtained by the procedure outlined above will be maximum ones.

In using this information to relate feeding activity to relevant biomass and availability of food, the minimum numbers of food organisms required for survival and growth of a larval population would have to be corrected by the per cent feeding success (shown in Fig. 5). Feeding success is lower in young larvae, hence the numbers of food organisms would have to be relatively higher to compensate for the greater number of feeding movements which fail.

A number of authors have taken the calculations further. From the metabolic requirements of juvenile herring in the Baltic and the density of prey species there in August, Ivlev (1960) concluded that they needed to feed 15 h/day, and that young bleak must feed 9.6 h/day to account for the ingestion of their maintenance and growth requirements. Blaxter (1966) concluded, from estimates of volume searched, calorific requirements and available relevant biomass of plankton, that the weight of food available for herring larvae was, perhaps, at least one order below the required level in the Firth of Clyde in the early spring. Later, in the North Sea, the weight of food available for juveniles was greatly in excess of the required level. Rosenthal & Hempel (1969) made somewhat similar estimates for larval herring and found that at an age of 6–9 days a food concentration of 21–42 nauplii/l was required, dropping to 13–25 nauplii/l some days later.

Mention should be made of the theoretical considerations of Cushing (1968) on the grazing of herbivorous copepods. Here the volume searched or swept clear can be estimated from the decrease in algal density with time. Presumably a similar approach could be made to the problems described in this paper. Cushing estimates the volume swept clear might range from 0.28 to 5.7 l/day in species of *Calanus* of length from 3.50 to 7.38 mm with perception distances from 0.38 to 1.0 mm. Harris (1968)

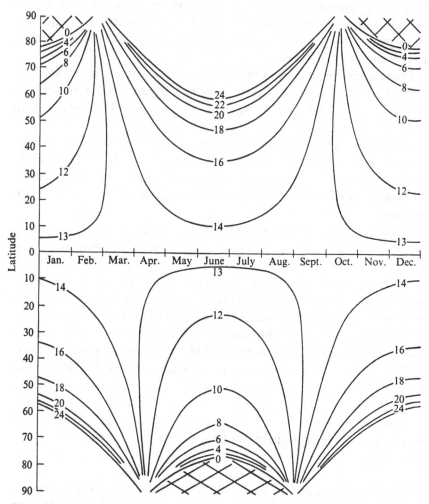

Fig. 9. Nomogram showing number of hours per day that the surface light intensity is above the visual threshold of 0.1 lx, for different latitudes and months.

elaborates this approach by taking into account the changing course of a copepod during searching, differences between searching and attack velocities and the influence of algal density on searching. These authors do not take into account the possible importance of visual thresholds nor any possible vertical asymmetry in the visual field of the copepods.

SUMMARY

The ability of marine fish larvae (herring *Clupea harengus*, pilchard *Sardina pilchardus*, plaice *Pleuronectes platessa* and sole *Solea solea*) to search for food during the weeks following yolk resorption was assessed by making the following observations: (a) distance covered per unit time, (b) per cent time spent in activity, (c) perception distance for food, (d) per cent success in predation on zooplankton.

Distance covered increased from about 20 to 80 cm/min over 8 weeks in herring, 10 to 30 cm/min over 3 weeks in pilchard, 10 to 60 cm/min in plaice and 5 to 40 cm/min in sole, both over 7 weeks. There was a sharp drop at metamorphosis in the flatfish.

Percentage activity increased from 30 to 100% in herring during 4 weeks, 5 to 100% in pilchard in 2 weeks and in sole 25 to about 80% in 5 weeks (never reaching 100%). In plaice it remained from 80 to 100% throughout the rearing period. In the flatfish activity decreased markedly at metamorphosis.

Perception distances increased gradually in herring from 3.5 to 5 mm, in pilchard from 1 to 2.5 mm and in plaice from 3.5 to 5.5 mm. It was not possible to measure these in sole larvae.

Feeding efficiency was initially low in herring and pilchard larvae, rather less than 5% of feeding movements being successful at yolk resorption. Efficiency in both species reached about 40% 2 weeks later and about 70% in herring after 5 weeks. Feeding success was much more constant in plaice, 40–70% throughout early larval life. It was not measurable in sole.

Calculations were made of volume searched per unit time in the first three species. It increased from 0.1 to 2.4 l/h in herring over 8 weeks, 0.01 to 0.2 l/h in pilchard over 3 weeks, and 0.1 to 2.3 l/h in plaice over 7 weeks.

These values of searching ability may be related in visual feeders to the hours of suprathreshold light available per day at different seasons and latitudes.

The authors are most grateful for material and facilities provided by the Marine Laboratory, Aberdeen, New Hatchery and Marine Station, Port Erin, Isle of Man and The Laboratory, Citadel Hill, Plymouth. The research was, in part, supported by a grant from the Natural Environment Research Council.

REFERENCES

BAINBRIDGE, V. & FORSYTH, D. C. T. (1971). The feeding of herring larvae in the Clyde. *Rapp. P.-v. Réun. Cons. perm. int. Explor. Mer.* (In Press).

BLAXTER, J. H. S. (1966). The effect of light intensity on the feeding ecology of herring. In: *Light as an Ecological Factor*, pp. 393–409 (ed. R. Bainbridge, G. C. Evans & O. Rackham). Oxford: Blackwell Scientific Publications.

BLAXTER, J. H. S. (1968a). Rearing herring larvae to metamorphosis and beyond. *J. mar. biol. Ass. U.K.* 48, 17–28.

BLAXTER, J. H. S. (1968b). Light intensity, vision and feeding in young plaice. *J. exp. mar. Biol. Ecol.* 2, 293–307.

BLAXTER, J. H. S. (1969a). Experimental rearing of pilchard larvae. *J. mar. biol. Ass. U.K.* 49, 557–75.

BLAXTER, J. H. S. (1969b). 2. Light. 2.32. Pisces. In: *Marine Ecology* Vol. 1, pp. 223–320 (ed O. Kinne). Chichester, Sussex: John Wiley and Sons.

BLAXTER, J. H. S. (1969c). Visual thresholds and spectral sensitivity of flatfish larvae. *J. exp. Biol.* 51, 221–30.

BLAXTER, J. H. S. (1971). Feeding and condition of Clyde herring larvae. *Rapp. P.-v. Réun. Cons. perm. int. Explor. Mer.* 160, 125–36.

BRAUM, E. (1964). Experimentelle Untersuchungen zur ersten Nahrungsaufnahme und Biologie an Jungfischen von Blaufelchen (*Coregonus wartmanni* Bloch) Weissfelchen (*Coregonus fera* Jurine) und Hechten *Esox lucius* L. *Arch. Hydrobiol. Suppl.* 28, 183–244.

BROWN, D. R. E. (1952). *Natural Illumination Charts*. Research and Development Project NS 714-100. US Dept. of the Navy, Bureau of Ships, Washington 25 D.C.

CUSHING, D. H. (1968). Grazing by herbivorous copepods in the sea. *J. Cons. perm. int. Explor. Mer.* 32, 70–82.

FRASER, J. H. (1969). Experimental feeding of some medusae and Chaetognatha. *J. Fish. Res. Bd Can.* 26, 1743–62.

HARRIS, J. G. K. (1968). A mathematical model describing the possible behaviour of a copepod feeding continuously in a relatively dense randomly distributed population of algal cells. *J. Cons. perm. int. Explor. Mer.* 32, 83–92.

HEMPEL, G. & BLAXTER, J. H. S. (1963). On the condition of herring larvae. *Rapp. P.-v. Réun. Cons. perm. int. Explor. Mer.* 154, 35–40.

IVLEV, V. S. (1960). On the utilisation of food by planktonophage fishes. *Bull. math. Biophys.* 22, 371–89.

ROSENTHAL, H. (1966). Beobachtungen über das Verhalten der Seezungenbrut. *Helgoländer wiss. Meeresunters.* 13, 213–28.

ROSENTHAL, H. & HEMPEL, G. (1969). Experimental studies in feeding and food requirements of herring larvae. *Symp. Marine Food Chains, Univ. of Aarhus, Denmark 1968*, pp. 344–64 (ed. J. H. Steele) Edinburgh: Oliver and Boyd.

SHELBOURNE, J. E. (1957). The feeding and condition of plaice larvae in good and bad plankton patches. *J. mar. biol. Ass. U.K.* 36, 539–52.

SHELBOURNE, J. E. (1964). The artificial propagation of marine fish. *Adv. mar. Biol.* 2, 1–83.

A NEW LIGHT TRAP FOR PLANKTON

D. A. JONES

Marine Science Laboratories, Menai Bridge

INTRODUCTION

The use of light at night to attract zooplankton is a well-known technique (Fage, 1923; Sheard, 1941; Hale, 1953) and has resulted in the capture of many animals which, because of their fast swimming ability or relatively sparse distribution in the plankton, would otherwise have only rarely been captured in townets. Previous apparatus has usually consisted of a water-tight light source, electric or acetylene, held submerged from a floating raft (Grein, 1912; Fage, 1923); the plankton was then fished by hand as it concentrated around the light source.

Successful trials were reported by the Liverpool Marine Biological Society (Herdman 1888–9), using a second type of apparatus. Submarine incandescent lamps of 100 candle power were mounted in the mouths of townets which were then lowered to various depths. The catches depended largely on the quick recovery of the townets which swept the congregating plankton into the nets, unlit nets being used as a comparison.

However, previous methods have always been time-consuming, requiring the constant attention of an operator, either to collect plankton as it con-centrated around the light source, or to haul up a townet. Previous methods also caught a large proportion of the general phyto- and zooplankton which is present in the water but which may not be specifically attracted to the light.

The present apparatus has been designed to run automatically during the night, retaining the catch, which can be removed the next day at the convenience of the operator. Preliminary trials indicate that only a small proportion of the catch is composed of plankton which is not specifically attracted to the light.

DESCRIPTION AND METHODS

The trap (Fig. 1) consists essentially of an electric light source enclosed in a wire frame covered with plankton netting. It operates on the principle that photopositive zooplankton, guided by a light beam, enters via the small conical entrance in the base, but cannot escape as it has difficulty in finding the exit as it attempts to sink downwards with the onset of daylight.

[487]

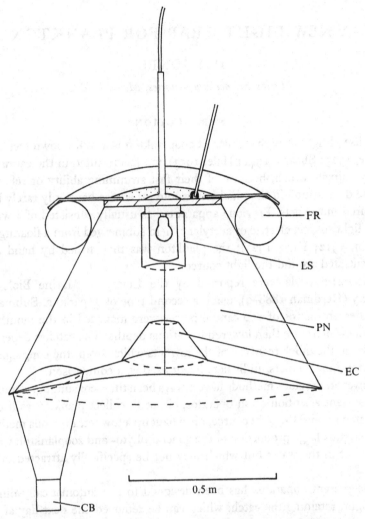

Fig. 1. The plankton light trap; EC, entrance cone; FR, floatation ring; LS, light source 100 W bulb; PN, plankton mesh 190 m; CB, collecting bucket.

The bulb (100 W at present) is mounted in a water-tight housing screwed to a wooden base, from which a cone-shaped frame of 0.6 cm diameter steel rod is suspended, surrounding the bulb housing. The frame terminates in a circular concave base 125 cm in diameter, from which a second internal cone 10 cm in diameter arises directly beneath the light source, giving access to the inside of the light trap. The frame is covered with plankton netting with 28 meshes/cm and a pore size of 190 μm; thus the earlier copepodite stages of copepods are not retained.

Initially, the light was left unshaded, but as this resulted in the attraction of plankton to the sides of the trap a shade was added to direct the light vertically downwards through the entrance cone. The trap was positioned at the end of Menai Bridge Pier in approximately 5 m of water during present investigations.

Removal of the catch is facilitated by the provision of a sleeve and collecting bucket on the lower rim of the trap so that the contents may be concentrated as the apparatus is lifted from the water. The whole apparatus weighs approximately 12 kg and is supported by an inflatable rubber tube, attached so that the trap floats with the light submerged.

Each sample was preserved in 5% formalin on collection and passed through a 1 mm sieve to separate off the macroplankton which was identified and counted. The microplankton was diluted to a standard volume (1.5 l), and subsampled with a Stempel pipette to estimate total numbers. Animals were identified in each sample until the first hundred holoplanktonic animals were reached according to the method adopted by Winsor & Walford (1936); repeats were used to obtain percentage abundance of species.

RESULTS

Total catch

A mean total of 43 700 micro- and 426 macroplanktonic animals were taken per night as a result of 15 trials in May and June 1969 (Table 1). The mean number of hours of darkness per night was approximately 6, with a range of 6.40 h on 9 May to 5.25 h on 14 June. A mean total of 1207 animals/m³ was recorded by Kenchington (1968) from the Menai Straits in May and June, but accurate estimations could not be obtained during present investigations due to the presence of a *Phaeocystis* bloom (Fig. 2*b*), which caused clogging of townets.

Table 1. *Comparison of catches of light trap and townets*

(Mean total catches per night in light trap compared with standard 8.3 m medium and coarse twonet hauls. Figures for total plankton/m³ in Menai Straits from Kenchington 1968.)

	Mean total/ night in trap. Light on	Mean total/ night in trap. Light off	Mean total standard net hauls (20 min)	Nos/m³ May/June 1968
Microplankton	43 700	3650	3600⎱	1207
Macroplankton	426	23	133⎰	

On three occasions the trap was left switched off all night and the mean totals of plankton retained were only 3650 micro- and 23 macroplankton

490 D. A. JONES

organisms. Finally, a comparison with total numbers of zooplankton caught in conventional standard 0.5 m nets (coarse and medium) towed at night during the operation of the trap, before the *Phaeocystis* bloom, indicates that the light trap catch is approximately proportional to 4 h townetting.

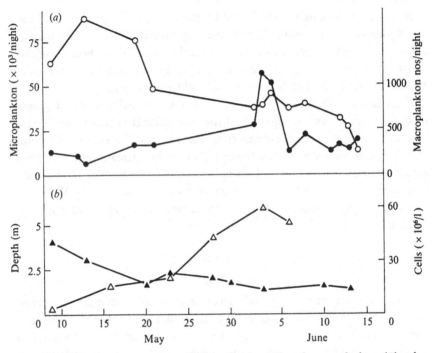

Fig. 2.(a) Results of night catches of microplankton (O) and macroplankton (●), taken in the light trap during May and June 1969. (b) Secci disc readings (▲) taken from Menai Bridge pier together with algal cell counts (△) for May and June 1969.

Composition of samples

Table 2 shows that a close correlation exists between the range of species of microplankton caught during these preliminary trials with the light trap (10 species) and those caught using standard net tows (10 species). The only notable exceptions were fish ova and the ova of *Littorina littorea* which did not appear in the light trap, but which were present in the townet hauls. All the microplankton species recorded from the light trap have, however, been taken in net tows (Kenchington, 1968).

The total number of macroplankton species caught in the light trap was 39 as opposed to 26 species caught by tow netting during the same period. Several species of amphipod, *Calliopius rathkei*, *Orchomenella commensalis*, and the mysids, *Praunus flexuosus* and *Schistomysis spiritus* were absent from townets, as was the euphausid *Nyctiphanes couchii*, and the heteronereid annelids. Only two species, *Corophium volutator* and *Palaemon serratus* zoea, were restricted to the townet samples. The percentage abundance of the microplankton species again appears similar in both the light trap and townets, reflecting the species natural abundance in the Menai Straits. This variation of species, one with another, does not appear to have altered greatly during the May–June period in recent years (Table 2).

Table 2. *Percentage abundance of micro- and macrozooplankton species in light trap compared with ratios obtained in standard net hauls*

(Column 1, mean of 15 catches with light trap. Column 2, mean of 3 standard net night hauls. Column 3, data from Kenchington (1968).)

	1 Abundance in light trap (%)	2 Abundance in standard net hauls (%)	3 Abundance in daylight hauls 1968 (%)
MICROPLANKTON			
Temora longicornis	86.8	85.4	71.8
Paracalanus/Pseudocalanus	4.4	4.1	7.2
Acartia clausii	2.1	1.4	3.8
Centropages hamatus	0.9	0.9	6.9
Carcinus meanus zoea	1.5	1.8	0.50
Annelid Larvae	1.1	1.7	0.65
Balanus Cyprids	1.5	1.7	3.3
Other species	2.8	2.9	8.8
MACROPLANKTON (species forming 1 %)			
Sagitta elegans	8.8	16.7	—
Pleurobrachia pileus	21.0	50.2	—
Atylus swammerdami	50.5	10.0	—
Orchestia gammerella	0.0	2.5	—
Bathyporeia guilliamsoniana	0.0	3.0	—
Fish larvae	2.0	2.7	—
Meganictyphanes norvegica, furcilia	2.9	6.3	—
Pagurus bernhardus zoea	1.0	2.5	—
Other species	13.8	6.1	—

In the light trap catches of macroplankton, the amphipod *Atylus swammerdami* constituted over 50% on average of the total catch, a phenomenon also reported by Herdman (1889); this amphipod formed only 10% in the townet samples. The abundance of this species in the light trap is thus the major contributory factor to the discrepancies between percentage abundances of other common species in the light trap and townet hauls.

Variation between nightly samples

Although the trap has only been operated on 15 occasions, Fig. 2(*a*) shows that a considerable variation exists between nightly catches of both micro- and macroplankton during these trials. The microplankton catches dropped from a peak of 88 000 per night on 13 May to 13 000 on 14 June, whilst the macroplankton totals remained fairly constant except for a sharp rise on 3 and 4 June.

The decline in microplankton numbers may perhaps be explained by reference to Fig. 2(*b*) which includes a series of Secci disc readings taken from the pier during May and June 1969 (Tyler, unpublished observations). These show that water transparency dropped from 4 m on 9 May to about 1.5 m during the period 1–14 June. This might be expected to reduce the effectiveness of the light penetration from the trap, resulting in reduced zooplankton catches. However, whilst transparency remained relatively constant throughout June, the microplankton numbers continued to drop, indicating perhaps that factors other than loss of transparency may be responsible.

The loss in water transparency appears to be linked to increased algal cell counts during this period (Fig. 2*b*) (Tyler, unpublished observations), with *Phaeocystis* (Chrysophyceae) predominating and undergoing a seasonal bloom. It is suggested, therefore, that the continued decline in microzoo-plankton numbers may be related to the dense *Phaeocystis* growth in surface waters of the Straits.

It is interesting to note that the numbers of macroplankton are not similarly affected by this algal bloom. This may possibly be due to the dominance of *Atylus swammerdami*, a benthic form which does not feed in surface waters during the night, although more observations are required before this can be confirmed.

DISCUSSION

The difference in zooplankton numbers in catches obtained when the light trap was switched on and off indicates an increased efficiency of 94% when operational. The mean catch of 44 000 zooplankton individuals per night (for 6 h fishing), possibly higher but for the presence of *Phaeocystis*, com-pares favourably with results obtained using a 0.5 m townet. In the present position, anchored to the pier, the trap is subject to strong tidal currents which have been observed to sweep away the plankton as it congregates around the light. Thus the totals may well represent plankton caught during periods of slack tide only. It is hoped to test this experimentally from a drifting boat.

Whilst the light trap cannot be used to obtain accurate quantitative samples of the zooplankton in surrounding waters, it has been shown that it will sample the major elements of the plankton. It thus provides an economical method for the collection of zooplankton, especially with regard to larger species. Furthermore, examination of catches show that the animals remain alive and in good condition, unlike townet samples which are often damaged as a result of overcrowding and forcing against the net meshes during towing. The relative lack of phytoplankton in samples, especially striking during the recent *Phaeocystis* bloom when normal tow-netting was virtually impossible, renders samples particularly suitable for calorimetry as they do not require laborious hand sorting.

Finally, it was hoped at first that the light trap would sample only the photopositive members of the zooplankton. However, the trials without light show that a small percentage (4%) of other plankton organisms are captured. However, as the trap clearly retains a vast majority of photo-positive animals, the apparatus may prove useful as a quantitative method for testing the effectiveness of different light intensities and regimes on zooplankton in the field.

I am grateful to Mr J. Byng for building the apparatus, to Mr P. Tyler for access to his data on *Phaeocystis*, and to Professor Crisp for his criticism of the manuscript.

REFERENCES

FAGE, L. (1923). Essais de pêche a la lumière dans la baie de Concarneau. *Bull. Inst. océanogr. Monaco* **431**, 1–20.

GREIN, K. (1912). Eine Elektrische Lampe zum Anlocken positiv phototaktischer Seetieve. *Bull. Inst. océanogr. Monaco* **242**, 1–15.

HALE, H. M. (1953). Notes on distribution and night collecting with artificial light. *Trans. R. Soc. S. Aust.* **76**, 70–6.

HERDMAN, W. A. (1888). Annual report of Liverpool biological station on Puffin Island. *Proc. Trans. Lpool biol. Soc.* **2**, 1–20.

HERDMAN, W. A. (1889). Third annual report of Liverpool biological station on Puffin Island. *Proc. Trans. Lpool. biol. Soc.* **3**, 1–47.

KENCHINGTON, R. A. (1968). *M.Sc. Thesis (University of Wales)* 1–61.

SHEARD, K. (1941). Improved methods of collecting Marine Organisms. *Rec. S. Aust. Mus.* **7**, 47–134.

WINSOR, C. P. & WALFORD, L. A. (1936). Sampling variations in the use of plankton nets. *J. Cons. perm. int. Explor. Mer.* 190–204.

RESPONSES TO LIGHT OF
ASTERIAS RUBENS L.

J. C. CASTILLA*

Marine Science Laboratories, Menai Bridge, Anglesey

INTRODUCTION

Reports on the responses to light of *Asterias rubens* appear rather discordant. Hyman (1955) and Yoshida (1966) summarized the state of the problem. A positive response to all light intensities was found by Plessner (1913). Romanes & Ewart (1881) found in *A. rubens* a strong disposition to crawl towards, and to remain under the lights, and Just (1927) found a similar positive response to a single light. Diebschlag (1938) also found that freshly collected *A. rubens* are nearly always positive to light and may be led around, for about 15 min, by shining a light on the tip of an arm, after which they become indifferent to light. Bohn (1908) claimed that the species has a general 'phototropisme négatif'. Although he found that starfish which had collected under stones moved directly away from the sunlight, those found on sandbanks did not. Van Weel (1935) working with eyeless animals found that they were photopositive when the light was dim and photonegative when it was bright but he did not measure the critical intensity. Smith (1950), experimenting with inverted starfish, found that the arms nearest the light source dominated the locomotory pattern, those which were furthest away from the light only infrequently came to lead.

The experiments described below were designed to investigate how the past experience of the starfish might influence the behaviour to light.

MATERIALS AND GENERAL METHODS

Individuals of *A. rubens* were collected under stones at Church Island, situated on the western side of the Menai Strait, close to the Marine Science Laboratories, Menai Bridge. All the starfish had the normal five arms which measured from 2.7 to 5.5 cm in length. They were kept in glass tanks with continuous supplies of air and running sea water. No food was offered to the animals except on one occasion. The starfish were conditioned to light in three ways. The first group, dark–light adapted (DLA), was kept in the

* Present address: Laboratorio de Zoología, Instituto de Ciencias Biológicas, Universidad Católica de Chile, Casilla 114-D, Santiago, Chile.

dark during the night and illuminated during the day for at least 10 h. The second group, light adapted (LA), was kept for at least 20 h/day, and often longer, under an illumination of approximately 1600 lx (100 W tungsten bulb suspended approximately 34 cm over the aquarium). The third group, dark adapted (DA), was kept in a dark box, under absolutely dark conditions, for at least 20 h/day. Starfish were taken out of the box only at the time of the experiment. Experiments were also conducted with freshly collected starfish (FC).

All the experiments were carried out in a shallow perspex tank, 100 cm in length and 5 cm in depth, divided into two equal channels by means of

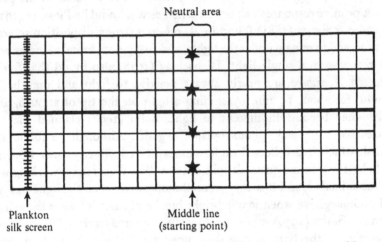

Fig. 1. Diagrammatic representation of the divided perspex tank used in all the experiments.

a longitudinal partition, each channel being about 18 cm wide (Fig. 1). A 5 cm square grid was drawn on the floor of the tank to enable the position of the starfish to be recorded; it was possible for either or both channels to be supplied with a flow of sea water along its length. When flowing sea water was used, which could be introduced from either end, a fine screen of plankton silk was placed 5 cm from the same end of the tank. This screen served two purposes: it prevented debris from the water entering the channels and it reduced turbulence along the channels. In all the experiments with flowing sea water, the water level was maintained by siphons to a constant depth of between 2.0 and 2.5 cm which was sufficient to cover the starfish completely but shallow enough to discourage them from climbing up the sides. In experiments carried out with static sea water, the same depth of water was used. Two starfish selected at random were placed on the middle line of each channel, with the oral side touching the bottom of the tank. An area, 5 cm to the right and 5 cm to the left of the middle line, was defined as the 'neutral area'. A trial was completed whenever a starfish left the neutral area. If the starfish remained inside the neutral area for the 7 min allowed for the experiment it was counted as an incomplete trial. During each trial the position of the starfish was recorded every minute. Temperatures were recorded in each experiment. Light intensities were measured with an E.E.L. Lightmaster photometer having a peak response at approximately 580 nm. In all the experiments light intensities were measured by placing the photocell in a horizontal position on the bottom of the tank. All light intensities are given in lux.

EXPERIMENTS

Experiments in which starfish were differentially illuminated from above
Method

The experimental situation A, illustrated in Fig. 2, was mainly used in these experiments. A lamp fitted with a 150 W tungsten filament was held 90 cm above the centre of the trough. A piece of wood covered with black material was placed over one-half of the tank. Light intensity over the illuminated first 20 cm, from the middle line, was approximately 240 lx. The intensity over the first covered 5 cm, from the middle line, was approximately 10 lx; the second covered 5 cm, 8 lx; the third covered 5 cm, 4 lx and the fourth covered 5 cm, 2 lx. Beyond this the light intensity was of approximately 1 lx. By varying the distance between the lamp and the trough and using lamps of different wattage, seven other experimental situations were used (Fig. 2, B–H). All the intensities were recorded at the bottom of the empty tank. By placing the starfish on the middle line of the tank it was possible to observe the responses to light when half of the body was exposed to a high, or relatively high, light intensity and the other half to a low intensity.

Results

Two batches of 36 starfish were collected on 19 February 1969, one batch kept in the dark and the other in the light. Each animal was tested in the apparatus described above (Fig. 2A) immediately after collection and at intervals over a period of 30 days. Static sea water was used. As Fig. 3 shows, nearly all the freshly collected starfish responded negatively, moving out of the light. The dark-adapted animals very quickly became indifferent to light; about 30–40% moved towards the light during the first

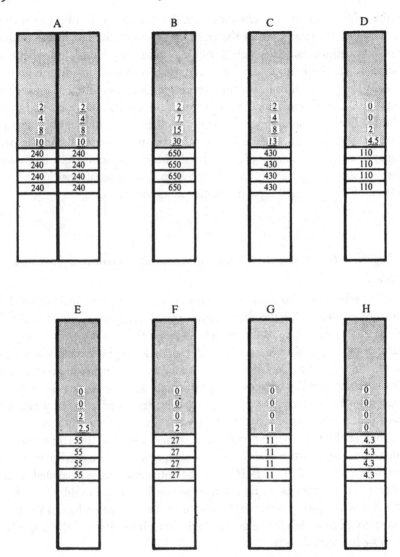

Fig. 2. Diagrammatic representation of the eight experimental situations (A–H) used in experiments where starfish were differentially illuminated from above. Light intensities in lux and measured at the bottom of the empty tank.

few days of dark adaptation but subsequently, after 2 weeks, lost almost all reaction to light, the proportion stabilized at about 40–50 %. In the light-adapted animals, the percentage moving towards the light increased steadily, the animals gradually becoming photopositive. By the 13th day, the light response was completely reversed. A small proportion of the animals appeared to be photonegative on the 30th day; nevertheless, there was still a significant difference from the initial values.

A further experiment was carried out with the light-adapted animals on the 40th day. On the 33rd day of light adaptation, 36 small mussels were offered to the starfish; every day during the following week, the number of mussels eaten were replaced by a similar number of fresh living mussels. The total number of mussels eaten during the week was 51. However, the

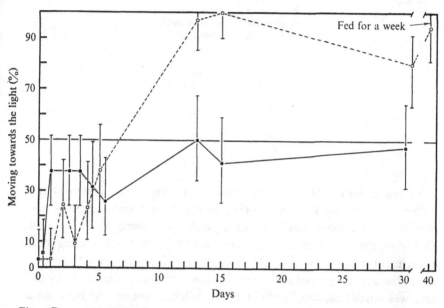

Fig. 3. Responses to light of two groups of 36 starfish collected on 19 February 1969, when differentially illuminated from above. Temperature ranging from 3.5 to 8.7 °C. Vertical bars represent 2.5 % limit of the expectation (Fisher & Yates, 1953, table VIII 1). Starfish freshly collected (◑), light adapted (○) and dark adapted (●).

starfish fed during this period did not show any difference in the pattern of response to light. The light-adapted animals were evidently photopositive as a result of some difference in the conditions in the laboratory, but not on account of starvation. The dark-adapted animals appeared to lose their sensitivity to light or, alternatively, were inhibited from displaying any response to light.

A third batch of 36 starfish was collected on 17 March 1969. The experiments were carried out in the same apparatus, using freshly collected animals and animals kept light adapted for 22 days. Table 1 shows the results. The initial negative response of this group of starfish and of those collected on 19 February 1969 (see Fig. 3) was the same, but this group required almost twice the time to lose their negative response and they did not become completely positive. Nevertheless, the same basic change in response was shown in both experiments.

Table 1. *Responses of* Asterias rubens *when differentially illuminated from above*

(36 *A. rubens*, collected on 17 March 1969; sea-water temperature ranging from 5.2 to 11.0 °C.)

Number of days under illumination	Per cent moving away from light
0	91.5
1	85.0
3	68.5
4	69.3
6	78.0
8	64.8
10	63.0
14	41.4
16	33.5
22	17.7

In the experiments conducted with light- and dark-adapted starfish collected on the 19 February 1969 (see Fig. 3), the time taken by each starfish to leave the neutral area was recorded and averaged. Fig. 4 shows that when the light response was clearly defined, as at the beginning of the experiments and also for the light-adapted animals at the end of the experiment, the animals moved rapidly so that the time taken to leave the neutral area was much less, usually about 2 min. When, however, the response to light was indefinite, as during the first week for both light- and dark-adapted animals, they moved more lethargically. The dark-adapted animals never showed a definite light response and thereafter moved slowly out of the neutral area.

Table 2 shows the results of seven groups of experiments in which the light intensity was varied, as shown in Fig. 2 B–H. As long as the animals were not kept for more than a few days in the laboratory, freshly collected and light-adapted animals moved away from the light whatever the range of intensity and so confirmed the previous experiments. Two experiments carried out with dark-adapted animals showed a corresponding indifference to light.

Method   *Experiments using a light gradient*

A lamp fitted with a 150 W tungsten bulb was held 45 cm above the central longitudinal partition and its light directed towards the centre of one end

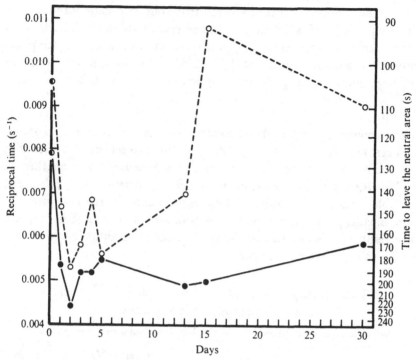

Fig. 4. Average time to leave the neutral area taken by starfish collected on 19 February 1969, when differentially illuminated from above. Each point represents average time of a mean of 35 completed trials. All values extracted from experiments of Fig. 3. Symbols as in Fig. 3.

Table 2. *Responses of* Asterias rubens *when differentially illuminated from above*

(All experiments conducted between 16 November 1968 and 29 April 1969 with starfish kept for about a week in captivity. Mean temperature 10.0 °C) (see Fig. 2B–H).

Light conditioning	Experimental situations (Fig. 2)	No. of individuals used	No. of trials completed	Light intensity on unshaded part (lx)	Per cent moving away from the light
FC	B	36	36	650	69·4
FC	C	36	33	430	75·7
FC	D	39	37	110	81.0
FC	E	40	40	55	95.0
FC	F	20	20	27	75.0
FC	G	54	53	11	81.1
FC	H	50	48	4·3	89.5
				Mean	80.9
LA	B	36	35	650	77.1
LA	C	36	34	430	67.6
LA	D	40	40	110	77·5
LA	E	33	33	55	72·7
LA	F	38	38	27	65·7
LA	G	55	50	11	66.0
LA	H	24	20	4·3	100.0
				Mean	75.2
DA	D	56	56	110	46.4
DA	E	36	35	55	51.4
				Mean	4.89

of the tank. All the experiments were carried out with static sea water and the lamp was used alternately at either end of the tank. The level of illumination was approximately 750 lx at the end over which the lamp was situated; 220 lx at the centre of the tank where the starfish were placed at the beginning of the experiment; and 60 lx at the opposite end.

Results

The responses to light of *Asterias rubens* were tested, as described above, between 14 and 25 November 1968, except for two sets of experiments at low temperature conducted on the 4 and 5 February 1969, another on 17 March 1969, and one experiment at high temperature on 24 July 1969. Table 3 shows the results of these experiments. In the first group of experiments, group A, normal sea temperatures were used (8.6–11.9 °C). The result of this experiment shows an effect of conditioning behaviour similar to that already discussed.

Table 3. *Responses to light of* Asterias rubens *at different temperatures using a light gradient*

(All experiments carried out with starfish kept for about a week in captivity.)

Light conditioning	Date of experiment	Experimental temperature (°C)	No. individuals used	No. of trials completed	No. moving towards light	No. moving away from light	Per cent moving away from light
Group A, mean temperature 10.5 °C							
FC	19. xi. 68	10.2–11.3	40	40	5	35	87.5
LA	14–20. xi. 68	9.7–11.4	60	60	20	40	66.6
DA	14–21. xi. 68	8.6–11.9	68	68	33	35	51.5
Group B, mean temperature 16.4 °C							
FC	24. vii. 69	17.5	42	42	9	33	78.5
LA	21. xi. 68	16.2–17.4	38	27	8	19	70.4
DA	23. xi. 68	14.5–17.5	38	38	19	19	50.0
Group C, mean temperature 6.5 °C							
FC	17. iii. 69	6.0–7.0	42	42	14	28	66.6
LA	22–25. xi. 68	5.3–7.2	102	82	34	48	58.5
LA	4–5. ii. 69	6.5–7.7	84	78	31	47	60.2
DA	22–23. xi. 68	5.3–7.3	80	77	34	43	55.8

Comparison by χ^2, 1 degree of freedom, using Yates correction for continuity: LA and DA groups A and B, 4.93; LA and DA group C, 0.14. Comparison by χ^2, 2 degrees of freedom: LA and DA and FC, all groups, 18.82.

In the second group, group B, temperatures higher than ambient were used (14.5–17.5 °C). Light-adapted animals continued to move away from the light as before. A rather large percentage (29 %) of the animals placed at the starting point remained inside the neutral area. Dark-adapted animals showed no response to light and all completed the trials. One experiment with freshly collected animals carried out on 24 July 1969 at ambient temperature, 17.5 °C, showed a clear negative response to light of *A. rubens*.

A third group of experiments, group C, were carried out with water of temperature lower than ambient (5.3–7.3 °C). Freshly collected and light-adapted starfish behaved as before but differed far less from the dark-adapted ones. In spite of a large number of light-adapted animals used (102) during the experiments conducted during November 1968, only a weak negative response to light was found. The same results were found in a new set of experiments using 84 light-adapted animals carried out during February 1969. It appears, as will be more clearly demonstrated later, that low temperatures prevent light-adapted starfish from exhibiting the clear negative response to light shown at higher temperatures.

The time taken to leave the neutral area by each starfish in the experiments recorded in Table 3, was also measured and is shown in Fig. 5. Freshly collected animals moved faster than light- or dark-adapted animals and were the most clearly photonegative. As the experimental temperature was increased, the rate of movement, recorded as reciprocal time to vacate the neutral area, increased steadily. The Q_{10} values for change in rate of locomotion calculated from the data in Fig. 5 were similar for all three groups between 6.5 and 11.0 °C, giving a mean value of 2.33. At 16.0 °C the dark-adapted animals reduced their rate of movement relative to the other groups resulting in a fall in the Q_{10} value to 1.21; no explanation is offered for this anomalous behaviour.

In order to confirm the effect of low temperature on the light response of light-adapted starfish (see Table 3, group C), a further experiment was carried out on 17 March 1969 with 36 freshly collected starfish and repeated at intervals after keeping them under illumination for up to 23 days. The lower curve of Fig. 6 shows the results. The average percentage of light-adapted starfish moving away from the light was only 53.7% at a mean temperature of 7.5 °C. This confirms the lowered response to light at low temperature demonstrated in Table 3, group C.

Experiments under various types of illumination and sea water flow
Method

A sea water current of about 0.6 cm/s was produced in one or both channels by a constant flow of 2 l/min into one end of the tank, the source. The same amount of water was siphoned off at the opposite end, the sink, the

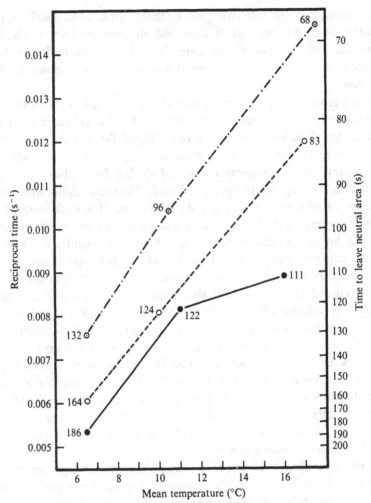

Fig. 5. Average time to leave the neutral area taken by *Asterias rubens* using a light gradient at different temperatures. Each point represents average time of a mean of 43 completed trials. All values extracted from experiments of Table 3. ⊙ Freshly collected. ○ Light adapted. ● Dark adapted.

direction of flow alternating. The rate of flow was checked by volume–time measurements. The velocity of the current is a mean value between readings of surface velocity of small particles and the theoretical value. One channel used with flowing sea water, was the flowing channel, and the other channel with no flow is referred to as the static channel. The various types of illumination will be described below. In this account, positive rheotaxis implies a movement of the animal upstream, i.e. towards the current source.

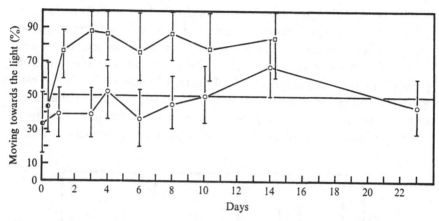

Fig. 6. Responses to light of a group of 36 light adapted starfish collected on 17 March 1969, using a light gradient and static sea water (○), mean temperature 7.5 °C; and using a light gradient and flowing sea water from the same direction as the light source (□), mean temperature 5.7 °C. Vertical bars represent 2.5 % limit of the expectation (Fisher & Yates, 1953, table VIII 1). ◑, Freshly collected starfish.

Results

Experiments with flowing sea water and uniform light. Two types of experiments were carried out. In the first a 150 W tungsten lamp suspended 90 cm over the centre of the trough provided a uniform illumination of about 240 lx over both channels. Twenty-four starfish were collected on 17 March 1969. They were tested immediately after collection and then at intervals over a period of 14 days, being kept continuously in the light. Fig. 7 shows that a positive rheotaxis was displayed throughout the experiment.

In the second type of experiment two lamps, each fitted with 150 W tungsten lamps were held 45 cm above the central partition at each end of the tank and provided a field of illumination of approximately 750 lx at each end of the tank and a rather lower intensity of 430 lx at the centre of the channels. As Table 4, group A (flowing channel) shows, light, dark and light–dark adapted animals show positive rheotaxis; those kept in the static channel moved equally in each direction, indicating no bias in the apparatus.

This upstream behaviour is in fact normal whenever *A. rubens* is placed in running water in our laboratory.

Experiments with flowing and static sea water using a light gradient. Three experimental conditions were tested, the same methods of illumination described under *Experiments using a light gradient* being followed. The first condition (A) with lights at both ends has already been described (Table 4, group A). The starfish show a simple and positive rheotactic response. Because of the direction of flow of the sea water when the lamp was placed at one end of the apparatus it could either be held over the flow sink (condition B) or over the flow source of the tank (condition C). Table 4 also gives the results for conditions B and C. In these conditions the animals in the static channel are subjected only to a light stimulus. The great majority of those starfish with previous experience of illumination (DLA,

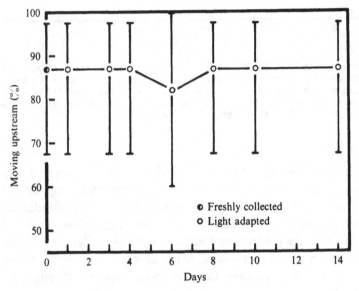

Fig. 7. Rheotactic behaviour of a group of 24 starfish collected on 17 March 1969. Mean temperature 7.5 °C. Vertical bars represent 2.5 % limit of the expectation (Fisher & Yates, 1953, table VIII 1).

LA) clearly moved away from the light. Those previously kept in the dark (DA), however, showed no response to light, moving in approximately equal numbers in each direction. These results are in agreement with those already discussed in Table 3, group B.

When the current was superimposed on the directional light field, the starfish in the flowing channel of experiment B, previously exposed to light, received a reinforcement of two responses, a rheotactic response driving them upstream and a photonegative response driving them away from the light; consequently they moved strongly towards the dimly lit end. In the experimental condition C, where the two stimuli were opposed their response was evidently a compromise, and they moved in roughly equal

(A, B and C are experimental conditions illustrated in the diagrams below. Flowing (i) and Static (ii) sea water is illuminated uniformly (A) or by a light gradient with the light source either at the water current sink (B) or at the water current source (C). All experiments carried out with starfish kept for about a week in captivity, between 30 August and 30 September 1968, at water temperatures ranging from 13.0–17.5 °C.)

Light conditioning of starfish	No. of individuals used	Condition of illumination and current	(i) Flowing channel			(ii) Static channel		
			No. of trials	Upstream (%)	Downstream (%)	No. of trials	To the left (%)	To the right (%)
Group A								
DLA	24	A	44	81.9	18.1	41	44.0	56.0
LA	24	A	60	73.5	26.5	54	53.8	46.2
DA	20	A	30	76.5	23.5	30	43.5	56.5
Group B				(To dim area)	(To bright area)		(To dim area)	(To bright area)
DLA	24	B	48	89.5	10.5	43	79.0	21.0
LA	24	B	23	95.5	4.5	23	78.0	22.0
DA	20	B	32	84.5	15.5	32	46.9	53.1
Group C				(To bright area)	(To dim area)		(To bright area)	(To dim area)
DLA	24	C	39	48.8	51.2	46	15.0	85.0
LA	24	C	33	51.5	48.5	31	35.5	64.5
DA	20	C	30	83.4	16.6	30	46.7	53.3

A Light ii i Light

B Light ii i

C ii i Light

numbers to each end of the tank. However, the dark-adapted starfish again behaved differently, following the upstream behaviour and ignoring the light stimulus. Thus, if we compare the proportion of trials of dark-adapted animals which resulted in a movement towards the current source (76.5%) in the absence of directional light (condition A) with the corresponding proportion in the case of dark-adapted individuals moving towards the current source in a light gradient (conditions B, C), 84.5 and 83.4% respectively, the results are not statistically distinguishable.

In the experimental condition C, the mean temperature used with light-adapted animals in the flowing channel was 14.6 °C and, as discussed, approximately equal numbers of starfish navigated upstream (photopositive) and downstream (photonegative). In view of the results obtained with light-adapted starfish using a light gradient at a low temperature (see Fig. 6, lower curve), a further experiment with the experimental condition C (flowing channel) was carried out, at a mean temperature of 5.7 °C, with the same 36 starfish collected on 17 March 1969. Fig. 6, upper curve, shows that at low temperature the results are quite different, the rheotactic response predominates over the negative response to light shown at higher temperatures. Evidently the negative response to light of light-adapted animals is much weaker at low temperatures.

Table 5. *Effect of water current on the responses of*
Asterias rubens *to light*

(36 light adapted *A. rubens* collected on 17 March 1969. Differentially illuminated from above. Temperature ranging from 5.7–7.5 °C.)

Number of days under illumination	Per cent moving towards light		Effect of current on the negative response to light(%)
	Without current stimulus (%)	With current stimulus (%)	
0	8	25	−17
1	15	50	−35
3	31	50	−29
4	31	46	−15
6	22	47	−25
8	35	40	−5
10	36	58	−22
14	58	71	−13
16	68	—	—
20	81	—	—

Experiments with flowing sea water using differential illumination from above. The apparatus used in the experiments where starfish were differentially illuminated from above (Fig. 2A) was again used in these experiments. The source end of the flowing channel was situated in the illuminated half of the tank, so that the positive response to current and the initial negative response to illumination, of light-adapted animals, were opposed. The 36 starfish collected on 17 March 1969 and used in the experiments illustrated in Table 1 were tested at intervals. Table 5 illustrates the results. In all the experiments, the movement of sea water reduced the negative response to light, as in the experiments using a light gradient. Both groups showed a tendency to switch from photonegative to photopositive behaviour during light conditioning.

DISCUSSION

The results of this investigation explain, to some extent, the contradictory reports on the reponses of *A. rubens* to light that are to be found in the literature. It clearly appears that individuals of *A. rubens* vary their response according to their physiological condition. Furthermore, the animal's behaviour is influenced not only by its previous experience of illumination, but also by the prevailing temperature, as was shown in experiments using a light gradient. As far as these experiments go, it appears that starvation does not play any part in the light response of these animals. The intensity of light over a wide range did not appear critical; experiments in which starfish were differentially illuminated from above show that they are able to exhibit a constant pattern of response when exposed to high intensities varying from 650 lx down to 4.3 lx. In order to evaluate and confirm our results dealing with the responses to light of *A. rubens*, under apparently natural conditions, we have had to take into account the very clear positive rheotaxis of the species, reported for the first time in this paper. It is possible that some attribute of the laboratory water supply may be involved, since many predators (i.e. *Nassarius, Buccinum, Carcinus*) move upstream only in response to odoriferous substances.

A. rubens has been regarded, as have starfish in general (Smith, 1965), as being positive to light, that is, moving into a lighted area or towards a light source, although some long established work has hinted at changes in the response to light. More recently (Millot, 1954; Yoshida, 1966; Yoshida & Ohtsuki, 1968) have demonstrated how behaviour in echinoids and asteroids depends generally on adaptative states. According to our experience, several factors must also be taken into account before assigning a positive or negative response to light to *A. rubens*. The photophysiological condition of the animal, the time kept in the laboratory, the temperature of the water used, the kind and intensity of illumination to which the animal has previously been exposed and the part of the animal illuminated, seem to be the most relevant.

510 J. C. CASTILLA

A typical confusion in the literature can be cited in connection with Bohn's (1908) paper. Hyman (1955, p. 354), reviewing light responses of starfish, says 'variable results have been reported for *Asterias rubens*, found by Bohn (1908) to be generally negative to light, although readily altering its response'. Hyman's statement suggests that Bohn demonstrated a change in the response of *Asterias* to light, from negative to positive. However, according to Bohn's (1908) original paper, all populations of *Asterias rubens* avoid light; those collected from under stones avoid light by translation ('fuite'), while populations of *Asterias* collected on sand banks avoid light by adopting 'attitudes phototropiques', illustrated on p. 634 fig. 6. The change Bohn refers to is in the mode of avoidance. Van Weel (1935) working with an artificial light beam of approximately 1.5–2.0 mm reported one animal as having taken an upside down position after the beam had passed all along the radial nerve of one arm and reached the nervous ring. He did not give any interpretation to such behaviour. In some experiments using a light gradient, we have observed *Asterias* behaving in the way observed by Van Weel (1935). They twisted and curved the arms up towards the aboral side and finally adopted completely reversed positions.

Thus, starfish behaviour to light may not only include a reversible taxis, but also a variety of other responses according to the actual situation and experience of the animal. To classify species as photopositive or photonegative seems an unjustifiable oversimplification.

SUMMARY

(1) Responses to light of *Asterias rubens* were found to be different in freshly collected animals and those which had experienced periods of light or dark adaptation in the laboratory.

(2) Freshly collected animals showed a very strong negative response to light over a wide range of intensities.

(3) Light-adapted animals showed a strong negative response to light at first but, if adapted to continuous light, the negative response steadily reversed to a positive response. Feeding the animals did not cause any change in behaviour to light.

(4) *A. rubens* adapted to continuous darkness lost its photonegative behaviour, its response to light becoming indeterminate.

(5) At low temperatures in the range of 5.0–7.0 °C, there was less difference between light- and dark-adapted animals, neither showing a very clear response to light.

(6) The rate of movement under a light gradient increased with in-

creasing temperature of the water. At all temperatures the rate of movement was highest for freshly collected animals and lowest for dark-adapted animals.

(7) A clear positive rheotaxis (upstream behaviour) was found in *Asterias rubens* when placed in a current of ordinary laboratory sea water.

This work was carried out during the tenure of a British Council Scholarship. I am grateful to Professor D. J. Crisp for his encouragement and for reading and correcting the manuscript.

REFERENCES

BOHN, G. (1908). De l'acquisition des habitudes chez les étoiles de mer. *C. r. Séanc. Soc. Biol.* **64**, 633–5.

DIEBSCHLAG, E. (1938). Ganzheitliches Verhalten und Lernen bei Echinodermen. *Z. vergl. Physiol.* **25**, 612–54.

FISHER, R. A. & YATES, F. (1953). *Statistical Tables for Biological, Agricultural and Medical Research*, 4th ed. Edinburgh: Oliver and Boyd.

HYMAN, L. H. (1955). *The Invertebrates*, Vol. IV, Echinodermata. New York: McGraw-Hill.

JUST, G. (1927). Untersuchungen über Orstbewegungsreaktionen. I. Das Wesen der phototaktischen Reaktionen von *Asterias rubens*. *Z. vergl. Physiol.* **5**, 247–82.

MILLOT, N. (1954). Sensitivity to light and the reactions to changes in light intensity of the echinoid *Diadema antillarum* Phillipi. *Phil. Trans. R. Soc. Ser.* B **238**, 187–220.

PLESSNER, H. (1913). Untersuchungen über die Physiologie der Seesterne. I. Lichtsinn. *Zool. Jb.* (Abt. 3) **33**, 361–86. From Yoshida, M. (1966). In: *Physiology of Echinodermata*, pp. 435–64 (ed. R. A. Boolootian). New York: Interscience.

ROMANES, G. & EWART, J. (1881). Observations on the locomotor system of Echinodermata. *Phil. Trans. R. Soc. Ser.* B **172**, 829–85.

SMITH, J. E. (1950). Some observations on the nervous mechanisms underlying the behaviour of starfishes. *Symp. Soc. exp. Biol.* **4**, 196–220.

SMITH, J. E. (1965). Echinodermata. In *Structure and Function in the Nervous Systems of Invertebrates*, Vol. II, pp. 1519–58 (ed. T. H. Bullock & G. A. Horridge). San Francisco: W. H. Freeman and Co.

VAN WEEL, P. B. (1935). Über die Lichtempfindlichkeit der Ambulakralfüsschen des Seesterns (*Asterias rubens*). *Archs néerl. Zool.* **1**, 347–53.

YOSHIDA, M. (1966). Photosensitivity. In: *Physiology of Echinodermata*, pp. 435–64 (ed. R. A. Boolootian). New York: Interscience.

YOSHIDA, M. & OHTSUKI, H. (1968). The phototactic behavior of the starfish, *Asterias amurensis* Lütken. *Biol. Bull. mar. biol. Lab., Woods Hole* **134**, 516–32.

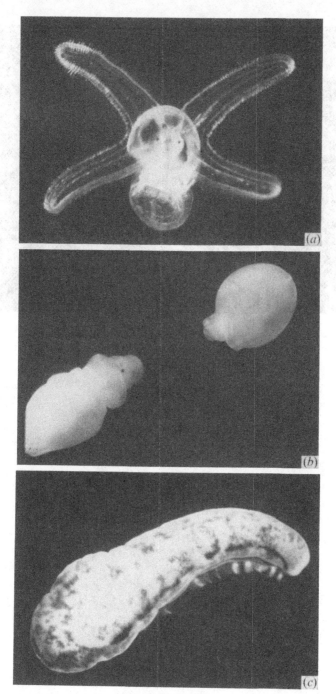

Scheltema: Plate 1

Teleplanic larvae from the warm-temperate and tropical Atlantic Ocean.

(a) Larva of *Philippia krebsii* (Archtectonicidae) with velum extended. The velar lobes increase in length at a later stage (after Robertson, Scheltema & Adams, 1970).

(b) Pelagosphaera larvae of the sipunculid designated 'smooth' by Jägersten (1963). The larvae are not swimming here and the introvert on one specimen is withdrawn.

(c) Semper's larva of *Zoanthella henseni* belonging to the genus *Palythoa*. Shown here is the stage at which the larvae were held for three months after capture.

(*Facing p.* 512)

Scheltema: PLATE 2

Veliger larva of *Cymatium parthenopeum* (Cymatiidae) taken from the Gulf Stream off the eastern coast of North America. The propodium of the foot is already partly developed (after Scheltema, 1966*b*).

This plate is available in colour for download from www.cambridge.org/9780521178259

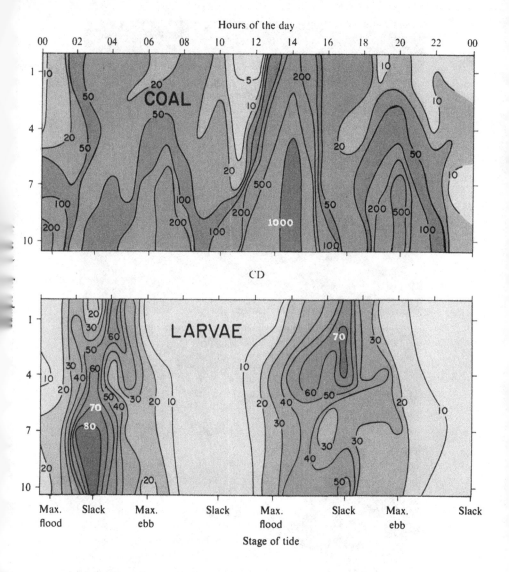

Wood and Hargis: PLATE I

Distribution of coal particles (upper diagram) and larvae (lower diagram) in relation to depth over a 24 h period at station CD. Contours show the numbers of larvae or particles per 100 l of water. The numbers of particles show four maxima, corresponding to flood and ebb tide, but the larvae show only two maxima, corresponding to periods of rising salinity. Simultaneous data from other two stations were markedly similar.

This plate is available in colour for download from www.cambridge.org/9780521178259

Moyse: Plate 1

(a) Live *Pyrgoma anglicum* on *Caryophyllia smithi*. A layer of coral flesh including acontia completely surrounds the two barnacles. As seen in this photograph the tentacles are only partly extended. (The coral had been kept in the laboratory for several weeks and was regenerating a second set of tentacles from its broken base.)

(b) Piece of *C. smithi* (flesh removed) with a large *P. anglicum* attached. Two cyprids have metamorphosed on the barnacle cone just beyond the adventitious scleroseptal plates.

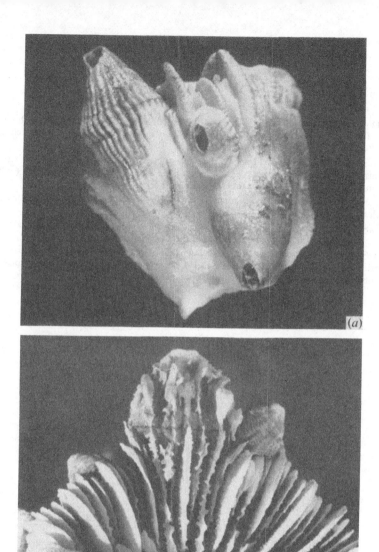

Moyse: PLATE 2

(a) *Pyrgoma* on *Caryophyllia*. The barnacle on the right had settled in the reversed position. Heterogonic growth of the rostral side of the basis has carried the top of the barnacle away from the coral rim. This is an extreme condition which would usually prove fatal but this particular coral had coenosarc extending well down the column thus protecting the barnacle.

(b) Close up of *Caryophyllia* with several attached *Pyrgoma*. Note the growth of coral sclerosepta on the cone of the oldest barnacles. These are attenuated or broken where they overlie the barnacle suture line.

(a) Distal region of the antennule showing the fourth segment (IV) arising from the third (III). There are sensory setae on the distal end of the fourth segment and on the attachment disc. Scanning electron micrograph. × 750.

(b) L.S. distal region of fourth segment. The cuticle of the segment is thinner around the base of terminal seta A (at 'x') than around the base of the subterminal seta (at 'y'). Within the segment can be seen tubes containing postciliary processes. × 3750.

(c) T.S. proximal region of the segment, showing dendrites containing mitochondria and vesicles. × 30 000.

(d) L.S. scolopidium showing two cilia, each arising from a dendrite; the extracellular space surrounding the cilia extends between the dendrites. × 20 000.

(e) T.S. distal region of two scolopidia. Each tube consists of an electron dense layer bounded by the sheath cell membrane and the membranes of the postciliary processes. The processes contain microtubules. × 30 000.

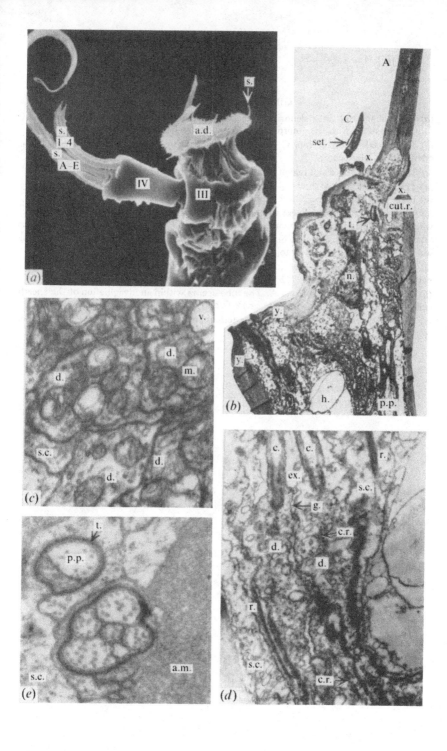

Gibson and Nott: PLATE 2

(a) L.S. ciliary root in a dendrite. Projections (some are arrowed) from the root may form connections with the amorphous material on the left of the micrograph. × 20 000.

(b) L.S. distal region of seta C showing postciliary processes; the seta appears to be open-ended. × 20 000.

(c) L.S. seta C. The distal region is out of the plane of the section. × 5 000.

(d) L.S. base of seta A showing the sheath cell, tube and postciliary processes entering the lumen of the seta. The cuticle around the base of the seta is thin ('x'). × 5 000.

(e) L.S. seta D. The thickness of the cuticle decreases distally in the region (arrow) where the regular external, cuticular ridges change to an irregular arrangement. × 5 000.

(f) L.S. subterminal seta showing circular bands of thickening within the thin cuticular wall. T.S. subterminal seta, at top right of micrograph. × 5 000.

(g) T.S. subterminal seta. The tube containing postciliary processes occurs eccentrically within the lumen. × 10 000.

(h) T.S. proximal region of seta A. The tube occurs within an invagination of the sheath cell membrane (arrowed) as in Plate 1 e. × 15 000.

(i) T.S. distal region of seta B. The postciliary processes are not enclosed in a tube. × 5 000.

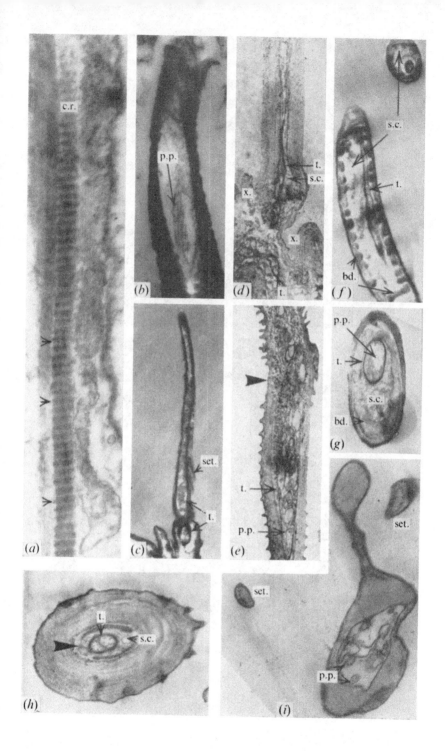

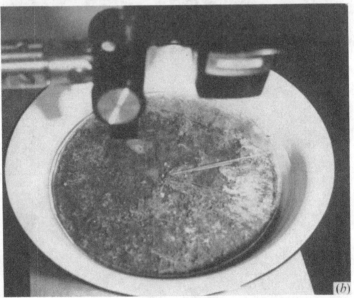

Pearce and Chess: PLATE I

(a) Prototype MDSA under construction; without supports and anchors. Disc materials include concrete, wood, glass, aluminum, steel and rubber.

(b) Subsampling a disc using a sector aliquot one-tenth the total upper surface of disc.

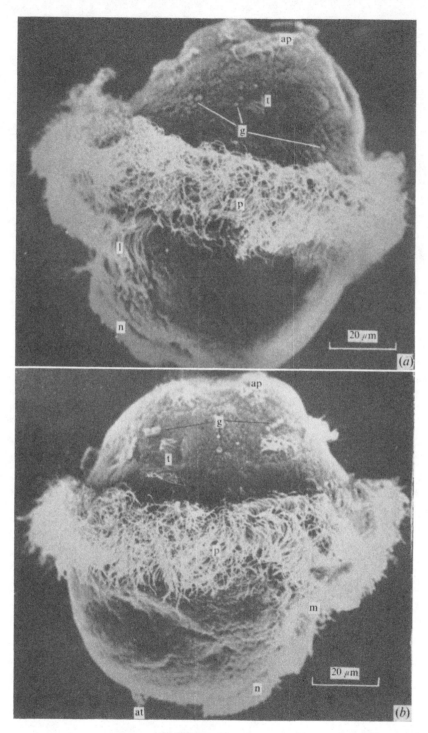

Holborow: PLATE I

Scanning electron micrographs of a trochophore. (*a*) Left-hand side view. (*b*) Right-hand side view. ap, apical cilia; t, small lateral tufts of cilia, probably those associated with the eyes; g, projecting pores of glands; p, prototroch; l, tongue of long cilia on the left-hand side of the mouth; m, position of mouth; n, neurotroch; at, anal tuft to the right posterior of the anus.

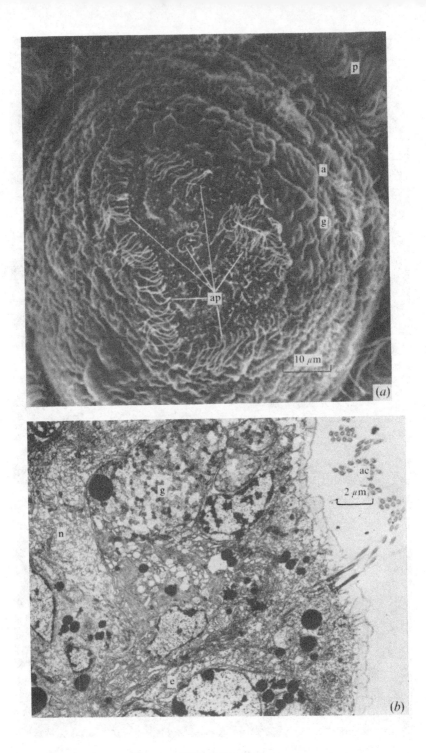

Holborow: PLATE 2

(*a*) Scanning electron micrograph of the upper hemisphere. ap, apical cilia; g, set of four gland pores at right-hand end of akrotroch; a, akrotroch; p, prototroch.

(*b*) Transmission electron micrograph of apical cells. ac, apical cilia; e, tapering extension of apical cell; g, glandular tissue; n, bundle of nerves.

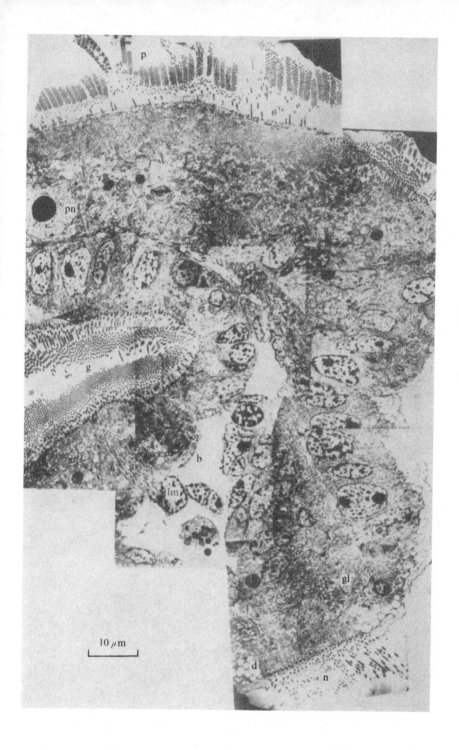

Holborow: PLATE 3

Panorama to show overall organization. p, prototroch cilia in orderly aggregations; pn, prototroch nucleus; b, blastocoel; g, gullet lined with cilia; lm, larval mesoderm cell; gl, glandular cell; n, neurotroch cilia; d, group of droplets on either side of neurotroch.

Holborow: PLATE 4

(*a*) Panorama of gullet (g) and stomach (s). t, area of thin tissue in stomach wall; c, cilia; mv, microvilli; ec, expanded cilia of gullet; r, oriented rootlets of normal gullet cilia.

(*b*) Expanded cilia of the gullet (ec); ax, axoneme with normal 9+2 configuration; f, extra filaments; m, fused membranes; nc, normal cilia.

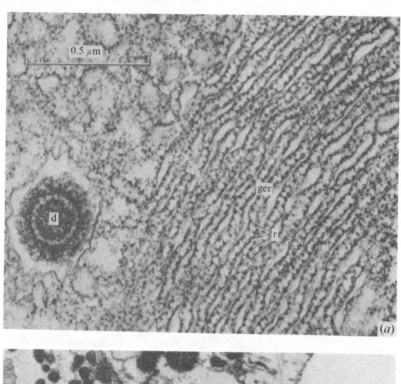

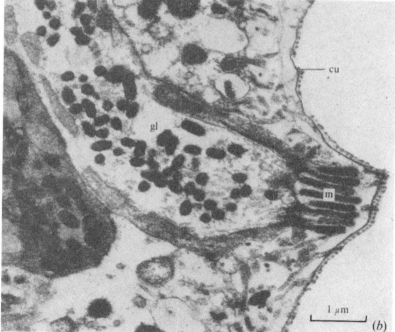

Holborow: PLATE 5

(*a*) A stomach secretory cell. ger, granular endoplasmic reticulum; r, ribosomes; d, secretion droplet.

(*b*) Superficial secretory cell (gl); m, microvilli supporting pore; cu, cuticle.

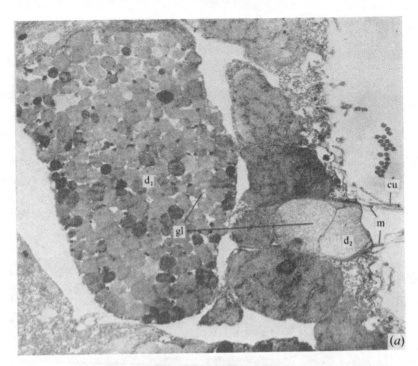

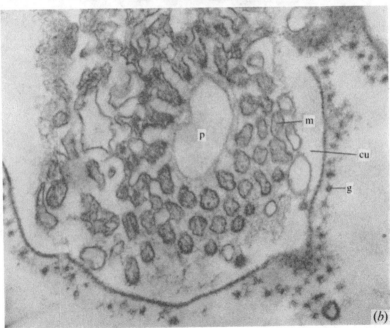

Holborow: PLATE 6

(a) Another type of superficial secretory cell (gl). d_1, immature droplets; d_2, mature droplets about to be released; m, microvilli supporting pore; cu, cuticle extending pore.

(b) Transverse section of pore of superficial gland. p, pore; m, microvilli; cu, cuticle; g, granules on external side of cuticle.

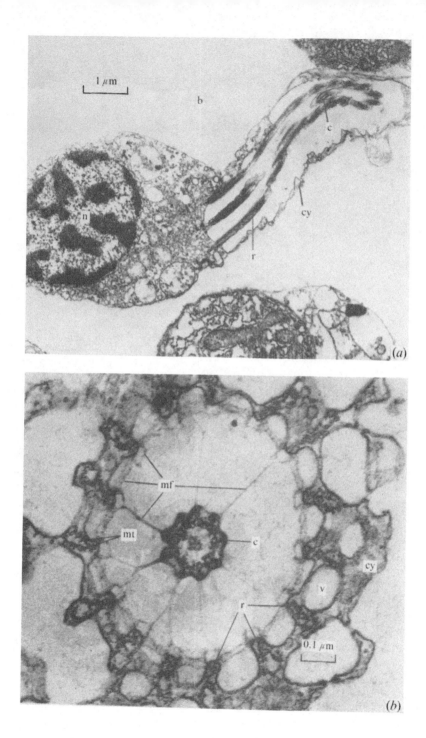

Holborow: PLATE 7

(a) Longitudinal section of part of solenocyte. b, blastocoel; c, cilium; r, rods of solenocyte tube; cy, cytoplasm surrounding tube; n, nucleus.

(b) Transverse section of solenocyte. c, cilium; mf, microfilaments; r, rods; mt, microtubules in rods; v, vacuoles; cy, cytoplasm.

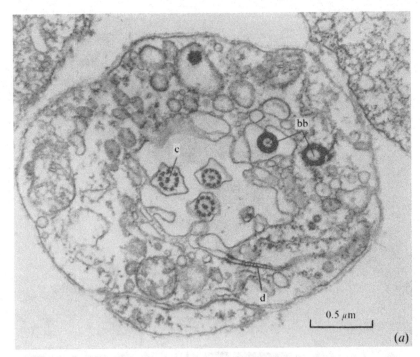

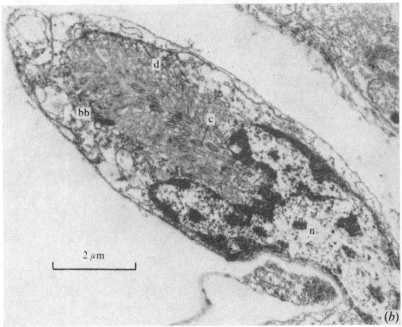

Holborow: PLATE 8

(*a*) Transverse section of duct of protonephridium. c, cilium; bb, basal bodies; d, septate desmosome.

(*b*) Transverse section of a lower portion of the protonephridial duct. c, cilium; bb, basal body; d, desmosome; n, nucleus.

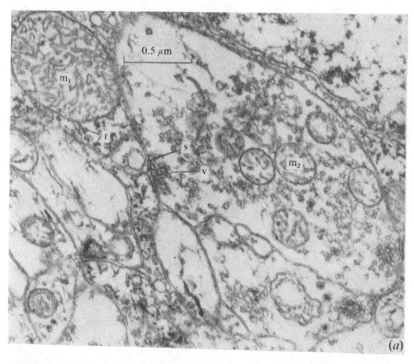

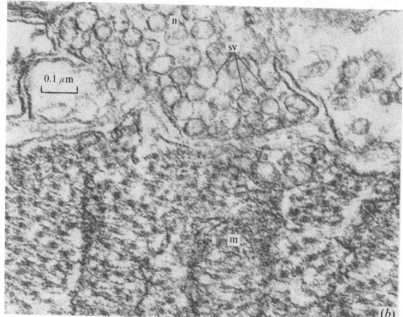

Holborow: Plate 9

(a) Neurociliary synapse (s). v, synaptic vesicles; m_1, mitochondrion of prototroch cell; r, ribosomes (in prototroch cell but not nerve cells); m_2, mitochondrion of nerve cell.

(b) Neuromuscular junction. n, nerve; m, muscle; sv, synaptic vesicles.

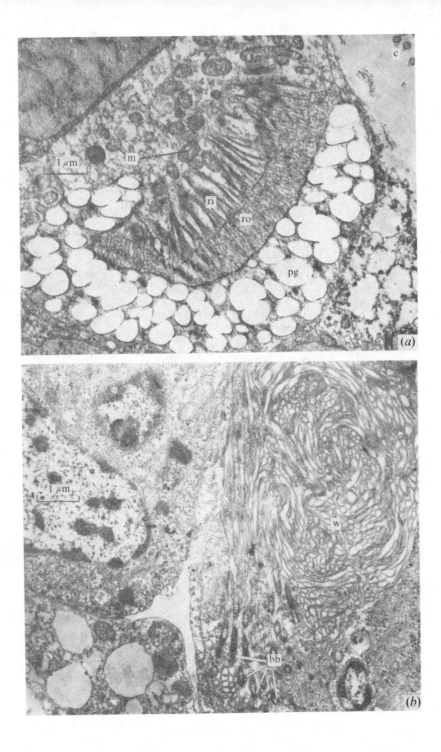

Holborow: PLATE 10

(a) Larval eye. pg, pigment granules of pigment cup; ri, inner layer of retina; ro, outer layer of retina; m, mitochondria of light sensitive cell; c, cilia.

(b) Problematic body, possibly developing adult eye. w, whorl of membranes; bb, basal bodies of cilia from which the whorl of membranes arise.

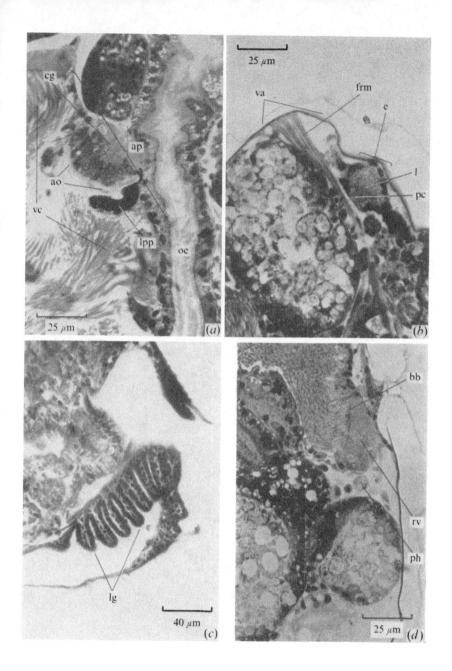

Hickman and Gruffydd: Plate 1

(a) Pediveliger. Sagittal section of the retracted velum and apical organ.

(b) Pediveliger. Vertical section of the eye and surrounding tissue.

(c) Well developed left gill at 48 h.

(d) Degenerating velum at 6 days.

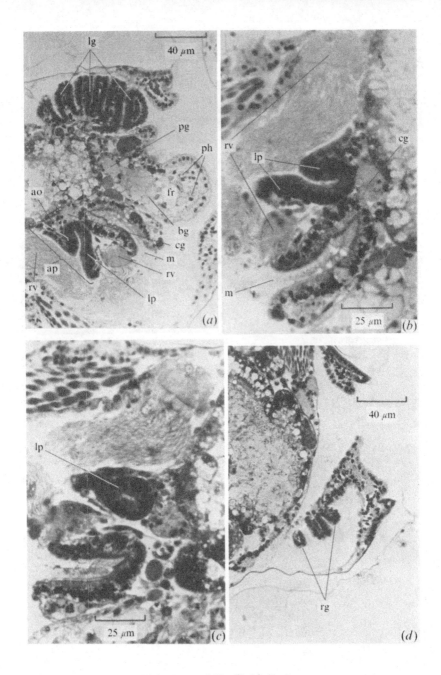

Hickman and Gruffydd: PLATE 2

(a) Vertical oblique section of 6-day-old spat.

(b) 6-day-old spat; sagittal section of the degenerating velum and the labial palp.

(c) 6-day-old spat. Vertical section showing the tubular nature of the distal portion of the labial palp.

(d) Right gill at 6 days.

 Key to abbreviations: ao, apical organ; ap, apical plate; bb, basal bodies; bg, byssus complex; cg, cerebral ganglion; dd, digestive diverticulum; e, eye; fr, remnants of foot; frm, foot retractor muscle; l, lens; lg, left gill; lpp, labial palp precursor; pc, pigment cup; pg, pedal ganglion; ph, phagocytes; rg, right gill; rv, remnants of ciliated crown of velum; va, shell valve; vc, large velar cilia.

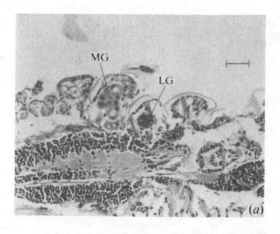

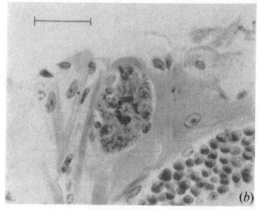

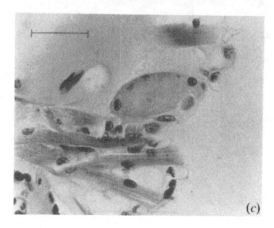

Hubschman: Plate I

(a) Frontal section of *Palaemonetes kadiakensis*. Form I zoea showing left side at level of the abdominal ganglia. LG, larval gland in base of maxilla I, MG, maxillary gland in base of maxilla II. Scale 100 μm, Haematoxylin–Eosin.

(b) Larval gland in *Palaemonetes pugio* Form I zoea. Scale 100 μm, Paraldehyde Fuchsin.

(c) Larval gland in *Palaemonetes intermedius* Form II zoea, Scale 100 μm. Azan.

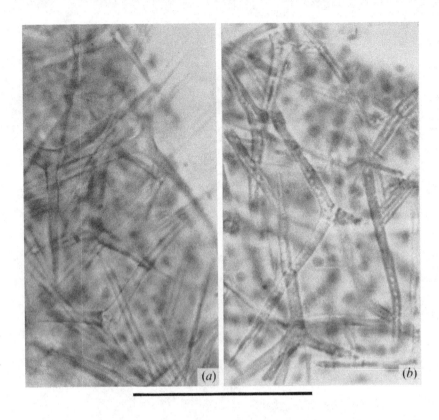

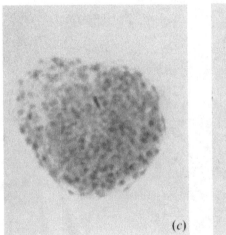
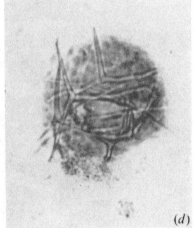

Jones: PLATE 1

(*a*, *b*) Spicule corrosion in living olynthi after 11.75 h immersion at 14 °C in, respectively, solutions 2 and 7. Note the coarser corrosion in the latter, bicarbonate-deficient, solution. The same two types were observed when alcohol-fixed olynthi were left in the same solutions. The line indicates 50 μm. (*c*, *d*) Recovery of spicule production on return to sea water. The depicted spongelets were attached to a pair of coverslips which were immersed in a bicarbonate-deficient solution for 24 h to remove the existing spicules, then in S 8 for 48 h at 14 °C, after which one (*c*) was fixed and the other (*d*) transferred to sea water for a further 48 h. Note that spicule formation does not require the presence of a functional water-conducting system. The line indicates 50 μm.

DIURNAL RHYTHMS IN SNAILS
AND STARFISH

VIVIEN M. THAIN

School of Biological Sciences, University of East Anglia, Norwich

INTRODUCTION

This work is concerned with three species of animals – *Gibbula cineraria* (L.), *Littorina littorea* (L.) and *Asterias rubens* L. – which live in the shallow sublittoral and intertidal zones. These animals were observed to undergo diurnal migrations and my intentions were to analyse the behavioural mechanisms of these migrations and to investigate their ecological significance.

Gibbula cineraria, a trochid gastropod, occurs in large numbers in the shallow sublittoral zone in parts of Lough Ine, County Cork, Ireland, where all the investigations on *Gibbula* were carried out. The animals were found to undergo a diurnal migration, moving to the tops of the rocks at dawn and to their undersides at dusk. Much work has been carried out on the ecology of Lough Ine and this work has recently been reviewed by Kitching & Ebling (1967). Lough Ine, because of the narrow tidal range, clear water and frequent calm conditions, is an ideal place for the continuous observation of sublittoral animals in their natural habitat.

Work on the gastropod *Littorina littorea*, and the asteroid *Asterias rubens*, was carried out on the shore at West Runton, Norfolk, and under laboratory conditions at the University of East Anglia. These animals could be observed on the shore only at low tide, owing to the turbidity of the water. Both *Littorina* and *Asterias* performed diurnal migrations on the shore, moving to the tops of the rocks at dusk and to their undersides at dawn. These migrations were in the opposite direction to the migration of *Gibbula* at Lough Ine.

Previous work on the behaviour of shore molluscs has been reviewed recently by Allen (1963) and by Newell (1964). Much of this work has been concerned with *Littorina* spp. and a fuller review of the work on these species has been given by Newell (1958 b).

Work on the responses of echinoderms to light and gravity has been reviewed by Yoshida (1966) and Reese (1966) respectively, but reports are conflicting.

The diurnal migrations described in the present work for all three species of animals have not previously been reported.

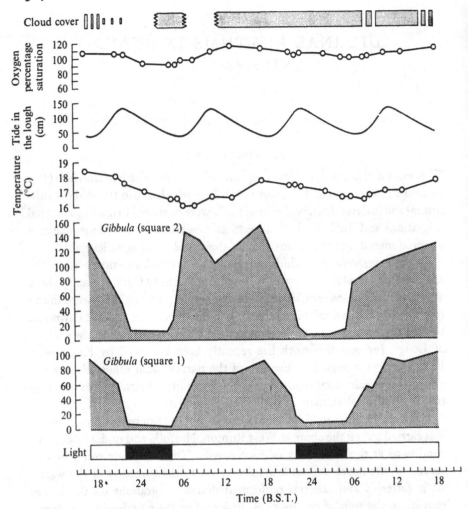

Fig. 1. Diurnal migration of *Gibbula* on the sea bed. Numbers of *Gibbula* visible in two separate 1 m squares from 19 to 21 July 1965, with parallel records of cloud cover, percentage saturation with dissolved oxygen, tide level, water temperature and light conditions. In this figure, and in others where the lighting conditions are represented by a strip of black and white areas, the black areas indicate darkness and the white areas indicate illumination.

GIBBULA CINERARIA

Diurnal migration of Gibbula *on the sea bed*

The behaviour of *Gibbula* was investigated under natural conditions on the sea bed by counting the number of animals visible in a marked area (1 metre square) of the sea bed, over periods from 24 to 48 h.

Gibbula showed a rhythmic diurnal migration, most animals being on top

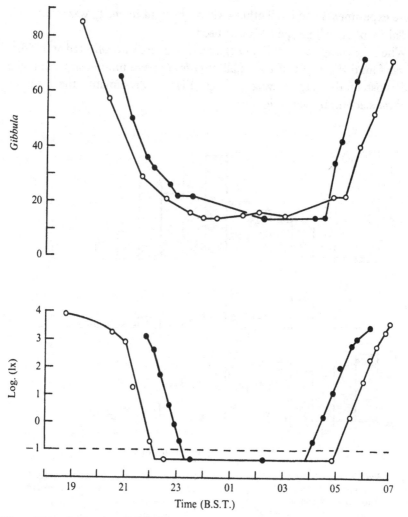

Fig. 2. Diurnal migration of *Gibbula* on the sea bed at different times of year. Numbers of *Gibbula* visible in two separate 1 m squares with a parallel record of light intensity: ●, 1–2 July 1966; ○, 15–16 August 1966. The horizontal dashed lines on the light intensity plots in this figure, and in Figs. 4, 7, 8, 9 and 10, indicate the limit of sensitivity of the photocell.

of the rocks by day and underneath them at night (Fig. 1). The movements correlated exactly with the increase and decrease of light intensity and not with changes in oxygen concentration, temperature or tide level. Two experiments conducted a week apart showed that the movements were unaffected by the change of the times of high and low tide.

At different times of the year the timing of the migrations varied so that they always occured at exactly the same time as dawn and dusk (Fig. 2). In these experiments and in all others where the light intensity was measured, a Clairex photocell (Type 5 H) was used.

When the complete population of a metre-square was collected at midday it was found that most of the smallest *Gibbula* were underneath the rocks and most of the largest were on top (Fig. 3). Presumably the smallest *Gibbula* seldom, if ever, migrate.

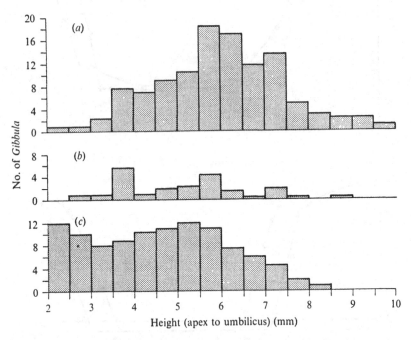

Fig. 3. Size distribution (numbers/m²) of complete populations of *Gibbula* averaged over two 1 m squares, determined by day on 17 August 1965. (*a*) tops of rocks, (*b*) sides of rocks, (*c*) underneath rocks.

Activity of Gibbula

The activity of *Gibbula* on the sea bed was recorded by watching ten marked animals on an area of the sea bed which consisted of small stones where the animals could not get out of sight. The positions of these animals were recorded at hourly intervals over 24 h. The results (Fig. 4) show that besides the rhythmic migration, *Gibbula* showed an activity rhythm, the animals being more active by day and less active by night.

In the above experiment torch light was used to observe the animals at night. To eliminate this possible source of error, the positions of individually marked *Gibbula* in a laboratory tank were recorded at 5 min inter-

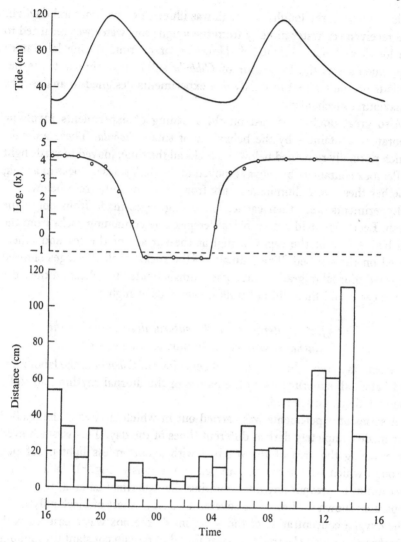

Fig. 4. Activity of *Gibbula* on the sea bed. Total distance (cm) moved in each hour by 9 marked *Gibbula* from 4 to 5 July 1966, with parallel records of tide and light intensity.

vals during the night, using only infra-red illumination and Leitz infra-red viewing equipment. This experiment confirmed that most *Gibbula* were stationary at night and moving by day.

Behaviour of Gibbula *in laboratory aquaria and in cages in the sea*

In an attempt to study the diurnal migration of *Gibbula* under controlled conditions, a laboratory tank was set up with conditions as similar as possible to those in the lough. The tank was illuminated with natural daylight and received sea water directly from the lough; this water was returned to the lough and not recirculated. However, at no time during laboratory experiments was the behaviour of *Gibbula* sufficiently similar to its behaviour on the sea bed to justify any experiments designed to analyse the behavioural mechanisms.

Also great doubt was cast on the meaning of experiments involving laboratory containers by the behaviour of small *Gibbula*. These animals, which normally remained below the rocks all the time, moved towards light in Perspex containers but always moved to the dark sides of rocks in a tank, whether these were illuminated only from above or only from below.

Experiments were then carried out using cages made from polythene mesh. Each cage held a deep black Perspex tray containing rocks from the sea bed. Some of the cages floated at the surface of the sea and others rested on the sea bed. The animals on the rocks in all the cages showed a normal diurnal migration but some animals tended to collect on the sides of the cages and then did not walk downwards at night.

The effect of alterations in the natural illumination on the diurnal migration of Gibbula *on the sea bed*

As attempts to reproduce the natural behaviour of *Gibbula* in the laboratory had failed, all experiments on the nature of the diurnal rhythm had to be conducted on the sea bed.

A series of experiments was carried out in which an area of the sea bed was made completely dark at different times of the day. This was achieved by covering the area of the sea bed with a bottomless light-proof box through which sea water was pumped continuously (Fig. 5). As a further precaution to prevent any light entering the experimental area, the whole apparatus shown in Fig. 5 was covered by a large sheet of black polythene. The oxygen concentration of the water inside the box was measured with a Makareth oxygen electrode and was found to remain constant throughout the experiments.

When darkness was imposed from before the first light of dawn until after dawn (Fig. 6B), *Gibbula* showed an upward migration similar to that in the control square. That this upward migration was not caused by the presence of the box can be concluded from the result of the experiment shown in Fig. 6A, which shows that when the experimental square was covered by the box during the night, no upward migration occurred. Since the temperature and oxygen concentration were constant in the experiment at dawn (Fig. 6B), the upward migration appears to be endogenous.

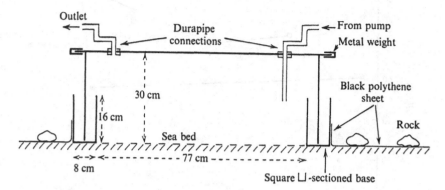

Fig. 5. Diagram of the apparatus used to make an area of the sea bed completely dark during the day.

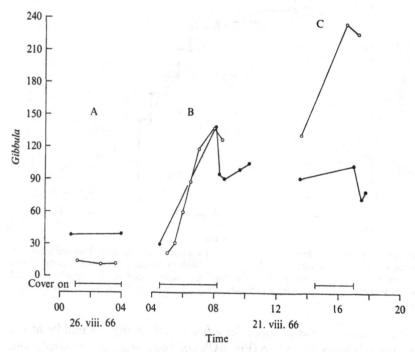

Fig. 6. Imposition of darkness on *Gibbula* on the sea bed. Numbers of *Gibbula* visible in two separate 1 m squares: ●, experimental square; ○, control square.

(A) Experimental square covered during darkness
(B) Experimental square covered over dawn
(C) Experimental square covered during full daylight.

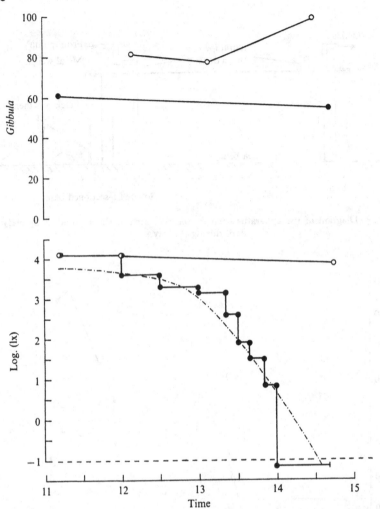

Fig. 7. Imposition of darkness on *Gibbula* on the sea bed. Light intensity lowered gradually during full daylight, 5 July 1967. Numbers of *Gibbula* visible and light intensity in two separate 1 m squares: ●, experimental square; ○, control square; ◐, experimental and control squares; —·—·, fall in light intensity at natural dusk, displaced for comparison with artificial dusk.

When darkness was imposed for a few hours during the middle of the day, *Gibbula* did not go down (Fig. 6C). *Gibbula* still did not go down even when the decreasing light of dusk was imitated in the middle of the day by means of neutral density Cinemoid filters fitted in a cumulative series to the top of a second box (Fig. 7). However, *Gibbula* did go down below the rocks when the sea bed was darkened over the period of dusk (Fig. 8).

In some further experiments an area of the sea bed was illuminated with

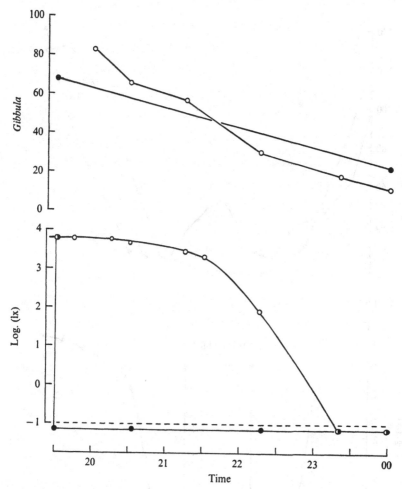

Fig. 8. Imposition of darkness on *Gibbula* on the sea bed. Darkness imposed over dusk, 1 July 1967. Numbers of *Gibbula* visible and light intensity in two separate 1 m squares: ●, experimental square; ○, control square; ◑, experimental and control squares.

blended mercury-tungsten lamps. Under artificial illumination in the middle of the night only very few *Gibbula* came up and most of these went down again before the lights were switched off (Fig. 9). However, when the sea bed was illuminated just before dawn, upward migration occurred earlier than the migration in the control square (Fig. 10).

Thus the results shown in Figs. 6C to 10 suggest that artificial changes in the conditions of illumination will evoke the upward and downward migrations only when these changes occur at or near the 'correct' time of day.

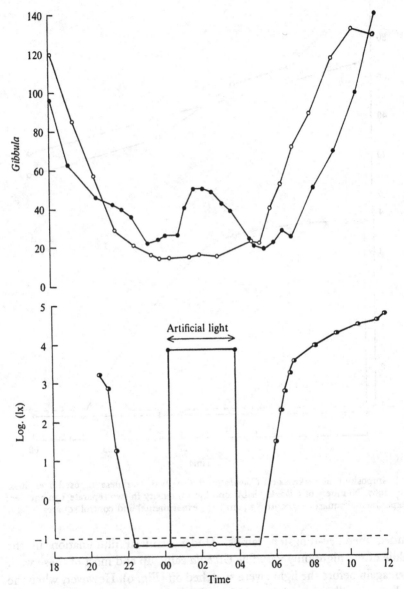

Fig. 9. Illumination during the night of *Gibbula* on the sea bed. Illumination in middle of night on 16 August 1966. Numbers of *Gibbula* and light intensity in two separate 1 m squares: ●, experimental square; ○, control square; ◑, experimental and control squares.

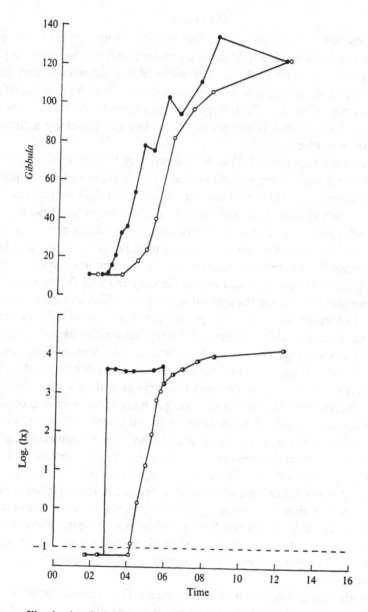

Fig. 10. Illumination during the night of *Gibbula* on the sea bed. Illumination just before dawn on 11 July 1966. Numbers of *Gibbula* and light intensity in two separate 1 m squares ●, experimental square; ○, control square; ◑, experimental and control squares.

Discussion

It appears that the normal stimuli for both the migratory and activity rhythms are the increasing light intensity at dawn and the decreasing light intensity at dusk. The evidence supporting this conclusion is that the movements of the animals coincided exactly with these light changes at different times of the year (July, August and November), and that the onset of activity and upward migration could be brought about by artificial illumination just before dawn.

The upward migration of *Gibbula* at dawn may be just a result of the increase in activity of the animals at that time. This would cause many animals to move on to the tops of the rocks where they might collect because the food supply is more abundant there than below the rocks. However, as it was not possible to analyse the movement in the laboratory it is not known whether a positive phototaxis or a negative geotaxis are also involved. Since the upward movement can occur at dawn in complete darkness, a positive phototaxis could not be the only cause of the movement.

It was argued earlier that the upward migration of *Gibbula* might be endogenous. In view of the mechanism proposed above to explain the upward migration it was probably only the activity rhythm which was endogenous.

The downward migration is very difficult to explain. It cannot be a result of the cessation of activity; some directional response must be involved to cause all the animals to go underneath the rocks at dusk. Two possibilities are that the movement is either a negative phototaxis or a positive geotaxis, but the response could not be analysed in the laboratory. However, when the sea bed was darkened over the period of dusk, all the downward migration occurred in total darkness and this suggests a positive geotaxis rather than a negative phototaxis. Therefore to explain the results one has to postulate that the animals are stimulated to become positively geotactic by a sudden or a gradual decrease in light intensity but only when this occurs at the time of dusk. It has not been possible to investigate whether the animals would still become positively geotactic at dusk under conditions of constant light or constant darkness (i.e. whether the downward migration is endogenous).

One other interesting result from this work is the difference between the behaviour of *Gibbula* under natural conditions on the sea bed and the behaviour of *Gibbula* in Perspex containers and on the sides of cages. One possible cause is the absence of food on the sides of the containers. When *Gibbula* was kept in a cage floating in the sea, without food for 1 day and then returned to the sea bed just before dusk, they did not go below the rocks. By the second or third night their behaviour had returned to normal. It could be that the animals do not walk downwards at dusk when they are starved.

LITTORINA LITTOREA

Diurnal migration and activity of Littorina *on the shore and in the laboratory*

Littorina on the Norfolk shore were on top of the rocks at night and underneath the rocks by day and they showed this diurnal migration throughout the year.

The migration continued in a laboratory aquarium under a diurnal light regime, the upward and downward movements occurring simultaneously with the changes in light intensity (Fig. 11a). Since the temperature and

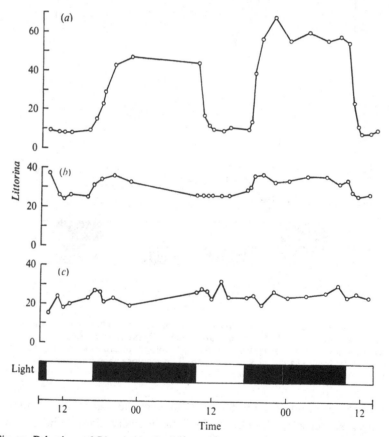

Fig. 11. Behaviour of *Littorina* in the laboratory tank. Number of *Littorina* visible out of a total of 140 large animals and 33 small animals. (*a*) Large *Littorina* on top of rocks. (*b*) Large *Littorina* on sides of tank. (*c*) Small *Littorina* on top of rocks. From 30 January to 1 February 1968 with a parallel record of light conditions. Temperature 10.2±0.1 °C.

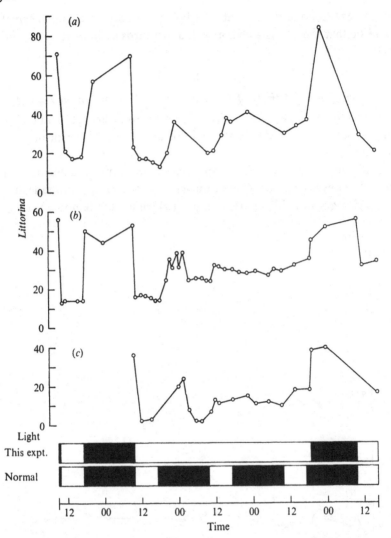

Fig. 12. Continuous illumination of *Littorina* in the laboratory. (*a*) Number of *Littorina* on top of rocks out of a total of 160 (6–10 March 1968). (*b*) Number of *Littorina* on top of rocks out of a total of 150 (27–31 March 1968). (*c*) Number of *Littorina* moving on top of and underneath the rocks out of a total of 150. The readings shown in plots (*b*) and (*c*) were taken during the same experiment. Temperature 11.0±0.2 °C in both experiments.

oxygen concentration of the water were constant the migration was almost certainly a response to the change in light intensity alone. The animals resting on the sides of the aquarium did not show a diurnal migration (Fig. 11*b*) nor did the very small animals of height less than 7 mm (Fig. 11*c*). During the night the animals were observed with the aid of

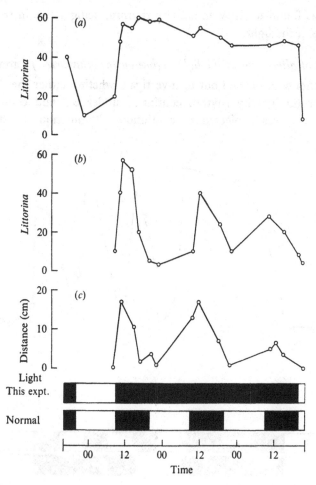

Fig. 13. Exposure of *Littorina* in the laboratory to continuous darkness. (*a*) Number of *Littorina* on top of rocks out of a total of 115. (*b*) Number of *Littorina* moving on top of and underneath rocks out of a total of 115. (*c*) Average distance moved in periods of 1 h by animals on the rocks out of a total of 12 marked individuals. From 8 to 11 April 1968, with a parallel record of light conditions, and at a temperature of 10.6 ± 0.2 °C. All plots refer to the same experiment.

infra-red light and Leitz infra-red viewing equipment. In all laboratory experiments the temperature of the sea water was between 8 and 12 °C with a maximum variation of ± 0.5 °C during any one experiment. Artificial illumination was provided by a combination of fluorescent and tungsten strip lamps.

The activity of *Littorina* was recorded by plotting at intervals the positions of individual animals on plans of the rocks in the laboratory tank.

Littorina was found to show an activity rhythm, being active in the dark and inactive in the light.

The effect of alterations in the light regime on the behaviour of Littorina

Experiments were carried out to investigate whether either the diurnal migration or the activity rhythm continued under constant conditions. When *Littorina* was subjected to continuous illumination the diurnal

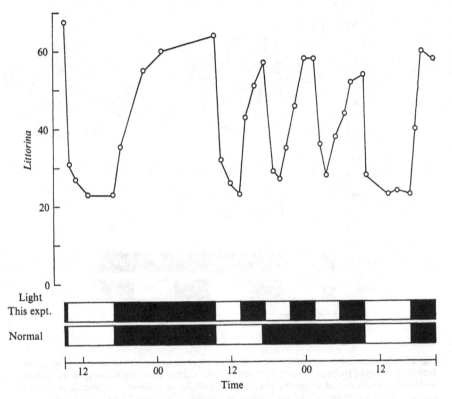

Fig. 14. Exposure of *Littorina* in the laboratory to alternating periods of 4 h light and 4 h darkness. Number of *Littorina* on top of rocks out of a total of 140 with a parallel record of light conditions, from 13 to 15 February 1968.

migration appeared to continue for at least one cycle (Fig. 12*a*, *b*). The activity rhythm, recorded as the number of animals moving at each observation, also showed this pattern (Fig. 12*c*). Under constant darkness the migration did not continue (Fig. 13*a*) but the activity rhythm did appear to continue (Fig. 13*b*, *c*). When the snails were subjected to periods of 4 h of light followed by 4 h of darkness, the animals showed a migration coincident with this new light regime (Fig. 14).

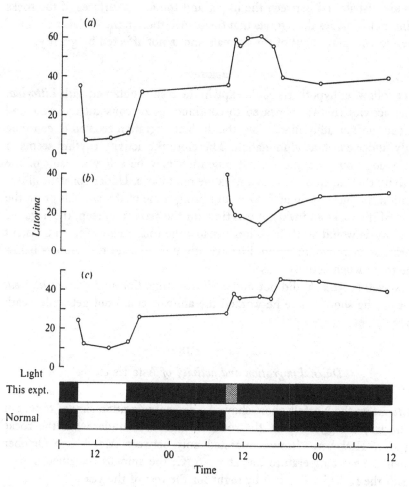

Fig. 15. Exposure of *Littorina* in the laboratory to illumination from below the tank. (*a*) Number of *Littorina* on top of rocks. (*b*) Number of *Littorina* underneath rocks. (*c*) Number of *Littorina* on sides of tank out of a total of 115 animals, from 4 to 6 April 1968, with a parallel record of light conditions. All plots refer to the same experiment. Stippled area represents illumination from below the tank.

The responses of Littorina to light and gravity

To investigate whether the downward movement of *Littorina* in the light was a response to light or gravity, experiments were carried out in which the rocks in tanks were illuminated from either above or below. It was found that the animals moved rapidly either upwards or downwards away from light (Fig. 15). Therefore the movement is presumably a negative phototaxis unaffected by gravity. In the dark the animals became approximately evenly distributed between the upper and the lower surfaces of the rocks (Fig. 15). This result suggests that the distribution in the dark is a result of the random movement of the animals and is not affected by gravity.

Discussion

The following hypothesis could explain the results obtained with *Littorina*. The activity rhythm appeared to continue under constant darkness and under constant illumination but the diurnal migration seemed to continue only under constant illumination. Therefore the activity rhythm seems to be endogenous and the diurnal migration may be solely a result of the activity rhythm coupled with a negative phototaxis. Under constant illumination the increase in activity would bring some of the animals on to the tops of the rocks and when it was time for the activity to stop, the negative phototaxis would send the animals below the rocks. However, in constant darkness there would be no directive stimulus to send the animals below the rocks when activity ceased.

Newell (1958 a, b) did not find a diurnal migration in *Littorina littorea*, but on the shore where he worked the animals could not get underneath the pebbles.

ASTERIAS RUBENS

Diurnal migration and activity of Asterias *on the shore and in the laboratory*

Asterias on the Norfolk shore showed a diurnal migration similar to that of *Littorina*, being on top of the rocks at night and underneath the rocks by day. However, *Asterias* migrated only between March and October when the sea temperature was above 6 °C; the animals remained underneath the rocks by day and by night for the rest of the year.

Asterias continued to show this diurnal migration in a laboratory aquarium with a diurnal light regime and, as with *Littorina*, the upward and downward migrations were found to be responses to changes in light intensity only (Fig. 16). Even when the animals were collected between November and February, at which time they were inactive on the shore, they performed a normal diurnal migration when placed in the laboratory at a temperature between 8 and 12 °C.

Infra-red light was used to observe the animals at night and was shown to have no effect on their behaviour. The temperature of the sea water was kept constant during all the laboratory experiments and the artificial illumination was the same as that used in experiments with *Littorina*.

The activity of individual *Asterias* was recorded by plotting the positions of animals on plans of the rocks in a laboratory tank. During a normal diurnal light regime nearly all the starfish were stationary underneath the rocks during the light period and as soon as the lamps were switched off most of the animals started to move. The animals continued to move during the dark period except when they were feeding.

The effect of alterations in the light regime on the behaviour of Asterias
When, after being under a normal diurnal light regime, *Asterias* was subjected to a period of 64 h of continuous darkness, the animals did not show any continuation of the rhythmic migration (Fig. 16). There was also no

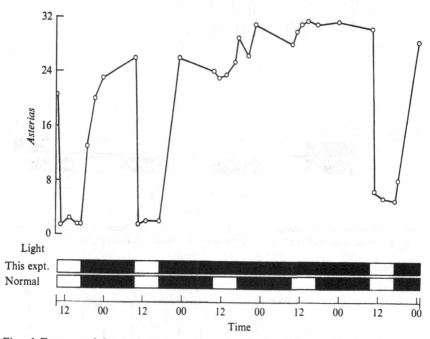

Fig. 16. Exposure of *Asterias* in the laboratory to continuous darkness. Number of *Asterias* visible out of a total of 60 with a parallel record of light conditions, from 2 to 5 March 1968.

apparent continuation of the activity rhythm under conditions of constant darkness. Conditions of continuous illumination also caused the rhythmic migration to stop (Fig. 17), but the number of *Asterias* on top of the rocks gradually increased during the light period. When *Asterias* was subjected to a light regime of 4 h of illumination followed by 4 h of darkness, the upward and downward migrations followed the new light regime and there was no evidence of any continuation of the original *diurnal* migration (Fig. 18).

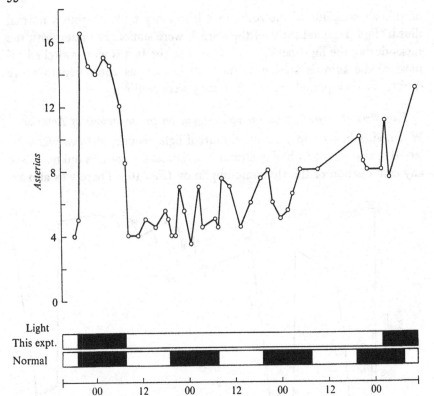

Fig. 17. Continuous illumination of *Asterias* in the laboratory. Number of *Asterias* visible out of a total of 25 with a parallel record of light conditions, from 13 to 16 June 1967.

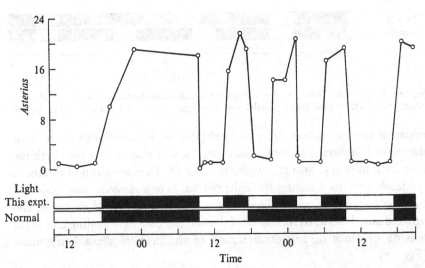

Fig. 18. Exposure of *Asterias* in the laboratory to alternating periods of 4 h light and 4 h darkness. Number of *Asterias* visible out of a total of 47 with a parallel record of light conditions, from 8 to 9 February 1968.

The responses of Asterias *to light and gravity*

A series of experiments, similar to those described for *Littorina*, was carried out to investigate whether the downward movement of *Asterias* in the light was a response to light or gravity. It was found that the animals moved rapidly either upwards or downwards away from light irrespective of their initial distribution; in the dark the animals became approximately evenly distributed between the upper and the lower surfaces of the rocks (Fig. 19). Therefore the movements appear to be a negative phototaxis in the light and a random movement in the dark, both being unaffected by gravity.

Feeding and the diurnal migration of Asterias

The effect of food on the diurnal migration of *Asterias* was investigated in two experiments. In the first experiment (Fig. 20), after the starfish had been starved for a month, large numbers continued to move around on top of the rocks by day. When food was given to these animals in the form of barnacles on only the tops of the rocks, three-quarters of the starfish remained feeding continuously on top of the rocks for 48 h. After this time the number of starfish remaining on top of the rocks by day decreased gradually. Thus starvation seems to decrease the photonegativity of the animals presumably because they are then moving around all the time, searching for food.

In the second experiment (Fig. 21), when barnacles covered all surfaces of the rocks about half the starfish were on top of the rocks at night. When barnacles were present on only the upper surfaces of the rocks nearly all the starfish were on top of the rocks at night, and when barnacles were present on only the lower surfaces of the rocks only one-third of the starfish were visible at night. Therefore the distribution of the starfish at night was directly dependent on the distribution of the food on the rocks.

Discussion

Neither the rhythmic migration nor the activity rhythm of *Asterias* showed any signs of being endogenous. The behaviour of this starfish appears to be the result of a negative phototaxis combined with a random search for food in the dark.

Just (1927) and Diebschlag (1938) found in laboratory experiments that *A. rubens* showed a positive response to light. This disagrees with the results presented here but many factors can affect the behaviour of *Asterias*. Castilla (1970) showed that freshly collected *Asterias* were negatively phototactic but that their response to light altered when they were

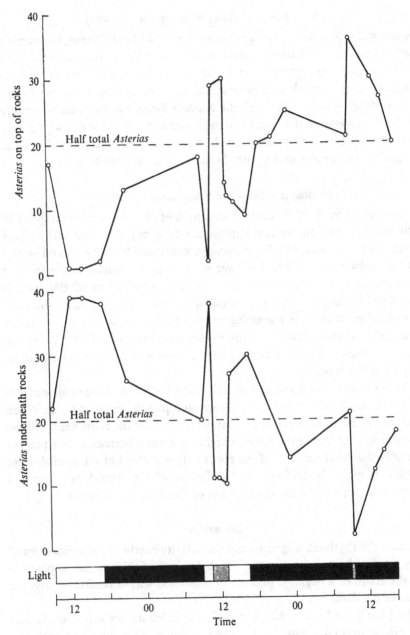

Fig. 19. Exposure of *Asterias* in the laboratory to illumination from below the tank. Numbers of *Asterias* on top of and underneath the rocks with a parallel record of light conditions, from 13 to 15 March 1968. Stippled areas represent illumination from below the tank.

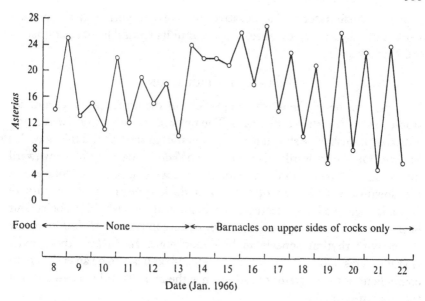

Fig. 20. Starvation of *Asterias* in the laboratory. A normal diurnal light regime was imposed and one reading of the number of *Asterias* visible out of a total of 28 was taken during each light period and each dark period.

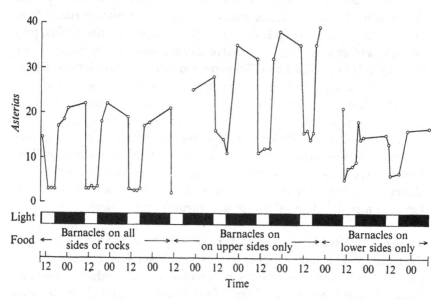

Fig. 21. Exposure of *Asterias* in the laboratory to different distributions of food. Number of *Asterias* visible out of a total of 45 with parallel records of light conditions and food distribution, from 15 to 24 February 1968.

kept in the laboratory. The negative phototaxis reported in the present work is the significant response of *Asterias* in its natural habitat, at least on the Norfolk shore.

GENERAL DISCUSSION

The mechanisms of the diurnal migrations of the three species of animals studied showed certain differences. The upward migrations of *Asterias* and *Littorina* at dusk were shown to be due to random activity and this may also be true for the upward migration of *Gibbula* at dawn. The downward migration of *Asterias*, and *Littorina*, at dawn were negative phototaxes but the downward movement of *Gibbula* at dusk appeared to be a positive geotaxis triggered by the decrease in light intensity at dusk. In the behaviour of *Asterias* there appears to be no endogenous component. In *Littorina* only the activity rhythm appears to be endogenous. In *Gibbula* the activity rhythm seems to be endogenous and there also appears to be an endogenous component in the migratory rhythm since the animals will only go down at the time of normal dusk.

The phenomenon of diurnal migration is obviously widespread amongst intertidal and shallow water animals and one would expect this behaviour to confer some advantage on the animals. Some of the food chains of the shallow sublittoral community at Lough Ine have been studied and it has been proposed that the diurnal migrations serve to separate predator from prey (Ebling *et al.* 1966). At Lough Ine, birds, which are active by day, prey on crabs and starfish which are active and exposed only at night. These latter animals in turn feed on *Gibbula cineraria* and *Paracentrotus* which are active and on top of the rocks by day. *Paracentrotus* and *Gibbula* are protected from birds by spines and hard camouflaged shells respectively. Thus it seems possible that the system of diurnal migrations helps to limit predation so that populations of both the predator and the prey may survive. The food chains operating in the intertidal community at West Runton have not yet been investigated, but sea birds have frequently been observed feeding there during the day. The migrations of *Asterias* and *Littorina* may therefore help to protect them from the birds.

The work reported here was carried out with the help of an S.R.C. Research Studentship. The author wishes to thank Professor J. A. Kitching for his continual help and encouragement. Figures 1 and 3 which previously appeared in the *Journal of Animal Ecology* are reproduced by the kind permission of the Editors and of Blackwell Scientific Publications.

REFERENCES

ALLEN, J. A. (1963). Ecology and functional morphology of molluscs. In *Oceanogr. mar. Biol. Ann. Rev.* 1, pp. 253–88 (ed. H. Barnes). London: George Allen and Unwin.

CASTILLA, J. C. (1970). Responses to light of *Asterias rubens* L. (This volume p. 495).

DIEBSCHLAG, E. (1938). Ganzheitliches Verhalten und Lernen bei Echinodermen. *Z. vergl. Physiol.* 25, 612–54.

EBLING, F. J., HAWKINS, A. D., KITCHING, J. A., MUNTZ, L. & PRATT, V. M. (1966). The ecology of Lough Ine. XVI. Predation and diurnal migration in the *Paracentrotus* community. *J. Anim. Ecol.* 35, 559–66.

JUST, G. (1927). Untersuchungen über Orstbewegungsreaktionen. I. Das Wesen der phototaktischen Reaktionen von *Asterias rubens*. *Z. vergl. Physiol.* 5, 247–82.

KITCHING, J. A. & EBLING, F. J. (1967). Ecological studies at Lough Ine. In *Adv. ecol. Res.* 4, pp. 197–291 (ed. J. B. Cragg). London: Academic Press.

NEWELL, G. E. (1958a). The behaviour of *Littorina littorea* (L.) under natural conditions and its relation to position on the shore. *J. mar. biol. Ass. U.K.* 37, 229–39.

NEWELL, G. E. (1958b). An experimental analysis of the behaviour of *Littorina littorea* (L.) under natural conditions and in the laboratory. *J. mar. biol. Ass. U.K.* 37, 241–66.

NEWELL, G. E. (1964). Physiological aspects of the ecology of intertidal molluscs. In *Physiology of Mollusca*, Vol. I, pp. 59–81 (ed. K. M. Wilbur & C. M. Yonge). London: Academic Press.

REESE, E. S. (1966). The complex behaviour of echinoderms. In *Physiology of Echinodermata*, pp. 157–218 (ed. R. A. Boolootian). New York: Interscience.

YOSHIDA, M. (1966). Photosensitivity. In *Physiology of Echinodermata*, pp. 435–64 (ed. R. A. Boolootian). New York: Interscience.

ORIENTATION TO LIGHT AND THE SHADING RESPONSE IN BARNACLES

L. FORBES, M. J. B. SEWARD AND D. J. CRISP

Marine Science Laboratories, Menai Bridge

INTRODUCTION

Visscher (1928) stated that the cyprids of *Balanus amphitrite* Darwin, *B. improvisus* Darwin and *Chthamalus fragilis* Darwin were negatively phototropic at the time of settlement, orienting the anterior end away from the light source. This view was disputed by McDougall (1943) working with *B. improvisus* Darwin, *C. fragilis* Darwin, and *B. eburneus* Gould, and by Pyefinch (1948) working with *B. balanoides* (L.) and *B. crenatus* Bruguière. These authors claimed that settlement occurred with the paired eye-spots facing the incident light. In 1951 the observations and experiments of Barnes, Crisp & Powell proved conclusively that barnacle cyprids remained photopositive up to and including the time of settlement, and that they settled with the anterior end towards the light. The contrary conclusions of Visscher may have resulted from confusion concerning the orientation of the metamorphosed barnacle. After metamorphosis the eyes are clearly visible beneath the rostral compartment in the position originally occupied by the posterior end of the settled cyprid and facing away from the source of light (Fig. 1).

Orientation to light is a clearly defined behaviour pattern in which the cyprid adjusts itself to the direction of light just prior to the discharge of cement which irrevocably fixes the animal in position (Knight-Jones & Crisp, 1953). Three factors appear to be important in determining the orientation: contour, light and water current; they form a hierarchy of stimuli in that order. The cyprid responds most clearly to the direction of grooves in the surface, its long axis being oriented in line with the groove (Crisp & Barnes, 1954). In the absence of a strong contour stimulus it responds to light, the anterior end of the cyprid pointing towards the source (Fig. 1). In the absence of contour stimuli and directional light, a very weak orientation has been detected which results in the posterior end of the cyprid being directed towards the current source. Orientation to water current is not detectable under normal circumstances (Crisp & Stubbings, 1957), but has been demonstrated by D. J. Crisp & P. S. Meadows (unpublished observations) in cyprids of *B. balanoides* and *B. crenatus* when settling in the dark on a smooth surface.

[539]

540   L. FORBES, M. J. B. SEWARD AND D. J. CRISP

Whereas the orientation to contour has an evident protective function enabling the cyprid to metamorphose in the shelter of a groove or crack, and the orientation to current is of clear benefit in allowing the concave side of the cirral net to face the oncoming stream of water, the reaction to light is less obviously adaptive. Some possibilities concerning its function have already been investigated. Barnes *et al.* (1951) found no significant differences in growth rates of *Elminius modestus* Darwin placed at different angles to the incident light, and individuals of cross-fertilizing species lying at different angles to the light and to each other have been found to contain fertilized eggs indicating that the reaction to light is not important in reproduction.

The nauplius eye, which is retained throughout the larval and adult life, comes to lie beneath the rostral plates as a result of the half forward somersault which the cyprid executes during metamorphosis (Fig. 1). In the newly metamorphosed barnacle the nauplius eye has separated into two parts which can be seen beneath the transparent rostral compartment, though when the compartment becomes calcified they are no longer visible. Nevertheless the adult eye may receive a certain amount of light transmitted through the semi-opaque shell, but when the barnacle opens its valves it is possible for light to enter the mantle space directly, passing between the terga and scuta and on to the light sensitive area below the rostrum as indicated in Fig. 1. The animal might then receive a greater proportion of light coming from the direction of the carina than from other directions in which the light has to pass through the semi-opaque wall plates. On this hypothesis, the passage of an opaque object between the animal and the light source would interrupt a greater proportion of the light incident on the eye if the source of light were situated at the carinal end of the animal rather than in any other position. Hence by orienting the carinal end to the light the barnacle might acquire a maximum sensitivity to shading by a predator large enough to interrupt part of the light from this quarter.

In this paper the above hypothesis is tested in two ways. First, barnacles orientated at different angles to the light were exposed to natural attack in aquaria by blennies. Secondly, barnacles were shaded from various directions about the rostrocarinal axis and the response to shading related to the orientation of the animal.

Throughout this investigation the anterior end of the barnacle will be defined as the direction of the anterior end of the cyprid before metamorphosis. It will therefore coincide after metamorphosis with the carinal (tergal) end of the barnacle. This direction will be assigned the orientation 0°. The rostral (scutal) direction will be taken as ± 180°. Deviations from 0°

in a clockwise direction will be read as positive, in an anticlockwise direction as negative (Fig. 1).

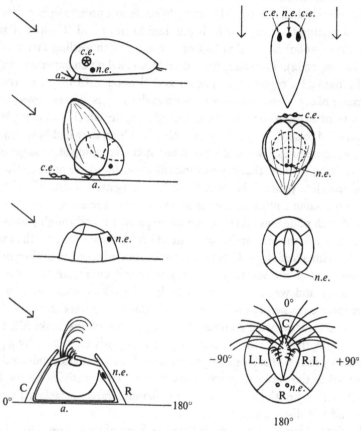

Fig. 1. The orientation of the cyprid to light (top) and the subsequent appearance of the barnacle during and after metamorphosis as viewed from the side (left) and from above (right). The conventional designation of the angle of orientation to light is also shown. The barnacle shown in the figure has its rostro-carinal axis oriented typically at 0° to the light source. *a*, antennule; *c.e.*, compound eye; *n.e.*, nauplius eye; C, Carinal; R, Rostral; R.L., L.L., right and left lateral compartments.

RELATIONSHIP BETWEEN ORIENTATION TO LIGHT
AND PREDATION BY *BLENNIUS PHOLIS* L.

Blennius pholis, a fish common on European rocky coasts, feeds largely on barnacles (Day, 1880–4; Smitt, 1892; McIntosh, 1904, 1905; Hartley, 1949; Qasim, 1957). They bite off the cirri of active barnacles (Soljan, 1932) and, according to Hartley, quoted by Wilson (1951), they can do so even in heavy surf.

Material and methods

Cyprids of *B. balanoides*, collected in quantity with a medium mesh plankton net, were induced in the laboratory to settle to a prearranged pattern on slates measuring 8 × 5 in. in which pits had been drilled (Crisp & Barnes, 1954; Crisp, 1960) and which had been treated with settling factor (Crisp & Meadows, 1962). The slates were then immersed on an experimental raft and the barnacles regularly inspected and cleaned until they had reached a diameter of 0.5–1 cm and were large enough to be used in the experiment. Two sets of experiments were conducted: one in which the cyprids had been allowed to settle at random, the other in which they had been exposed during settlement to a beam of light of about 500 lx falling at an angle of 45° to the settling surface, thereby causing the majority to orientate to the long axis of the slate panel. The procedure adopted ensured that all barnacles were in the same condition and of approximately the same size. A photograph of each plate was taken before the experiment and rough prints were marked to show the orientation and fate of each barnacle during the experiment. Specimens of *B. pholis* between 5 and 9 cm in length, including mature and juvenile individuals (Qasim, 1957), were collected from the west coast of Anglesey and were given access to live barnacles while in captivity, except for 2 days before each experiment when they were given no food.

The experiments were conducted in large black plastic tanks filled with water to a depth of 18 cm, the water being vigorously circulated by a pump and steadily renewed at a rate of 350 ml/min. Fresh plankton collected daily was added to the water to promote continuous cirral activity by the barnacles (Crisp & Southward, 1961). The blennies rarely attacked barnacles which did not display cirral activity.

The slates were examined daily for the effects of predation. Occasionally it was found that whole barnacles had been removed without trace; more frequently the opercular valves and the attached soft parts were torn away leaving the six parietal plates. Sometimes the soft parts of the barnacles were damaged with no external evidence of attack. Any individual which was clearly moribund, and every barnacle that was not observed to be active and which, when tested with a pin, did not respond by a downward jerk of the operculum, was removed and examined for damage indicative of fish predation.

Experiment 1. Randomly oriented barnacles

In the first set of experiments the lack of orientation corresponded more closely to the natural arrangement of barnacles on the shore where the direction of light continually changes and other factors than light influence orientation (Moore, 1933; Crisp, 1953; Gregg, 1948; Crisp & Barnes, 1954; Crisp & Stubbings, 1957). These slates, each with some 200 barnacles equally oriented in all directions, were placed in a tank with several blennies at a temperature of 17 °C. The tank was illuminated at one end by tungsten lamps producing an intensity, as measured by an E.E.L. light meter, varying from 560 to 280 lx. The orientation of each slate to the light was kept unaltered throughout the experiment. Because of the random orientation of the barnacles, the fixed light source gave a wide range of barnacle–light orientations.

Table 1. *Predation on randomly orientated barnacles*

(Ten slates bearing over 2000 randomly orientated *Balanus balanoides* presented to several *Blennius pholis* in the same tank.)

	I	II	III	IV	V	VI
Quadrant ...		$+45°$ $-45°$	$+45°$ $+135°$	$-45°$ $-135°$	$+135°$ $-135°$	
Equivalent light direction ...		From carina to rostrum	From right to left side	From left to right side	From rostrum to carina	Total
Number eaten		44	26	29	31	130
Expectation		32.5	32.5	32.5	32.5	130
Deviation		$+11.5$	-6.5	-3.5	-1.5	—

χ^2 between all quadrants $= 5.81$ for 3 degrees of freedom.
χ^2 between carinal and other quadrants $= 5.97$ for 1 degree of freedom.

Rather a small proportion of barnacles were attacked, the numbers eaten within each quadrant being shown in Table 1. More were eaten when the light fell on the carinal (Column II) than when the light fell on the rostral (Column V) or on the lateral compartments (Columns III, IV). However the values of χ^2 did not differ significantly from random expectation over the four quadrants as a whole. A comparison of the number attacked with the carinal compartment towards the light (44) against the numbers attacked with the light incident on to the other three quadrants taken together (86) differed significantly from the expected $1:3$ ratio with a value of χ^2 of 5.97 (with correction for continuity). Hence there is some evidence from this experiment that when the light is received from the direction of the carina the barnacle is slightly more prone to fall victim to predators than when it is received from other directions, and thus contradicts the hypothesis advanced above.

Experiment 2. Barnacles oriented towards and away from the light

In this experiment 18 slates with the majority of barnacles oriented in the same sense were placed in pairs in nine separate tanks, one slate with the barnacles' carinal ends predominantly oriented towards the light and the other with the barnacles' rostral ends towards the light. Only those barnacles oriented within $\pm 30°$ of the long axis of the slate were taken into account in the results. The illumination was provided by tungsten lamps at one end giving an average intensity of approximately 500 lx in all nine tanks. The temperature was $17 \pm 0.5\,°C$, and the water exchange 250 ml/min. Only one blenny at a time was placed in each tank but the fish were moved from tank to tank every 2 days in order to equalize predation rates as far as possible. The experiment was continued until about a quarter of the barnacles were eaten. Table 2 shows that in this experiment almost identical proportions of the experimental population were eaten regardless of orientation, the differences being quite insignificant.

Table 2. *Predation on orientated barnacles*

(Pairs of slates with *Balanus balanoides* orientated at $\pm 30°$ and $\pm 150°$ exposed simultaneously to attack by individual *Blennius pholis*.)

Direction of barnacles	Number eaten	Number left	Percentage eaten
Carina facing light ($+30°$ to $-30°$)	337	999	25.2
Rostrum facing light ($+150°$ to $-150°$)	383	1049	26.8

$\chi^2 = 0.75$ for 1 degree of freedom

Thus the orientation to light appears to confer little or no adaptive value to barnacles against attack by blennies. In view of these results the feeding behaviour of the blenny was investigated to determine how far it was likely to give warning of its approach by casting a shadow.

Observations on the feeding behaviour of Blennius pholis
in relation to light

One fish at a time was placed in a clear glass tank illuminated from one side so that the light intensity at the centre of the tank measured 1500 lx. The remainder of the room was completely dark, enabling observations to be made without disturbing the fish. Other conditions were maintained as before, the barnacles actively feeding and apparently in good condition.

Barnacles oriented at various angles to the direction of light were placed in the tank. Nine fish between 5 and 9 cm long were observed in turn and the number and direction of actual attacks recorded. The convention used for determining the direction of the fish's attack in relation to the incident light is made clear from Table 3. When the fish attacked against the light no shadow was cast on the barnacle before the attack. If the fish attempted to attack from the same direction as the incident light, its body and pectoral fins cast a shadow. It was difficult to determine whether a shadow was cast when the fish attacked from either side. The number and direction of actual attacks for the nine fish is shown in Table 3. The results, when analysed by the χ^2 test, were highly significant showing that the blennies attack mainly against the light and so do not usually cast a shadow on their prey until the very last moment. Out of a total of 631 attacks in 9.5 h, only 11 were successful. In only one case were all the soft parts of the barnacle removed, together with the valves, leaving the six shell plates. In the other ten cases, the cirri were bitten off only.

Table 3. *Orientation of* Blennius pholis *in relation to the direction of light when feeding on barnacles*

Angle of B. pholis to light

	+45°--45° Looking into light, no shadow	+45°-+135° Light on left side	-45°--135° Light on right side	+135°--135° Looking away from light, casting shadow	Total
Attacks delivered	253	147	145	86	631
Attacks expected	157.75	157.75	157.75	157.75	631
Deviation from random expectation	+95.25	-10.75	-12.75	-71.75	—

χ^2 for 3 degrees of freedom = 91.6.

These observations show that in the majority of attacks the barnacle has little time to withdraw, but the low proportion of attacks that are successful indicates that withdrawal is nevertheless effective in preventing the barnacle from being caught. Even when a grown barnacle is caught is it often no more than mutilated by the loss of the cirri. These may possibly regenerate. Darwin (1854) mentions 'monstrous individuals...with the posterior cirri distorted, unequal on the opposite sides and in an almost rudimentary condition...'.

There is no evidence from the blenny's hunting behaviour that it arranges itself deliberately to avoid shading the prey. Rather it attacks any barnacle whose movement catches its eye. It appears first to move forward, approaching within range, and then suddenly grabs at the prey. In the 'approach' stance the fish, moving forward in small leaps rather than swimming, is supported by the pelvic fins with the pectoral fins outstretched on either side. On noticing an active barnacle, the fish's head is bent slightly forward and usually to one side. The pectoral fins are flattened against the body at the moment of attack, and on grasping the barnacle the body twists in an attempt to swim backwards and remove the barnacle. Small barnacles are sometimes dislodged completely in this way. When the fish is moving into the light, most of the barnacles remain beating during its approach and it attacks them in turn. When the fish reaches the end of a row of barnacles, it turns and moves back. The shadow cast by the fish in its approach then causes the barnacles to close before any attack can be made, though active barnacles on either side are sometimes spotted and attacked. If the blenny remains in one position for a time the barnacles in its shadow may resume activity and they also are then attacked.

RELATION BETWEEN ORIENTATION TO LIGHT
AND RESPONSE TO SHADING

The shadow reflex is a well-known response enabling many sessile organisms, including barnacles, to withdraw quickly into the safety of their shells, tubes or burrows. Visual acuity is clearly unimportant to such a reflex. Since a shadow may fall over any part of the solid angle of the potential field of vision, the eye should possess a very wide angle of acceptance as well as having a sensitive 'off' response.

The sensitivity of the response may vary over different parts of the visual field. In order to map out variations in sensitivity of the eye from different directions it would be necessary to measure the minimum reduction in light necessary to evoke a response by shading different parts of the visual field. Our experiments were concerned only with variations in response to shadows falling at different angles to the rostro-carinal axis of the barnacle and therefore it was found necessary to consider only two of the many possible directions from which the shadow might fall. First, the animals were illuminated by nearly parallel light from above and a variable proportion of this light was extinguished to simulate a shadow. This system will be referred to as vertical shading. Secondly, while the animals were illuminated from above, an additional variable source of light was allowed to fall on them from an angle of 48° to the horizontal; the latter light could be extinguished to test the shading response at different angles to the rostro–carinal axis. This system will be referred to as lateral shading.

Method

Fig. 2 illustrates the apparatus used. The barnacles were placed in a glass dish fitted with air jets to circulate the water. Light from a tungsten bulb (L_1) placed about a metre above the dish fell on the animal at a constant intensity of 30 lx. For vertical shading light from a second point source (L_2) was reflected on to the barnacle by a small mirror (M) (Fig. 2b). The mirror was slightly offset from the vertical to prevent a shadow being cast from the first light source (L_1); the reflected light fell almost vertically on the animal. The source L_2 could be moved along an optical bench. The intensity of illumination received by the barnacle was measured by means of an E.E.L. photometer for a number of positions of L_2 along the bench

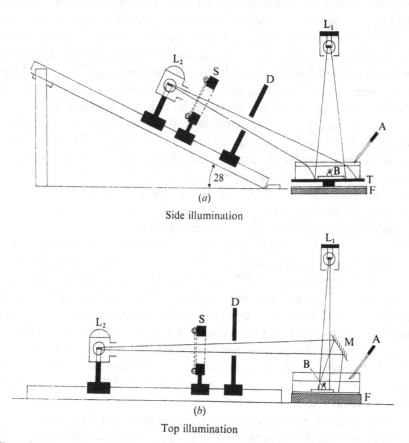

(*a*)

Side illumination

(*b*)

Top illumination

Fig. 2. Systems of illumination used for measuring shading response. A, air jet; B, barnacle; D, diaphragm; F, foam rubber pad; L_1, reference light source; L_2, variable light source on optical bench; M, mirror; S, shutter; T, turntable.

with the light L_1 switched off. The fraction of the total light intensity attributable to the second source L_2 was plotted graphically against the position along the optical bench so that the relative shading when L_1 was kept on and L_2 was extinguished could be read off from the graph. A foot operated shutter (S) was used to produce sudden extinction of the second light source, and a large diaphragm (D) was inserted to protect the experiment from any stray light coming from the second source except by reflection from the mirror. Both lights were battery operated to eliminate intensity fluctuations caused by mains surges and the current was adjusted to a constant value by means of a rheostat to ensure uniform luminous output.

When conditions of lateral shading were required the light used for shading (L_2) was allowed to fall directly on the animals from the side by tilting the optical bench to an angle of 28° to the horizontal. The light from L_2 then passed directly to the animal after refraction through the water surface instead of being reflected by the mirror (Fig. 2a). The angle of the light beam to the horizontal after refraction became 48°. When lateral shading was used, the dish containing the animal was placed on a turntable so that any required orientation to the lateral light beam could be obtained. The light from L_1 was maintained as before. Unlike the simpler situation with vertical shading, in calculating the percentage drop in illumination with lateral shading the actual value of percentage reduction in light intensity experienced is a function of the angle of acceptance of the light by the receiving surface or sense organ. It is necessary therefore to postulate the orientation to the horizontal of the receiving surface. This has been assumed to be in the horizontal plane. The variable component of the laterally directed light (which is periodically extinguished during experiments) falling on this plane will therefore be $I_2 \cos 48°$. The constant light component falling on this plane from above will be $I_1 \cos 90°$. The value of I_2 was measured at the point occupied by the barnacle but with the dish of sea water removed and the illumination falling normally on to the photometer cell. A small correction had to be applied (Jerlov, 1968) for the loss of reflected light on passing through the water surface at an angle of incidence of 62°, the complement of the angle subtended by the light beam to the horizontal.

Had the light receiving surface been chosen as the plane tangential to the barnacle shell at the point facing the direction of lateral illumination, the component of illumination received from the lamp L_1 would have been reduced and that received from the lamp L_2 would have been increased, hence the apparent sensitivity of the response to shadow would have been less. Fortunately, whatever the angle of acceptance of light by the eye, it will remain the same when the animal is rotated about a vertical axis, and will not therefore affect the measurement of the orientation at which the barnacle is most sensitive to shading.

In all the experiments the method of determining the level of sensitivity to shading was the same. The second light source L_2 was varied by moving the lamp along the optical bench until the position was found at which the barnacle just gave a perceptible response when the light source L_2 was extinguished by the shutter. The proportionate reduction in illumination was then recorded. Thus a score of 0.10 indicated that the reduction in illumination was 0.10 and the value of the light intensity was brought down to 90% of the reference level. Great care had to be taken to avoid vibrations being transmitted when the shutter closed. The dish containing the animals was mounted on a rubber cushion standing on a mechanically separated part of the laboratory bench. Opening and closing the shutter with the light L_2 extinguished produced no effect on the barnacle. Sufficient time was allowed between trials to prevent habituation to the stimulus.

The results are strictly valid for the reference level of illumination employed, but probably the response approximately follows the Weber–Fechner law and the proportionate reduction in light intensity, $\Delta I/I$, would be applicable to the response over a range of reference light intensities.

Influence of laboratory conditioning

Tests were made to establish the range of individual variation in the sensitivity to shadow, and changes resulting from laboratory storage. The mean values obtained under various conditions together with the standard errors are given in Table 4. Evidently the sooner the barnacle is tested after collection and the more closely the water movement during storage and experimentation resemble those experienced under natural conditions, the more

Table 4. *Effect of laboratory conditions on shading response*

Species	Minimum light reduction to induce shadow reflex in freshly collected animals in moving water	Modification of experimental condition during measurement	Modification of condition of storage	Minimum light reduction to induce shadow reflex under modified condition
Balanus crenatus	0.119 ± 0.016	Still water	Fresh	0.158 ± 0.019
		Still water	24 h in running water	0.170 ± 0.02
		Still water	24 h in still water	0.306 ± 0.02
		None	24 h in running water	0.132 ± 0.011
Elminius modestus	0.227 ± 0.01	Still water	Fresh	0.221 ± 0.018
		None	1 week in running water	0.268 ± 0.023
Balanus balanoides	0.178 ± 0.01	None	1 week in running water	0.243 ± 0.031
Chthamalus stellatus	0.537 ± 0.103 (Only 60% gave responses)	None	1 day in running water	0.98 (Only 2% gave responses)

18-2

readily does the animal display a response to shadow. *B. crenatus*, a mainly subtidal form which does not necessarily experience intertidal wave action, was least affected by laboratory conditions. *C. stellatus*, a high water form favouring wave exposed habitats, almost failed to respond after being kept in the laboratory. For all critical experiments therefore, unless otherwise stated, freshly collected specimens were used, but for some of the work on *B. crenatus*, specimens maintained in fast running sea water were found to give satisfactorily consistent results.

Influence of size and tidal level

Table 5 gives the minimum stimulus for the shadow reflex to be elicited for samples of large and small *B. balanoides* collected from high and low tidal levels. The difference in response is small and insignificant in relation to the standard error. Evidently there is no need to select individuals of any particular size or from any particular level on the shore for use in these experiments.

Table 5. *Effect of tidal level and size on shading response*
of Balanus balanoides *at c. 30 lx*

(Stimulus: fractional reduction of vertically incident light.)

Tidal level	Size	Minimum light reduction to induce shadow reflex
H.W.	< 0.5 cm	0.174 ± 0.019
	> 0.5 cm	0.184 ± 0.030
L.W.	< 0.5 cm	0.162 ± 0.011
	> 0.5 cm	0.168 ± 0.015

H.W. = high water; L.W. = low water.

Differences between species

A number of individuals of different species were compared using freshly collected animals where possible. The results are shown in Table 6.

Three species of *Balanus*, and *Elminius modestus*, all of which have white translucent shells, were the most sensitive. Nearly all the individuals gave a response when a light intensity of 30 lx was reduced appreciably, and the average reduction in light required to produce a response lay between a tenth and a quarter of the total illumination. The most sensitive individuals of *B. crenatus* required only a twentieth reduction in light intensity, the most sensitive individuals of the other three species responded to a reduction of approximately one-tenth. The other species tested were considerably less sensitive, perhaps because they have more opaque shells.

The shell of *C. stellatus* is lined with a darkly pigmented epidermis and the shell of *Pyrgoma anglicum* and of *B. perforatus* is very thick.

Table 6. *Shading stimulus as fraction of incident light for minimal response*

(Temperature 17–18 °C. Water stirred during observations. Barnacles freshly collected except where stated. Significant differences exist between *B. crenatus, B. balanoides* and *Elminius modestus*, but not between *E. modestus* and *B. balanus*. There is a significant difference between *B. balanus* and the succeeding species.)

Species	Number tested	Minimum light reduction to induce shadow reflex	Mean light reduction to induce shadow reflex	Approx. light intensity (lx)	Percentage giving shading response
Balanus crenatus	11	0.05	0.12 ± 0.016	30	100
B. balanoides	49	0.08	0.18 ± 0.010	30	100
Elminius modestus	7	0.10	0.23 ± 0.010	30	93
Balanus balanus	10	0.12	0.25 ± 0.040	30	100
Chthamalus stellatus	10	0.25	0.54 ± 0.10	30–40	55
Pyrgoma anglicum*	3	0.65	0.81 ± 0.03	300–600	75
Balanus perforatus*	21	c. 0.98	Few responses	1000	49

* Specimens not freshly collected.

Influence of orientation on shading responses

When the illumination received directly on to the compartments from a direction of 48° from the horizontal was reduced (lateral shading, see Fig. 2a) a clear difference in sensitivity was recorded according to the orientation of the barnacle to the direction of light. The results for *E. modestus* and *B. crenatus* are shown in Table 7, together with an analysis of variance for each species. The analyses show that the sensitivity of the barnacle varied significantly with its orientation, being greatest when the rostral compartment was shaded and least when the carinal was shaded. This result is at variance with the hypothesis given on p. 540. There was no difference between right and left lateral compartments.

Since the rostral compartment is the thinnest, and therefore probably the most translucent, and the carinal is the thickest, the above result strongly suggests that light reaches the eye by passing directly through the shell as well as, or instead of, through the opercular opening. To test this the compartments were made opaque with tin foil.

Effect of covering the compartments

Specimens of *B. crenatus* were fitted with tin foil shields to prevent light striking the compartments but allowing light to enter the opercular aperture. They were illuminated both vertically (Fig. 2b) and laterally at various orientations to light (Fig. 2*a*). The minimum reduction in intensity of light to elicit a response was measured with the covers on and after they had been removed.

Table 7. *Effect of shading compartments*

(The table gives the minimal reduction of illumination striking the side of the barnacle at an angle of 48° to the horizontal to elicit a shading response. The proportional reduction in intensity of light was resolved on to the horizontal surface to which the barnacle was attached. The variance ratio indicates in both species: (1) No significant difference between values of minimum stimulus when either the right or left lateral compartments were shaded. (2) A significant difference between the minimum shading stimulus when (*a*) the carinal and lateral, and (*b*) lateral and rostral compartments were shaded.)

Mean reduction of illumination in

Compartment shaded	E. modestus	B. crenatus
Carinal	0.275	0.198
Left lateral	0.158	0.163
Right lateral	0.168	0.152
Rostral	0.133	0.108

Analysis of variance

	Source of variation	Degrees freedom	Variance ratio
Elminius modestus	Between individuals	11	5.25
	Between compartments	3	21.09
	Residual	33	(1.00)
Balanus crenatus	Between individuals	8	6.37
	Between compartments	3	5.88
	Residual	24	(1.00)

Table 8. *Reduction in sensitivity of barnacles to shadow*

(The reduction in sensitivity to shadow when the parietal plates were covered with tin foil under conditions of vertical shading, using the same specimens of *B. crenatus* throughout, at illumination of *c.* 30 lx.)

	Mean reduction in light intensity for response
Compartments not covered	0.124
Compartments covered	0.199

Mean difference = 0.075; S.E. of difference = 0.024. Student's *t* test for 10 degrees of freedom, *t* = 3.1; *P* = 0.01.

The results for the experiments in which the barnacles were illuminated directly from above are given in Table 8. The application of the *t* test to these results shows that a highly significant reduction in sensitivity ($P < 0.01$) took place as a result of applying the covers, but the animals remained still fairly sensitive. Light is therefore received both through the opercular aperture, perhaps through the translucent valves, and also through the substance of the parietes. Table 9 shows the effect of covering the compartments when the shadow was applied laterally at various orientations. The results without covers confirm the previous finding that the rostral compartment is the most sensitive. When the covers were applied

Table 9. *Effect of covering compartments on the response of* Balanus crenatus *to lateral shading at different orientations*

(The calculations of light reduction are based on illumination resolved on to the horizontal plane. This population of *B. crenatus* was, on average, less sensitive than that recorded in a similar experiment in Table 7. Differences between individuals, compartments, and between shaded and unshaded specimens were significant. No significant difference was found between compartments in shaded individuals.)

	Mean reduction in illumination for response	
	Compartments normal	Compartments covered
Carinal	0.279	0.348
Left lateral	0.216	0.363
Right lateral	0.220	0.355
Rostral	0.148	0.340
Grand mean	0.216	0.352

Analysis of variance

	Source of variation	Degrees of freedom	Variance ratio
Compartments normal	Individuals	8	5.66
	Compartments	3	17.32
	Residual	24	(1.00)
Compartments covered	Individuals	8	242
	Compartments	3	0.9
	Residual	24	(1.00)
All compartments together	Between normal and shaded animals	1	10.74
	Residual	70	(1.00)

there was no significant difference in sensitivity between any of the compartments and the animals were rendered distinctly less sensitive. This result therefore discredits the view suggested in the introduction that light might pass unimpeded from the direction of the carina through the opercular aperture on to the light sensitive areas beneath the rostrum. In fact the rostral rather than the carinal end is the more sensitive to light, and its sensitivity seems likely to result from the eyes being closely applied to the inner surface of the translucent shell of the rostrum.

DISCUSSION AND CONCLUSIONS

The experiments on predation by *Blennius pholis* indicated an almost uniform susceptibility to predation whatever the orientation of the barnacle, though perhaps with a slightly greater hazard when the attack was delivered from the direction of the carina. These experiments, however, failed to indicate any definite effect of orientation on the response to shadow. However, the experiments to determine the degree of shading necessary to evoke a response, using controlled illumination, clearly indicated that the rostral compartment, which overlies the light sensitive organs, was the most receptive area.

The discrepancy between these two approaches may partly be explained by the tendency of the blenny to attack while moving towards the light. Under these circumstances the barnacle would be shaded only at the last moment before the attack and successful predation might depend more on the rate of the response to shadow than on the sensitivity to a particular degree of shading.

It might appear disadvantageous for the barnacle to be so oriented that the least sensitive area pointed towards the predominant source of light since its maximum sensitivity to shadow would be reduced. The initial hypothesis presented in the Introduction was based on this view. However, if it is correct to assume that the shading response is a defence against predation, the fact that the animal appears deliberately to orient its least sensitive area, the carinal region, towards the light at settlement must be explained adaptively.

Since the barnacle is sessile it must be capable of responding to shadows from any quarter above the level of the substratum to which it is attached. Otherwise, it would always fall an easy victim to attack from the particular direction in which it was blind – a catastrophe which no amount of increased sensitivity in another direction could possibly avert. It follows that the ideal defence for such an animal is not a highly developed sensitivity in one direction; it needs rather to be equally alert to danger in all directions.

An object approaching from the less illuminated side of the animal will cause a smaller degree of shading than one casting a definite shadow on the better illuminated side. The approach of a predator from the darker side therefore presents a greater danger and calls for a greater degree of watchfulness on the part of the prey. Herein probably lies the advantage of a differential sensitivity to light, the more sensitive area facing the direction from which the shadow of a predator is least readily perceived. The requirements of a sessile organism are therefore quite different from those of a mobile species: the latter can afford a limited field of acute perception since it can bring the region of greatest visual acuity or sensitivity to bear on a hazard from any direction at will. The sessile organism must normally possess a minimum sensitivity in all directions simultaneously.

A test of the above hypothesis would be to consider the orientation of sessile species which are likely to be attacked only from a given direction irrespective of the direction of light. Two examples could be suggested: the epizoic barnacles, *Chelonibia patula* and *Pyrgoma anglicum*, which are attached respectively to crabs and solitary corals. Attack on these species is most likely to come from the periphery of the host. Both species orientate predominantly with the rostral compartment, presumably the most sensitive, facing outwards (Fig. 3).

SUMMARY

At settlement the cyprid responds to three orienting stimuli; to contour, to light and to current. The strength of the response is such that contour takes precedence over light, and current orientation can be observed only if there is no directional stimulus caused by contour or light. Whereas the advantage of lying in the protection of a groove and of having the cirral net beating into the prevailing current is obvious, the significance of orientation to light is less clear. It has been suggested, on account of the position of the eye, that the carinal end might be more sensitive to light and shade when the valves are opened and, if so, the preferred orientation would increase the sensitivity of the response to shadow and thereby the chance of avoiding predation.

Experiments in which blennies (*Blennius pholis*) were allowed to attack barnacles oriented at various angles towards a source of light, failed to reveal any effect of orientation on predation rate. However, the behaviour of the blenny may have precluded any manifestation of differences in shading response because the fish usually attacked against the direction of the light and rarely snapped at barnacles lying within its own shadow. Experiments were therefore carried out on several barnacle species to measure directly the threshold response to shadow in relation to the direction of light incident on the barnacles.

C. stellatus and *B. perforatus* were either insensitive to, or quickly adapted to shading and were therefore unsatisfactory for this purpose. *B. balanus*, *E. modestus* and *B. balanoides* gave good results when fresh but became less sensitive after a day or two in the laboratory. *B. crenatus*, if kept in running water, gave the most reproducible results.

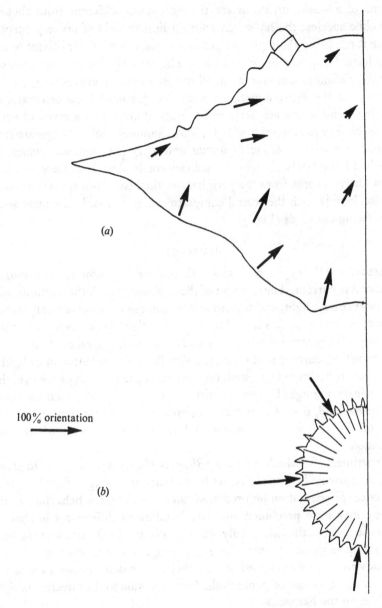

Fig. 3. Orientation (a) of *Chelonibia patula* on the carapace of the portunid, *Callinectes sapidus* and (b) of *Pyrgoma anglicum* on the solitary coral *Caryophyllia smithi*. Only the left half of the host is shown. The direction of the arrow represents the predominant orientation of the commensal barnacles; the length of the arrow indicates the degree of conformity of the orientation of the barnacles at the position shown. The arrow points from the rostral towards the carinal compartment.

In all the species tested, the rostral compartment was found to be the most sensitive to shading, and the carinal compartment the least sensitive. If the compartments were covered with foil, so that light could enter only via the operculum, not only was the sensitivity of the animal reduced but the response became independent of its orientation to light. Evidently light reaches the eye by passing directly through the rostral plate as well as via the operculum.

These experiments indicate that the barnacle orientates itself with its most insensitive area directed towards the light. The preferred orientation will thus have the effect of increasing the responsiveness of the barnacle in the direction that is likely to receive the least shading when a predator approaches from that quarter. It is perhaps an advantage to a sessile animal to be equally alert to danger in all directions. It is interesting to note that species that settle around the edge of other animals, such as *Pyrgoma* on *Caryophyllia* or *Chelonibia* on *Callinectes*, orientate themselves with the rostrum outwards, the direction from which attack is most likely to come.

REFERENCES

BARNES, H., CRISP, D. J. & POWELL, H. T. (1951). Observations on the orientation of some species of barnacles. *J. Anim. Ecol.* **20**, 227–41.

CRISP, D. J. (1953). Changes in orientation of barnacles of certain species in relation to water currents. *J. Anim. Ecol.* **22**, 331–43.

CRISP, D. J. (1960). Factors influencing the growth rate in *Balanus balanoides*. *J. Anim. Ecol.* **29**, 95–116.

CRISP, D. J. & BARNES, H. (1954). The orientation and distribution of barnacles at settlement with particular reference to surface contour. *J. Anim. Ecol.* **23**, 142–62.

CRISP, D. J. & MEADOWS, P. S. (1962). The chemical basis of gregariousness in Cirripedes. *Proc. Roy. Soc. Lond.* B **156**, 500–20.

CRISP, D. J. & SOUTHWARD, A. J. (1961). Different types of cirral activity of barnacles. *Phil. Trans. R. Soc. ser.* B **243**, 271–308.

CRISP, D. J. & STUBBINGS, H. G. (1957). The orientation of barnacles to water current. *J. Anim. Ecol.* **26**, 179–96.

DARWIN, C. (1854). *A Monograph on the sub-class Cirripedia.* II. *The Balanidae, the Verrudicae, etc.*, 684 pp. London: Ray Society.

DAY, F. (1880–4). *The Fishes of Great Britain and Ireland*, 1, 336 pp. London and Edinburgh: Williams and Norgate.

GREGG, J. H. (1948). Replication of substrate detail by barnacles and some other marine organisms. *Biol. Bull. mar. biol. Lab.*, *Woods Hole* **94**, 161–8.

HARTLEY, P. H. J. (1949). Blennies and the ecology of littoral fishes. *Rep. Challenger Soc.* **31**, 19 (Abstract).

JERLOV, N. G. (1968). *Optical oceanography*, 194 pp. Amsterdam: Elsevier.

558   L. FORBES, M. J. B. SEWARD AND D. J. CRISP

KNIGHT-JONES, E. W. & CRISP, D. J. (1953). Gregariousness in barnacles in relation to the fouling of ships and to anti-fouling research. *Nature, Lond.* **171**, 1109.

MCDOUGALL, K. D. (1943). Sessile Marine Invertebrates of Beaufort, North Carolina: A study of settlement, growth, and seasonal fluctuations among pile-dwelling organisms. *Ecol. Monogr.* **13**, 321–74.

MCINTOSH, W. C. (1904). Notes on the eggs of the shanny (*Blennius pholis* L.). *Rep. Br. Ass. Advmt Sci.* (1903), 697.

MCINTOSH, W. C. (1905). On the life history of the shanny (*Blennius pholis* L.). *Z. wiss. Zool.* **82**, 368.

MOORE, H. B. (1933). The change of orientation of a barnacle after metamorphosis. *Nature, Lond.* **132**, 969–70.

PYEFINCH, K. A. (1948). Notes on the biology of Cirripedes. *J. mar. biol. Ass. U.K.* **27**, 464–503.

QASIM, S. Z. (1957). The biology of *Blennius pholis* L. (Teleostei). *Proc. zool. Soc. Lond.* **128**, 161–208.

SMITT, F. A. (1892). *A History of Scandinavian Fishes*, Part 1, 2nd ed., 566 pp. Stockholm: Norstedt and Söner.

SOLJAN, T. (1932). *Blennius galerita* L., poisson amphibien des zones supra-littorale et littorale exposées de l'Adriatique. *Acta adriat.* **2**, 1–14.

VISSCHER, J. P. (1928). Reactions of the cyprid larvae of barnacles at the time of attachment. *Biol. Bull. mar. biol. Lab., Woods Hole* **54**, 327–35.

WILSON, D. P. (1951). *Life of the Shore and Shallow Sea*, 213 pp. London: Nicholson and Watson.

THE LIGHT SENSITIVITY
AND LIGHT ENVIRONMENT OF
COROPHIUM VOLUTATOR

A. R. GIDNEY

Department of Natural History, University of Aberdeen

ABSTRACT

An experimental study of the sensitivity to light of the mud-dwelling amphipod, *Corophium volutator*, has been carried out. Using horizontal light rays, the extinction of the positive phototactic response has been used to determine a visual threshold.

Dark adaptation increases the response threshold from 10^{-3} to 4×10^{-6} lx. Measurements of the position of masking pigment in the compound eye have shown that dark adaptation takes place between 10^{-2} and 10^{-3} lx, and that light adaptation occurs faster than dark adaptation.

The spectral sensitivity curve of *Corophium* has a broad maximum from 520 to 570 nm, sensitivity being 10 times greater in the green–yellow region than in the violet, and 400 times greater than in the red. Maximum sensitivity in this region would suit an animal in estuarine water where transmission of longer wavelengths is favoured.

Underwater light measurements on the river bed show that the animal is subjected to large changes in light intensity depending on the time of day, season and state of tide. Light intensity here is usually 10^{-3} lx or lower at high water during darkness, but may be as high as 10^4 lx at high water, and over 10^5 lx at low water, in daytime. Light attenuation is greater on the ebb than on the flood tide. Samples taken during tidal cycles show few animals occurring above the mud during the day, but large numbers occurring on the ebb tide at night. This suggests that the tide and light intensity together act as a stimulus for emergence from the mud.

Measurements made at different depths in mud show that light is no longer measurable below 1.5 mm. Light will, however, penetrate burrows but its intensity is probably reduced by at least 100 times.

It is suggested that the intensity experienced by *Corophium* at most times of day is such that the eye remains light adapted, and that the onset of dark adaptation may contribute to the stimulus for emergence.

LA LUMIERE ET LE DECLENCHEMENT
DE LA PONTE CHEZ *CIONA INTESTINALIS*

D. GEORGES

Laboratoire de Zoologie, Cedex 53. 38 Grenoble-gare, France

ENGLISH SUMMARY

In normal daily alternation of light and darkness the removal of the neural complex or of the neural gland alone of *Ciona intestinalis* L. (Tunicata, Ascidiacea) increases the spawning rate on the first day after operation.

Permanent lighting strongly inhibits the spawning and this inhibitory action prevails over the stimulating effect of the removal of the neural gland.

The inversion of light and darkness with regard to the normal day and night cycle results in a shifting of the time of spawning, which occurs with regularity shortly before the end of a period of darkness.

INTRODUCTION

Les Tuniciers présentent au cours de leur développement des caractères de Cordés et de nombreuses recherches ont voulu établir des analogies entre certains de leurs organes et ceux des Vertébrés. Ainsi, la glande neurale des Ascidies, située entre les siphons buccal et cloacal, ventralement par rapport au ganglion nerveux, fut longtemps considérée comme l'homologue de l'hypophyse, particulièrement par Julin (1881). Cependant Willey (1894), et, plus récemment, Lender & Bouchard-Madrelle (1964) ont reconnu et prouvé l'origine nerveuse unique de tout le complexe neural.

Le but des recherches entreprises est d'élucider le rôle joué par le complexe neural formé par le ganglion nerveux et la glande neurale, ainsi que l'influence des conditions d'éclairement dans le déclenchement de la ponte de *Ciona intestinalis*.

Les expériences ont été réalisées pendant les mois d'été à la Station Biologique de Roscoff (France). Les animaux récoltés à faible profondeur, à l'abri de la lumière, sont maintenus en eau courante à une température de 18 °C. Pendant les expériences, les Ascidies sont élevées dans des coupelles individuelles dont l'eau est renouvelée une ou deux fois par jour. Chaque matin le nombre des animaux ayant pondu est noté. Avant les opérations les animaux sont anesthésiés pendant 0.5 h dans une solution de MS 222 dans de l'eau de mer filtrée.

D. GEORGES

RÔLE DU COMPLEXE NEURAL DANS LE DÉCLENCHEMENT DE LA PONTE

Dans les conditions d'expérience précitées, les sujets adultes mâtures, non opérés, pondent de façon régulière tous les deux ou trois jours, entre 03.00 et 04.00 h; certains pondent plusieurs jours consécutifs.

Si l'on réalise l'ablation du complexe neural ou de l'une de ses parties les résultats sont les suivants. L'ablation du complexe neural déclenche la ponte chez un grand nombre d'animaux dès le lendemain de l'opération (87% des cas). Ceci n'est pas dû au choc opératoire car les témoins ayant subi une opération factive ne pondent que dans 40% des cas. L'ablation de la glande neurale et des ocelles, opération qui détruit les mêmes voies nerveuses et humorales que dans le cas précédent, donne des résultats semblables: un jour après l'opération 79% des animaux pondent.

Afin de préciser les rôles respectifs du ganglion nerveux et de la glande neurale, deux autres séries d'expériences ont été faites. La première consiste à exciser la glande neurale seule, ce qui provoque la ponte de 79% des animaux. La seconde concerne l'ablation du ganglion nerveux; les animaux ne pondent alors que dans 31% des cas et l'on peut constater que la périodicité de la ponte ne reprend qu'à partir du 6ème jour après l'opération. C'est à cette période que les moignons des nerfs commencent à régénérer (cf. Lender & Bouchard-Madrelle, 1964).

Ces expériences suggèrent que, dans les conditions normales d'alternance du jour et de la nuit, la glande neurale a un rôle inhibiteur dans le déclenchement de la ponte et que le ganglion nerveux n'intervient que dans la périodicité de la libération des produits génitaux.

INFLUENCE DES CONDITIONS D'ÉCLAIREMENT

Pérès (1943) avait pensé que l'alternance normale de lumière et d'obscurité était l'un des facteurs essentiels du déclenchement initial de la ponte, mais que le cycle de ponte une fois amorcé les émissions ultérieures ne pouvaient plus être entravées par l'expérimentation. Lambert & Brandt (1967) utilisant de la lumière blanche et de la lumière monochromatique obtiennent des pontes en moyenne 27 min après un éclair lumineux; d'autres expériences confirment que l'excitation lumineuse n'a pas besoin d'être continue pour être efficace. Whittingham (1967) a constaté une réduction de la ponte lorsque l'intensité lumineuse diminue et même une inhibition de la libération des produits génitaux à partir d'une intensité minimum (2.25 lux).

Nous avons tenté de répondre aux questions suivantes. (1) Des variations des conditions d'éclairement sont-elles capables de perturber l'activité de ponte de *Ciona intestinalis* intactes ayant déjà effectué plusieurs cycles de pontes? (2) Etant donné l'effet stimulateur de l'ablation de la glande neurale sur la libération des ovules, quel est le rôle joué par la lumière chez des animaux privés des différentes parties du complexe neural?

Animaux non opérés

Quatre lots d'animaux ont été placés dans des conditions d'éclairement différentes.

Lot 1: éclairement nycthéméral normal (En).

Lot 2: obscurité permanente (Op).

Lot 3: éclairement permanent à l'aide de tubes fluorescents de 20 W, type lumière du jour, situés à 60 cm au-dessus des coupelles d'élevage (Ep).

Lot 4: éclairement normal pendant 2 jours puis éclairement permanent.

Les résultats obtenus sont les suivants: (voir Fig. 1). Les animaux du lot 2 pondent plus ou moins régulièrement, comme les animaux maintenus en éclairement normal, ce qui n'est pas étonnant car dans la nature, ils sont généralement fixés à l'abri de la lumière.

Parmi les animaux du lot 3, quelques uns pondent au cours des 4 premiers jours; ensuite, la lumière agit de façon inhibitrice: en effet, bien que leur oviducte soit plein d'oeufs, les animaux ne pondent pas.

Cette action inhibitrice de la lumière est encore mise en évidence par l'arrêt de la ponte chez des animaux maintenus en éclairement nycthéméral normal puis placés en éclairement permanent (lot 4).

Animaux opérés

Des expériences analogues à celles qui ont été réalisées en éclairement nycthéméral normal ont été reprises en éclairement continu: ablation du ganglion nerveux, et ablation de la glande neurale et des ocelles.

Lors de l'ablation du ganglion nerveux, les résultats obtenus sont semblables à ceux des expériences en éclairement nycthéméral normal: les pourcentages des animaux ayant pondu sont peu différents entre les témoins (8.9%) et les opérés (14%) un jour après l'opération. Seule la périodicité de la ponte est supprimée pendant les 6 premiers jours (Fig. 2).

L'ablation de la glande neurale et des ocelles donne au contraire des résultats différents si les animaux sont maintenus en éclairement permanent: elle n'augmente pas le pourcentage des animaux ayant pondu (0% chez les opérés pendant les deux premiers jours) (Fig. 3).

Les expériences réalisées en éclairement continu montrent donc une forte action inhibitrice de la lumière sur la ponte des animaux opérés aussi bien que chez les témoins (voir Fig. 4) et cette action n'est pas limitée au déclenchement initial de la ponte. Cependant, la lumière n'affecte pas le rythme périodique de la libération des produits génitaux.

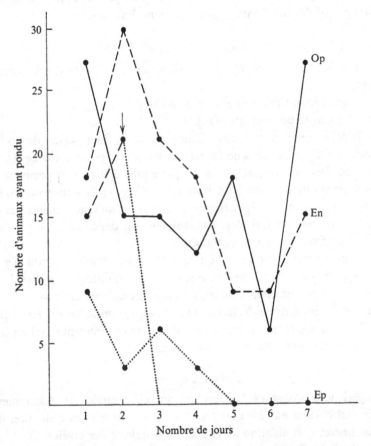

Fig. 1. Comparaison de l'activité de ponte chez des *Ciona intestinalis* placées dans différentes conditions d'éclairement. En (trait interrompu) éclairement nycthéméral normal; Op (trait continu) obscurité permanente; Ep (pointillé) éclairement permanent. La flèche indique la mise en éclairement permanent des animaux du lot 4.

Il semble, comme l'avait pensé Pérès (1943), que l'alternance des phases de lumière et d'obscurité soit un facteur important dans le comportement de ces Ascidies, en particulier dans leur rythme de ponte: en effet, deux lots d'animaux sexuellement mûrs sont mis en expérience; le premier, servant de témoin, est élevé dans les conditions d'éclairement normal: 13 heures de lumière (de 06.00 à 19.00 h) et 11 h d'obscurité. Le second lot subit un cycle diurne inversé (exposition à la lumière de 19.30 à 08.30 h, obscurité de 08.30 à 19.30 h). Dans les deux lots la ponte se produit *dès la fin de la première* période d'obscurité, entre 03.30 et 04.00 h chez les témoins et entre 17 et 18 h pour les animaux dont le cycle diurne a été inversé. Pour ces derniers, la première ponte est retardée de 14 à 15 h par rapport à celle des témoins (Fig. 5).

Une période de 9 heures d'obscurité environ semble nécessaire pour que la ponte soit déclenchée chez des animaux mâtures non opérés.

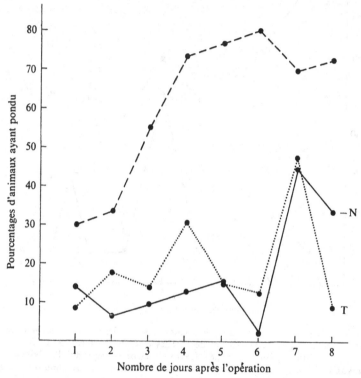

Fig. 2. Activité de ponte chez des *Ciona intestinalis* maintenues en éclairement permanent et ayant subi l'ablation du ganglion nerveux (– N, trait continu). T (pointillé) témoins ayant subi une opération simulée. Le trait interrompu reproduit la courbe correspondant à l'activité de ponte des animaux ayant subi l'ablation du ganglion nerveux et élevés dans des conditions d'éclairement normal.

DISCUSSION

Ces résultats semblent être en contradiction avec ceux d'autres auteurs, qui pour la plupart, attribuent un rôle stimulant à la lumière.

Quel que soit son rôle (stimulant ou inhibiteur) la lumière a une influence certaine: déjà en 1936, Warren & Scott avaient pu inverser l'heure de ponte des oiseaux domestiques par rapport à l'heure normale en inversant le cycle diurne. Pour ces animaux qui pondent le jour, il était possible de déclencher la ponte le jour et la nuit en les maintenant en éclairement permanent. Ce résultat était obtenu après une adaptation de 14 jours.

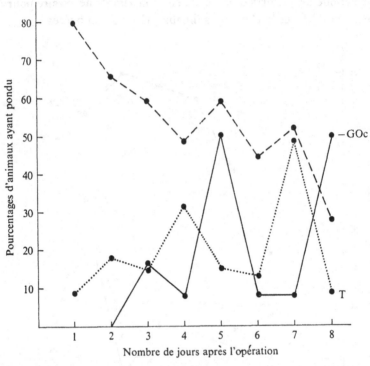

Fig. 3. Activité de ponte chez des animaux maintenus en éclairement permanent et ayant subi l'ablation de la glande neurale et des ocelles (– GOc, trait continu). T (pointillé) témoins ayant subi une opération simulée. Le trait interrompu reproduit l'activité de ponte des animaux élevés en éclairement normal et ayant subi la même opération (cf. Sengel & Georges, 1966).

Plusieurs auteurs se sont intéressés à l'influence de la lumière sur la gamétogenèse de différents Vertébrés: Burger en 1939 et 1947 a mis en évidence, chez l'étourneau, *Sturnus vulgaris*, la nécessité d'une période lumineuse de durée suffisante pour accélérer la spermatogenèse: des périodes lumineuses trop longues (l'été) ou trop brèves (l'automne) sont inopérantes dans la nature: c'est ce qui explique un cycle sexuel au cours de l'année. En ce qui concerne l'ovogenèse, Burger a constaté que, malgré l'accroissement des ovaires, dû au traitement lumineux, la lumière n'avait jamais causé la ponte chez les femelles traitées (Burger, 1940). Dans le cas

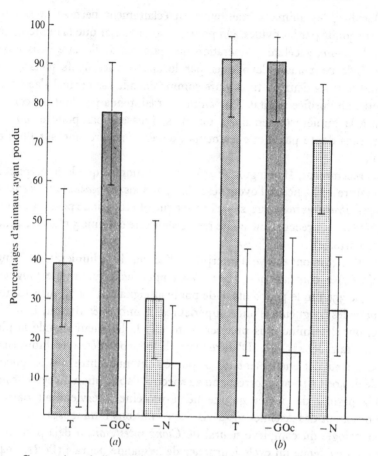

Fig. 4. Comparaison de l'activité de ponte chez des animaux maintenus en éclairement nycthéméral normal (pointillé) et en éclairement permanent (blanc); (a) au cours du premier jour après l'opération; (b) au cours des trois jours suivant l'opération. T, témoins; – GOc, animaux privés de glande neurale et des ocelles. – N, animaux privés de ganglion nerveux. (Les traits verticaux représentent les limites de l'intervalle de confiance correspondant au coefficient de sécurité de 95 %.)

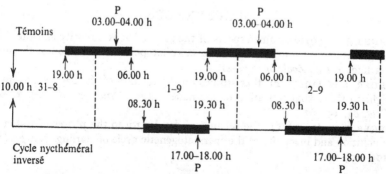

Fig. 5. Comparaison de l'heure de ponte chez *Ciona intestinalis* élevées selon un cycle nycthéméral inverse de celui d'animaux témoins maintenus dans des conditions normales d'éclairement, au cours des trois premiers jours de l'expérience. P, ponte. Les zones noires représentent les périodes d'obscurité.

des Ascidies, les animaux maintenus en éclairement permanent ont un oviducte gonflé par les ovules. On pourrait donc penser que la lumière, qui inhibe la ponte, accélère la libération des gamètes de l'ovaire dans l'oviducte. Cela peut aussi s'expliquer par le fait que les oeufs n'étant pas éliminés tous les deux ou trois jours comme dans le cas normal s'accumulent près de l'orifice génital. Des animaux maintenus pendant un certain temps à la lumière (3 semaines, un mois) finissent par pondre tout de même, sans doute parce que les oeufs exercent une pression sur l'orifice génital.

Plus récemment, Harrington (1956, 1957) a montré que la lumière et la température influencent l'ovogenèse des poissons *Enneacanthus obesus* et *Notropis bifrenatus* mais que, seule une longue photopériode peut l'induire à un achèvement précoce (une ponte peut même être obtenue 5 mois 1/2 avant la date normale).

Peu d'auteurs ont essayé d'interpréter l'action de la lumière: Whittingham (1967) a constaté que, chez *Ciona intestinalis*, pour une longueur d'onde de 390 nm le pourcentage de ponte atteignait 81.84%, mais diminuait pour les longueurs d'onde supérieures. Lambert & Brandt, la même année, ont déterminé plus précisément que la longueur d'onde la plus efficace était de 415 nm. D'autre part, la courbe d'énergie lumineuse, nécessaire au déclenchement de la ponte, obtenue entre des longueurs d'onde de 400 à 610 nm, correspond au spectre d'absorption du cytochrome *c*. Cela permet de penser qu'une hème-protéine interviendrait dans le comportement des Ascidies vis-à-vis des conditions lumineuses.

L'histologie du complexe neural de *Ciona intestinalis* a déjà permis de mettre en évidence un cycle journalier de la glande neurale (Pérès, 1943; Georges, 1967). Des expériences actuellement en cours permettront de compléter, à l'aide de la microscopie électronique et de la culture *in vitro*, les données des auteurs précités.

REFERENCES

BURGER, J. W. (1939). Some aspects of the roles of light intensity and the daily length of exposure to light in the sexual photoperiodic activation of the male starling. *J. exp. Zool.* **81**, 333–41.

BURGER, J. W. (1940). Further studies on the relation of the daily exposure to light to the sexual activation of the male starling (*Sturnus vulgaris*). *J. exp. Zool.* **84**, 351–62.

BURGER, J. W. (1947). On the relation of day-length to the phases of testicular involution and inactivity of the spermatogenetic cycle of starling. *J. exp. Zool.* **105**, 259–67.

GEORGES, D. (1967). La glande neurale de *Ciona intestinalis* (Tunicier, Ascidiacé) observée aux microscopes photonique et électronique. *C. r. hebd. Séanc. Acad. Sci.*, Paris **265**, 1984–7.

HARRINGTON, R. W. (1956). An experiment on the effects of daily photoperiods on gametogenesis and reproduction in the centrarchid fish, *Enneacanthus obesus* (Girard). *J. exp. Zool.* **131**, 203–23.

HARRINGTON, R. W. (1957). Sexual photoperiodicity of the cyprinid fish *Notropis bifrenatus* (Cope), in relation to the phases of its annual reproductive cycle. *J. exp. Zool.* **135**, 529–56.

JULIN, CH. (1881). Recherches sur l'organisation des Ascidies simples. Sur l'hypophyse et quelques organes qui s'y rattachent, dans les genres *Corella, Phallusia, Ascidia. Archs Biol.*, Paris **2**, 59–126 et 211–32.

LAMBERT, CH.C. & BRANDT, CH.L. (1967). The effect of light on the spawning of *Ciona intestinalis. Biol. Bull. mar. biol. Lab.*, Woods Hole **132**, 222–8.

LENDER, TH. & BOUCHARD-MADRELLE, C. (1964). Etude expérimentale de la régénération du complexe neural de *Ciona intestinalis* (Prochordé). *Bull. Soc. zool. Paris* **89**, 546–54.

PÉRÈS, J. M. (1943). Recherches sur le sang et les organes neuraux des Tuniciers. *Annls Inst. océanogr.*, Monaco **21**, 229–359.

SENGEL, P. & GEORGES, D. (1966). Effets de l'éclairement et de l'ablation du complexe neural sur la ponte de *Ciona intestinalis* L. (Tunicier Ascidiacé). *C. r. hebd. Séanc. Acad. Sci.*, Paris **263**, 1876–9.

WARREN, D. C. & SCOTT, H. M. (1936). Influence of the light on ovulation in the fowl. *J. exp. Zool.* **74**, 137–56.

WHITTINGHAM, D. G. (1967). Light-induction of shedding of gametes in *Ciona intestinalis* and *Molgula manhattensis. Biol. Bull. mar. biol. Lab.*, Woods Hole **132**, 292–8.

WILLEY, A. (1894). Studies on the Protochordata. *Q. Jl microsc. Sci.* **35**, 295–334.

LIGHT CONTROLLED SWARMING IN THE POLYCHAETE *AUTOLYTUS*

By L. GIDHOLM

ABSTRACT

Many polychaetes migrate to the surface of the sea for spawning, but very little is known about the environmental factors controlling these migrations. The sexually mature epitokes of the syllid *Autolytus edwardsi* are suitable for experimental studies on this question. Laboratory observations on the swimming activity of males or females during natural light conditions reveal that there are periods of swimming at dusk and dawn, and that the animals rest during the day and the night. If the daylight is prolonged artificially in the evening, or the darkness in the morning, distinct swimming periods fail to appear. This demonstrates that light changes stimulate swimming, and an endogenous rhythm plays little or no role.

The swimming response to photic stimulation was studied further in a simple 'test box', in which the light intensity was varied by a transformer and controlled by a photometer. In samples of 8–16 animals the activity was recorded at intervals of 1–10 min by estimating the number of worms swimming at each observation. This study revealed the presence of a summation mechanism, and showed that the stimulatory factor was the *change* in the light intensity.

It is suggested that, in nature, swarming is initiated by photic stimulation at dusk and dawn. Since the animals are positively phototactic and swim upwards towards the light, the result will be that they gather near the surface. This means that one spatial dimension is eliminated – the males have only to seek the females along a surface.

Provided the light changes at sunset or sunrise are rapid enough, this mechanism would be important for the co-ordination in time and space of swarming in populations living in the photic layers of the sea.

LIGHT AND GONAD DEVELOPMENT
IN *PONTOPOREIA AFFINIS*

S. G. SEGERSTRÅLE

*Institute of Marine Research, Biological Laboratory,
Helsinki/Helsingfors, Finland*

INTRODUCTION

For many animals living at high latitudes where there are marked seasonal fluctuations in temperature, day length, food and so forth, the timing of vital activities in such a way that they are limited to those seasons when conditions are especially favourable is essential for survival. Two well-known examples of this phenomenon are the reproduction of birds in spring and summer and the arrest of development, or diapause, in insects during late autumn and winter. In the sea, the appearance of planktotrophic larvae in many boreo-arctic species during spring, when the phytoplankton is abundant, is a corresponding phenomenon (Thorson 1936, 1946; Grainger 1959; Bogorov, 1958; Barnes, 1957, 1963).

Among the factors responsible for this timing of events in the life of animals, not only temperature but also light has proved to be of special importance. As a landmark in this field of study will remain the experimental work on birds which was started by Rowan in the 1920s and which opened up the entire field of research on photoperiodism in animals. As far as plants are concerned, it had already been shown that flowering could be induced out of season by changes in day length. Today, we know more about the photoperiodic responses of birds than of any other group of animals except perhaps the insects. Surveys of the experimental work performed on photoperiodism in animals are given by Bullough, 1951; Withrow, 1959; Wolfson, 1964 and Hoar, 1966.

As far as aquatic invertebrates are concerned, there is comparatively little information on light as a timing factor (for surveys see Giese, 1959; Steele, 1964). On this occasion, I shall restrict myself to reviewing the data on crustaceans, the group to which the species to be discussed belong. In fact, more experimental work on the effect of light has been performed on crustaceans than on most other aquatic invertebrates.

The influence of illumination on gametogenesis has been demonstrated in some freshwater Crustacea. In the crayfish *Cambarus virilis* the ovarian cycle, development of secondary sexual characteristics and moult can be

[573]

manipulated by changing the photoperiod, extended illumination having a stimulating effect (Stephens, 1952; Stephens, 1955). The stimulating influence of light was also established in similar studies on *Cambarellus shufeldti* (Lowe, 1961) and the freshwater shrimp *Palaemonetes paludosus* (Paris & Jenner, 1952).

By contrast, other cases are known in which illumination has proved to inhibit the maturation process in crustaceans. If in juvenile or non-breeding females of the shrimp *Leander* the eyestalks are removed and perception of light is thus rendered impossible, a rapid increase in ovarian weight results through stimulation of vitellogenesis (Panouse, 1944). This effect is concluded to be due to the secretion of a gonad-inhibiting hormone by the neurosecretory cells of the eyestalk. The extensive and important series of studies on barnacles which has been carried out in recent years by Crisp, Barnes and others deserves special mention (Crisp, 1959; Crisp & Clegg, 1960; Barnes, 1963; Crisp & Patel, 1969; see also Tighe Ford, 1967). These studies have mainly been concerned with the species *Balanus balanoides*. In this boreo-arctic barnacle constant illumination has proved to inhibit maturation of the gonads, although perhaps not indefinitely. Thus, a certain degree of darkness seems to be necessary for gametogenesis in the species concerned.

Finally, as regards the role of light as a regulator of the life-cycle in crustaceans, some observations referring to the copepod genus *Cyclops* should be mentioned. In 1959 Elgmork showed that in *C. strenuus strenuus* diapause started at the same time in habitats with very different conditions as regards temperature and oxygen supply. This apparent independence of the local environment led to the hypothesis that, in the crustacean concerned as in many insects, diapause is initiated primarily by changes in the length of day. This possibility was later considered also by Smyly (1961) and also by Einsle (1964a), who has recently demonstrated by laboratory experiments that photoperiodicity can induce dormancy in species of *Cyclops* (Einsle, 1964b).

In those areas of Northern Europe, Asia and North America which were covered with ice during the Pleistocene, the lakes and certain brackish areas are inhabited by a number of crustaceans, whose presence there is due to this particular prehistory. Among the species belonging to the group concerned, the so-called glacial relicts, the amphipod *Pontoporeia affinis* and the mysid *Mysis relicta*, are especially common and widespread.

P. affinis Lindstr., one of the characteristic benthic animals of the Baltic Sea, has for a long time been studied in connection with my work on the bottom fauna of the coastal waters of Finland. In recent years my interest has focused on the factors responsible for the timing of reproduction in the species.

BREEDING SEASONS OF PONTOPOREIA AFFINIS

In the study area, situated near the Zoological Station at Tvärminne on the south coast of Finland, *Pontoporeia affinis* breeds in the winter, a feature that is, on the whole, typical of this glacial relict. Reproduction in the summer, too, has been observed at depths below about 100 m in the Baltic, as well as in the deeper parts of some North American lakes (see Segerstråle, 1967; Green, 1968). In the Tvärminne region, fertilization and oviposition take place in November–December. Owing to the low temperature in winter, embryonic development is slow, most of the young not leaving the brood-pouch of the mother until 4–5 months later, in March and April.

Sampling work, repeated during several years in the Tvärminne area, revealed the unexpected fact that spawning commences at practically the same time even in situations with very different temperature regimes. Observations in two intensely studied localities are especially illustrative (Segerstråle, 1937). One of these localities is situated at 3 m depth near the Zoological Station, the other is an open area where the depth is about 35 m. In successive years (1930, 1931) the onset of spawning in both localities was ascertained by sampling every 10 days during the season concerned. In spite of striking differences in the water temperature in the two localities during the preceding months (Segerstråle, 1933) fertilization was practically synchronous; the time at which 50% of the females of the two parental stocks had laid their eggs fell within the same 10 day period. Furthermore, this period was identical in the 2 years concerned (21 November to 1 December).

The conclusion that temperature is not the factor responsible for the timing of breeding of *P. affinis* is also supported by the data given by Green (1965, 1968) for Cayuga Lake, N.Y., America.

In the circumstances, it was only natural to suspect that light might be the operative factor. As mentioned above, decrease in illumination in autumn has proved to control vital functions in many other animals. Special interest, of course, attaches to those studies which have established that light exerts an inhibitory effect in crustaceans. It should be mentioned that the idea that illumination might control the reproductive cycle in *P. affinis* was tentatively put forward as long ago as 1920 by the Swedish zoologist Sven Ekman (Ekman 1920).

EXPERIMENTAL OBSERVATIONS ON THE EFFECT
OF DAY LENGTH

Some preliminary experiments performed in 1967 at the Tvärminne Station, showed that the amphipod under discussion could be kept in the laboratory without difficulty for several months. This was primarily due to the fact that *Pontoporeia* is a substrate feeder and therefore does not need special food, provided that the bottom of the aquarium is covered with a layer of mud from the normal environment of the animal. This point requires to be stressed, as our poor knowledge of the influence of light on aquatic animals as compared with the terrestrial fauna is no doubt chiefly caused by the difficulty of keeping them alive for long periods under laboratory conditions.

On 24 June 1968, two aquaria measuring 30 × 50 cm were each stocked with about 150 large *Pontoporeia*. The material was obtained at the aforementioned locality from a depth of 35 m. The aquaria had been provided in advance with a 2–3 cm bed of mud taken from the nearby bay, Krogarviken, and passed through a 1 mm mesh sieve. During the experiment both aquaria were kept in a tank containing cooled water and continuously aerated; their temperatures, as expected, proved to be practically identical. One of the aquaria was placed near the window of the room and so subjected to daily and seasonal fluctuations of light, while the other was isolated and exposed to continuous illumination emanating from a fluorescent tube (Philips no. 32, 20 W) placed 2 m above the aquarium. The former aquarium will be referred to as the 'window aquarium'.

The experiment was continued until the middle of November, the time of year when the first sign of spawning of the species is normally observed in the study area. By that time the first adult males with their characteristic strongly elongated antennae appear, and sometimes egg-bearing females are also found. All females of the parental stock have at this time strongly developed ovaries.

On 14 November, the two aquaria were emptied. In both a considerable proportion of the amphipods had survived: in the window aquarium 87 (58%), and in the one with constant illumination 110 (73%).

Twenty-five females from each aquarium were subsequently dissected and the developmental stage of the ovarian eggs assessed. The results are shown in Fig. 1.

As emerges from Fig. 1, a very clear-cut difference in ovarian development was found between the populations of the two aquaria. In the window aquarium the average volume of the eggs had increased about ten times

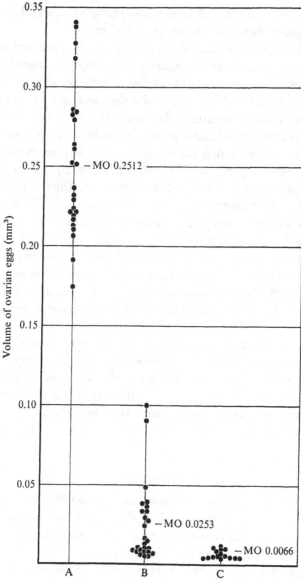

Fig. 1. Volumes of ovarian eggs in *Pontoporeia affinis*. A and B: material from the experimental aquaria, emptied 14 November 1969 (25 specimens from each aquarium, each dot
refers to one specimen and represents the mean volume of 3 eggs). A, 'window aquarium',
B, constantly illuminated aquarium. C, volumes of ovarian eggs before the onset of maturation (15 specimens, volumes calculated on the basis of size data given in Segerstråle, 1937,
p. 80). For practical reasons (irregularity of egg form), the egg volume was estimated by
simply multiplying the length, width and thickness of the egg (thickness, on the basis of
experience, taken, on the average, as 0.6 of the width). This method, which does not take
into consideration the more or less rounded form of the egg, gives absolute values that are
somewhat too high, but the relations between the volume values are not affected. The mean
volume of the ovarian eggs (MO) is shown for each group.

578 S. G. SEGERSTRÅLE

more than in the aquarium with continuous illumination, and there was no overlap whatsoever between the two groups of values.

Sampling in the locality from which the experimental material had been taken on the same day that the aquaria were emptied showed that the ovaries, as was to be expected, had reached a late stage of maturation. In fact, the eggs proved to be of practically the same size as those of the females in the window aquarium. It should also be noted that in this aquarium, and in the field locality as well, a number of mature males were found – a clear indication that the breeding season has commenced.

A note on the temperature in the experimental aquaria and in this sampling locality should be inserted. Conditions in August are of special interest, earlier observations having suggested that it is in this month that maturation sets in. If the temperature measurements made in August in the aquaria are compared with those in the sea where the natural population was studied, it turns out that the sea was much colder than the water in the aquaria. The fact that the development of the gonads was nevertheless synchronous in the two situations lends additional support to the conclusion that the induction of maturation in P. affinis is not connected with temperature.

On inspection of Fig. 1 it also emerges that in a considerable proportion of the females that had lived in the constantly illuminated aquarium the size of the eggs did not exceed that typical of ovaries before the onset of maturation (C). It should also be noted that the males of the aquarium concerned exhibited comparatively early stages of antennal development, certainly stages prior to the penultimate one as described by Segerstråle (1937).

DISCUSSION

The experiments described clearly show that constant illumination inhibits the normal development of the gonads in P. affinis. As maturation seems to begin in late summer, the relative darkness of this season and of autumn no doubt creates favourable conditions for this process.

It is tempting to assume that a decrease in illumination is not only a prerequisite for the normal maturation in the amphipod investigated but also triggers the development of the gonads and thus times the breeding period. However, further experimental work is required on this point. Caution in drawing conclusions from the results obtained so far is prompted by certain observations on Balanus balanoides. As was mentioned earlier, in this barnacle, as in P. affinis, maturation is inhibited by constant illumination. On the other hand, however, Barnes has demonstrated that breeding takes place even in continuous darkness; in other words, maturation sets in without any change in the light conditions.

The fact that a proportion of those *Pontoporeia* that were subjected to constant illumination nevertheless exhibited some degree of ovarian development seems, at first sight, to exclude the possibility that light functions as an inductor of this process. However, it should be taken into consideration that in this case illumination was not, strictly speaking, constant, since, for technical reasons beyond control, some breaks occurred between 22 August and 16 September, with a duration of up to 10 h. It seems not unlikely that even such short variations in the light conditions may have a considerable physiological effect. For instance, experiments with the insect *Chironomus salinarius* have proved that development from the fourth-instar larva to the adult requires light, but that exposure to daylight for as little as 5 min is sufficient to bring about this effect (Koskinen, 1968).

In order to elucidate this and other related problems, further work with *Pontoporeia* is planned and some new experiments are at present in progress at the Tvärminne Station. Continuation of the studies is prompted not only by our present ignorance of the effect of light on reproduction in the group of amphipods – it is also hoped that the case may contribute to our general knowledge of the photoperiodic phenomena in aquatic animals.

The present scarcity of experimentation with marine invertebrates aimed at clearing up these problems is no doubt largely due to the difficulty of keeping most of these animals alive for long periods under laboratory conditions, not least because of the feeding problems (cf. Giese, 1959). In this respect *Pontoporeia* provides the experimental biologist with an almost ideal material.

REFERENCES

BARNES, H. (1957). Processes of restoration and synchronization in marine ecology. The spring diatom increase and the 'spawning' of the common barnacle, *Balanus balanoides* (L.). *Année biol.* **33**, 67–85.

BARNES, H. (1963). Light, temperature and the breeding of *Balanus balanoides*. *J. mar. biol. Ass. U.K.* **43**, 717–27.

BOGOROV, B. G. (1958). Perspectives in the study of seasonal changes of plankton and of the number of generations at different latitudes. In *Perspectives in marine biology* (ed. A. A. Buzzati-Traverso). Berkeley and Los Angeles: University of California Press.

BULLOUGH, W. S. (1951). *Vertebrate Sexual Cycles*. London: Methuen.

CRISP, D. J. (1959). Factors influencing the time of breeding of *Balanus balanoides*. *Oikos* **10**, 275–89.

580   S. G. SEGERSTRÅLE

CRISP, D. J. & CLEGG, D. J. (1960). The induction of the breeding condition in *Balanus balanoides*. *Oikos* 11, 265–75.

CRISP, D. J. & PATEL, B. (1969). Experimental control of the breeding of three boreo-arctic cirripedes. *Mar. Biol.* 2, 283–95.

EINSLE, U. (1964a). Die Gattung Cyclops s.str. im Bodensee. *Arch. Hydrobiol.* 60, 133–99.

EINSLE, U. (1964b). Larvalentwicklung von Cyclopiden und Photoperiodik. *Naturwissenschaften*, 345.

EKMAN, S. (1920). Studien über die marinen Relikte der nordeuropäischen Binnengewässer. VIII. Fortpflanzung und Lebenslauf der marin-glazialen Relikte und ihrer marinen Stammformen. *Int. Revue ges. Hydrobiol.* 8, 543–89.

ELGMORK, K. (1959). Seasonal occurrence of *Cyclops strenuus strenuus*. *Folia limnol. scand.* 12, 1–83.

GIESE, A. C. (1959). Comparative physiology: annual reproductive cycles of marine invertebrates. *A. Rev. Physiol.* 21, 547–76.

GRAINGER, W. H. (1959). The annual oceanographic cycle at Igloolik in the Canadian Arctic. 1. The zooplankton and physical and chemical observations. *J. Fish. Res. Bd Can.* 16, 453–501.

GREEN, R. H. (1965). *The Population Ecology of the Glacial Relict Ampiphod Pontoporeia affinis Lindström in Cayuga, New York*. Doctoral thesis, Cornell Univ.

GREEN, R. H. (1968). A summer-breeding population of the relict amphipod *Pontoporeia affinis* Lindström. *Oikos* 19, 191–7.

HOAR, W. S. (1966). *General and Comparative Physiology*. Englewood Cliff, N.J.: Prentice-Hall.

KOSKINEN, R. (1968). Seasonal and diel emergence of *Chironomus salinarius* Kieff. (Dipt., Chironomidae) near Bergen, Western Norway. *Ann. zool. Fenn.* 5, 65–70.

LOWE, M. E. (1961). The female reproductive cycle of the crayfish *Cambarellus shufeldti*: the influence of environmental factors. *Tulane Stud. Zool.* 8, 157–76.

PANOUSE, M. J. (1944). L'action de la glande du sinus sur l'ovaire chez la Crevette *Leander*. *C. r. hebd. Séanc. Acad. Sci., Paris* 218, 293–4.

PARIS, O. H. & JENNER, C. E. (1952). Photoperiodism in the fresh-water shrimp *Palaemonetes paludosus* (Gibbes). *J. Elisha Mitchell Scient. Soc.* 68, 144.

SEGERSTRÅLE, S. G. (1933). Studien über die Tierwelt in südfinnländischen Küstengewäsusern. I. Untersuchungsgebiete, Methodik und Material. *Commentat. biol.* IV 8, 1–62.

SEGERSTRÅLE, S. G. (1937). Studien über die Tierwelt etc. III. Zur Morphologie und Biologie des Amphipoden *Pontoporeia affinis*, nebst einer Revision der *Pontoporeia*-Systematik. *Ibid*, VII: 1, 1–183.

SEGERSTRÅLE, S. G. (1967). Observations of summer-breeding in populations of the glacial relict *Pontoporeia affinis* Lindstr. (Crustacea Amphipoda), living at greater depths in the Baltic Sea, with notes on the reproduction of *P. femorata* Kröyer. *J. exp. mar. Biol. Ecol.* 1, 55–64.

SMYLY, W. J. P. (1961). The life-cycle of the freshwater copepod *Cyclops leuckarti* Claus in Easthwaite Water. *J. Anim. Ecol.* 30, 153–71.

STEELE, V. J. (1964). *Reproduction and Metabolism in Gammarus oceanicus Segerstråle and Gammarus setosus Dementieva.* Doctoral thesis, McGill University, Montreal, 1–165.

STEPHENS, G. C. (1955). Induction of molting in the crayfish, *Cambarus*, by modification of daily photoperiod. *Biol. Bull. mar. biol. Lab.*, Woods Hole **108**, 235–41.

STEPHENS, G. J. (1952). Mechanisms regulating the reproductive cycle in the crayfish *Cambarus*. I. The female cycle. *Physiol. Zoöl.* **25**, 70–84.

THORSON, G. (1936). The larval development, growth and metabolism of Arctic marine bottom invertebrates compared with those of other seas. *Meddr Grønland* **100**, 1–155.

THORSON, G. (1946). Reproduction and larval ecology of Danish marine bottom invertebrates. *Meddr. Kommn Danm. Fisk.-og Havunders.*, Ser. Plankton **4**, 1–523.

TIGHE FORD, D. J. (1967). Possible mechanisms for the endocrine control of breeding in a cirripede. *Nature, Lond.* **216**, 920–1.

WITHROW, R. B. (ed.) (1959). Photoperiodism and related phenomena in plants and animals. *Publs Am. Ass. Advmt. Sci.* **55**, 1–903.

WOLFSON, A. (1964). Animal photoperiodism. In *Photophysiology* (ed. A. C. Giese), Vol. II, 1–49. New York and London: Academic Press.

AUTHOR INDEX

Bold numbers refer to the reference lists

584 AUTHOR INDEX

588 AUTHOR INDEX

SUBJECT INDEX

Abietinaria abietina, Spirorbis on, 100
Acanthocyclops viridis, temperature and development of, 218
Acartia clausii, in light-trap and tow-net, 491
Acetabularia, photomorphogenetic responses of, 379, 384, 388–9
Aclididae (Prosobranchs), larvae of, 223
Actinaria, 10
  planktonic larvae of, 18, 20
adaptation to light or dark
  in *Corophium*, 559
  and responses of *Asterias* to light, 495–511 *passim*
Adriatic Sea, larvae of bivalves in plankton of, 45–53
aggregation
  of cell macerates of *Ophlitaspongia* from different sources, 161–3
  of metamorphosing Ophlitaspongia, 167–8, 172, 174–5; and mortality, 168–71
Alburnus alburnus (bleak) larvae, food-searching potential of, 474, 482
Alcyonidium polyoum, on *Fucus serratus*, 121
algae
  choice of, by *Spirorbis*, 89–104
  culture of, under irradiation regimes, 323–5
  distribution of, in cave, 420
  light, and growth of spores of, 363–74
  light quality, and photomorphogenesis of, 375–92
  mussel larvae on, 65, 68
  obscurity values for littoral species of, 436
  respiration and photosynthesis of, 347, 349–59
Alima hylina, larval stage of *Squilla alba*, 20, 21
alimentary system
  of *Mytilus* larvae, 266–8, 278; after metamorphosis, 274
  of *Ostrea* larvae, 286; after settlement, 288, 289–90
  of Prosobranch larvae and adults, 225
allophycocyanin, photoreceptor in *Nostoc*, 377, 378
Amaroucium spp., responses of larvae of, to light, 443, and gravity, 463
Amphipoda
  caprellid, associated with *Tubularia*, 57, 60
  as hosts of Digenea, 185
  quantitative recording of, 58
ancestrula, primary zooid of Bryozoan colony, 106
Anglesey, measurement of submarine

irradiation on east and west coasts of, 321–33
animal species, in cave, 431
Annelids, larvae of, in light-trap and tow-net, 491
Anomia sp., planktonic larvae of, 48, 51, 52
Anomura, planktonic larvae of, 20
antennae
  of *Balanus* larva, sense-organ in, 227–36
  of Crustacea, glands in, 296
Aporrhaidae (Prosobranchs), larvae of, 223, 224
Archidoris pseudoargus, in cave, 424
Architectonicidae (sun dials), planktonic larvae of, 12, 13, 14, 15
ascidians, responses of larvae of, to light and gravity, 443–4
Ascophyllum, in cave, 419
Aspidosiphon (Sipunculid), duration of development of, 24
Asterias forbesii, predator on epibenthic communities, 60
Asterias rubens
  diurnal rhythms in, 513, 530–6
  responses of, to light, 495–511
Asteroidea, in cave, 432
Atlanta lesueuri, larvae of, 221, 223, 224
Atlantic Ocean
  currents in, 10–12; and distribution of planktonic larvae, 20–2; velocities of, 22–4
  distribution of planktonic larvae in, 12–20
Atlantidae (Prosobranchs), larvae of, 223
Atylus swammerdami, in light-trap and tow-net, 491, 492
Autolytus edwardsi (Polychaete), swarming of, 571

Baccaria citrinella (Sipunculid), planktonic larvae of, 17
Balanophyllia regia (coral), 125
Balanus balanoides
  antennular sense-organ of larvae of, 227–36
  cross-fertilization of, 173
  light, and reproduction in, 574, 578–9
  orientation of, to light, and predation by *Blennius*, 542–4, 555
  settlement of larvae of, 539
  settlement-inducing factor from, 143–53, 227, 542
  shadow response in, 551, 555
Balanus balanus, shadow response in, 551, 555

salinity, in estuary, 30
  association of movements of bivalve
    larvae with increase of, in flood tide,
    36–7, 39, 41
  sampling apparatus, multiple-disc, for
    epibenthic organisms, 55–7
Sargassum fulvellum, *Clytia* on, 121
Scenedesmus acuminatus, response of, to
    blue light, 380, 389
  scioconstant zone of algal settlement, 434,
    435
  resemblance of communities in, to those
    in caves, 438–9, 440
Scrupocellaria reptans
  assessment of growth of, 109
  on *Flustra* and *Laminaria*, 108, 120, 122
  larvae of, 105, 107; settlement and
    metamorphosis of, 108
  orientation of, 113–14, 115, 116, 117, 118,
    119
Scyllaridae (Decapods), planktonic larvae
    of, 20, 24
sea water
  sediment in: attenuation of submarine
    irradiation with load of, 327–8; nature
    of, 330–2
  volume of, searched by fish larvae, per
    hour, 479–81; per day, 482–3
Semper's larvae of Zoanthidia, 18–20
Sesarma cinereum, larvae of, 218
setae, of antennular sense-organ of *Balanus*
    larva, 230–5
settlement
  of Bryozoans, 105–23
  contour, light, and water flow as factors
    in orientation of barnacles at, 539, 540,
    555
  delay in, 23–4
  of *Diplosoma* larvae on different surfaces,
    458–61
  of *Mytilus* larvae, 63–9
  of *Ophlitaspongia* larvae on different
    surfaces, 164–8
  of *Pyrgoma* on *Caryophillia*, 125–41
  of *Spirorbis* larvae on algae, 89–104
settlement activity, assay of, 146
settlement-inducing factor, from *Balanus*,
    143–53, 227, 542
shadow response in barnacles, 139, 540,
    554–5
  influence of conditioning on, 549–50
  minimum reduction in illumination for,
    in different species, 550–1
  relation between orientation of barnacles
    to light and, 546–9, 551–3
shell
  in *Mytilus* larva, 264; after metamorpho-
    sis, 273–4

in *Ostrea* larva, after settlement, 291
in Prosobranch larvae, 221, 222
Simnia spelta (Prosobranch), larvae of, 221,
    222, 223, 224
Sipunculida, planktonic larvae of, 16–17, 24
sole (*Solea solea*) larvae, 467, 468
  food-searching potential of, 469, 471–5
    passim, 481
Solen sp., planktonic larvae of, 48, 52
Solenacea, planktonic larvae of, 49
solenocytes, in *Harmothoë* larva, 241, 242
spawning
  in *Ciona*, after removal of neural gland,
    562–5
  season of: for bivalves in Adriatic,
    46–51; for mussels in Oslo Fjord, 64
speciation, in *Spirorbis*, 101, 102
spermatophores, of *Microspio*, 249
spicules, formation and erosion of, in
    Sycon, 301–20
Spiochaetopterus (Polychaete), duration of
    development of, 24
Spionidae
  development of, 247
  planktonic larvae of, 15
Spirorbinae, speciation in, 101, 102
Spirorbis spirorbis, choice between species
    of *Fucus* by larvae of, 89–104
Spirorbis spp., 100, 101, 102
sponges
  in cave, 422, 430
  reproductive processes of, 173–4, 175
  see also individual species
Spongilla lacustris, larvae of, 166
spores of algae, light and growth of, 363–74
Squilla alba, planktonic larvae of (*Alima
    hyalina*), 20, 21
starvation
  and diurnal rhythms of *Asterias*, 533; of
    Gibbula, 524–5
  and responses of *Asterias* to light, 499,
    510
Sternaspididae, planktonic larvae of, 15
Stomatopoda, planktonic larvae of, 20
surfaces
  contour of: and orientation of *Balanus*
    larvae at settlement, 539, 540; and site
    of settlement of *Pyrgoma*, 127, 129
  roughness of, and settlement of *Balanus*
    larvae, 235
  settling preferences for different arti-
    ficial: of *Diplosoma* larvae, 458–61; of
    invertebrates, 56–60
Sycon ciliatum (calcareous sponge), spicule
    formation and corrosion in, 301–20
Sycon raphanus, 301
Sycon setosa, 301, 310–11
Synedra (diatom), mussel larvae on, 65